Universitext

Editors

F.W. Gehring
P.R. Halmos
C.C. Moore

A.I. Kostrikin

Introduction to Algebra

Translated from the Russian by
Neal Koblitz

Springer-Verlag
New York Heidelberg Berlin

A.I. Kostrikin
Moscow State University
Department of Mathematics
Moscow, U.S.S.R. 117234

Neal Koblitz
Department of Mathematics
University of Washington
Seattle, Washington 98195, U.S.A.

AMS Classifications: 13-01, 16-01, 20-01

Library of Congress Cataloging in Publication Data

Kostrikin, A. I. (Alekseĭ Ivanovich)
Introduction to algebra.

(Universitext)
Translation of: Vvedenie v algebru.
Bibliography: p.
Includes index.
1. Algebra. I. Title.
QA154.2.K6713 512 82-5534
AACR2

Printed in the United States of America

9 8 7 6 5 4 3 2 1

ISBN 0-387-90711-4 Springer-Verlag New York Heidelberg Berlin
ISBN 3-540-90711-4 Springer-Verlag Berlin Heidelberg New York

A Note on the English Edition

Every textbook is written taking into account the traditions in a given university or, more generally, in the universities of a given country. My algebra textbook is no exception. At the same time, the exchange of ideas in the area of mathematics teaching in different countries is no less important than the exchange of ideas in research. The Soviet Union, in particular, has accumulated a rich experience in translating into Russian monographs and textbooks published in other countries.

It will be a pleasure, as well as an honor, for me to see this textbook translated into English, under the auspices of Springer publishing house, which is famous throughout the mathematical world. I would like to express my deep thanks to Neal Koblitz for his excellent translation.

Moscow, November 1981 A. Kostrikin

Translator's Preface

This textbook, written by a dedicated and successful pedagogue who developed the present undergraduate algebra course at Moscow State University, differs in several respects from other algebra textbooks available in English. The book reflects the Soviet approach to teaching mathematics with its emphasis on applications and problem-solving -- note that the mathematics department in Moscow is called the "Mechanics-Mathematics" Faculty. In the first place, Kostrikin's textbook motivates many of the algebraic concepts by practical examples, for instance, the heated plate problem used to introduce linear equations in Chapter 1. In the second place, there are a large number of exercises, so that the student can convert a vague passive understanding to active mastery of the new ideas. These problems are intended to be challenging but doable by the student; the harder ones have hints at the back of the book. This feature also makes the book ideally suited for learning algebra on one's own outside of the framework of an organized course. In the third place, the author treats material which is usually not part of an elementary course but which is fundamental in applications. Thus, Part II includes an introduction to the classical groups and to representation theory. With many American colleges now trying to bring their undergraduate mathematics curriculum closer to applications, it seems worthwhile to translate Soviet textbooks which reflect their greater experience in this area of mathematical pedagogy.

I would like to express my deep appreciation to Robert Cornell for his careful reading of Part I and his many suggestions for increasing clarity and readability. I would also like to thank Barbara Moody for her meticulous typing of the manuscript.

Neal Koblitz, Seattle, February 1982

Contents

Foreword

This book was written to give a systematic exposition of the course in algebra for students of the Mechanics-Mathematics Faculty of Moscow University that has developed in recent years. The natural evolution of the standard syllabus necessitated at least a partial re-working and modernization of the textual material in algebra.

Formally, the book is divided into two parts, in rough correspondence to the algebra courses taught in the first and third semesters at Moscow University. In Part II we assume that the reader is well grounded in the theory of abstract vector spaces and linear operators -- material which is studied in the second semester course in linear algebra and geometry. However, real vector spaces are presented in Chapter 2 of Part I, several concepts of linear algebra are developed along the way in the text, and a small appendix contains the geometric theory of matrix reduction to Jordan normal form. The book can therefore be studied independently of any other sources.

A significant role is played by the problems at the end of most of the sections. Because of the availability of excellent exercise books in algebra, it seemed pointless here to emphasize numerical calculations; therefore, the problems have a more substantive character, and help to develop the basic ideas. In several cases they are referred to in the body of the text, but all such exercises are supplied with detailed hints so that there will be no difficulty in solving them. We recommend that the reader look at these hints as little as possible, only after persistent attempts on his own to solve the problem.

It is probably unrealistic to expect that a small number of lecture hours will be sufficient to cover the contents of the entire book. This is especially true of Part II, the material of which is not completely traditional. This material includes a fair amount of intuitive motivation, but certain "delicacies" (such as the Sylow theorems, invariants of linear groups, representations of rotation groups, and non-associative algebras) are more advanced, and are consciously directed toward enthusiasts who may be stimulated to further study.

After studying the fairly difficult 7-th chapter, one should decide whether to concentrate on elements of representation theory (Chapter 8) or on the general theory of rings, modules and fields, which is touched upon in Chapter 9 (where it was not possible, however, to go deeply into structural questions). The first choice seems preferable not only because of its connection with geometry and the material of the second semester course in linear algebra, but also because knowledge of the basic facts about group representations is very useful to mathematicians who specialize in fields other than algebra. It is extremely desirable to solidify one's understanding of group representations, which are illustrated in the book using only a few basic examples, by studying further applications. Examples of themes for further study are Galois theory, groups generated by reflections (including crystallographic groups), representations of compact groups, and so on. On the other hand, Chapter 9, with its number theoretic slant, more closely corresponds to the usual syllabus in algebra. In any case, either choice of emphasis provides a foundation for further work in algebra.

The beginning of each part of the book contains a small list of supplementary literature, which makes no claim to completeness.

One point should be clearly stated, since it may not be obvious to the beginning student. A course in higher algebra, despite its name, can in no way reflect the full gamut of modern algebra. It is for this reason that the book is called an "introduction". A further purpose of an introduction is to be a sourcebook of concepts and results needed for the study of other fields of mathematics. The importance of learning the language of algebra will become immediately apparent to anyone who attempts independent study of mathematics without first acquiring this knowledge.

Despite its elementary character, the traditional course in algebra presented difficulties to the student because of the inherently formal nature of algebraic thinking. The author had this constantly in mind, and for this reason attempted to emphasize connections between algebra and other areas of the mathematical sciences. It is unfortunate that elements of category theory and partially ordered systems did not find a place in the book. However, it would have been pointless to overload an introductory course with a conglomerate of abstract notions which would tend to kill interest in those subjects because of the inevitable superficiality of their exposition.

Many different variants of a required algebra course were put into practice in the Mechanics-Mathematics Faculty of Moscow University over the last ten or fifteen years. It is reasonable to hope that the present realization in book form of the recently adopted version of the course will be useful for students and instructors at other colleges as well, and also for those who would like to begin independent study of algebra. Of course, the order and the degree of completeness with which the material in the book is presented in lectures will depend strongly on the concrete circumstances and pedagogical traditions in each college.

The author is very grateful to the experienced teaching staff of the Department of Higher Algebra of Moscow University, and wishes to thank those who gave much useful advice for presenting the course. All constructive suggestions and remarks concerning errors and misprints will be gratefully accepted.

A. Kostrikin

Zvenigorod
July, 1976

Advice to the Reader

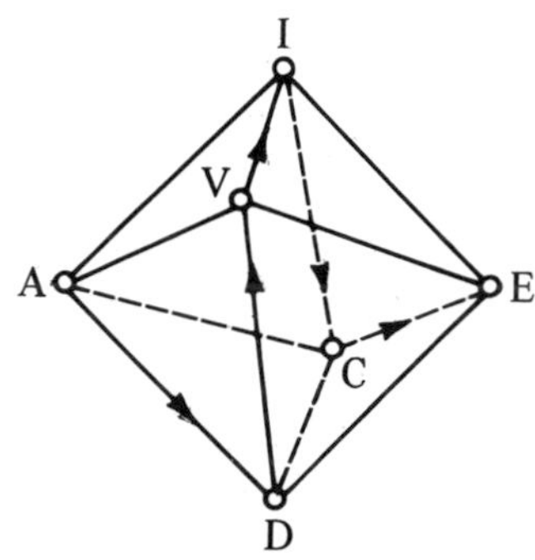

As explained in the Foreword, the interdependence of the chapters is as follows:

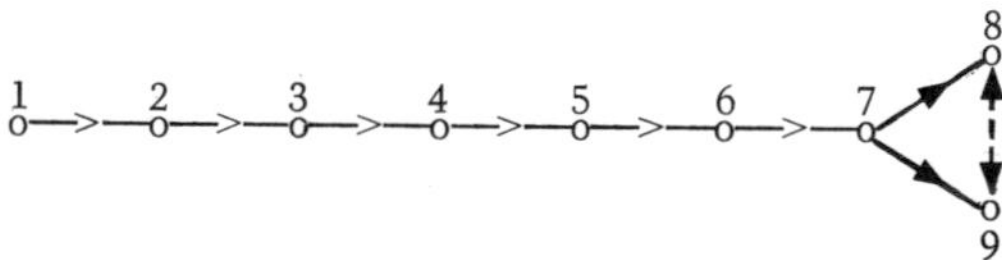

(the broken arrow indicates weak dependence). Of course, an experienced reader (such as an instructor or an advanced student) will have no trouble beginning to read from almost any place, if he is willing from time to time to turn back to the definitions in the earlier sections and chapters. Not all new concepts are introduced in paragraphs beginning with the word "Definition". The detailed Table of Contents and the Index can be used to find the needed place in the book.

Each chapter is divided into several sections, and each section is divided into several subsections with appropriate headings. The theorems, propositions, lemmas, and corollaries within each section have their own numbering: Theorem 1, Theorem 2,...; Lemma 1, Lemma 2,... . With this primitive but simple system of numbering, when referring to assertions in another section we must write, for example, Theorem i §j , or even Theorem i §j Ch. k; but this will not cause any difficulties.

The end (or absence) of a proof is indicated by the sign □ .

For brevity, we use the simplest logical symbols. The implication sign ==> in A ==> B has the simple meaning "A implies B" or "B follows from A", while "A <==> B" means that assertions A and B are equivalent (A if and only if B). The general quantifier ∀ replaces the expression "for all". The other notation will be clear from the context.

Below we give the full Greek alphabet, indicating the pronounciation of the letters. Any confusion here can be annoying, since the letters of the Greek alphabet are very widely used in mathematics.

GREEK ALPHABET

Α α	Β β	Γ γ	Δ δ	Ε ε	Ζ ζ	Η η	Θ θ	Ι ι	Κ κ	Λ λ
alpha	beta	gamma	delta	epsilon	zeta	eta	theta	iota	kappa	lambda

Μ μ	Ν ν	Ξ ξ	Ο ο	Π π	Ρ ρ	Σ σ	Τ τ	Υ υ	Φ φ	Χ χ
mu(myu)	nu(nyu)	xi(ksee)	omicron	pi	rho	sigma	tau	upsilon	phi	chi(ki)

Ψ ψ	Ω ω
psi	omega

"L'algèbre est généreuse, elle donne souvent plus qu'on liu demande".

-- d'Alembert

Part One
Foundations of Algebra

This part can be considered "algebra in miniature". The fundamental concepts of groups, rings and fields, which are unfamiliar to the beginning student, are introduced informally and in small doses, although the total number of interrelated ideas presented to the reader turns out to be quite large. The definitions and theorems should not be memorized: they will become familiar after working independently on the problems and exercises. It is helpful to concentrate on a few of the most widely used algebraic systems (the groups $(\mathbb{Z}, +)$, S_n, A_n, $GL(n)$, $SL(n)$; polynomial rings; the fields $\mathbb{Q}$, $\mathbb{R}$, $\mathbb{C}$, and $\mathbb{Z}_p$) which serve to illustrate the language of algebra. In accordance with tradition and considerations of compatibility between high school and college, we first present matrices and determinants, which are used to find and study the solutions of systems of linear equations. Along the way, basic algebraic structures arise in a natural way.

Further Reading

1. Z. I. Borevich and I. R. Shafarevich, Number Theory, Academic Press (New York), 1966.

2. H. Davenport, The Higher Arithmetic, Hutchinson's Univ. Library (London), 1952.

3. D. K. Faddeev and I. S. Sominskii, Problems in Higher Algebra, W. H. Freeman (San Francisco), 1965.

4. I. Herstein, Topics in Algebra, Xerox College Publishing (Lexington, Mass.), 1975.

5. K. Ireland and M. Rosen, Elements of Number Theory, Bogden and Quigley, 1972.

6. S. Lang, Algebra, Addison-Wesley, 1971.

7. B. L. van der Waerden, Algebra, Frederick Ungar (New York), 1970.

Chapter 1. Foundations of Algebra

What does algebra start from? To a certain extent, one can say that the sources of algebra are implicit in the art of adding and multiplying integers and raising them to powers. If one formally replaces integers with letters -- a step which is far from obvious and can be carried out in many ways -- one can then proceed according to similar rules in much more general algebraic systems. In fact, an attempt to give an exhaustive answer to our question would take us not only far back through the ages, but also into the mysteries of the emergence of mathematical thought. A difficult part of the answer to this question would consist of describing the basic structures of the algebra of our day: groups, rings, fields, modules, and so on. But the entire book is devoted precisely to this, so that the goal of Chapter I seems at this point to be out of reach.

Fortunately, under the abstract shell of most axiomatic theories of algebra one finds very concrete problems of a theoretical or practical nature, whose solution once served as a fortuitous and sometimes indispensable stimulus to far-reaching generalizations. The development of a general theory, in turn, gave an impulse and a technique for the solution of new problems. The complicated interaction between the theoretical and practical aspects, which is inherent in all mathematics, takes an especially pronounced form in algebra and to some extent provides a justification for the concentric style of presentation adopted in this book.

After some brief general remarks on the history of the subject, we shall formulate several problems which motivate the material in the chapters which follow. One of these problems is the point of departure for our study of systems of linear equations and the theory of matrices and determinants. We shall give Gauss's method and thus obtain our first facts about the solutions of linear systems.

At this stage it will already be useful to introduce some standard notation and terminology, which will be done in a brief survey of set theory. We shall introduce the important concepts of an equivalence relation and a quotient map. Further, in order to explain the principle of mathematical induction, we establish some elementary combinatoric relations. Finally, the simple arithmetic properties of the integers which are given in the last section are not only used throughout the subsequent chapters, but are also the prototype for constructing similar rules of arithmetic in more complicated algebraic systems.

The material in this chapter does not go far beyond high school mathematics. The reader is only required to adopt a more general point of view.

The student may begin reading in §3.

§1. Algebra in brief

For good reason one often hears these days about the "algebraization" of mathematics, i.e., the penetration of algebraic ideas and methods into both theoretical and practical fields of mathematics. This state of affairs, which became completely apparent in the middle of the twentieth century, has by no means always been with us. As in every area of human endeavor, mathematics is subject to the influence of fashion. The fashion for algebraic methods had substantive causes, but sometimes enthusiasm for these methods exceeds reasonable boundaries. And since an algebraic shell which obscures the content is no less of a disaster than basic ignorance of algebra, it has become customary (justifiably so) for books to be praised whenever the author manages to avoid an overloading of algebraic formalism.

While avoiding extremes, one should realize that algebra has from time immemorial made up an essential part of mathematics. The same could just as well be said about geometry, but here we should cite the opinion of Sophie Germain (nineteenth century): "Algebra is nothing but geometry in symbols, and geometry is nothing but algebra in pictures". The situation has changed somewhat since then, but it still seems that "the 'true nature' of mathematical objects is really of secondary importance, and it does not much matter, for example, whether we describe a result as a theorem in 'pure' geometry or, using analytic geometry, as an algebraic theorem" (N. Bourbaki).

According to the principle that "it is not the mathematical objects which are important, but the relationship between them", algebra can be defined (in a way that is somewhat tautological and is completely incomprehensible to the uninitiated) as the science of algebraic operations performed on elements of various sets. The algebraic operations themselves grew out of elementary arithmetic. Algebraic ideas, in turn, give the most natural proofs of many facts of "higher arithmetic", number theory.

But the significance of algebraic structures -- sets with algebraic operations -- goes far beyond number-theoretic applications. Many mathematical objects (topological spaces, differential equations, functions of several complex variables, etc.) are studied by first constructing suitable algebraic structures; even if these structures do not tell the whole story about the objects under consideration, they often reflect their most important properties. The same can be said about applications of algebra to the real world.

The definitive opinion on this subject was given more than 45 years ago by P. Dirac, who was one of the founders of quantum mechanics: "Modern physics increasingly requires abstract mathematics and the development of its foundations. Thus, non-Euclidean geometry and non-commutative algebra, which were once considered to be merely the fruit of imagination or fascination with logical reasoning, are now recognized to be very necessary to describe the general picture of the physical world".

Algebraic techniques are very useful in studying elementary particles in quantum mechanics, investigating the properties of rigid solids and crystals (here the theory of group representations is especially important), analyzing models in economics, constructing modern computers, and so on and so forth.

Algebra, in turn, is nourished by the life-blood of other disciplines, including the other mathematical disciplines. For example, homological methods in algebra grew out of topology and algebraic number theory.

It is not surprising that the appearance of algebra and the way it is viewed change over time. Here we cannot give a detailed account of these changes, not only because of lack of space, but, even more, because such a historical discussion must be concrete, and this is only possible after a basic knowledge of the subject has been acquired.

We shall only give a schematic list of names and periods.

The ancient civilizations of Babylonia and Egypt. Greek civilization. The "arithmetic" of Diophantus (3rd century B. C.).		Arithmetic operations on integers and positive rational numbers. Algebraic formulas in geometry and astronomy. Formulation of construction problems (doubling the cube and trisecting an angle) which occupied algebraic minds at a much later time.
Eastern civilizations of the Middle ages. The work "ilm al-jabr wa'l muqabalah" by Mohammed ibn Mûsâ al-Khowârizmî (approx. 825).		Algebraic equations of degree one and two. Introduction of the term "algebra".
Renaissance		Solution of general algebraic equations of degree three and four.
Fibonacci (Leonardo of Pisa)	(approx. 1170-1250)	
S. Ferro	(1465-1526)	
N. Tartaglia	(1500-1557)	
G. Cardano	(1501-1576)	
L. Gerrari	(1522-1565)	
F. Vieta	(1540-1603)	Creation of modern algebraic symbolism.
R. Bombelli	(1530-1572)	
XVII - XVIII centuries		
R. Descartes	(1596-1650)	Emergence of analytic geometry -- a solid
P. Fermat	(1601-1665)	bridge between geometry and algebra.
I. Newton	(1643-1727)	Increasing activity in number theory.
G. Leibniz	(1646-1716)	Development of the algebra of polynomials.
L. Euler	(1707-1783)	Intensive search for general formulas for

J. d'Alembert	(1717-1783)	solutions of algebraic equations. The first
J.-L. Lagrange	(1736-1813)	approaches to proving the existence of a root
G. Cramer	(1704-1752)	of an equation with numerical coefficients.
P. Laplace	(1749-1827)	The beginnings of the theory of determinants.
A. Vandermonde	(1735-1796)	

XIX - early XX centuries

K. F. Gauss	(1777-1855)	Proof of the basic existence theorem for roots
P. Dirichlet	(1805-1859)	of an equation with numerical coefficients.
E. Kummer	(1810-1893)	Intensive development of algebraic number
L. Kronecker	(1823-1891)	theory.
R. Dedekind	(1831-1916)	
E. I. Zolotarev	(1847-1878)	Search for methods of approximate solution
G. F. Voronoi	(1868-1908)	of algebraic equations. Conditions on the
A. A. Markov	(1856-1922)	coefficients which ensure a certain location
P. L. Chebyshev	(1821-1894)	for the roots.
C. Hermite	(1822-1901)	
N. I. Lobachevskii	(1792-1856)	Proof of the unsolvability in radicals of the
A. Hurwitz	(1859-1919)	general equation of degree $n \geq 5$. Devel-
P. Ruffini	(1765-1822)	opment of the theory of algebraic functions.
N. H. Abel	(1802-1829)	Creation of Galois theory. The beginnings
C. Jacobi	(1804-1851)	of the theory of finite groups, mainly based
E. Galois	(1811-1832)	on permutation groups.
G. Riemann	(1826-1866)	
A. L. Cauchy	(1789-1857)	Intensive development of methods of linear
C. Jordan	(1838-1922)	algebra. The emergence, after the discovery
L. Sylow	(1832-1918)	of quaternions, of the theory of hypercomplex
H. Grassmann	(1809-1877)	systems (such systems are now called
J. Sylvester	(1814-1897)	algebras). In particular, in connection with
A. Cayley	(1821-1895)	the development of the theory of continuous
W. Hamilton	(1805-1865)	groups (Lie groups), the foundations were laid
G. Boole	(1815-1864)	for the theory of Lie algebras. Algebraic
S. Lie	(1842-1899)	geometry and the theory of invariants became
F. Frobenius	(1849-1918)	important branches of mathematics. In the
J. Serret	(1819-1885)	XIX century, mathematics had not yet become
M. Noether	(1844-1922)	highly specialized, and many leading scientists
D. A. Gravier	(1863-1939)	worked successfully in several areas.
H. Poincaré	(1854-1912)	
F. Klein	(1849-1925)	The first half of the XX century saw a radical
W. Burnside	(1852-1927)	reconstruction of the entire edifice of
I. Schur	(1885-1941)	mathematics. Algebra gave up the title of
H. Weyl	(1885-1955)	the science of algebraic equations and took a
F. Enriques	(1871-1946)	decisive step along an axiomatic and much
J. von Neumann	(1903-1957)	more abstract path of development. The
D. Hilbert	(1862-1943)	language of rings, modules, categories,
E. Cartan	(1869-1951)	homology came into wide use. Many diverse
K. Hensel	(1861-1941)	theories fit into the general scheme of
E. Steinitz	(1871-1928)	universal algebra. The theory of models
E. Noether	(1882-1935)	arose in the overlap between algebra and
E. Artin	(1898-1962)	mathematical logic. Old theories were
N. Bourbaki, "Elements of Mathematics"		rejuvenated, broadening their applications.
		Examples are modern algebraic geometry,
		algebraic topology, algebraic K-theory ,

the theory of algebraic groups. The theory of finite groups had many bright moments.

All of algebra is now in a state of dynamic development. Among the Soviet mathematicians who have made great contributions to this research are N. G. Chebotarev (1894-1947), O. Ju. Shmidt (1891-1956), A. I. Mal'tsev (1909-1967), A. G. Kurosh (1908-1971), P. S. Novikov (1901-1975).

§2. Some model problems

The four problems below are at different levels of difficulty. The first three, which themselves are not all at the same level, are designed exclusively to motivate the study of different types of fields, vector spaces, groups, and group representations, i.e., the algebraic theories which will be discussed later in the book. Many specialized monographs are devoted to "solving" these problems. The fourth problem, which motivates the study of linear systems, is worthwhile for the reader to try to solve right now, without looking at the next section, which contains the necessary steps.

1. Solvability of equations in radicals. The formula

$$x_1, x_2 = \frac{-b \pm \sqrt{b^2 - 4ac}}{2a} \tag{1}$$

for the solutions x_1, x_2 of the quadratic equation $ax^2 + bx + c = 0$ is well known from elementary algebra.

A cubic equation $x^3 + ax^2 + bx + c = 0$ takes the form $x^3 + px + q = 0$ after performing the substitution $x \mapsto x - \frac{1}{3}a$. Let x_1, x_2, x_3 be the three roots of the equation $x^3 + px + q = 0$. If we set

$$D = -4p^3 - 27q^2, \qquad \varepsilon = \frac{-1 + \sqrt{-3}}{2},$$

$$u = \sqrt[3]{-\frac{27}{2}q + \frac{3}{2}\sqrt{-3D}}, \qquad v = \sqrt[3]{-\frac{27}{2}q - \frac{3}{2}\sqrt{-3D}} \tag{2}$$

(where the cube roots must be chosen so that $uv = -3p$), then it is possible to show that

$$x_1 = \frac{1}{3}(u+v) \ , \ x_2 = \frac{1}{3}(\varepsilon^2 u + \varepsilon v) \ , \ x_3 = \frac{1}{3}(\varepsilon u + \varepsilon^2 v) \ . \tag{3}$$

Formulas (2) and (3), which are known as Cardano's formulas (1545) and are also associated with the names of other Italian Renaissance mathematicians (Ferro, Tartaglia), are valid, just like formula (1), for absolutely any values of the letters a, b, c, p, q, for example, for any rational values. Similar formulas were found for the roots of a fourth degree equation. Then for almost three hundred years mathematicians attempted unsuccessfully to "solve in radicals" the general fifth degree equation. It was only in 1813 that Ruffini (in rough form) and in 1827 that Abel (independently and completely rigorously) proved the theorem that the general equation $x^n + a_1 x^{n-1} + \cdots + a_n = 0$ cannot be solved in radicals if $n > 4$. The fundamental discovery in this field was made by twenty-year-old Evariste Galois in 1831 (his work only became known in 1846), when he gave a general criterion for any equation (say, with rational coefficients), not just the general n-th degree equation, to be solvable in radicals.

To every polynomial (or equation) of degree n Galois associated a "splitting field" and a finite family (of cardinality no greater than $n!$) of so-called "automorphisms" of this field. These automorphisms are now called the "Galois group" of the field (or of the original polynomial). Although we shall not dwell on Galois theory in detail, Chapter 7 contains an intrinsic characterization of the special class of so-called "solvable" groups. It turns out that an equation of degree n with rational ceofficients is solvable in radicals if and only if the corresponding Galois group is a solvable group. For example, suppose that we are given the fifth degree equation $x^5 - ax - 1 = 0$, where a is some integer. This equation corresponds to a Galois group G_a, which depends in some complicated way on a. G_0 is the cyclic group of order 4 (and all cyclic groups are solvable, by definition), and the equation $x^5 - 1 = 0$ is, of course, solvable in radicals. On the other hand, G_1 has the same structure as the symmetric group S_5, which has order 120, and, as we show in Chapter 7, this group is not solvable. Hence, the equation $x^5 - x - 1 = 0$ is not solvable in radicals.

In conclusion, we note that the possibility of expressing a root of an algebraic equation explicitly in terms of radicals is not very important from a practical standpoint; approximation methods are more relevant for computations. But this does not diminish the beauty of Galois' achievement, which had a profound conceptual influence on the subsequent development of mathematics. To begin with, it was Galois theory that set the stage for group theory. In the XX century, Galois' one-to-one correspondence between subfields of the splitting field and subgroups of its Galois group has been generalized and enriched with new abstract constructions, so that now this correspondence provides an indispensable tool for studying mathematical objects.

2. The states of a molecule. Every molecule can be considered as a system of particles, i.e., atomic nuclei (surrounded by electrons). If the system's configuration at the initial moment of time is close to an equilibrium configuration, then, under certain conditions, the particles in the system will always remain close to equilibrium positions, and will not acquire large velocities. Motion of this type is called oscillation relative to the equilibrium configuration, and such a system is called stable. It is known that any small oscillation of the molecule near a position of stable equilibrium is a superposition of so-called "normal" oscillations. In many cases it is possible to determine the potential energy of the molecule and its normal frequencies by taking into account the internal symmetries of the molecule. The symmetry of the molecular structure is described by the "point group" of the molecule. Different realizations of this finite group (its irreducible representations) and functions on the group which are associated to these realizations (characters of the representations) give parameters of the oscillations of the molecule.

For example, the water molecule H_2O (Fig. 1) corresponds to the Klein four-group (the direct product of two cyclic groups of order two); and the phosphorus molecule P_4 (Fig. 2), which has the form of a right tetrahedron with phosphorus atoms at the vertices, corresponds to the symmetric group S_4, which has order 24. The irreducible representations of these groups will be studied in Chapter 8. Nowadays it is hard to imagine how the structure theory of molecules could have developed without the use of group theory.

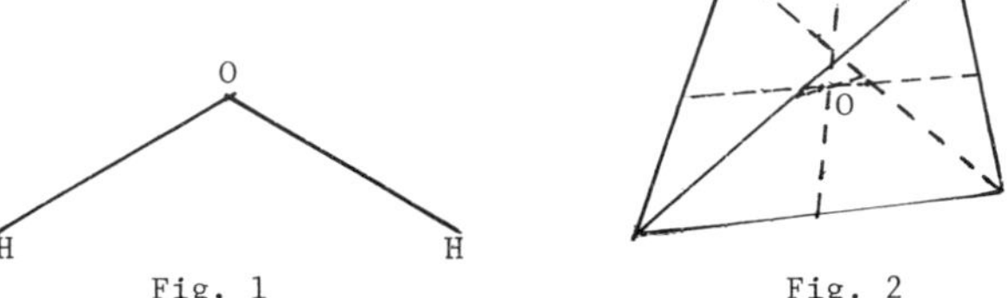

Fig. 1 Fig. 2

Much earlier applications of group theory are found in crystallography. As early as 1891, the great Russian crystallographer Fedorov, and then the German scientist Schoenflies found the 230 crystallographic space-groups which describe all crystal symmetries which are found in nature. Ever since then, group theory has been continually used to study the influence of symmetry on the physical properties of crystals.

3. Coding information. In constructing automatic communication systems, on earth or in the cosmos, one usually takes the basic message to be an ordered sequence, which we call a row (or word): $a = (a_1, a_2, \cdots, a_n)$ of length n, where $a_i = 0$ or 1. Since the usual operations of addition and multiplication modulo 2 are well suited for execution on an electronic machine, and the symbols 0 and 1 are themselves easily transmitted as electronic signals (1 and 0 are distinguished by how successive signals are separated, or else one corresponds to a signal and the other to its absence), it is not surprising that the field GF(2) (see §4 of Chapter 4) is used constantly by the specialist in information theory. It is sometimes convenient to take the a_i to be elements of other finite fields.

If one wants to minimize the influence of static (atmospheric and cosmic interference), which can turn 0's into 1's and vice-versa, one must take a to be sufficiently long and use a special coding system, i.e., a choice of a subset S_0 of admissible rows from among the set S of all possible words. S_0 is called the code, and its elements are called code-words. In that way it is possible to reconstruct a from a distorted word a', provided that there weren't too many erroreous signals.

In this way, error-correcting codes arise. Algebraic coding theory has been developing rapidly in recent years, and now includes many clever methods of coding. This theory is largely concerned with the special linear codes which are obtained when the choice of S_0 is connected with the construction of special rectangular matrices and the solution of systems of linear equations whose coefficients belong to a given finite field. A simple example of such a code will be given in Chapter 5.

4. The heated plate problem. A flat rectangular plate with three holes (Fig. 3) is used as a valve in an imaginary set-up for obtaining low temperatures. It is covered with

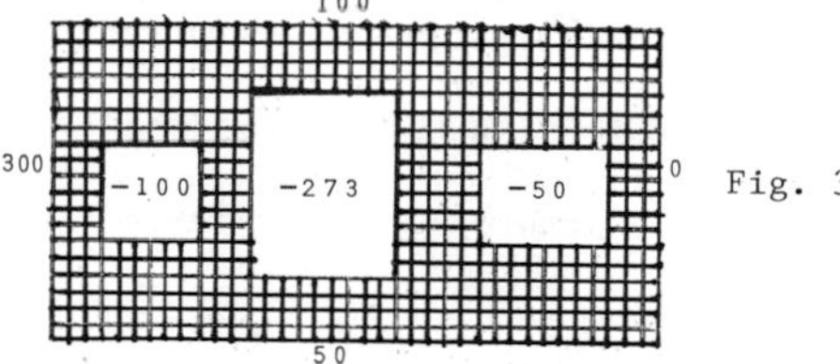

Fig. 3

a square net (grid). The vertices of the grid which lie on the four contours are called boundary vertices, and all the other ones are called interior vertices. Experiments show that, during any heating or cooling, the temperature at any interior vertex is the arithmetic mean of the temperatures at the four nearest vertices (interior or boundary). We would like the temperature at the vertices along the contours to have the values indicated in Fig. 3. Is this possible, and, if it is possible, is the distribution of temperatures at the interior vertices uniquely determined?

§3. Systems of linear equations. The first steps

Linear equations $ax = b$ and systems of the type

$$\begin{aligned} ax + by &= e \quad , \\ cx + dy &= f \end{aligned} \tag{1}$$

with real coefficients a, b, c, d, e, f are "solved" in high school. Our purpose is to learn how to work with a system of linear algebraic equations (or briefly: a linear system) of the most general type:

$$\begin{aligned} a_{11}x_1 + a_{12}x_2 + \cdots + a_{1n}x_n &= b_1 , \\ a_{21}x_1 + a_{22}x_2 + \cdots + a_{2n}x_n &= b_2 , \\ \cdots\cdots\cdots\cdots\cdots\cdots\cdots\cdots \\ a_{m1}x_1 + a_{m2}x_2 + \cdots + a_{mn}x_n &= b_m . \end{aligned} \qquad (2)$$

Here m and n are arbitrary positive integers. m is the number of equations and n is the number of unknowns. The simple step of letting m and n be greater than two in passing from (1) to (2) is of major importance. Systems of type (2) occur literally in every branch of mathematics, and so-called "linear methods", whose end products are often the solutions of linear systems, constitute the most developed parts of mathematics. For example, at the end of the XIX century the theory of systems of type (2) served as a prototype for the creation of a theory of integral equations which plays a vital role in mechanics and physics. A large number of practical problems which are handled by computer also reduce to systems of type (2).

1. Terminology. Note the following efficient and convenient notation for the coefficients of (2): the a_{ij} coefficient (read "a-i-j", so that, for example, a_{12} is a one-two, never a-twelve) is the coefficient of the j-th unknown x_j in the i-th equation. The number b_i is called the free term (or constant term) of the i-th equation. The system (2) is called homogeneous if $b_i = 0$ for $i = 1, 2, \cdots, m$. Given any system (2), the linear system

$$\begin{aligned} a_{11}x_1 + a_{12}x_2 + \cdots + a_{1n}x_n &= 0 , \\ a_{21}x_1 + a_{22}x_2 + \cdots + a_{2n}x_n &= 0 , \\ \cdots\cdots\cdots\cdots\cdots\cdots\cdots\cdots \\ a_{m1}x_1 + a_{m2}x_2 + \cdots + a_{mn}x_n &= 0 \end{aligned} \qquad (2_0)$$

is called the homogeneous system associated to (2).

The coefficients of the unknowns make up a rectangular table

$$\left\|\begin{matrix} a_{11} & a_{12} & \cdots & a_{1n} \\ a_{21} & a_{22} & \cdots & a_{2n} \\ \cdot & \cdot & \cdot & \cdot \\ a_{m1} & a_{m2} & \cdots & a_{mn} \end{matrix}\right\| \tag{3}$$

which is called an $m \times n$ matrix (a square matrix if $m = n$). Such a matrix is written in abbreviated form as (a_{ij}), or else is denoted simply by the letter A. It is natural to call $(a_{i1}, a_{i2}, \cdots, a_{in})$ the i-th row of the matrix (3), and to call

$$\left\|\begin{matrix} a_{1j} \\ a_{2j} \\ \vdots \\ a_{mj} \end{matrix}\right\|$$

the j-th column. To economize on space, we shall denote a column by writing a row in brackets: $[a_{1j}, a_{2j}, \cdots, a_{mj}]$. For a square matrix we speak of its main diagonal, which consists of the elements $a_{11}, a_{22}, \cdots, a_{nn}$. A matrix (a_{ij}) all of whose elements not on the main diagonal are zero is sometimes denoted $\operatorname{diag}(a_{11}, a_{22}, \cdots, a_{nn})$ and is called a diagonal matrix. A diagonal matrix with $a_{11} = a_{22} = \cdots = a_{nn} = a$ is denoted $\operatorname{diag}_n(a)$ and is called a scalar matrix. The matrix $\operatorname{diag}_n(1)$, which has 1's on the main diagonal and zeros elsewhere, is called the identity matrix and is usually denoted E_n or simply E, when the dimension of the matrix is fixed in the discussion.

Besides the matrix (3), we also consider the extended matrix $(a_{ij} | b_i)$ of the system (2), which is obtained from (3) by adding on the column of constant terms $[b_1, b_2, \cdots, b_m]$; for clarity, this column is separated from the other columns by a vertical line.

If each of the equations in (2) becomes an identity when the unknowns x_i are replaced by numbers x_i°, then we call the set of n numbers $x_1^\circ, x_2^\circ, \cdots, x_n^\circ$ a solution of the system (2), and we call x_i° the i-th component of the solution. We also say that the n-tuple $x_1^\circ, x_2^\circ, \cdots, x_n^\circ$ of numbers satisfies all of the equations in

(2). A system which does not have any solution is called <u>incompatible</u>. If the system has a solution, it is called <u>compatible</u>, and it is called a <u>determined</u> system if it has one and only one solution. It is possible for there to be more than one solution, in which case the system is called <u>under-determined</u>. The problem of deciding when a given system is compatible, and, if it is, then what are all of its solutions, is the first series of questions we must answer.

Now let us once again look at the fourth problem in §2. Suppose we first number all of the interior vertices of the plate from 1 to 416 (the number of such vertices in Fig. 3) in an arbitrary way. We then add 204 indices for the boundary vertices, and, following the rule for computing the temperature t_i at the i-th interior vertex, we write down 416 equations of the type

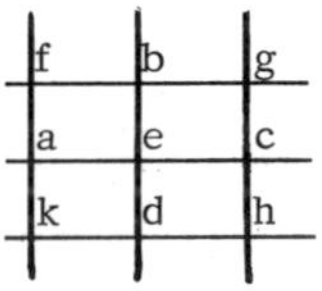

f	b	g
a	e	c
k	d	h

$$t_e = \frac{t_a + t_b + t_c + t_d}{4} .$$

Suppose, for example, that $a, b, c \leq 416$, and $d > 416$. Then this equality can be rewritten as the linear equation

$$-t_a - t_b - t_c + 4t_e = t_d ,$$

with right side $t_d = -273, -100, -50, 0, 50, 100,$ or 300. If, on the other hand, $a, b, c, d \leq 416$, then we obtain a similar equation with five t's with indices ≤ 416 on the left and 0 on the right. All of these equations taken together give a square linear system of the form (2) with $n = m = 416$. All of the coefficients a_{ij} are equal either to 0 (most of them), -1, or 4. Is this system compatible and determined? We have obtained a new, mathematically precise formulation of a qualitative problem. The question of existence and uniqueness (in this case, of a solution to the linear equations) is very typical of the questions that arise in many areas of mathematics connected with physical phenomena.

2. <u>Equivalence of linear systems.</u> Suppose that we are given another linear system having "the same size" as (2) :

$$\begin{aligned} a'_{11}x_1 + a'_{12}x_2 + \cdots + a'_{1n}x_n &= b'_1 , \\ \cdots\cdots\cdots\cdots\cdots &\cdots \\ a'_{m1}x_1 + a'_{m2}x_2 + \cdots + a'_{mn}x_n &= b'_m . \end{aligned} \tag{2'}$$

We say that the system (2') is obtained from (2) by an elementary transformation of type (I) if all of the equations in (2) except for the i-th and k-th remain the same, while the i-th and k-th equations interchange places. On the other hand, if all of the equations in (2') except the i-th are the same as in (2), while the i-th equation in (2') has the form

$$(a_{i1} + c\,a_{k1})x_1 + \cdots + (a_{in} + c\,a_{kn})x_n = b_i + c\,b_k , \tag{*}$$

where c is any number (in other words, $a'_{ij} = a_{ij} + c\,a_{kj}$, $b'_i = b_i + c\,b_k$), then we say that the system (2') is obtained from (2) by an elementary transformation of type (II).

We call two linear systems (2) and (2') equivalent if either both are incompatible, or else both are compatible and have the same solutions. Let us denote equivalence of two systems (a) and (b) as follows: (a) $\sim$ (b). Note the following properties of equivalence of linear systems: (a) $\sim$ (a), (a) $\sim$ (b) implies (b) $\sim$ (a), and (a) $\sim$ (b) and (b) $\sim$ (c) together imply (a) $\sim$ (c). The following theorem gives a sufficient condition for equivalence.

THEOREM 1. Two linear systems are equivalent if one is obtained from the other by applying a finite sequence of elementary transformations.

To prove this, it suffices to prove that two systems (2) and (2') are equivalent if (2') is obtained from (2) by applying one elementary transformation. Note that in this case (2) is also obtained from (2') by applying a single elementary transformation, since each elementary transformation has an inverse elementary transformation. In other words, in the case of type (I), if we again interchange the i-th and k-th equations, we return to the original system; and in type (II), if we add (-c) times the k-th equation in (2') to the i-th equation in (2'), we obtain the i-th equation of (2).

We now prove that any solution $(x_1^\circ, x_2^\circ, \cdots, x_n^\circ)$ of the system (2) is also a solution of the system (2'). If the elementary transformation used to obtain (2') was of type (I), then the equations have not changed at all; only the order in which they are written has changed. Hence, the numbers $x_1^\circ, x_2^\circ, \cdots, x_n^\circ$ which satisfied them before will satisfy them after the elementary transformation. Next, if the elementary transformation used to obtain (2') was of type (II), then all of the equations except for the i-th remain the same, and so the solution $(x_1^\circ, x_2^\circ, \cdots, x_n^\circ)$ satisfies these equations. As for the i-th equation, in (2') it has the form (*). Since our solution satisfies the i-th and k-th equations of (2), we have

$$a_{i1}x_1^\circ + \cdots + a_{in}x_n^\circ = b_i \quad , \quad a_{k1}x_1^\circ + \cdots + a_{kn}x_n^\circ = b_k \quad .$$

Multiplying both sides of the second equation by c and adding it to the first equation, and grouping terms as in (*), we find that (*) holds with $x_i = x_i^\circ$.

Because, as noted above, the elementary transformations are invertible, it follows that the same reasoning shows that any solution of (2') is also a solution of (2).

It remains to observe that incompatibility of one system implies incompatibility of the other. (Use proof by contradiction.) □

3. Reducing to step form. By successively applying elementary transformations, we can change a given system of equations to a system having a simpler form.

First of all, we may assume that there is at least one non-zero coefficient a_{i1} in the first column of coefficients; otherwise there would be no point in referring to the unknown x_1. If $a_{11} = 0$, use a transformation of type (I) to interchange the first equation with a j-th equation for which $a_{j1} \neq 0$. Now the coefficient of the first unknown in the first equation is non-zero. Let a'_{11} denote this coefficient. Now for each $i = 2, 3, \cdots, m$, we subtract c_i times the first equation from the i-th equation, where c_i is chosen so that, after we subtract, the coefficient of x_1 becomes 0. Obviously, the value of c_i which will do this is $c_i = a_{i1}/a'_{11}$. We thus apply $m-1$

elementary transformations of type (II). We now have a system in which x_1 only appears in the first equation.

It can sometimes happen that the second unknown x_2 also appears only in the first equation of our new system. Let x_k be the unknown with the lowest index which appears in some equation other than the first. We obtain the system

$$\begin{aligned}
a'_{11}x_1 + \cdots\cdots\cdots + a'_{1n}x_n &= b'_1 \;, \\
a'_{2k}x_k + \cdots + a'_{2n}x_n &= b'_2 \;, \\
\cdots\cdots\cdots\cdots\cdots & \\
a'_{mk}x_k + \cdots + a'_{mn}x_n &= b'_m \;, \quad k > 1 \,, \quad a'_{11} \neq 0 \;.
\end{aligned}$$

Ignoring the first equation, we now apply the same reasoning as before to the remaining equations. After several more elementary transformations, our system takes the form

$$\begin{aligned}
a''_{11}x_1 + \cdots\cdots\cdots\cdots\cdots + a''_{1n}x_n &= b''_1 \;, \\
a''_{2k}x_k + \cdots\cdots\cdots + a''_{2n}x_n &= b''_2 \;, \\
a''_{3\ell}x_\ell + \cdots + a''_{3n}x_n &= b''_3 \;, \\
\cdots\cdots\cdots\cdots\cdots & \\
a''_{m\ell}x_\ell + \cdots + a''_{mn}x_n &= b''_m
\end{aligned}$$

$$\ell > k > 1 \;, \quad a''_{11} \neq 0 \;, \quad a''_{2k} \neq 0 \;.$$

Of course, here $a''_{1j} = a'_{1j}$ and $b''_1 = b'_1$, since the first equation was not touched.

We continue to apply this procedure as long as possible. Clearly, we will have to stop when all the coefficients in the remaining equations of all the remaining unknowns up through the n-th are zero. We then finally have the system (2) in the form

$$
\begin{aligned}
\bar{a}_{11}x_1 + \cdots\cdots\cdots\cdots + \bar{a}_{1n}x_n &= \bar{b}_1 \ , \\
\bar{a}_{2k}x_k + \cdots\cdots\cdots + \bar{a}_{2n}x_n &= \bar{b}_2 \ , \\
\bar{a}_{3\ell}x_\ell + \cdots\cdots + \bar{a}_{3n}x_n &= \bar{b}_3 \ , \\
\cdots\cdots\cdots\cdots\cdots\cdots\cdots & \\
\bar{a}_{rs}x_s + \cdots + \bar{a}_{rn}x_n &= \bar{b}_r \ , \\
0 &= \bar{b}_{r+1} \ , \\
\cdots\cdots & \\
0 &= \bar{b}_m \ .
\end{aligned}
\tag{4}
$$

Here $\bar{a}_{11}, \bar{a}_{2k}, \bar{a}_{3\ell}, \cdots, \bar{a}_{rs}$ are all nonzero, $1 < k < \ell < \cdots < s$. It may happen that $r = m$, in which case the system (4) has no equations of the form $0 = \bar{b}_i$. We say that a system of equations in the form (4) has <u>step</u> form. (This is not the only common terminology: such a system is sometimes said to be in <u>trapezoidal</u> form or in <u>quasi-triangular</u> form.)

THEOREM 2. <u>Every system of linear equations is equivalent to a system in step form.</u>

The proof follows immediately from the above procedure. □

It is sometimes useful to think of the elementary transformations as applied not to the system but to its extended matrix $(a_{ij}|b_i)$. In the same way as Theorem 2 we can prove

THEOREM 2'. <u>Every matrix can be reduced to step form using elementary transformations.</u> □

4. <u>Studying a system of linear equations.</u> By virtue of Theorems 1 and 2, the questions of compatibility and determinacy need only be investigated for systems in the step form (4).

We begin with the question of compatibility. It is obvious that, if the system (4) contains an equation of the form $0 = \bar{b}_t$ with $\bar{b}_t \neq 0$, then this system is incompatible, since the equation $0 = \bar{b}_t$ cannot be satisfied by any choice of values of the unknowns. We now prove that, if there are no such equations in (4), then the system is compatible.

Thus, suppose $\bar{b}_t = 0$ for $t > r$. We call the unknowns $x_1, x_k, x_\ell, \cdots, x_s$ with which the first, second, $\cdots$, r-th equations begin principal (or pivotal) variables, and we call the remaining unknowns, if there are any, free variables. There are r principal variables in all.

We prescribe arbitrary values to the free variables and substitute these values in the equations in (4). We then obtain a single equation for x_s (the r-th) of the form $ax_s = b$ with $a = \bar{a}_{rs} \neq 0$; such an equation has a unique solution. Substituting this value $x_s = x_s^{\circ}$ in the first $r-1$ equations and continuing in this way from the bottom to the top in (4), we see that values for the principal variables are uniquely determined once we have chosen an arbitrary set of values for the free variables. We have proved

THEOREM 3. <u>A system of linear equations is compatible if and only if, after reduction to step form, it includes no equations of the form $0 = \bar{b}_t$ with $\bar{b}_t \neq 0$. If this condition holds, then the free variables can be given arbitrary values, and the values of the principal variables are uniquely determined by the system once the values of the free variables are chosen.</u> □

Assuming now that this compatibility condition holds, we explain when a system is determined. If the system (4) has free variables, then the system is automatically undetermined: we can give any values at all to the free variables, and then express the principal variables in terms of these values, by Theorem 3. But if there are no free variables -- i.e., all of the unknowns are principal variables -- then, by Theorem 3, the values of the unknowns are uniquely determined by the system; hence, the system is determined. Finally, we note that the condition that there be no free variables is equivalent to: $r = n$. We have proved the following assertion.

THEOREM 4. <u>A compatible linear system</u> (2) <u>is determined if and only if</u> $r = n$

in the system (4) in step form that is obtained from (2). □

A square linear system, i.e., for which $m = n$, after being reduced to step form, can also be written in the following triangular form:

$$\begin{aligned} \bar{a}_{11}x_1 + \bar{a}_{12}x_2 + \cdots + \bar{a}_{1n}x_n &= \bar{b}_1 , \\ \bar{a}_{22}x_2 + \cdots + \bar{a}_{2n}x_n &= \bar{b}_2 , \\ \cdots\cdots\cdots\cdots\cdots \\ \bar{a}_{nn}x_n &= \bar{b}_n , \end{aligned} \tag{5}$$

if we do not insist that $\bar{a}_{ii} \neq 0$ for all i. In fact, the form (5) merely means that the k-th equation in the system does not contain unknowns x_i with $i < k$, and this is automatically true for systems in step form.

A matrix (a_{ij}) whose elements a_{ij} are zero whenever $i > j$ is called upper triangular. We similarly define a lower triangular matrix.

Theorems 3 and 4 have some useful corollaries.

COROLLARY 1. *A linear system (2) in which $m = n$ is compatible and determined if and only if, after reduction to the step form (5), all of the $\bar{a}_{ii}$ are non-zero.* □

Notice that the condition in Corollary 1 does not depend on the right side of the system of equations. Thus, when $m = n$, the system (2) is compatible and determined if and only if the corresponding homogeneous system (2_0) is compatible and determined. But a homogeneous system is always compatible; for example, it always has the zero solution $x_1^\circ = 0$, $x_2^\circ = 0, \cdots, x_n^\circ = 0$.

The condition that all of the $\bar{a}_{ii}$ are non-zero means that the homogeneous system only has the zero solution. We thereby obtain another form of Corollary 1 not involving the step form of the system.

COROLLARY 1'. *A linear system (2) in which $m = n$ is compatible and determined if and only if the associated homogeneous system (2_0) has only the zero solution.* □

Special attention should also be given to the case $n > m$.

COROLLARY 2. A compatible system (2) with $n > m$ is never determined. In particular, a homogeneous system with $n > m$ always has a non-zero solution.

In fact, we always have $r \leq m$, since the system (4) does not have more equations than the system (2) from which it was derived. Hence, if $n > m$, it follows that $n > r$, and so, by Theorem 4, the system (2) is undetermined. It remains to note that in the case of a homogeneous system, it is undetermined if and only if it has a non-zero solution. □

Some of our results are summarized in the following table.

Type of linear system

	general	homogeneous	non-homogeneous, $n > m$	homogeneous, $n > m$
Number of solutions	$0, 1, \infty$	$1, \infty$	$0, \infty$	∞

5. Some remarks and examples. The method just given for solving systems of linear equations is called Gauss's method or the method of successive elimination. The method is very convenient for small n, and also for computer solution in the case of large n (although for a variety of reasons it is often more practical to use other methods, for example, iteration methods). This method is especially useful when the coefficients are fixed, and we are looking for a solution with a specified degree of accuracy. However, in theoretical investigations, it is often of greater importance to find compatibility or determinacy conditions for a linear system and also to find general formulas for the solutions in terms of the coefficients and the constant terms -- without reducing the system to step form. To some extent Corollary 1' is of this type (i.e., not requiring reduction to step form).

Example 1. We again return to the heated plate problem of §2. As we saw in the first subsection of §3, the question that interests us can be stated in terms of the properties of a certain very concrete linear system (which we denote the HP system), which has a rather large number of unknowns t_i. Following the criterion in Corollary 1', we consider the homogeneous linear system HHP associated to HP. In other words, we now take the temperature of all boundary vertices to be identically zero. Let e be the index of an interior vertex having <u>maximal</u> value $|t_e|$. Then the condition

$$t_e = \frac{t_a + t_b + t_c + t_d}{4}$$

implies that $|t_e| = |t_a| = |t_b| = |t_c| = |t_d|$. Moving one vertex at a time in each of the four directions, we similarly find that each vertex we pass through has $|t_i| = |t_e|$. Eventually we reach a boundary vertex, having temperature zero. Hence $t_e = 0$, and so $t_i = 0$ for all i. Thus, the system HHP has only the zero solution, and so the system HP is compatible and determined. This solves the heated plate problem: there is one and only one possible distribution of temperatures.

<u>Example 2.</u> Consider the linear system

$$\begin{array}{rcl} x_1 \ldots\ldots\ldots\ldots\ldots & = & 1 \,, \\ x_2 \ldots\ldots\ldots\ldots & = & 1 \,, \\ -x_1 - x_2 + x_3 \ldots\ldots\ldots & = & 0 \,, \\ \ldots\ldots\ldots\ldots\ldots\ldots\ldots & & \\ -x_{n-2} - x_{n-1} + x_n & = & 0 \,. \end{array}$$

This is obviously a compatible and determined system, which already has a step (triangular) form, except that it must be solved from top to bottom, instead of from bottom to top as with (5). By definition, its solution is the first n numbers in the sequence of <u>Fibonacci numbers</u> $f_1, f_2, \cdots, f_n$. These numbers are connected with a certain botanical phenomenon, called phyllotaxis (the arrangement of leaves on a stem). It would be nice to have an expression (an analytic formula) for the n-th Fibonacci number when $n = 1000$, or even for arbitrary n. You might object that, with patience, even f_{1000} can be computed using the inductive definition of these numbers. But this is not what we mean. In Chapters 2 and 3 we shall give two expressions for f_n (although, in the case of this specific problem we could proceed more directly, without waiting for the general techniques).

Remark. It is sometimes more convenient to find a solution to a linear system without reducing it to step form. This is especially the case when the matrix of the system contains many zeros. Here some practice in doing this is more useful than reading lengthy explanations.

§4. Determinants of small order

When presenting Gauss's method, we did not much care about the values of the coefficients of the principal variables. It was only important for these coefficients to be non-zero. We now do a more careful job of eliminating unknowns, at least in the case of square linear systems of small size. This will give us some food for thought, and a starting point for constructing a more general theory of determinants in Chapter 3.

As in §3, we consider a system of two equations in two unknowns

$$\begin{aligned} a_{11}x_1 + a_{12}x_2 &= b_1 , \\ a_{21}x_1 + a_{22}x_2 &= b_2 \end{aligned} \tag{1}$$

and we try to find general formulas for the components x_1°, x_2° of its solution. By the determinant of the matrix

$$\begin{Vmatrix} a_{11} & a_{12} \\ a_{21} & a_{22} \end{Vmatrix}$$

we mean the expression $a_{11}a_{22} - a_{21}a_{12}$; we denote the determinant as follows: $\begin{vmatrix} a_{11} & a_{12} \\ a_{21} & a_{22} \end{vmatrix}$. To every square 2×2 matrix we thereby associate a number

$$\begin{vmatrix} a_{11} & a_{12} \\ a_{21} & a_{22} \end{vmatrix} = a_{11}a_{22} - a_{21}a_{12} . \tag{2}$$

If we try to eliminate x_2 from the system (1) by multiplying the first equation by a_{22}

and adding it to $(-a_{12})$ times the second equation, we obtain

$$\begin{vmatrix} a_{11} & a_{12} \\ a_{21} & a_{22} \end{vmatrix} x_1 = b_1 a_{22} - b_2 a_{12} \quad .$$

The right side is nothing other than the determinant of the matrix $\begin{Vmatrix} b_1 & a_{12} \\ b_2 & a_{22} \end{Vmatrix}$. We suppose that $\begin{vmatrix} a_{11} & a_{12} \\ a_{21} & a_{22} \end{vmatrix} \neq 0$. We then have

$$x_1 = \frac{\begin{vmatrix} b_1 & a_{12} \\ b_2 & a_{22} \end{vmatrix}}{\begin{vmatrix} a_{11} & a_{12} \\ a_{21} & a_{22} \end{vmatrix}} \quad \text{and similarly} \quad x_2 = \frac{\begin{vmatrix} a_{11} & b_1 \\ a_{21} & b_2 \end{vmatrix}}{\begin{vmatrix} a_{11} & a_{12} \\ a_{21} & a_{22} \end{vmatrix}} \quad . \tag{3}$$

Once we have formulas for finding the solutions of a system of two equations with two unknowns, we can also solve certain other systems. For example, consider a system of two homogeneous equations with three unknowns:

$$\begin{aligned} a_{11}x_1 + a_{12}x_2 + a_{13}x_3 &= 0 \quad , \\ a_{21}x_1 + a_{22}x_2 + a_{23}x_3 &= 0 \quad . \end{aligned} \tag{4}$$

We are interested in finding a non-zero solution of this system, i.e., a solution for which at least one $x_i \neq 0$. Suppose, for example, that $x_3 \neq 0$. Dividing both sides of the two equations by $-x_3$ and setting $y_1 = -x_1/x_3$ and $y_2 = -x_2/x_3$, we rewrite (4) in the same form as (1):

$$\begin{aligned} a_{11}y_1 + a_{12}y_2 &= a_{13} \quad , \\ a_{21}y_1 + a_{22}y_2 &= a_{23} \quad . \end{aligned}$$

If we assume that $\begin{vmatrix} a_{11} & a_{12} \\ a_{21} & a_{22} \end{vmatrix} \neq 0$, then the formulas (3) give

$$y_1 = -\frac{x_1}{x_3} = \frac{\begin{vmatrix} a_{13} & a_{12} \\ a_{23} & a_{22} \end{vmatrix}}{\begin{vmatrix} a_{11} & a_{12} \\ a_{21} & a_{22} \end{vmatrix}} \quad , \quad y_2 = -\frac{x_2}{x_3} = \frac{\begin{vmatrix} a_{11} & a_{13} \\ a_{21} & a_{23} \end{vmatrix}}{\begin{vmatrix} a_{11} & a_{12} \\ a_{21} & a_{22} \end{vmatrix}} \quad .$$

It is not surprising that, starting with (4), we determined not x_1, x_2, x_3 themselves but rather their ratios. We immediately see from the homogeneity of the system that, if $(x_1^\circ, x_2^\circ, x_3^\circ)$ is a solution and c is any number, then $(cx_1^\circ, cx_2^\circ, cx_3^\circ)$ is also a solution. Thus, we can set

$$x_1 = -\begin{vmatrix} a_{13} & a_{12} \\ a_{23} & a_{22} \end{vmatrix}, \quad x_2 = -\begin{vmatrix} a_{11} & a_{13} \\ a_{21} & a_{23} \end{vmatrix}, \quad x_3 = \begin{vmatrix} a_{11} & a_{12} \\ a_{21} & a_{22} \end{vmatrix} \tag{5}$$

and say that any solution is obtained from this solution by multiplying all of the x_i by some number c. We can give these formulas a more symmetric appearance if we note that always

$$\begin{vmatrix} a & b \\ c & d \end{vmatrix} = -\begin{vmatrix} b & a \\ d & c \end{vmatrix},$$

as is clear from (2). Hence, (5) can be written in the form

$$x_1 = \begin{vmatrix} a_{12} & a_{13} \\ a_{22} & a_{23} \end{vmatrix}, \quad x_2 = -\begin{vmatrix} a_{11} & a_{13} \\ a_{21} & a_{23} \end{vmatrix}, \quad x_3 = \begin{vmatrix} a_{11} & a_{12} \\ a_{21} & a_{22} \end{vmatrix} \tag{6}$$

These formulas were derived under the assumption that $\begin{vmatrix} a_{11} & a_{12} \\ a_{21} & a_{22} \end{vmatrix} \neq 0$. But it is not hard to see that, as long as at least one of the determinants in (6) is non-zero, it is still true that the solutions of (4) are precisely the multiples of the triple in (6). However, if all three determinants are zero, then, while (6) still gives a solution of (4) (the zero solution), it is no longer the case that all solutions can be obtained from (6) by multiplying the three determinants by some number. For example, consider the system consisting of two identical equations $x_1 + x_2 + x_3 = 0$.

We now proceed to the case of a system of three equations with three unknowns:

$$\begin{aligned} a_{11}x_1 + a_{12}x_2 + a_{13}x_3 &= b_1, \\ a_{21}x_1 + a_{22}x_2 + a_{23}x_3 &= b_2, \\ a_{31}x_1 + a_{32}x_2 + a_{33}x_3 &= b_3. \end{aligned}$$

We would like to eliminate x_2 and x_3 from this system, in order to obtain a value for

x_1. To do this we multiply the first equation by c_1, the second by c_2, and the third by c_3, and add them. We choose c_1, c_2, c_3 in such a way that the resulting equation has zero coefficient of x_2 and x_3. Setting these two coefficients equal to zero gives the following system of equations for the unknowns c_1, c_2, c_3:

$$a_{12}c_1 + a_{22}c_2 + a_{32}c_3 = 0 ,$$
$$a_{13}c_1 + a_{23}c_2 + a_{33}c_3 = 0 .$$

These equations are of the same type as (4). Hence we can take

$$c_1 = \begin{vmatrix} a_{22} & a_{32} \\ a_{23} & a_{33} \end{vmatrix} , \quad c_2 = - \begin{vmatrix} a_{12} & a_{32} \\ a_{13} & a_{33} \end{vmatrix} , \quad c_3 = \begin{vmatrix} a_{12} & a_{22} \\ a_{13} & a_{23} \end{vmatrix} .$$

After using these values of c_1, c_2, c_3 to combine the three equations, we obtain the following equation for x_1:

$$\left(a_{11} \begin{vmatrix} a_{22} & a_{23} \\ a_{32} & a_{33} \end{vmatrix} - a_{21} \begin{vmatrix} a_{12} & a_{13} \\ a_{32} & a_{33} \end{vmatrix} + a_{31} \begin{vmatrix} a_{12} & a_{13} \\ a_{22} & a_{23} \end{vmatrix} \right) x_1 = \tag{7}$$
$$= b_1 \begin{vmatrix} a_{22} & a_{23} \\ a_{32} & a_{33} \end{vmatrix} - b_2 \begin{vmatrix} a_{12} & a_{13} \\ a_{32} & a_{33} \end{vmatrix} + b_3 \begin{vmatrix} a_{12} & a_{13} \\ a_{22} & a_{23} \end{vmatrix} .$$

The coefficient of x_1 in (7) is called the determinant of the matrix $\begin{Vmatrix} a_{11} & a_{12} & a_{13} \\ a_{21} & a_{22} & a_{23} \\ a_{31} & a_{32} & a_{33} \end{Vmatrix}$ and is denoted $\begin{vmatrix} a_{11} & a_{12} & a_{13} \\ a_{21} & a_{22} & a_{23} \\ a_{31} & a_{32} & a_{33} \end{vmatrix}$.

Thus, we take the third order determinant to be the expression

$$\begin{vmatrix} a_{11} & a_{12} & a_{13} \\ a_{21} & a_{22} & a_{23} \\ a_{31} & a_{32} & a_{33} \end{vmatrix} = a_{11} \begin{vmatrix} a_{22} & a_{23} \\ a_{32} & a_{33} \end{vmatrix} - a_{21} \begin{vmatrix} a_{12} & a_{13} \\ a_{32} & a_{33} \end{vmatrix} + a_{31} \begin{vmatrix} a_{12} & a_{13} \\ a_{22} & a_{23} \end{vmatrix} = \tag{8}$$
$$= a_{11}a_{22}a_{33} + a_{12}a_{23}a_{31} + a_{13}a_{21}a_{32} - a_{11}a_{23}a_{32} - a_{12}a_{21}a_{33} - a_{13}a_{22}a_{31} ,$$

which we have defined using second order determinants. Now notice that the right side in

(7) can be obtained from the coefficient of x_1 on the left by replacing a_{11} by b_1, a_{21} by b_2, and a_{31} by b_3. Hence, equation (7) can be written in the form

$$\begin{vmatrix} a_{11} & a_{12} & a_{13} \\ a_{21} & a_{22} & a_{23} \\ a_{31} & a_{32} & a_{33} \end{vmatrix} x_1 = \begin{vmatrix} b_1 & a_{12} & a_{13} \\ b_2 & a_{22} & a_{23} \\ b_3 & a_{32} & a_{33} \end{vmatrix} .$$

Suppose that the coefficient of x_1 here is non-zero. Then, if we carry out analogous computations for x_2 and x_3, we arrive at the formulas

$$x_1 = \frac{\begin{vmatrix} b_1 & a_{12} & a_{13} \\ b_2 & a_{22} & a_{23} \\ b_3 & a_{32} & a_{33} \end{vmatrix}}{\begin{vmatrix} a_{11} & a_{12} & a_{13} \\ a_{21} & a_{22} & a_{23} \\ a_{31} & a_{32} & a_{33} \end{vmatrix}}, \quad x_2 = \frac{\begin{vmatrix} a_{11} & b_1 & a_{13} \\ a_{21} & b_2 & a_{23} \\ a_{31} & b_3 & a_{33} \end{vmatrix}}{\begin{vmatrix} a_{11} & a_{12} & a_{13} \\ a_{21} & a_{22} & a_{23} \\ a_{31} & a_{32} & a_{33} \end{vmatrix}}, \quad x_3 = \frac{\begin{vmatrix} a_{11} & a_{12} & b_1 \\ a_{21} & a_{22} & b_2 \\ a_{31} & a_{32} & b_3 \end{vmatrix}}{\begin{vmatrix} a_{11} & a_{12} & a_{13} \\ a_{21} & a_{22} & a_{23} \\ a_{31} & a_{32} & a_{33} \end{vmatrix}} \qquad (9)$$

Clearly, the same reasoning can be applied to a system of four, five, and so on equations in an equal number of unknowns. To treat the case of four equations, we must first derive formulas similar to (6) for the solutions of a homogeneous system of three equations with four unknowns; then we eliminate x_2, x_3, x_4 in the system of four equations with four unknowns by multiplying the equations by c_1, c_2, c_3, c_4 and adding them. We find the values of the c_i $(i = 1, 2, 3, 4)$ by solving a system of three homogeneous equations.

By analogy with (8), we define the fourth order determinant to be the coefficient of x_1 in the resulting equation; it will be built up from third order determinants. Carrying out the same procedure for x_2, x_3, x_4, we find formulas analogous to (9) for the x_i.

We can continue in this way indefinitely. We can be sure that we will eventually be able to solve systems of n equations with n unknowns for any n, because of a principle that is widely used in mathematics: the principle of mathematical induction (see §7).

EXERCISES

1. Formula (8) can be easily remembered if one uses a visual device which gives the rule for the sign of the products which occur in the third order determinant (see Fig. 4). Find a similar visual rule for the sign in the fourth order determinant.

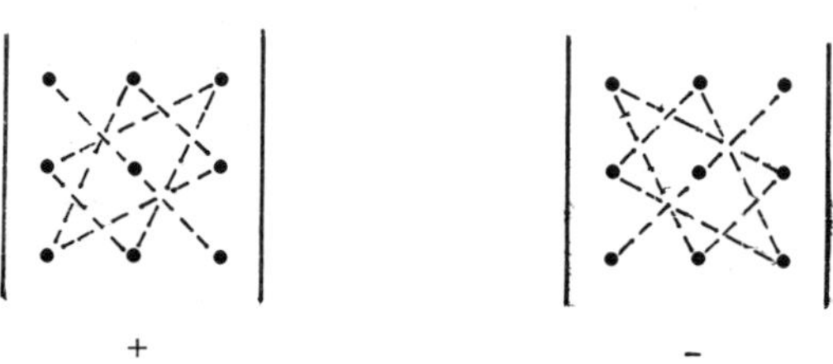

Fig. 4

2. Show that it is impossible for all six terms in the expansion of the third order determinant to be simultaneously positive.

3. The square of the area of the parallelogram which is constructed using the vectors from the origin to the points P, Q with rectangular coordinates (α, β) and (γ, δ) (see Fig. 5), is given by the formula

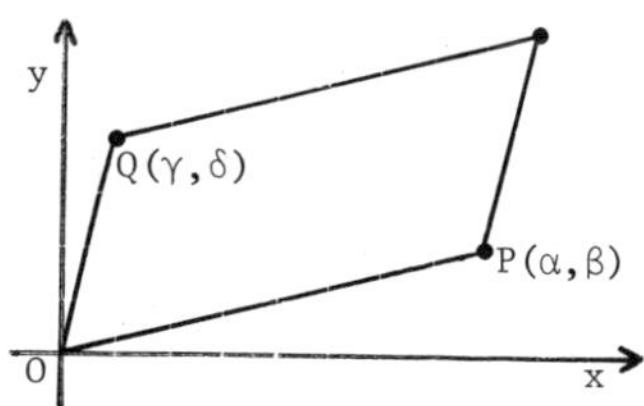

$$\Delta^2 = \begin{vmatrix} \alpha^2+\beta^2 & \alpha\gamma+\beta\delta \\ \alpha\gamma+\beta\delta & \gamma^2+\delta^2 \end{vmatrix}.$$

Fig. 5

(This is easy to see, if one changes to a coordinate system in which P lies on the x-axis; one must check that the determinant on the right does not change under a change of coordinates.) Find a similar expression for the square of the volume of a parallelopiped in three-dimensional space; use a third order determinant.

§5. Sets and mappings

In the preceding two sections we have encountered various sorts of sets of elements, and various sorts of mappings between sets. The set of solutions of a given system of linear equations, the rule which associates to every 2×2 matrix its determinant -- these are only special cases of certain formal notions with which it is important to become familiar, at least on an intuitive level, as soon as possible.

1. Sets. By a set we mean a collection of objects, which are called the elements of the set. A set with finitely many elements can be described by explicitly enumerating all of its elements; these elements are usually enclosed in braces. For example $\{1,2,4,8\}$ is the set of powers of two between 1 and 10. As a rule, a set is denoted by a capital letter in some alphabet, and an element in a set is denoted by a small letter in the same or another alphabet. Certain designations for some of the most important sets have become standard, and should be consistently used. Thus, the letters $\mathbb{N}$, $\mathbb{Z}$, $\mathbb{Q}$, $\mathbb{R}$ denote the set of positive integers (natural numbers), the set of all integers, the set of rational numbers, and the set of real numbers, respectively. For a given set S, the symbol $a \in S$ means that a is an element of the set S; if a is not an element of S, we write $a \notin S$. We say that S is a subset of a set T or write $S \subset T$ (S is contained in T), if we have the implication

$$\forall x \, , \; x \in S \Longrightarrow x \in T \quad .$$

(Concerning this notation, see "Advice to the Reader" at the beginning of the book.) Two sets S and T are said to coincide (to be equal) if they have the same elements. Symbolically:

$$S = T \Longleftrightarrow S \subset T \text{ and } T \subset S \quad .$$

($\Longleftrightarrow$ means "if and only if", i.e., "two-way implication".) By definition, the empty set ϕ, which is the set without any elements, is a subset of every set. If $S \subset T$, but $S \neq \phi$ and $S \neq T$, then S is called a proper subset of T. Subsets $S \subset T$ are often defined by giving a property which only elements of S possess. For example,

$$\{n \in \mathbb{Z} \,|\, n = 2m \text{ for some } m \in \mathbb{Z}\}$$

is the set of all even integers, and

$$\mathbb{N} = \{n \in \mathbb{Z} \mid n > 0\}$$

is the set of natural numbers.

By the <u>intersection</u> of two sets S and T we mean the set

$$S \cap T = \{x \mid x \in S \text{ and } x \in T\} \quad ,$$

and by their <u>union</u> we mean the set

$$S \cup T = \{x \mid x \in S \text{ or } x \in T\} \quad .$$

The intersection $S \cap T$ might be the empty set. In that case we say that S and T are <u>disjoint</u> sets. The operations of intersection and union satisfy the identities

$$R \cap (S \cup T) = (R \cap S) \cup (R \cap T) \quad ,$$

$$R \cup (S \cap T) = (R \cup S) \cap (R \cup T) \quad ,$$

the verification of which we leave to the reader as an exercise. The diagrams

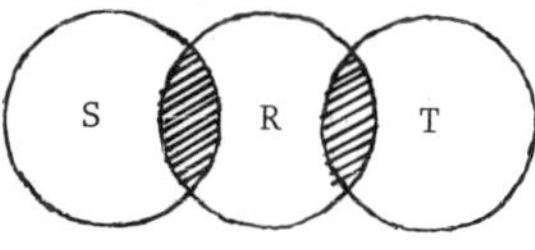

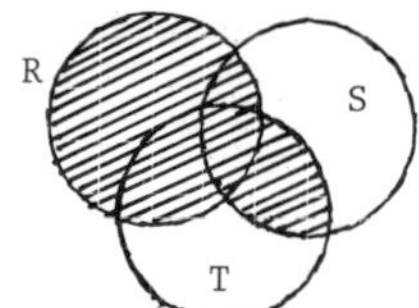

will help the reader think through the simple arguments.

By the <u>difference</u> $S \setminus T$ of the sets S and T we mean the set of all elements of S which are not elements of T. Here we do not require $T \subset S$. The notation $S - T$ is sometimes used instead of $S \setminus T$.

If T is a subset of S, then the difference $S \setminus T$ is also called the <u>complement</u> of T in S. If we set $R = S \setminus T$, then we have: $R \cap T = \phi$, $R \cup T = S$. Notice that there is a correspondence between the operations of intersection, union, and complement, and the logical connectives "and", "or", "not".

Now let X and Y be arbitrary sets. A pair of elements (x, y), where $x \in X$ and $y \in Y$, which is taken in a definite order, is called an <u>ordered pair.</u> We consider two ordered pairs (x_1, y_1) and (x_2, y_2) to be equal if and only if $x_1 = x_2$ and $y_1 = y_2$. The <u>cartesian product</u> of two sets X and Y is the set of all ordered pairs (x, y):

$$X \times Y = \{(x, y) \mid x \in X, \ y \in Y\} \quad .$$

For example, let $\mathbb{R}$ be the set of all real numbers. Then the cartesian product $\mathbb{R}^2 = \mathbb{R} \times \mathbb{R}$ is simply the set of all of the cartesian coordinates of the points on the plane relative to a fixed choice of coordinate axes.

In a similar way we can introduce the cartesian product $X_1 \times X_2 \times X_3$ of three sets (this is $(X_1 \times X_2) \times X_3$, or equivalently $X_1 \times (X_2 \times X_3)$), the cartesian product of four sets, and so on. If $X_1 = X_2 = \cdots = X_k$, we abbreviate $X^k = X \times X \times \cdots \times X$, and call this the k-th cartesian power of the set X. The elements of X^k are sequences (rows) of length k: $(x_1, x_2, \cdots, x_k)$.

In order to get a feeling for the difference between the sets $X \times Y$ and $X \cup Y$, we take the case when X and Y are sets with finitely many elements (of finite cardinality; the number of elements in a set is called its "cardinality" and is denoted Card or $|\ |$):

$$|X| = \text{Card } X = n, \quad |Y| = \text{Card } Y = m \quad .$$

Then

$$|X \times Y| = nm, \quad \text{while} \quad |X \cup Y| = n + m - |X \cap Y| \quad .$$

If these equalities are not immediately clear, the reader should carefully reread all of the definitions.

2. Mappings. The notion of a function or mapping (also: "map") plays a central role in mathematics. Given two sets X and Y, a mapping f with domain of definition X and range of values Y associates to every element $x \in X$ an element $f(x) \in Y$, which can also be denoted fx. In the case $Y = X$ we also call f a transformation of the set X to itself. A mapping is written symbolically in the form $f : X \to Y$ or $X \xrightarrow{f} Y$. The image of a mapping f is the set of all elements of the form $f(x)$:

$$\text{Im } f = \{f(x) \mid x \in X\} = f(X) \subset Y \quad .$$

The set

$$f^{-1}(y) = \{x \in X \mid f(x) = y\}$$

is called the preimage of the element $y \in Y$. More generally, for $Y_0 \subset Y$ we set

$$f^{-1}(Y_0) = \{x \in X \mid f(x) \in Y_0\} = \bigcup_{y \in Y_0} f^{-1}(y) \quad .$$

If $y \in Y \setminus \mathrm{Im}\, f$, then obviously $f^{-1}(y) = \phi$.

A mapping $f : X \to Y$ is called a surjective or an onto mapping, if $\mathrm{Im}\, f = Y$; it is called an injective mapping if $x \neq x'$ implies $f(x) \neq f(x')$. Finally, $f : X \to Y$ is called a bijective mapping or a one-to-one correspondence if it is both surjective and injective.

To say that two mappings f and g are equal means that their domains and ranges are the same: $X \xrightarrow{f} Y$, $X \xrightarrow{g} Y$, and that $f(x) = g(x)$, $\forall x$. The symbol $x \mapsto f(x)$ denotes the correspondence of a value $f(x) \in Y$ to the "argument" x, i.e., to an element $x \in X$.

To take an example, let f_n be the n-th Fibonacci number (see §4). The correspondence $n \mapsto f_n$ gives a mapping $\mathbb{N} \to \mathbb{N}$. The mapping is obviously not surjective, and is also not injective, since $f_1 = f_2 = 1$. Another example: if $\mathbb{R}_+$ is the set of non-negative real numbers, then the mappings $f : \mathbb{R} \to \mathbb{R}$, $g : \mathbb{R} \to \mathbb{R}_+$, $h : \mathbb{R}_+ \to \mathbb{R}_+$ defined by the same rule $x \mapsto x^2$, are all different mappings. Here f is neither surjective nor injective; g is surjective but not injective; and h is bijective. Thus, the specification of the domain of definition and range of values is an essential part of defining a mapping (function).

The identity mapping $e_X : X \to X$ is the mapping which takes every element $x \in X$ to itself. If X is a subset of $Y : X \subset Y$, it is sometimes useful to consider the inclusion mapping $I : X \to Y$, which takes every element $x \in X$ to the same element, now regarded as an element of Y. A mapping $f : X \to Y$ is called a restriction of the map $g : X' \to Y'$ if $X \subset X'$, $Y \subset Y'$, and $f(x) = g(x)$, $\forall x \in X$. In this situation g is called an extension of f. For example, the inclusion $I : X \to Y$ is a restriction of the identity mapping $e_Y : Y \to Y$.

We shall also have occasion to speak of functions of several variables. It is worthwhile to convince oneself that, if we use the notion of a cartesian power X^n of a set X

(see above), we can then treat a function $f(x_1, \cdots, x_n)$ of several variables $x_i \in X$, $i = 1, \cdots, n$, as an ordinary function $f : X^n \to Y$ of one variable $x \in X^n$.

The <u>product</u> (<u>composition</u>) of two mappings $g : U \to V$ and $f : V \to W$ is the mapping $f \circ g : U \to W$ which is defined by

$$(f \circ g)(u) = f(g(u)) \quad , \quad \forall u \in U \quad .$$

This definition can be depicted visually by means of the <u>triangular diagram</u>

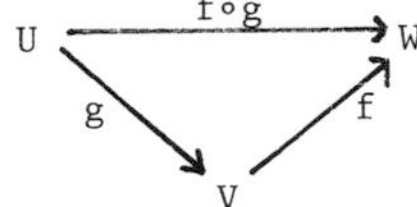

We say that this diagram "commutes" (or "is commutative"), i.e., the result of going from U to W does not depend on whether we go directly using $f \circ g$ or via V, using f and g. Note that the composition is not defined for just any mappings f and g. In the above notation, it is necessary that the same set V be both the range of g and the domain of f. The composition of two mappings from a set X to itself always makes sense.

We shall henceforth write simply fg instead of $f \circ g$.

An obvious verification shows that for any mapping $f : X \to Y$ we have

$$f e_X = f \quad , \quad e_Y f = f \quad .$$

An important property of the composition of mappings is given in the following

THEOREM 1. <u>Composition obeys the associative law. This means that, if</u> $h : U \to V$, $g : V \to W$, <u>and</u> $f : W \to T$ <u>are three mappings, then</u>

$$f(gh) = (fg)h \quad .$$

<u>Proof.</u> The necessary argument is expressed in the following diagram:

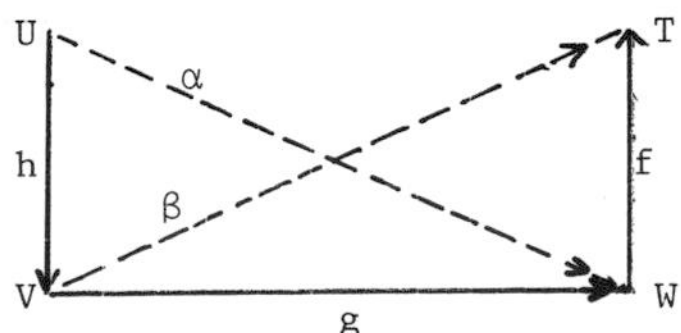

where $\alpha = gh$, $\beta = fg$. According to the definition of equality of mappings, we need only compare the values of the mappings $f(gh) : U \to T$ and $(fg)h : U \to T$ at an arbitrary element $u \in U$. But, by definition of composition, we have

$$(f(gh))u = f((gh)u) = f(g(hu)) = (fg)(hu) = ((fg)h)u \quad . \qquad \square$$

In general, the composition of mappings $X \to X$ is not commutative, i.e., $fg \neq gf$. One can see right away from the example of a two-element set $X = \{a,b\}$ with $f(a) = b$, $f(b) = a$, $g(a) = a$, $g(b) = a$. Another example: let f and g be constant mappings from X to X , i.e., the values $f(x)$ and $g(x)$ do not depend on x . Then $f \neq g \Rightarrow fg \neq gf$.

Some functions have inverses. Suppose that $f : X \to Y$ and $g : Y \to X$ are any two mappings; then the compositions fg and gf are defined. If $fg = e_Y$, then f is called a left inverse of g , and g is called a right inverse of f . If the product in either order is the identity map:

$$fg = e_Y \quad , \quad gf = e_X \quad , \tag{1}$$

then we call g the two-sided inverse (or, simply, the inverse) of f (in which case f is the inverse of g) , and we denote g by the symbol f^{-1} . Thus, $f(u) = v \Leftrightarrow f^{-1}(v) = u$.

If there were another mapping $g' : Y \to X$ for which

$$fg' = e_Y \quad , \quad g'f = e_X \quad , \tag{1'}$$

we could conclude, using (1), (1') , and Theorem 1, that

$$g' = e_X g' = (gf)g' = g(fg') = ge_Y = g \quad .$$

Thus, whenever a two-sided inverse of f exists, it is unique. Thus, the notation f^{-1} is unambiguous.

THEOREM 2. A mapping $f : X \to Y$ has an inverse if and only if it is bijective.

The proof of the theorem is based on the following lemma, which is useful in its own right.

LEMMA. *If* $f : X \to Y$ *and* $g : Y \to X$ *are any mappings for which* $gf = e_X$, *then* f *is injective and* g *is surjective.*

To prove the lemma, first suppose that $x, x' \in X$ and $f(x) = f(x')$. Then $x = e_X(x) = (gf)x = g(fx) = g(fx') = (gf)x' = e_X(x') = x'$. Thus, f is injective. Next, if x is any element of X, then $x = e_X(x) = (gf)x = g(fx)$, and this proves that g is surjective.

Returning to Theorem 2, first suppose that f has an inverse $g = f^{-1}$. Then the equations (1) and the lemma give both injectivity and surjectivity of f. In other words, f is bijective. Conversely, if we suppose that f is bijective, for any $y \in Y$ we can find a *unique* element $x \in X$ for which $f(x) = y$. Setting $g(y) = x$, we define a mapping $g : Y \to X$ having the properties (1). Thus, $f^{-1} = g$. □

COROLLARY. *If* $f : X \to Y$ *is bijective, then* f^{-1} *is also bijective, and*

$$(f^{-1})^{-1} = f \quad . \tag{2}$$

Further, suppose that $f : X \to Y$ *and* $h : Y \to Z$ *are bijective mappings. Then the composition* hf *is also bijective, and*

$$(hf)^{-1} = f^{-1}h^{-1} \quad . \tag{3}$$

Proof. By Theorem 2, the bijectivity of f implies the existence of f^{-1}. Then symmetry of the conditions in (1), written in the form $ff^{-1} = e_Y$, $f^{-1}f = e_X$, shows that f is the inverse of f^{-1}, which must then be bijective, by Theorem 2. Next, by Theorem 2 and the hypothesis, we have inverse mappings $f^{-1} : Y \to X$, $h^{-1} : Z \to Y$ and their composition $f^{-1}h^{-1} : Z \to X$. The equations

$$(hf)(f^{-1}h^{-1}) = ((hf)f^{-1})h^{-1} = (h(ff^{-1}))h^{-1} = hh^{-1} = e_Z \quad ,$$

$$(f^{-1}h^{-1})(hf) = f^{-1}(h^{-1}(hf)) = f^{-1}((h^{-1}h)f) = f^{-1}f = e_X$$

imply that $f^{-1}h^{-1}$ is the inverse of hf. □

The mapping $\sigma : \mathbb{N} \to \mathbb{N}$ defined by $\sigma(n) = n+1$ is injective but not surjective, since 1 does not belong to $\operatorname{Im}\sigma$. It is interesting that this cannot happen with finite sets.

THEOREM 3. If X is a finite set and the mapping $f : X \to X$ is injective, then it is bijective.

Proof. We need only show that f is surjective, i.e., for any element $x \in X$ we must find x' with $f(x') = x$. Set

$$f^k(x) = f(f \cdots (fx) \cdots) = f(f^{k-1}x) \quad , \quad k = 1, 2, \cdots \quad .$$

Since X is finite, there must be repetitions in this sequence of elements, say, $f^m(x) = f^n(x)$, $m > n$. If $n > 0$, then, because $f(f^{m-1}x) = f(f^{n-1}x)$ and f is injective, we must have $f^{m-1}(x) = f^{n-1}(x)$. Canceling f in this way n times, we obtain an element $x' = f^{m-n-1}(x)$ with the required property: $f(x') = x$. □

It is similarly easy to see that a surjective mapping of a finite set to itself must be bijective.

A few words on cardinality. We say that two sets X and Y have the same cardinality if and only if there exists a bijective mapping $f : X \to Y$. Sets with the same cardinality as $\mathbb{N}$ (or $\mathbb{Z}$) are called countable.

EXERCISES

1. Let $\Omega = \{+, -, ++, +-, -+, --, +++, \cdots\}$ be the set of all finite sequences of pluses and minuses, and let $f : \Omega \to \Omega$ be the mapping which takes an element $\omega = \omega_1\omega_2 \cdots \omega_n \in \Omega$ to the element $\omega' = \omega_1\bar{\omega}_1\omega_2\bar{\omega}_2 \cdots \omega_n\bar{\omega}_n$, where $\bar{\omega}_k = -$ if $\omega_k = +$ and $\bar{\omega}_k = +$ if $\omega_k = -$. Show that any interval of length > 4 in $f(f\omega)$ contains $++$ or $--$.

2. Does the mapping $f : \mathbb{N} \to \mathbb{N}$ given by $n \mapsto n^2$ have a right inverse? Find two mappings which are both left inverses for f.

3. Let $f : X \to Y$ be a mapping, and let S and T be subsets of X. Show that

$$f(S \cup T) = f(S) \cup f(T) \quad , \quad f(S \cap T) \subset f(S) \cap f(T) \quad .$$

Give an example showing that the second inclusion cannot, in general, be replaced by equality.

4. Let the symbol $\mathcal{P}(S) = \{T \mid T \subset S\}$ denote the set of all subsets of S. For example, if $S = \{s_1, s_2, \cdots, s_n\}$ is a finite set with n elements, then $\mathcal{P}(S)$ consists of the empty set ϕ, the n one-element sets $\{s_1\}, \{s_2\}, \cdots, \{s_n\}$, the $n(n-1)/2$ two-element sets $\{s_i, s_j\}$, $1 \leq i < j \leq n$, and so on, until we reach $T = S$. What is the cardinality of the set $\mathcal{P}(S)$?

5. Let $f : X \to Y$ be a mapping, and let $b = f(a)$ for some $a \in X$. The pre-image

$$f^{-1}(b) = f^{-1}(f(a)) = \{x \mid f(x) = f(a)\}$$

is sometimes called the <u>fibre</u> over the element $b \in \operatorname{Im} f$. Show that the set X is a disjoint union of fibres, i.e., that the fibres give a partition of X. (WARNING: the symbol $f^{-1}(b)$ should not be thought of as referring to an inverse mapping, since an inverse may not exist.)

6. Show that a finite cartesian power of a countable set is itself a countable set.

7. The symbol $S \Delta T$ denotes the symmetric difference of the two sets S and T:

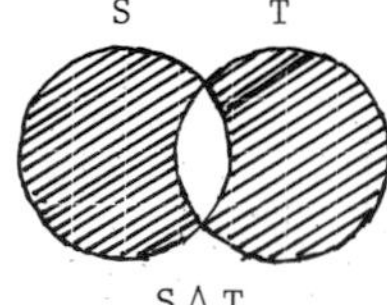

$$S \Delta T = (S \setminus T) \cup (T \setminus S) \quad .$$

Show that

$$S \Delta T = (S \cup T) \setminus (S \cap T) \quad .$$

§6. Equivalence relations. Quotient maps

The idea of equivalence of systems of linear equations, which we introduced in §3 , leads to the thought of introducing such a concept in our general setting, especially since various types of equivalence are used, often unconsciously, both in logical reasoning and in daily life.

1. Binary relations. Given two sets X and Y, any subset $O \subset X \times Y$ is called a binary relation between X and Y (or simply a binary relation on X, if $Y = X$). If an ordered pair (x,y) is an element of O, we use the notation xOy and say that x has the relation O to y. This notation is useful, since, for example, the ordering "<" on the set of real numbers $\mathbb{R}$ is the binary relation on $\mathbb{R}$ consisting of all points of the plane $\mathbb{R}^2$ which lie above the line $y = x$ (see Fig. 6); in this case the cumbersome notation

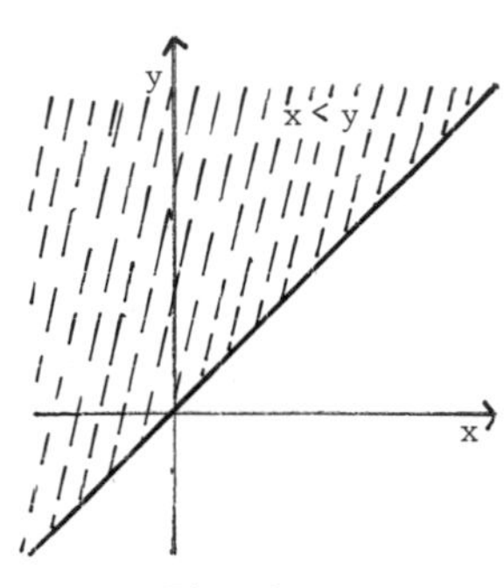

Fig. 6

$$(x,y) \in O \quad (O = <)$$

can be replaced by the usual inequality $x < y$.

To every function $f : X \to Y$ we associate its graph, which is the subset

$$\Gamma(f) = \{(x,y) | x \in X, y = f(x)\} \subset X \times Y \quad .$$

The graph $\Gamma(f)$ is a binary relation between X and Y. The graphs in $\mathbb{R}^2$ of functions $\mathbb{R} \to \mathbb{R}$ are studied in calculus courses. It is clear that not every binary relation O can be the graph of a function. A binary relation O is the graph of some function from X to Y if and only if for every $x \in X$ there is exactly one y with xOy.

Specifying X, Y, and the graph $\Gamma(f)$ is enough to reconstruct the function f.

2. Equivalence relations. A binary relation $\sim$ on X is called an equivalence relation if the following conditions hold for all $x, x', x'' \in X$:

(i) $x \sim x$ (reflexivity);

(ii) $x \sim x' \Rightarrow x' \sim x$ (symmetry);

(iii) $x \sim x'$ and $x' \sim x'' \Rightarrow x \sim x''$ (transitivity).

The notation $a \not\sim b$ means that the elements $a, b \in X$ are not equivalent. The subset

$$\bar{x} = \{x' \in X \mid x' \sim x\} \subset X$$

of all elements equivalent to a given x is called the <u>equivalence class</u> containing x. Since $x \sim x$ by (i), we do have $x \in \bar{x}$. Any element $x' \in \bar{x}$ is called a <u>representative</u> of the class $\bar{x}$. We have the following fact:

<u>Definition.</u> If a set X is a disjoint union of subsets X_i, we say that $\{X_i\}$ is a <u>partition</u> of X.

PROPOSITION. <u>The set of $\sim$ -equivalence classes is a partition of the set</u> X. (This partition can be denoted by the symbol $\pi_{\sim}(X)$.)

<u>Proof.</u> For any $x \in X$ we have $x \in \bar{x}$; hence, $X = \bigcup_{x \in X} \bar{x}$. Next, a class $\bar{x}$ is uniquely determined by any representative in it, i.e., $\bar{x} = \bar{x}' \iff x \sim x'$. In one direction: $x \sim x'$ and $x'' \in \bar{x} \Rightarrow x'' \sim x \Rightarrow x'' \sim x' \Rightarrow x'' \in \bar{x}' \Rightarrow \bar{x} \subset \bar{x}'$. But $x \sim x' \Rightarrow x' \sim x$ by (ii); hence, the reverse inclusion also holds: $\bar{x}' \subset x$. Thus, $\bar{x} = \bar{x}'$. In the other direction: since $x \in \bar{x}$, we have $\bar{x}' = \bar{x} \Rightarrow x \in \bar{x}' \Rightarrow x \sim x'$.

Now suppose $\bar{x}' \cap \bar{x}'' \neq \phi$. If $x \in \bar{x}' \cap \bar{x}''$, then $x \sim x'$ and $x \sim x''$, so that, by (ii) and (iii), we have $x' \sim x''$, hence $\bar{x}' = \bar{x}''$. Thus, distinct classes are disjoint. $\square$

Let $\Pi = \mathbb{R}^2$ be the real plane with rectangular coordinates. If we take $\sim$ between two points $P, P' \in \Pi$ to mean that P and P' lie on the same horizontal line, we obviously have an equivalence relation whose equivalence classes are the horizontal lines (Fig. 7). Similarly, the hyperbolas Γ_ρ (Fig. 8) of the form $xy = \rho$, where $\rho > 0$, determine an equivalence relation in the region $\Pi_+ \subset \Pi$ of points $P(x, y)$ with coordinates $x > 0$, $y > 0$. These

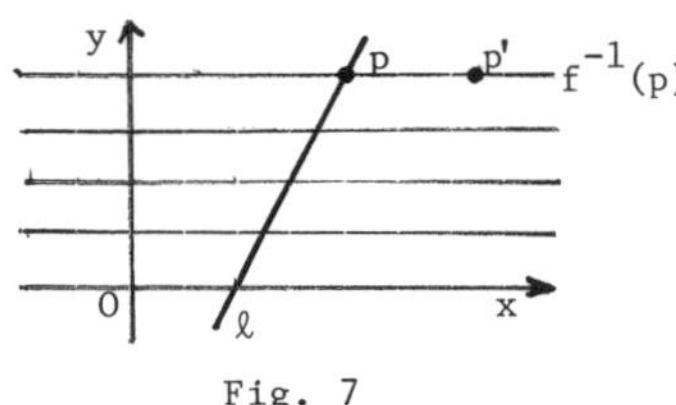

Fig. 7

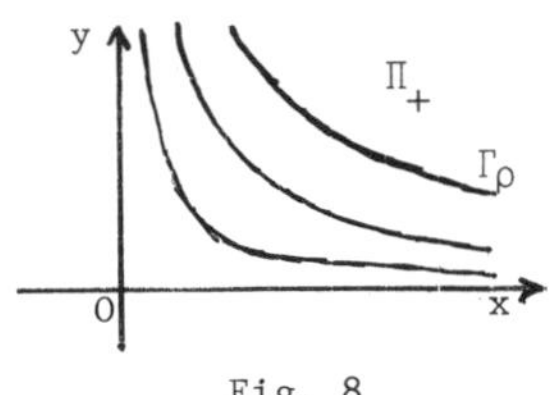

Fig. 8

geometrical examples visually illustrate the following assertion.

If $\pi(X)$ is a partition of a set X into disjoint subsets C_x, then the C_x are equivalence classes for some equivalence relation $\sim$ on X.

Proof. By assumption, each element $x \in X$ is contained in precisely one subset C_a. We define $\sim$ by saying that $x \sim x'$ if and only if x and x' lie in the same C_a. This relation $\sim$ is obviously reflexive, symmetric, and transitive, i.e., it is an equivalence relation. Furthermore, $x \in C_a \Rightarrow \bar{x} = C_a$, so that $\pi_\sim(X) = \pi(X)$. □

3. Quotient maps. We just saw that there is a one-to-one correspondence between equivalence relations and partitions of a set X. It is customary to let the symbol $X/\sim$ denote the partition of X corresponding to $\sim$; this set of equivalence classes is called the quotient set of X relative to $\sim$. The surjective mapping

$$p : x \mapsto p(x) = \bar{x} \tag{1}$$

is called the natural mapping (or the canonical projection) of X onto the quotient set $X/\sim$. In the example in Fig. 7, $X/\sim$ is the set of horizontal lines, and the canonical projection is the mapping that associates to each point in $\mathbb{R}^2$ the horizontal line through it.

Let X and Y be two sets, and let $f : X \to Y$ be a mapping. The binary relation O_f:

$$xO_f x' \iff f(x) = f(x') \quad , \quad \forall\, x, x' \in X \quad ,$$

is clearly reflexive $(f(x) = f(x))$, symmetric $(f(x') = f(x) \Rightarrow f(x) = f(x'))$, and transitive $(f(x) = f(x')$ and $f(x') = f(x'') \Rightarrow f(x) = f(x''))$. Thus, O_f is an equivalence relation on X. The equivalence classes $\bar{x}$ are the fibres (preimages) in the sense of Exercise 5 of §5. In other words,

$$\bar{x} = \{x' \mid f(x') = f(x)\} \quad .$$

The mapping $f : X \to Y$ <u>induces</u> a mapping $\bar{f} : X/O_f \to Y$ given by the rule

$$\bar{f}(\bar{x}) = f(x) \quad , \tag{2}$$

or, if we use our notation for the canonical projection (see (1)),

$$\bar{f}p(x) = f(x) \quad . \tag{2'}$$

Since $\bar{x} = \bar{x}' \iff f(x) = f(x')$, it follows that the equality (2) defining $\bar{f}$ does not depend on the choice of representative x of the equivalence class $\bar{x}$. (In this case, it is customary to say that $\bar{f}$ is <u>well-defined</u>, or that the definition (2) is <u>correct.</u>) In other words, by definition f has a fixed value on all x in an equivalence class; hence we can think in terms of a function $\bar{f}$ whose domain is a set of equivalence classes. For example, if $f : \mathbb{R}^2 \to \mathbb{R}$ is defined by $f(x,y) = y^3$, then we get the equivalence relation in Fig. 7, and f (the line $y = y_0) = y_0^3$.

The commutative diagram

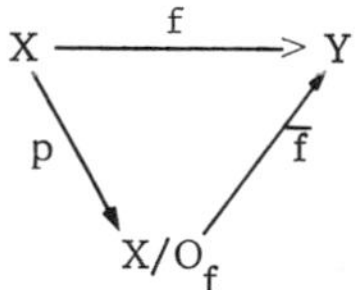

depicts the <u>factoring</u> (<u>decomposition</u>)

$$f = \bar{f} \cdot p \tag{3}$$

of f into a product of a surjective mapping p and an injective mapping $\bar{f}$. Notice that $\bar{f}$ is injective because

$$\bar{f}(\bar{x}_1) = \bar{f}(\bar{x}_2) \iff f(x_1) = f(x_2) \iff \bar{x}_1 = \bar{x}_2 \quad .$$

The mapping f is surjective if and only if $\bar{f}$ is surjective. Note that, if $f' : X/O_f \to Y$ is another mapping for which $f'p = f$, then since $f'(\bar{x}) = f'(px) = (f'p)x = f(x) = \bar{f}(\bar{x})$ (by (2)), it follows that in fact $f' = \bar{f}$. Thus, the mapping $\bar{f}$ which makes the above triangular diagram commute is unique.

4. Ordered sets. By an ordering on a set X we mean a binary relation $\leq$ on X which has the properties of reflexivity $(x \leq x)$, anti-symmetry $(x \leq y$ and $y \leq x \Longrightarrow x = y)$, and transitivity $(x \leq y$ and $y \leq z \Longrightarrow x \leq z)$. If $x \leq y$ and $x \neq y$, we write $x < y$. If $x \leq y$ we also write $y \geq x$. It is possible for a pair of elements $x, x' \in X$ not to be related in either direction by $\leq$. But if our X and $\leq$ are such that either $x \leq x'$ or $x' \leq x$ for every pair of elements in X, then X is said to be linearly ordered (or totally ordered). In the general case we speak of a partial ordering on X.

Some examples of partially ordered sets are: the set $X = \mathcal{P}(S)$ of subsets of a set S (see Exercise 4 of §5) with $\leq$ being the usual inclusion relation $R \subset T$ between subsets, and the set $\mathbb{N}$ of natural numbers with $\leq$ being the relation $d|n$ (n is divisible by d).

Let X be an arbitrary partially ordered set, and let x and y be elements of X. We say that y follows (or covers) x if $x < y$ and there does not exist z with $x < z < y$. If X is a finite set, then $x < y$ if and only if there is a chain of elements $x_1 = x, x_2, \cdots, x_{n-1}, x_n = y$ in which x_{i+1} follows x_i. The notion of following (covering) is useful for depicting a finite partially ordered set X by a plane diagram. The elements of X are represented by points. If y covers x, then y is placed higher than x and x is joined to y by a straight line. If y and x are related by $<$, they are joined by a "descending" polygonal line, and perhaps by several such lines. The first diagram in Fig. 9 depicts an interval of the natural numbers with the usual ordering $<$ (under which $\mathbb{N}$ is linearly ordered); the second diagram shows $\mathcal{P}(\{a,b,c\})$ with the inclusion ordering described above.

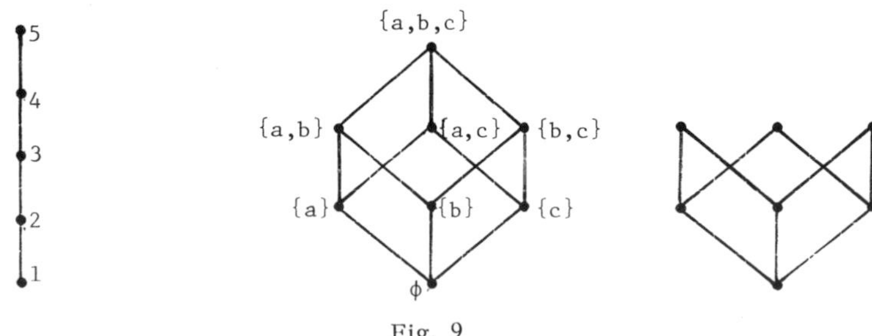

Fig. 9

A greatest element of a partially ordered set X is an element $n \in X$ such that $x \leq n$ for all $x \in X$, and a maximal element is an element $m \in X$ such that $m \leq x \in X$ implies that $x = m$. A greatest element is always maximal, but the converse is not true. There can be many maximal elements, but the greatest element, if it exists, is unique. There are analogous definitions and remarks for least and minimal elements. The first two diagrams in Fig. 9 each have a greatest and a least element. In the third diagram there are three maximal elements and one least element, but no greatest element.

The theory of partially ordered algebraic systems (Boolean algebras, lattices) is full of interesting results and occupies an important place in algebra, but we cannot go into it here. This section has the modest goal of acquainting the reader with a further example of a binary relation and giving some exposure to diagrams of the type that will later be of use in understanding, for example, the location and interrelation of subgroups in a group or subfields in a field.

EXERCISES

1. Give a one-to-one correspondence between the quotient set $\mathbb{R}^2/\sim$ obtained from the picture in Fig. 7 and the points of an arbitrary straight line ℓ which intersects the x-axis.

2. Let $P(x,y) \sim P(x',y')$ for points of $\mathbb{R}^2$ if and only if both $x' - x \in \mathbb{Z}$ and $y' - y \in \mathbb{Z}$. Prove that $\sim$ is an equivalence relation, and that the quotient set $\mathbb{R}^2/\sim$ is in one-to-one correspondence with the points on a torus (surface of a donut; see Fig. 10).

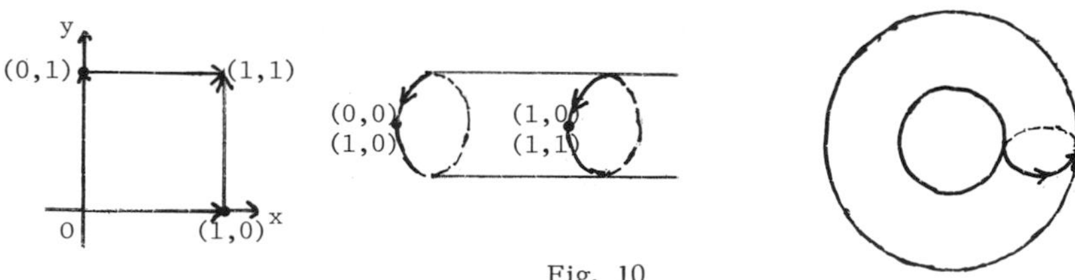

Fig. 10

3. Show that sets of two, three, and four elements have, respectively, 2, 5, and

15 different quotient sets (i.e., that there are 2, 5, 15 different equivalence relations that can be defined on them).

4. Let $\sim$ be an equivalence relation on a set X, and let $f: X \to Y$ be a mapping for which $x \sim x' \Rightarrow f(x) = f(x')$. Show that this compatibility condition of f with $\sim$ allows us to construct a well-defined mapping $\bar{f}: \bar{x} \mapsto f(x)$ from $X/\sim$ to Y which gives a factoring of $f: f = \bar{f}p$. However, note that $\bar{f}$ is no longer necessarily injective. What condition must be satisfied in order for $\bar{f}$ to be injective?

5. Draw diagrams for the following partially ordered sets: (1) $P(\{a,b,c,d\})$, (2) the set of all divisors of 24 (use the ordering by divisibility: $x \leq y \Leftrightarrow x|y$).

§7. The principle of mathematical induction

The set $\mathbb{N} = \{1, 2, 3, \cdots\}$ of all natural numbers (positive integers) is considered to be a very familiar set. Actually, the point of departure for studying $\mathbb{N}$ is the axioms of Peano (1858-1932). His three axioms (which we shall not list here) can be used to derive the properties of addition and multiplication and the linear ordering (see above) of the natural numbers (more precisely, of the non-negative integers $\mathbb{N} \cup \{0\}$). Those axioms imply the following intuitively clear assertion: every non-empty set $S \subset \mathbb{N}$ has a least element, i.e., there is a natural number $s \in S$ which is smaller than all of the other numbers in S. Using this assertion, one can deduce the following

Principle of Induction. Suppose that for every $n \in \mathbb{N}$ we are interested in an assertion $M(n)$. Further suppose that we know that $M(1)$ is true, and we have a procedure which tells us that $M(\ell)$ is true whenever $M(k)$ is true for all $k < \ell$. Then $M(n)$ is true for all $n \in \mathbb{N}$.

To prove this from the "least element principle" in the previous paragraph, let

$$S = \{s \in \mathbb{N} | M(s) \text{ is false}\} \subset \mathbb{N} .$$

Suppose S is non-empty. Then S has a least element s_0. This means that $M(s_0)$ is false, but $M(s)$ is true for all $s < s_0$. But this contradicts our assumption, which assures us that we know $M(s_0)$ to be true whenever $M(s)$ is true for $s < s_0$. □

This is not the place for a philosophical discussion of the principle of induction. Suffice it to say that, in some sense, it reflects the essence of what is meant by a natural sequence.

Notice that, when using the principle of induction, it is essential to establish a starting point for the induction, i. e., to verify that the assertion holds for a certain small n. If this step is neglected, one can "prove" such ludicrous assertions as "all students have the same height". Here is the argument: The empty set of students and the set of one student have this property. Proceeding by induction, we suppose that any set of $\leq n$ students has the property. Then, in a set of $n+1$ students, the first n and the last n students have the same height by the induction assumption. These sets intersect in a set of $n-1$ students, all with the same height. Hence, all $n+1$ students have the same height. The fallacy here is that the first use of the induction step relates to the set of any two students, and it is there that the induction step is unjustified. For how many low values of n must we verify an assertion before we can be sure that the induction step is valid? Usually this is clear from the proof. In our example the assumption implicit in the induction step is that the two sets of size n in the set of $n+1$ elements must have non-empty intersection; this means that $n \geq 2$.

In more complicated situations, especially when one defines or constructs an object by induction using recursion relations (as we shall do for the determinant of a matrix in Chapter 3), one must pay careful attention to establishing a basis, or starting point, for the proof or definition by induction. On the other hand, one must not go to the other extreme of erroneously concluding $M(n)$ for all $n \in \mathbb{N}$ on the basis of a case-by-case verification of $M(k)$ for all k in a very long sequence $1 \leq k \leq \ell$. Here are two unpleasant examples of what that can lead to.

1. Fermat conjectured that all numbers of the form $F_n = 2^{2^n} + 1$, $n = 0, 1, \cdots$, (the so-called "Fermat numbers") are primes. (Concerning prime numbers, see §8.) The

first five Fermat numbers are in fact prime, but Euler found that F_5 is composite: $F_5 = 4294967297 = 641 \cdot 6700417$. Persistent attempts to find at least one more prime Fermat number using the latest computers have not yet met with success. One of the most recent "accomplishments" in this area was the verification that F_{1945} is divisible by $5 \cdot 2^{1947} + 1$.

2. If one looks at numbers of the form $n^2 - n + 41$ (a polynomial studied by Euler) for $n = 1, 2, \cdots, 40$, one might think that this polynomial takes prime values for all n. However $41^2 - 41 + 41 = 41^2$.

Sometimes the most important part of proving a formula by induction is having the right form for the formula one is trying to prove. For example, suppose that we want to find the sum

$$p_k(n) = 1^k + 2^k + 3^k + \cdots + (n-1)^k + n^k \; ; \quad k = 1, 2, 3 \quad .$$

The problem becomes much easier when we are told that the answer is supposed to be the following expressions:

$$p_1(n) = \frac{n(n+1)}{2} \; , \quad p_2(n) = \frac{n(n+1)(2n+1)}{6} \; ,$$

$$p_3(n) = \left[\frac{n(n+1)}{2}\right]^2 \quad .$$

Although $p_1(n)$ is not hard to think of (Gauss did this as a young boy), the form of $p_2(n)$ and $p_3(n)$ is not quite so trivial, and the relation

$$p_5(n) + p_1(n) = 2\left[\frac{n(n+1)}{2}\right]^4$$

could probably only be thought of if one had developed a framework or theory to predict these expressions. In this case such general procedures can be found, but that is not our concern here. Once we know the formula we want to prove, all we have to do to prove it is make the trivial computation for $n = 1$, and then, also by direct computation, verify the induction step from n to $n + 1$. It would be a worthwhile exercise for the reader to carry out this procedure for the formulas given above.

To do the above exercise one uses the so-called binomial formula

$$(a+b)^n = a^n + \binom{n}{1} a^{n-1}b + \cdots + \binom{n}{k} a^{n-k}b^k + \cdots + b^n \quad . \tag{1}$$

Here a and b are arbitrary numbers, and the binomial coefficient $\binom{n}{k}$ of the monomial $a^{n-k}b^k$ has the form

$$\binom{n}{k} = \frac{n!}{k!\,(n-k)!} = \frac{n(n-1)\ldots(n-k+1)}{k(k-1)\ldots 2\cdot 1} \quad . \tag{2}$$

It is useful to adopt the convention that $0! = 1$ and $\binom{n}{k} = 0$ for $k < 0$. Note that

$$\binom{n}{n-k} = \binom{n}{k}$$

(the symmetry property for binomial coefficients).

We prove (1) by induction on n. The formula is obviously true for $n = 1, 2$. We assume it holds for all exponents $\leq n$, and then multiply both sides of (1) by $a+b$. We obtain

$$\begin{aligned}
(a+b)^{n+1} &= (a+b)^n (a+b) = \\
&= a^n(a+b) + \cdots + \binom{n}{k} a^{n-k}b^k(a+b) + \cdots + b^n(a+b) = \\
&= a^{n+1} + a^n b + \cdots + \binom{n}{k-1} a^{n+2-k}b^{k-1} + \binom{n}{k-1} a^{n+1-k}b^k + \\
&\qquad + \binom{n}{k} a^{n+1-k}b^k + \binom{n}{k} a^{n-k}b^{k+1} + \cdots + ab^n + b^{n+1} \quad .
\end{aligned}$$

Combining similar terms, we see that the coefficient of $a^{n+1-k}b^k$ is

$$\begin{aligned}
&\binom{n}{k-1} + \binom{n}{k} = \\
&= \frac{n!}{(k-1)!\,(n-k+1)!} + \frac{n!}{k!(n-k)!} = \frac{n!}{(k-1)!\,(n-k)!}\left[\frac{1}{n-k+1} + \frac{1}{k}\right] = \\
&\qquad = \frac{n!}{(k-1)!\,(n-k)!} \cdot \frac{n+1}{k(n-k+1)} = \frac{(n+1)!}{k!(n+1-k)!} = \binom{n+1}{k} \quad ,
\end{aligned}$$

i.e., the binomial coefficient of the form (2) with upper index increased by one. We have thereby proved (1) for all $n \in \mathbb{N}$.

If we write

$$(a+b)^n = (a+b)(a+b)\cdots(a+b) \quad ,$$

give each factor on the right an index from 1 to n, and look at all ways of choosing k b's to get $a^{n-k}b^k$, i.e., at all subsets of k indices $1 \leq i_1 < i_2 < \cdots < i_k \leq n$, we conclude that $\binom{n}{k}$ <u>is equal to the number of subsets of</u> k <u>elements in a set of</u> n <u>elements.</u> For this reason the alternate (and somewhat old-fashioned) notation and terminology $C_n^k = \binom{n}{k}$, the combination of n things taken k at a time, is sometimes used.

If we think of the binomial coefficients as the number of subsets of given cardinality, we see that the cardinality of $\mathcal{P}(\{s_1, \cdots, s_n\})$ (see Exercise 4 of §5) is equal to $\binom{n}{0} + \binom{n}{1} + \cdots + \binom{n}{n-1} + \binom{n}{n}$. But, setting $a = b = 1$ in (1), we obtain

$$2^n = \binom{n}{0} + \binom{n}{1} + \binom{n}{2} + \cdots + \binom{n}{n-1} + \binom{n}{n} \quad .$$

Thus, $\operatorname{Card} \mathcal{P}(\{s_1, s_2, \cdots, s_n\}) = 2^n$.

Sometimes a theorem can be proved or an object constructed using a more complicated form of induction. For example, we can use the following principle of "double induction". Suppose that for any pair of natural numbers m and n we are interested in an assertion $A(m,n)$. Suppose that: (i) $A(m,1)$ and $A(1,n)$ are true for all m and n; (ii) if $A(k-1,\ell)$ and $A(k,\ell-1)$ are true, then $A(k,\ell)$ is also true (equivalently: (ii') if $A(k',\ell')$ is true for all $k' \leq k$ and $\ell' \leq \ell$ for which $k' + \ell' < k + \ell$, then $A(k,\ell)$ is also true). Then the assertion $A(m,n)$ is true for all natural numbers m and n.

§8. Integer arithmetic

The purpose of this section is to give a brief description of the simplest divisibility properties of the integers, which we shall have occasion to refer to in various connections later in the book. Further facts will be given in Chapter 5, where the theory of divisibility is carried over to more general algebraic systems.

1. The fundamental theorem of arithmetic. An integer s is called a divisor (or factor) of an integer n if $n = st$ for some $t \in \mathbb{Z}$. In that case n is called a multiple of s. The notation $s \mid n$ means that s is a divisor of n, and $s \nmid n$ means that it is not. Divisibility is a transitive relation on $\mathbb{Z}$. In addition, if both $m \mid n$ and $n \mid m$, then we must have $n = \pm m$, in which case the integers n and m are called associated. An integer p whose only divisors are $\pm p$ and ± 1 (the improper divisors) is called prime. One usually agrees to consider prime numbers to be positive and > 1. The basic role played by prime numbers is brought out by the so-called

Fundamental Theorem of Arithmetic. Every positive integer $n \neq 1$ can be written as a product of prime numbers: $n = p_1 p_2 \cdots p_s$. This prime decomposition is unique except for the order of the factors.

The proof of the Fundamental Theorem will be postponed until Chapter 5. At first glance, it may seem so obvious that we should not have to prove it. But the proof is not so trivial. Although the theorem itself only refers to multiplicative properties of integers (divisibility), it turns out to be necessary to use both multiplication and addition in $\mathbb{Z}$ in the proof. To illustrate the non-triviality of the theorem, let us consider the subset $S = \{4k+1 \mid k = 0, 1, 2, \cdots\} \subset \mathbb{N}$. S is closed with respect to multiplication: $(4k_1 + 1)(4k_2 + 1) = 4k_3 + 1$. Using induction on $n \in S$, it is not hard to prove that any $n \in S$ can be written $n = q_1 \cdots q_\ell$, where q_i are elements of S which cannot be further factored into elements of S (this is analogous to the first part of the Fundamental Theorem). Examples of indecomposable elements of S are 5, 9, 13, 17, 21, 49. But the second part of the Fundamental Theorem is false for S, since, for example, the integer $441 \in S$ has two different decompositions as a product of indecomposable elements of S:

$$441 = 9 \cdot 49 = 21 \cdot 21 \quad .$$

In the Fundamental Theorem, if we combine identical prime factors and modify our notation, we can write n in the form $n = p_1^{\varepsilon_1} p_2^{\varepsilon_2} \cdots p_k^{\varepsilon_k}$, $\varepsilon_i > 0$, $1 \le i \le k$.

Every rational number $a = n/m \in \mathbb{Q}$ has a similar decomposition, except that the exponents ε_i can be either positive or negative. Note the following important fact (a theorem of Euclid):

The set $P = \{2, 3, 5, 7, 11, 13, \dots\}$ of all prime numbers is infinite.

To prove this, suppose that there were only finitely many primes, say $p_1, p_2, \cdots, p_t$. Then, by the Fundamental Theorem, the number $c = p_1 p_2 \cdots p_t + 1$ would be divisible by at least one of the p_i. Without loss of generality we may assume that $c = p_1 c'$. Then $p_1(c' - p_2 \cdots p_t) = 1$, which is impossible, since the only divisors of one in $\mathbb{Z}$ are ± 1. □

2. g.c.d. and l.c.m in $\mathbb{Z}$. If we agree to allow zero exponents for primes in a factorization (of course, taking $p_i^0 = 1$), any two integers n and m can be written as a product of the same primes:

$$n = \pm p_1^{\alpha_1} p_2^{\alpha_2} \cdots p_k^{\alpha_k}, \quad m = \pm p_1^{\beta_1} p_2^{\beta_2} \cdots p_k^{\beta_k}.$$

We introduce the two integers

$$\text{g.c.d.}(n,m) = p_1^{\gamma_1} p_2^{\gamma_2} \cdots p_k^{\gamma_k}, \quad \text{l.c.m.}(n,m) = p_1^{\delta_1} p_2^{\delta_2} \cdots p_k^{\delta_k}, \tag{1}$$

where $\gamma_i = \min(\alpha_i, \beta_i)$, $\delta_i = \max(\alpha_i, \beta_i)$, $i = 1, 2, \cdots k$. Since $d \mid n \Rightarrow d = \pm p_1^{\alpha'_1} \cdots p_k^{\alpha'_k}$, $0 \le \alpha'_i \le \alpha_i$, the following assertions follow from the definitions (1):

(i) $\text{g.c.d.}(n,m) \mid n$, $\text{g.c.d.}(n,m) \mid m$, and if $d \mid n$ and $d \mid m$, then $d \mid \text{g.c.d.}(n,m)$.

(ii) $n \mid \text{l.c.m.}(n,m)$, $m \mid \text{l.c.m.}(n,m)$, and if $n \mid u$ and $m \mid u$, then $\text{l.c.m.}(n,m) \mid u$.

It is properties (i) and (ii) which explain the terminology greatest common divisor (g.c.d.) and least common multiple (l.c.m). For $n > 0$, $m > 0$ we have the relation

$$\text{g.c.d.}(n,m) \cdot \text{l.c.m.}(n,m) = nm. \tag{2}$$

Two integers n, m are called <u>relatively prime</u> if $g.c.d.(n,m) = 1$. In this case (2) takes the form: $l.c.m.(n,m) = nm$.

3. <u>The division algorithm in</u> $\mathbb{Z}$. <u>Given</u> $a, b \in \mathbb{Z}$, $b > 0$, <u>there always exist</u> $q, r \in \mathbb{Z}$ <u>such that</u>

$$a = bq + r \quad , \quad 0 \leq r < b \quad .$$

(If we only require $b \neq 0$, then we have the same thing with $0 \leq r < |b|$.)

<u>Proof.</u> The set $S = \{a - bs \mid s \in \mathbb{Z}, a - bs \geq 0\}$ is clearly non-empty (for example, $a - b(-a^2) \geq 0$). Hence, S contains a least element; let us denote this element $r = a - bq$. By the definition of S, $r \geq 0$. If we had $r \geq b$, we would obtain an element $r - b = a - b(q+1) \in S$ which is less than r. This contradicts the definition of r, so we must have $r < b$. □

The simple proof also gives a prescription (<u>algorithm</u>) for finding the <u>quotient</u> q and the <u>remainder</u> r in a finite number of steps. This division algorithm can be used to give another definition of $g.c.d.$ (and hence of $l.c.m.$, because of (2)).

Namely, given integers n and m, not both zero, set

$$J = \{nu + mv \mid u, v \in Z\} \quad . \tag{3}$$

Choose the least positive element in $J: d = nu_0 + mv_0$. Using the division algorithm, write $n = dq + r$, $0 \leq r < d$. Because of our choice of d, the fact that

$$r = n - dq = n - (nu_0 + mv_0)q = n(1 - u_0q) + m(-v_0q) \in J$$

implies $r = 0$. Hence, $d \mid n$. We similarly prove that $d \mid m$. Now let d' be any divisor of the integers n and m. Then

$$d' \mid n \;,\; d' \mid m \Rightarrow d' \mid nu_0 \;,\; d' \mid mv_0 \Rightarrow d' \mid (nu_0 + mv_0) \Rightarrow d' \mid d \quad .$$

Thus, d has all of the properties of the greatest common divisor, and so $d = g.c.d.(n,m)$. (Note that there can be only one positive integer, the $g.c.d.$, with property (i) above, since if there were two, g and g', we would have $g \mid g'$ and $g' \mid g$, and so $g' = \pm g$.) We have proved the following assertion.

Given two integers n and m, not both zero, their greatest common divisor can always be written in the form

$$\text{g.c.d.}(n,m) = nu + mv ; \quad u,v \in \mathbb{Z} \quad . \tag{4}$$

In particular, two integers n and m are relatively prime if and only if

$$nu + mv = 1 \tag{4'}$$

for some $u,v \in \mathbb{Z}$.

The last part of this assertion follows because we have already verified that we can write $1 = \text{g.c.d.}(n,m) = nu + mv$; and, conversely, if (4') holds, then $d|n$, $d|m \Rightarrow d|nu$, $d|mv \Rightarrow d|(nu + mv) \Rightarrow d|1 \Rightarrow d = \pm 1$. □

The proof of (4) and (4') was effective. One takes an arbitrary positive element of J (see (3)), and then finds smaller and smaller elements of J using the division algorithm, until one obtains the least element, which will be the g.c.d.

EXERCISES

1. Every prime number > 2 has the form $4k + 1$ or $4k - 1$. Using the multiplicativity of the set S in subsection 1, prove that there are infinitely many primes of the form $4k - 1$.

2. It can be proved that if $n, m \in \mathbb{Z}$, $\text{g.c.d.}(m,n) = 1$, and p is an odd prime dividing $n^2 + m^2$, then p is of the form $4k + 1$. (See subsection 1 of §2, Ch. 9.) Use this fact to prove that there are infinitely many prime number of the form $4k + 1$.

3. If a natural number n is divisible by exactly r different prime numbers $p_1, \cdots, p_r$, then the number of positive integers less than n and relatively prime to n is equal to

$$\varphi(n) = n\left(1 - \frac{1}{p_1}\right) \cdots \left(1 - \frac{1}{p_r}\right) \quad .$$

The function $\varphi : \mathbb{N} \to \mathbb{N}$ is called Euler's function. Show that this formula holds for $n \leq 25$, and also for n of the form $n = p^m$ (see also subsection 4 of §1, Ch. 9).

4. Using the binomial formula and induction on n, prove that, if p is a prime, then $n^p - n$ is divisible by p for any $n \in \mathbb{Z}$.

Chapter 2. Sources of Algebra

The rectangular matrices introduced in §3 of Chapter 1 occur so often that an independent branch of mathematics called matrix theory has evolved. Although it arose in the middle of the last century, it acquired a complete and elegant form somewhat later, when linear algebra developed. To this day matrix theory remains a tool well-suited both to applied problems and to the abstract constructions of modern theoretical mathematics. Here we shall present the simplest results of matrix theory.

The title of the chapter may give rise to the illusion that we intend to rely upon geometry to describe our purely algebraic objects. But actually, it is only that we find it convenient and efficient to express the properties of matrices and solutions of linear systems in a language borrowed from geometry. The concepts of a space, a vector, linear dependence, the rank of a system, etc., are developed precisely to the extent that they are needed for our immediate purposes. Our approach is basically algebraic, and we do not give geometrical intuition the key role that it plays in some other treatments of the subject.

Linear spaces will be necessary to us in order to speak of linear mappings, the companion concept to matrices. It is composition of mappings (subsection 2, §5, Ch. 1) that gives the most natural explanation of matrix multiplication.

§1. Vector spaces

1. Motivation. When we studied systems of linear equations, we had to consider rows of length n in various contexts. They were the rows $(a_{i1}, a_{i2}, \cdots, a_{in})$, $1 \le i \le m$, of an $m \times n$ matrix $A = (a_{ij})$, and also the solutions $(x_1^\circ, x_2^\circ, \cdots x_n^\circ)$ of the linear system with matrix A. The elementary transformations of type (II) that were used in §3 of Chapter 1 to reduce a matrix to step form involved two basic operations: multiplying rows by a number, and adding two rows. The same operations can be performed upon the solutions of a homogeneous linear system. That is, if $(x_1', x_2', \cdots, x_n')$ and $(x_1'', x_2'', \cdots, x_n'')$ are two solutions of the system

$$a_{i1}x_1 + a_{i2}x_2 + \cdots + a_{in}x_n = 0 \quad , \quad i = 1, 2, \cdots, m \quad ,$$

and if α and β are any two real numbers, then the row

$$(\alpha x_1' + \beta x_1'', \alpha x_2' + \beta x_2'', \cdots, \alpha x_n' + \beta x_n'')$$

will also be a solution of our system:

$$a_{i1}(\alpha x_1' + \beta x_1'') + a_{i2}(\alpha x_2' + \beta x_2'') + \cdots + a_{in}(\alpha x_n' + \beta x_n'') =$$

$$= \alpha(a_{i1}x_1' + a_{i2}x_2' + \cdots + a_{in}x_n') + \beta(a_{i1}x_1'' + a_{i2}x_2'' + \cdots + a_{in}x_n'') = 0 .$$

On the other hand, any row of length n, no matter what it stands for, is an element of the "universal" set $\mathbb{R}^n$ -- the n-th cartesian power of the set $\mathbb{R}$ of real numbers. So it would be worthwhile to study this general object; its properties can then be carried over to matrices and to the solutions of homogeneous systems.

2. Basic definitions. Let n be a fixed natural number. The n-dimensional vector space over $\mathbb{R}$ is the set $\mathbb{R}^n$ (whose elements are called row-vectors, or simply vectors), considered along with the operations of adding vectors and multiplying vectors by real numbers (real numbers will be called scalars). We shall denote scalars by small Latin or Greek letters, and we shall denote vectors by capital Latin leters, like matrices. In fact, a vector $X = (x_1, x_2, \cdots, x_n)$ can be though of as a $1 \times n$ matrix.

Let $Y = (y_1, y_2, \cdots, y_n)$ be another vector, and let λ be a scalar. By definition

$$X + Y = (x_1 + y_1, x_2 + y_2, \cdots, x_n + y_n) \quad ,$$

$$\lambda X = (\lambda x_1, \lambda x_2, \cdots, \lambda x_n) \quad .$$

We shall let the usual symbol 0 denote both the real number 0 and also the zero vector $(0, 0, \cdots, 0)$. In addition, it is customary to identify $\mathbb{R}^1$ with $\mathbb{R}$.

The formal properties of operations with real numbers, which are well known to the reader, carry over to $\mathbb{R}^n$. Although it is boring to list them, doing so gives a precise idea of how one can define an abstract vector space; such abstract vector spaces (not necessarily having real number coordinates or finitely many dimensions) are important in many fields. Here is the list of properties satisfied by the operations of addition and scalar multiplication on a vector space:

VS_1 : $X + Y = Y + X$ for any vectors $X, Y \in \mathbb{R}^n$ (commutative law);

VS_2 : $(X + Y) + Z = X + (Y + Z)$ for any three vectors $X, Y, Z \in \mathbb{R}^n$ (associative law);

VS_3 : there exists a special vector 0 such that $X + 0 = X$ for all $X \in \mathbb{R}^n$;

VS_4 : every $X \in \mathbb{R}^n$ has a negative (additive inverse) vector $-X$ such that $X + (-X) = 0$;

VS_5 : $1 \cdot X = X$ for all $X \in \mathbb{R}^n$;

VS_6 : $(\alpha\beta) X = \alpha(\beta X)$ for all $\alpha, \beta \in \mathbb{R}$, $X \in \mathbb{R}^n$;

VS_7 : $(\alpha + \beta) X = \alpha X + \beta X$ (distributive law for scalars);

VS_8 : $\alpha(X + Y) = \alpha X + \alpha Y$ (distributive law for vectors).

The uniqueness of the vectors 0 and $-X$ in VS_3 and VS_4, and the other simple consequences of these rules (which are called axioms if we have in mind an abstract vector space), will not be derived here, since the derivations are very easy and can be left to the reader.

We referred to $\mathbb{R}^n$ as an n-dimensional space, but the notion of dimension will only acquire a precise meaning at the end of the section, after a little preliminary material. The origin of the term "vector space" is clear after studying analytic geometry, where one learns of the one-to-one correspondence between points (vectors) in the cartesian plane and ordered pairs (x,y). Adding vectors by the parallelogram rule and multiplying them by real numbers precisely correspond to the operations on row-vectors of $\mathbb{R}^2$, as defined above.

In addition to the vector space of row-vectors $(x_1, x_2, \cdots, x_n)$ of length n, one can consider the vector space of <u>column-vectors</u> of height n

$$\begin{Vmatrix} x_1 \\ x_2 \\ \vdots \\ x_n \end{Vmatrix} = [x_1, x_2, \cdots x_n] ,$$

as we agreed to denote them in §3 of Chapter 1. There is clearly no essential difference between these two vector spaces, but we shall soon see that it is useful to have both versions of a vector space. It is usually clear from the context whether one is talking about row- or column-vectors, so that we shall not introduce any further notation to distinguish the two types of vectors.

Let V be a non-empty subset of $\mathbb{R}^n$. We shall call V a <u>linear subspace</u> of $\mathbb{R}^n$ if

$$X, Y \in V \Longrightarrow \alpha X + \beta Y \in V \qquad (1)$$

for all $\alpha, \beta \in \mathbb{R}$. (This definition at first glance might not seem satisfactory, since it does not explain in what sense V is a "space", but we shall say some words in its defense at the end of the section.) Note that (1) implies that the zero vector always belongs to V.

For example, the set of all row-vectors $(x_1, \cdots, x_{n-1}, 0)$ with $x_n = 0$ is a linear subspace of $\mathbb{R}^n$; it is customary to identify this subspace with $\mathbb{R}^{n-1}$. We have the "chain" of imbedded subspaces

$$0 \subset \mathbb{R} \subset \mathbb{R}^2 \subset \cdots \subset \mathbb{R}^{n-1} \subset \mathbb{R}^n .$$

The solutions of the homogeneous equation $x_1 + x_2 + \cdots + x_n = 0$ make up a subspace in $\mathbb{R}^n$, $n > 1$, which is different from the zero subspace and from all the $\mathbb{R}^i$ in the above chain. Other examples will be given below.

3. Linear combinations. Linear span. Let $X_1, X_2, \cdots, X_k$ be vectors in $\mathbb{R}^n$, and let $\alpha_1, \alpha_2, \cdots, \alpha_k$ be scalars. The vector $X = \alpha_1 X_1 + \alpha_2 X_2 + \cdots + \alpha_k X_k$ is called a linear combination of the vectors X_i with coefficients α_i. For example, $(2,3,5,5) - 3(1,1,1,1) + 2(1,0,-1,-1) = (1,0,0,0)$. Now let $Y = \beta_1 X_1 + \beta_2 X_2 + \cdots + \beta_k X_k$ be a linear combination of the same vectors X_i with coefficients β_i, and let $\alpha, \beta \in \mathbb{R}$. Then

$$\begin{aligned}\alpha X + \beta Y &= \alpha(\alpha_1 X_1 + \alpha_2 X_2 + \cdots + \alpha_k X_k) + \beta(\beta_1 X_1 + \beta_2 X_2 + \cdots + \beta_k X_k) = \\ &= (\alpha\alpha_1 + \beta\beta_1) X_1 + (\alpha\alpha_2 + \beta\beta_2) X_2 + \cdots + (\alpha\alpha_k + \beta\beta_k) X_k\end{aligned}$$

is also a linear combination of the vectors X_i; its coefficients are $\alpha\alpha_i + \beta\beta_i$. We thus see that the set of all linear combinations of a given set of vectors $X_1, X_2, \cdots, X_k$ is a linear subspace of $\mathbb{R}^n$. We denote this subspace by the symbol $\langle X_1, X_2, \cdots, X_k \rangle$, and we call it the linear span of the set of vectors $X_1, X_2, \cdots, X_k$. It is also customary to say that the space $\langle X_1, X_2, \cdots, X_k \rangle$ is spanned or is generated by the vectors $X_1, X_2, \cdots, X_k$.

It is possible to define the linear span of any subset $S \subset \mathbb{R}^n$: $\langle S \rangle$ is the set of all linear combinations of finite sets of vectors in S. Clearly, if V is a subspace of $\mathbb{R}^n$, then $\langle V \rangle = V$, since any linear combination of vectors in V belongs to V. More generally, since $S \subset V \Rightarrow \langle S \rangle \subset V$, it follows that the linear span $\langle S \rangle$ can be defined as the intersection of all subspaces containing the given set S of vectors of $\mathbb{R}^n$:

$$\langle S \rangle = \bigcap_{S \subset V} V \quad . \tag{2}$$

It might not be obvious at first glance, but the intersection of any set of subspaces, as in (2), will always be a subspace. Namely, if $X, Y \in \cap V$, then $X, Y \in V$ for every V in the set of subspaces. Hence, $\alpha X + \beta Y \in V$ for all $\alpha, \beta \in \mathbb{R}$; since this is true for all V, we have $\alpha X + \beta Y \in \cap V$, as required.

Unlike intersections, the union of two subspaces U and V is not, in general, a subspace. For example, let $U = \{(\lambda,0)|\lambda \in \mathbb{R}\}$, $V = \{(0,\lambda)|\lambda \in \mathbb{R}\}$ in $\mathbb{R}^2$. The linear span $\langle U \cup V\rangle$ is called the sum of the subspaces U and V:

$$U + V = \langle U \cup V\rangle = \{u+v | u \in U, v \in V\} \quad .$$

If $U \cap V = 0$, we say that the sum $U+V$ is direct, and we write $U \oplus V$. Let $V = V_1 \oplus V_2$, and let $X = X_1 + X_2 = X_1' + X_2'$ be two expressions for a vector $X \in V$ as a linear combination of vectors $X_1, X_1' \in V_1$ and $X_2, X_2' \in V_2$. Then we have $X_1 - X_1' = X_2 - X_2' \in V_1 \cap V_2$; but since $V_1 \cap V_2 = 0$, we have $X_1 = X_1'$, $X_2 = X_2'$. Conversely, if every $X \in V = V_1 + V_2$ can be written as $X_1 + X_2$, $X_i \in V_i$, $i = 1, 2$, in a unique way, then the sum $V = V_1 + V_2$ is direct (we leave this as an exercise). More generally, if V is a sum of subspaces $V_1, \cdots, V_k \subset \mathbb{R}^n$, we call V the direct sum and write $V = V_1 \oplus \cdots \oplus V_k$ if every vector in V can be written uniquely as a sum of vectors in the V_i.

Example 1. Consider the following two sets in $\mathbb{R}^n$ $(m < n)$:

$$U_m = \{(\lambda_1, \cdots, \lambda_m, 0, \cdots, 0)|\lambda_i \in \mathbb{R}\}$$

and

$$V_m = \{(0, \cdots, 0, \lambda_{m+1}, \cdots, \lambda_n)|\lambda_i \in \mathbb{R}\} \quad .$$

One immediately checks that U_m and V_m are subspaces of $\mathbb{R}^n$, that $U_m + V_m = \mathbb{R}^n$, and that $U_m \cap V_m = 0$. Hence $\mathbb{R}^n = U_m \oplus V_m$.

Example 2. In $\mathbb{R}^n$ we consider the so-called unit row-vectors

$$E_1 = (1,0,\cdots,0), \quad E_2 = (0,1,\cdots,0), \cdots, \quad E_n = (0,0,\cdots,1) \quad . \tag{3}$$

Every vector $X = (x_1, x_2, \cdots, x_n)$ can be uniquely written in the form $X = x_1E_1 + x_2E_2 + \cdots + x_nE_n$. Hence,

$$\mathbb{R}^n = \langle E_1\rangle \oplus \langle E_2\rangle \oplus \cdots \oplus \langle E_n\rangle \quad .$$

In an analogous way, we shall denote the unit column-vectors by the sumbols

$$E^{(1)} = [1,0,\cdots,0],\ E^{(2)} = [0,1,\cdots,0],\ \cdots,\ E^{(n)} = [0,0,\cdots,1] \quad . \tag{3'}$$

We shall use the notation E_i and $E^{(i)}$ later.

4. Linear dependence. A set of vectors $X_1, \cdots, X_k$ in $\mathbb{R}^n$ is called linearly dependent if there exist k numbers $\alpha_1, \alpha_2, \cdots, \alpha_k$, not all zero, such that

$$\alpha_1 X_1 + \alpha_2 X_2 + \cdots + \alpha_k X_k = 0 \tag{4}$$

(the right side is the zero vector). We say that (4) is a non-trivial linear dependence relation. On the other hand, if $\alpha_1 X_1 + \alpha_2 X_2 + \cdots + \alpha_k X_k = 0 \Longrightarrow \alpha_1 = \alpha_2 = \cdots = \alpha_k = 0$, then the vectors $X_1, X_2, \cdots, X_k$ are called linearly independent.

Example 2 above shows that the unit vectors $E_1, E_2, \cdots, E_n$ are linearly independent. A single non-zero vector X is always linearly independent, since $\lambda X = 0$, $X \neq 0 \Longrightarrow \lambda = 0$. Also note that the property of linear independence of a set of vectors $X_1, \cdots, X_k$ does not depend on the order in which the vectors are taken, since the terms $\alpha_i X_i$ in (4) can be permuted in any way desired.

THEOREM 1. (i) If any subset of $\{X_1, \cdots, X_k\}$ is linearly dependent, then the entire set of vectors is linearly dependent.

(ii) Any subset of a linearly independent set of vectors $\{X_1, \cdots, X_k\}$ is linearly independent.

(iii) If the vectors $X_1, \cdots, X_k$ are linearly dependent, then at least one is a linear combination of the others.

(iv) If one of the vectors $X_1, \cdots, X_k$ is a linear combination of the others, then the vectors $X_1, \cdots, X_k$ are linearly dependent.

(v) If the vectors $X_1, \cdots, X_k$ are linearly independent, and the vectors $X_1, \cdots, X_k, X$ are linearly dependent, then X is a linear combination of the vectors $X_1, \cdots, X_k$.

(vi) If the vectors $X_1, \cdots, X_k$ are linearly independent, and the vector X_{k+1} cannot be expressed as a linear combination of $X_1, \cdots, X_k$, then the set

$X_1, \cdots, X_k, X_{k+1}$ is linearly independent.

Proof. (i) Suppose, for example, that the first s vectors $X_1, \cdots, X_s$, $s < k$, are linearly dependent, i.e.,

$$\alpha_1 X_1 + \cdots + \alpha_s X_s = 0 \quad ,$$

where the α_i are not all zero. If we then set $\alpha_{s+1} = \cdots = \alpha_k = 0$, we obtain a non-trivial linear dependence relation

$$\alpha_1 X_1 + \cdots + \alpha_s X_s + \alpha_{s+1} X_{s+1} + \cdots + \alpha_k X_k = 0 \quad .$$

(ii) follows immediately from (i) (proof by contradiction).

(iii) Suppose, for example, that $\alpha_k \neq 0$ in (4). Then

$$X_k = -\frac{\alpha_1}{\alpha_k} X_1 - \cdots - \frac{\alpha_{k-1}}{\alpha_k} X_{k-1} \quad .$$

(iv) Suppose, for example, that $X_k = \beta_1 X_1 + \cdots + \beta_{k-1} X_{k-1}$. If we set $\alpha_1 = \beta_1, \cdots, \alpha_{k-1} = \beta_{k-1}, \alpha_k = -1$, we arrive at the relation (4) with $\alpha_k \neq 0$.

(v) If we have a non-trivial relation

$$\beta_1 X_1 + \cdots + \beta_k X_k + \beta X = 0$$

in which $\beta \neq 0$, then we obtain what we want as in (iii). But we cannot have $\beta = 0$, since $X_1, \cdots, X_k$ were assumed to be linearly independent.

(vi) follows immediately from (v). □

5. Bases. Dimension. We now given an important

Definition. Let V be a subspace of $\mathbb{R}^n$. A set of vectors $X_1, \cdots, X_r \in V$ is called a basis for V if it is linearly independent and its linear span coincides with V:

$$\langle X_1, \cdots, X_r \rangle = V \quad .$$

From this definition and the definition of the linear span of a set of vectors it follows that every vector $X \in V$ can be expressed in a unique way in the form $X = \alpha_1 X_1 + \cdots + \alpha_r X_r$. The coefficients $\alpha_1, \cdots, \alpha_r \in \mathbb{R}$ are called the coordinates of

X in the basis $X_1, \cdots, X_r$.

We have seen that the linearly independent unit vectors (3) span $\mathbb{R}^n$. Hence, $\{E_1, E_2, \cdots, E_n\}$ is a basis of $\mathbb{R}^n$. This basis, which is far from the only basis of $\mathbb{R}^n$, is called the <u>standard basis.</u> The reader can verify that, for example, the vectors

$$E_1' = E_1 , \; E_2' = E_1 + E_2 , \; E_3' = E_1 + E_2 + E_3 , \cdots , \; E_n' = E_1 + E_2 + \cdots + E_n$$

also make up a basis of $\mathbb{R}^n$.

We have not yet answered the important questions: does every subspace of $\mathbb{R}^n$ have a basis, and, if so, is the number of basis vectors the same in each basis? Both questions have a positive answer. To see this we shall need the following lemma.

LEMMA. <u>Let</u> V <u>be a subspace of</u> $\mathbb{R}^n$ <u>with basis</u> $X_1, \cdots, X_r$, <u>and let</u> $Y_1, Y_2, \cdots, Y_s$ <u>be a linearly independent set of vectors in</u> V. <u>Then</u> $s \le r$.

<u>Proof.</u> $Y_1, \cdots, Y_s$, like all vectors in V, are linear combinations of the basis vectors. Let

$$\begin{aligned} Y_1 &= a_{11}X_1 + a_{21}X_2 + \cdots + a_{r1}X_r , \\ Y_2 &= a_{12}X_1 + a_{22}X_2 + \cdots + a_{r2}X_r , \\ &\cdots\cdots\cdots\cdots\cdots\cdots \\ Y_s &= a_{1s}X_1 + a_{2s}X_2 + \cdots + a_{rs}X_r , \end{aligned}$$

where a_{ij} are scalars (since they are the coordinates of the Y_j, they are uniquely determined, but at this point we are not concerned about that). We use proof by contradiction. Suppose $s > r$.

We form a linear combination of the vectors Y_j with coefficients x_j:

$$x_1Y_1 + \cdots + x_sY_s = (a_{11}x_1 + a_{12}x_2 + \cdots + a_{1s}x_s)\,X_1 + \cdots$$
$$\cdots + (a_{r1}x_1 + a_{r2}x_2 + \cdots + a_{rs}x_s)\,X_r ,$$

and we consider the following system of r linear equations with s unknowns:

$$a_{11}x_1 + a_{12}x_2 + \cdots + a_{1s}x_s = 0 ,$$
$$\cdots\cdots\cdots\cdots\cdots$$
$$a_{r1}x_1 + a_{r2}x_2 + \cdots + a_{rs}x_s = 0 .$$

Since we have supposed that $s > r$, Corollary 2 of §3, Ch. 1 applies. It says that our system has a non-zero solution $(x_1^\circ, x_2^\circ, \cdots, x_s^\circ)$. But this gives us a non-trivial linear dependence relation

$$x_1^\circ Y_1 + x_2^\circ Y_2 + \cdots + x_s^\circ Y_s = 0 ,$$

which contradicts the hypothesis of the lemma. Hence, $s \leq r$. □

THEOREM 2. <u>Every non-zero subspace</u> $V \subset \mathbb{R}^n$ <u>has a finite basis. All bases of</u> V <u>have the same number</u> $r \leq n$ <u>of vectors.</u> (This number r is called the <u>dimension</u> of V and is denoted $\dim_{\mathbb{R}} V$ or simply $\dim V$.)

<u>Proof.</u> By assumption, $V \neq 0$. Let X_1 be any non-zero vector in V. Suppose that we have found k linearly independent vectors in $V : X_1, \cdots, X_k$. If the linear span $\langle X_1, \cdots, X_k \rangle$ does not coincide with V, then we choose any vector X_{k+1} in V which is not in $\langle X_1, \cdots, X_k \rangle$. In other words, X_{k+1} is not a linear combination of the vectors $X_1, \cdots, X_k$. By Theorem 1 (vi), the set $X_1, \cdots, X_k, X_{k+1}$ is linearly independent. This process cannot go on indefinitely, since all of the vectors X_i lie in $\mathbb{R}^n = \langle E_1, \cdots, E_n \rangle$, and, by the above lemma, no linearly independent set of vectors in $\mathbb{R}^n$ can contain more than n vectors. Hence, for some $r \leq n$ the linearly independent set $X_1, \cdots, X_k, \cdots, X_r \in V$ becomes <u>maximal,</u> i.e., no matter what vector $X \in V$ we add to the set, $\{X_1, \cdots, X_r, X\}$ is linearly dependent. By Theorem 1 (v), this means that $X \in \langle X_1, \cdots, X_r \rangle$ for every $X \in V$. Hence, $V = \langle X_1, \cdots, X_r \rangle$ and the vectors $X_1, \cdots, X_r$ are a basis for V.

Now suppose that $Y_1, \cdots, Y_s$ is another vasis of V. By the lemma, we have: $s \leq r$. But if we interchange the roles of $X_1, \cdots, X_r$ and $Y_1, \cdots, Y_s$ in the lemma, we similarly obtain: $r \leq s$. Hence, $s = r$, and the theorem is proved. □

Note that all of the above discussion applies equally well to spaces of row-vectors or column-vectors.

Theorem 2 allows us to associate to every linear subspace V of $\mathbb{R}^n$ a positive integer $r \le n$, which we call the dimension of V: $r = \dim V$. In particular, $\dim \mathbb{R}^n = n$. This important number can be characterized in other ways (see the exercises). One possible definition of dimension is based on the notion of the rank of a set of vectors. Namely, if $\{X_1, X_2, \cdots\}$ is a set of vectors (possibly infinite) in $\mathbb{R}^n$, then we now know that the dimension of the linear span $\langle X_1, X_2, \cdots\rangle$ is no greater than n. We call this dimension the <u>rank</u> of $\{X_1, X_2, \cdots\}$:

$$\operatorname{rank}\{X_1, X_2, \cdots\} = \dim \langle X_1, X_2, \cdots\rangle \quad .$$

Finally, some words to justify the term "linear subspace". In a linear subspace $V \subset \mathbb{R}^n$ choose an arbitrary basis $X_1, \cdots, X_r$. Then $X = \alpha_1 X_1 + \cdots + \alpha_r X_r$ for every $X \in V$, and the set V is in one-to-one correspondence with the set of all rows $(\alpha_1, \cdots, \alpha_r)$ of length r. Under this correspondence, a linear combination of vectors corresponds to the same linear combination of the rows of coordinates. Hence, once we choose a basis of V, we can interpret V as the vector space $\mathbb{R}^r$ imbedded in a certain way in $\mathbb{R}^n$ $(n \ge r)$.

EXERCISES

1. Let V, V_1 and V_2 be subspaces of $\mathbb{R}^n$, where $V \subset V_1 + V_2$. It is always true that $V = V \cap V_1 + V \cap V_2$? What can be said about this in the special case when $V_1 \subset V$?

2. Let V be a subspace of $\mathbb{R}^n$. If $V = U \oplus W$, then the subspace W is called a <u>complement</u> of U in V, and U is called a complement of W in V. Does U have only one complement in V? Compare W with the set-theoretic notion of complement $V \backslash U$ (see §4, Ch. 1).

3. Show that the vectors $X_1 = (1, 2, 3)$, $X_2 = (3, 2, 1)$ are linearly independent;

consider the linear span $V = \langle X_1, X_2 \rangle$; show that the vector $X = (-5, 2, 9)$ belongs to V, and find its coordinates in the basis X_1, X_2; find a complement (any complement) of V in $\mathbb{R}^3$.

4. Show that a set of n vectors in $\mathbb{R}^n$ spans $\mathbb{R}^n$ if and only if it is linearly independent.

5. Show that every linearly independent set of vectors $X_1, \cdots, X_k$ in a subspace $V \subset \mathbb{R}^n$ can be included in (extended to) a basis for V.

6. Let U and V be subspaces of $\mathbb{R}^n$. Prove that if $U \cap V = 0$, then $\dim(U + V) = \dim U + \dim V$.

7. Find the rank of the set of vectors $(0,1,1)$, $(1,0,1)$, $(1,1,0)$.

§2. The rank of a matrix

1. Back to equations. In the vector space $\mathbb{R}^m$ of columns of height m, consider n vectors

$$A^{(j)} = [a_{1j}, a_{2j}, \cdots, a_{mj}], \quad j = 1, 2, \cdots, n,$$

and their linear span $V = \langle A^{(1)}, A^{(2)}, \cdots, A^{(n)} \rangle$. Suppose we are given another vector $B = [b_1, b_2, \cdots, b_m]$. We would like to know whether B belongs to the subspace $V \subset \mathbb{R}^m$, and, if so, how its coordinates $b_1, \cdots, b_m$ (in the standard basis, see (3') §1) can be expressed in terms of the coordinates of the vectors $A^{(j)}$. In the case $\dim V = n$, i.e., when the $A^{(j)}$ form a basis, the second part of the question asks for the coordinates of B in the basis $A^{(1)}, \cdots, A^{(n)}$.

To answer this question, we take a linear combination of the vectors $A^{(j)}$ with arbitrary coordinates x_j, and write the equation $x_1 A^{(1)} + \cdots + x_n A^{(n)} = B$. Writing this in the form

$$x_1 \begin{Vmatrix} a_{11} \\ a_{21} \\ \vdots \\ a_{m1} \end{Vmatrix} + x_2 \begin{Vmatrix} a_{12} \\ a_{22} \\ \vdots \\ a_{m2} \end{Vmatrix} + \cdots + x_n \begin{Vmatrix} a_{1n} \\ a_{2n} \\ \vdots \\ a_{mn} \end{Vmatrix} = \begin{Vmatrix} b_1 \\ b_2 \\ \vdots \\ b_m \end{Vmatrix} \tag{1}$$

we see that we have a system of m linear equations with n unknowns:

$$\begin{aligned} a_{11}x_1 + a_{12}x_2 + \cdots + a_{1n}x_n &= b_1 \ , \\ a_{21}x_1 + a_{22}x_2 + \cdots + a_{2n}x_n &= b_2 \ , \\ \cdots\cdots\cdots&\cdots\cdots \\ a_{m1}x_1 + a_{m2}x_2 + \cdots + a_{mn}x_n &= b_m \ . \end{aligned} \tag{2}$$

This is the type of system we first discussed in §3 of Chapter 1. There we introduced the matrix and the extended matrix of the linear system (2):

$$A = \begin{Vmatrix} a_{11} & a_{12} & \cdots & a_{1n} \\ a_{21} & a_{22} & \cdots & a_{2n} \\ \cdots & \cdots & \cdots & \cdots \\ a_{m1} & a_{m2} & \cdots & a_{mn} \end{Vmatrix} , \quad (A|B) = \left\Vert \begin{array}{cccc|c} a_{11} & a_{12} & \cdots & a_{1n} & b_1 \\ a_{21} & a_{22} & \cdots & a_{2n} & b_2 \\ \cdots & \cdots & \cdots & \cdots & \cdot \\ a_{m1} & a_{m2} & \cdots & a_{mn} & b_m \end{array} \right\Vert \ . \tag{3}$$

It may seem at first that we are back where we started, having wasted time and accomplished nothing. But, in fact, we now have several important concepts at our disposal. We need only become accustomed to using them.

At this point it is useful to agree on some notation. We shall often abbreviate a sum $s_1 + s_2 + \cdots + s_n$ by writing $\sum_{i=1}^{n} s_i$. Here $s_1, \cdots, s_n$ can be anything (numbers, row vectors, etc.) which satisfies the usual rules of addition of numbers or vectors. We have:

$$\sum_{i=1}^{n} ts_i = t \sum_{i=1}^{n} s_i \ , \quad \sum_{i=1}^{n} (s_i + t_i) = \sum_{i=1}^{n} s_i + \sum_{i=1}^{n} t_i \ .$$

We shall also consider <u>double sums</u>

$$\sum_{j=1}^{n} \sum_{i=1}^{m} a_{ij} = \sum_{j=1}^{n} \left(\sum_{i=1}^{m} a_{ij} \right) = \sum_{i=1}^{m} \left(\sum_{j=1}^{n} a_{ij} \right) = \sum_{i,j} a_{ij} \ ,$$

in which the order of summation can be chosen in whichever way is convenient. It is easy to see that the sum does not depend on the order of summation, if we arrange the a_{ij} in the form of an $m \times n$ rectangular matrix; it makes no difference whether we sum the entries along the rows or along the columns.

Other types of summation will be introduced when we need them.

2. The rank of a matrix. By the column space of a rectangular $m \times n$ matrix A (see (3)) we mean the space $V = \langle A^{(1)}, A^{(2)}, \cdots, A^{(n)} \rangle$ introduced above. We shall denote this space by the symbol $V_v(A)$ or simply V_v (v for vertical). We call its dimension $r_v(A) = \dim V_v$ the column rank of the matrix A. We similarly define the row rank of A: $r_h(A) = \dim V_h$, where $V_h = \langle A_1, A_2, \cdots, A_m \rangle$ is the subspace of $\mathbb{R}^n$ spanned by the row-vectors $A_i = (a_{i1}, a_{i2}, \cdots, a_{in})$, $i = 1, 2, \cdots, m$ (h is for horizontal). In other words,

$$r_v(A) = \operatorname{rank}\{A^{(1)}, A^{(2)}, \cdots, A^{(n)}\}$$

$$r_h(A) = \operatorname{rank}\{A_1, A_2, \cdots, A_m\}$$

are the ranks of the set of column-vectors and the set of row-vectors, respectively. By Theorem 2 of §1 the number $r_v(A)$ and $r_h(A)$ are well defined.

Following the definition in §3 of Chapter 1, we shall say that a matrix A' is obtained from A using an elementary transformation of type (I) if $A'_s = A_t$, $A'_t = A_s$ for some pair of indices $s \neq t$ and $A'_i = A_i$ for $i \neq s, t$. We shall say that A' is obtained from A using an elementary transformation of type (II) if $A'_i = A_i$ for all $i \neq s$ and $A'_s = A_s + \lambda A_t$, $s \neq t$, $\lambda \in \mathbb{R}$.

Note that both types of elementary transformations are invertible, i.e., the matrix A' obtained from A using an elementary transformation can be changed back into A using another elementary transformation (in fact, of the same type).

LEMMA. *If A' is obtained from A using a finite sequence of elementary transformations, then:*

(i) $r_h(A') = r_h(A)$;

(ii) $r_v(A') = r_v(A)$.

Proof. It is sufficient to consider the case when A' is obtained from A using a single elementary transformation.

(i) An elementary transformation of type (I) clearly does not change $r_h(A)$, since the linear span of a set of vectors does not depend on the order in which they are listed: $\langle A_1, \cdots, A_s, \cdots, A_t, \cdots, A_m \rangle = \langle A_1, \cdots, A_t, \cdots, A_s, \cdots, A_m \rangle$. Next, $A'_s = A_s + \lambda A_t \Rightarrow A_s = A'_s - \lambda A_t$, and hence: $\langle A_1, \cdots, A_s + \lambda A_t, \cdots, A_t, \cdots, A_m \rangle = \langle A_1, \cdots, A_s, \cdots, A_t, \cdots, A_m \rangle$. Thus, $r_h(A)$ does not change when an elementary transformation of type (II) is applied.

(ii) Let $A'^{(j)}$, $j = 1, \cdots, n$, be the columns of A'. It is enough to prove that

$$\sum_{j=1}^{n} \lambda_j A^{(j)} = 0 \iff \sum_{j=1}^{n} \lambda_j A'^{(j)} = 0 \quad .$$

In that case every independent set of columns of one matrix corresponds to an independent set of columns (the ones with the same indices) of the other matrix; in particular, maximal independent sets of columns (i.e., bases) correspond, and so $r_v(A') = r_v(A)$. Further note that, since elementary transformations are invertible, it suffices to prove the implication in one direction. Suppose, for example, that $\sum_{j=1}^{n} \lambda_j A^{(j)} = 0$. Then, if we replace x_j by λ_j and all of the b_i by 0 in (1), we see that $(\lambda_1, \lambda_2, \cdots, \lambda_n)$ is a solution of the homogeneous systems HS associated with the system (2). By Theorem 1 of Chapter 1, this solution is also a solution of the homogeneous system HS' which has A' as its matrix and is obtained from HS using an elementary transformation of type (I) or (II). Since the system HS' can be written in the form $\Sigma x_j A'^{(j)} = 0$, we arrive at the relation: $\Sigma \lambda_j A'^{(j)} = 0$. □

The basic result of this section is the following fact:

THEOREM 1. $r_v(A) = r_h(A)$ <u>for any rectangular</u> $m \times n$ <u>matrix</u> A. (This number is called simply the <u>rank</u> of A, and is denoted $\operatorname{rank} A$.)

Proof. By Theorem 2 of §3, Ch. 1, the matrix A can be reduced to the following step form by applying a finite number of elementary transformations to the rows of A:

$$\overline{A} = \left\| \begin{matrix} \overline{a}_{11} \cdots \overline{a}_{1k} \cdots \overline{a}_{1\ell} \cdots \overline{a}_{1s} \cdots \overline{a}_{1n} \\ 0 \;\cdots\; \overline{a}_{2k} \cdots \overline{a}_{2\ell} \cdots \overline{a}_{2s} \cdots \overline{a}_{2n} \\ 0 \;\cdots\; 0 \;\cdots\; \overline{a}_{3\ell} \cdots \overline{a}_{3s} \cdots \overline{a}_{3n} \\ \cdot\;\cdot\;\cdot\;\cdot\;\cdot\;\cdot\;\cdot\;\cdot\;\cdot\;\cdot\;\cdot\;\cdot\;\cdot\;\cdot\;\cdot \\ 0 \;\cdots\; 0 \;\cdots\; 0 \;\cdots\; \overline{a}_{rs} \cdots \overline{a}_{rn} \\ 0 \;\cdots\; 0 \;\cdots\; 0 \;\cdots\; 0 \;\cdots\; 0 \\ \cdot\;\cdot\;\cdot\;\cdot\;\cdot\;\cdot\;\cdot\;\cdot\;\cdot\;\cdot\;\cdot\;\cdot\;\cdot\;\cdot\;\cdot \\ 0 \;\cdots\; 0 \;\cdots\; 0 \;\cdots\; 0 \;\cdots\; 0 \end{matrix} \right\| \qquad (4)$$

where $\overline{a}_{11}\overline{a}_{2k}\overline{a}_{3\ell}\cdots\overline{a}_{rs} \neq 0$. According to the lemma,

$$r_v(A) = r_v(\overline{A}) \quad , \quad r_h(A) = r_h(\overline{A}) \quad ,$$

so that it is sufficient to prove that $r_v(\overline{A}) = r_h(\overline{A})$.

The columns of A and $\overline{A}$ whose indices $1, k, \ell, \cdots, s$ correspond to the principal variables in (2) are called basis columns. We now show that this terminology is justified. Suppose we have a linear dependence relation

$$\lambda_1 \overline{A}^{(1)} + \lambda_k \overline{A}^{(k)} + \lambda_\ell \overline{A}^{(\ell)} + \cdots + \lambda_s \overline{A}^{(s)} = 0 \quad ,$$

for the column-vectors $\overline{A}^{(1)} = [\overline{a}_{11}, 0, \cdots, 0]$, $\overline{A}^{(k)} = [\overline{a}_{1k}, \overline{a}_{2k}, 0, \cdots, 0], \cdots, \overline{A}^{(s)} =$ $= [\overline{a}_{1s}, \overline{a}_{2s}, \cdots, \overline{a}_{rs}, 0, \cdots, 0]$ of the matrix (4). We then successively obtain: $\lambda_s \overline{a}_{rs} = 0, \cdots, \lambda_\ell \overline{a}_{3\ell} = 0$, $\lambda_k \overline{a}_{2k} = 0$, $\lambda_1 \overline{a}_{11} = 0$, and, since all of the $\overline{a}_{rs}, \cdots, \overline{a}_{3\ell}, \overline{a}_{2k}, \overline{a}_{11}$ are non-zero, it follows that $\lambda_1 = \lambda_k = \lambda_\ell = \cdots = \lambda_s = 0$. Hence, $\operatorname{rank}\{\overline{A}^{(1)}, \overline{A}^{(k)}, \overline{A}^{(\ell)}, \cdots, \overline{A}^{(s)}\} = r$, and so $r_v(\overline{A}) \geq r$. But the space $\overline{V}_v$ spanned by the columns of $\overline{A}$ can be identified with the space spanned by the columns of the matrix obtained by deleting the last $m-r$ zero rows from $\overline{A}$. Hence, $r_v(\overline{A}) = \dim \overline{V}_v \leq \dim \mathbb{R}^r = r$. Comparing the two inequalities shows that $r_v(\overline{A}) = r$. (The inequality $r_v(\overline{A}) \leq r$ also follows from the observation that all of the columns of $\overline{A}$

are linear combinations of the basis columns; we leave the details of this argument to the reader as an exercise.)

We now consider $r_h(\overline{A})$. We prove that all of the non-zero rows of $\overline{A}$ are linearly independent in the same way as we proved that the basis columns are linearly independent. Namely, if we had a relation

$$\lambda_1 \overline{A}_1 + \lambda_2 \overline{A}_2 + \cdots + \lambda_r \overline{A}_r = 0 \quad , \quad \lambda_i \in \mathbb{R} \quad ,$$

then we would obtain successively $\lambda_1 \overline{a}_{11} = 0$, $\lambda_2 \overline{a}_{2k} = 0, \cdots, \lambda_r \overline{a}_{rs} = 0$, and then $\lambda_1 = \lambda_2 = \cdots = \lambda_r = 0$. Thus, $r_h(\overline{A}) = r = r_v(\overline{A})$. □

3. <u>Solvability criterion.</u> The step form of a matrix A, which allows us to answer several questions about the linear system (see §3 of Chapter 1), involves an element of arbitrary choice, because of the flexibility in our choice of basis columns. In spite of this, our proof of Theorem 1 leads us to the following

COROLLARY. <u>The number of principal variables in a linear system</u> (2) <u>does not depend on the manner in which the system is reduced to step form. This number equals</u> rank A, <u>where</u> A <u>is the matrix of the system.</u>

<u>Proof.</u> We have seen that the number of principal variables is equal to the number of non-zero rows in $\overline{A}$ (see (4)), which coincides with the rank of A. The rank of a matrix was defined in an invariant way, i.e., it depends only on the matrix, not on $\overline{A}$. □

In the next chapter we shall find an effective method for computing the rank of a matrix A, which does not require reducing A to step form. Of course, such a method increases the value of any assertions based on the concept of rank. A simple but useful example of such an assertion is the following criterion for solvability of a linear system.

THEOREM 2 (Kronecker-Capelli). <u>A system of linear equations is compatible if and only if the rank of its matrix is equal to the rank of the extended matrix</u> (see (3)).

<u>Proof.</u> As explained at the beginning of this section, a system of linear equations (2) can be written in the form (1), and so it is compatible if and only if the column-vector B

of free terms can be written as a linear combination of the column-vectors $A^{(j)}$ of the matrix A. If it can be so written, then $B \in \langle A^{(1)}, \cdots, A^{(n)} \rangle$, and hence $\operatorname{rank}\{A^{(1)}, \cdots, A^{(n)}\} = \operatorname{rank}\{A^{(1)}, \cdots, A^{(n)}, B\}$, and $\operatorname{rank} A = r_v(A) = r_v(A|B) =$ $= \operatorname{rank}(A|B)$ (see Theorem 1).

Conversely, if the ranks of A and $(A|B)$ coincide, and $\{A^{(j_1)}, \cdots, A^{(j_r)}\}$ is a maximal linearly independent set of columns of A, then the set $\{A^{(j_1)}, \cdots, A^{(j_r)}, B\}$ is linearly dependent. By Theorem 1 (v) of §1, this means that B is a linear combination of the columns $A^{(j_i)}$. Thus, the system (2) is compatible. □

EXERCISES

1. Prove Theorem 1 without reducing the $m \times n$ matrix $A = (a_{ij})$ to step form.

2. As in the case of rows, interchanging the s-th and t-th columns of a matrix A is called an elementary transformation of type (I), and adding the t-th column multiplied by a scalar λ to the s-th column is called an elementary transformation of type (II). Describe the step form of A obtained by applying elementary transformations to the columns. Use elementary transformations of the columns to reduce the matrix $\overline{A}$ (see (4)) to the form

$$\tilde{A} = \left\| \begin{matrix} \tilde{a}_{11} & & & & & & \\ & \tilde{a}_{22} & & & & & \\ & & \ddots & & & & \\ & & & \tilde{a}_{rr} & & & \\ & & & & 0 & & \\ & & & & & \ddots & \\ & & & & & & 0 \end{matrix} \right\|,$$

where $\tilde{a}_{11} = \overline{a}_{11}$, $\tilde{a}_{22} = \overline{a}_{2k}$, $\tilde{a}_{33} = \overline{a}_{3\ell}, \cdots, \tilde{a}_{rr} = \overline{a}_{rs}$; $\prod_{i=1}^{r} \tilde{a}_{ii} \neq 0$.

3. Show that, if $a_0 \neq 0$, then the square matrix

$$A = \begin{Vmatrix} 0 & 0 & \dots & 0 & 0 & a_0 \\ 1 & 0 & \dots & 0 & 0 & a_1 \\ 0 & 1 & \dots & 0 & 0 & a_2 \\ \cdot & \cdot & \cdot & \cdot & \cdot & \cdot \\ 0 & 0 & \dots & 1 & 0 & a_{n-1} \\ 0 & 0 & \dots & 0 & 1 & a_n \end{Vmatrix}$$

has rank n.

4. Express the condition that the two matrices

$$A = \begin{Vmatrix} \alpha_1 & \alpha_2 & \dots & \alpha_n \\ & & & \\ \beta_1 & \beta_2 & \dots & \beta_n \end{Vmatrix}, \quad B = \begin{Vmatrix} \alpha_1 & \alpha_2 & \dots & \alpha_n \\ \beta_1 & \beta_2 & \dots & \beta_n \\ \gamma_1 & \gamma_2 & \dots & \gamma_n \end{Vmatrix}$$

have equal rank in terms of a geometrical property of a set of n lines in the plane.

§3. Linear maps. Matrix operations

1. Matrices and maps. Let $\mathbb{R}^n$ and $\mathbb{R}^m$ be the vector spaces of columns of height n and m, respectively. Let $A = (a_{ij})$ be an $m \times n$ matrix. We define a map $\varphi_A : \mathbb{R}^n \to \mathbb{R}^m$ as follows. For any $X = [x_1, x_2, \cdots, x_n] \in \mathbb{R}^n$, let

$$\varphi_A(X) = x_1 A^{(1)} + x_2 A^{(2)} + \cdots + x_n A^{(n)} , \tag{1}$$

where $A^{(1)}, \cdots, A^{(n)}$ are the columns of A (compare with (1) in §2). Since these columns have height m, the right side of (1) gives a column-vector $Y = [y_1, y_2, \cdots, y_m] \in \mathbb{R}^m$. Writing (1) in more detail, we have

$$y_i = \sum_{j=1}^{n} a_{ij} x_j , \quad i = 1, 2, \cdots, m . \tag{1'}$$

If $X = X' + X'' = [x_1' + x_1'', x_2' + x_2'', \cdots, x_n' + x_n'']$, then

$$\varphi_A(X' + X'') = \sum_{i=1}^{n} (x_i' + x_i'') A^{(i)} = \sum_{i=1}^{n} x_i' A^{(i)} + \sum_{i=1}^{n} x_i'' A^{(i)} = \varphi_A(X') + \varphi_A(X'') .$$

In addition,

$$\varphi_A(\lambda X) = \sum_{i=1}^{n} \lambda x_i A^{(i)} = \lambda \sum_{i=1}^{n} x_i A^{(i)} = \lambda \varphi_A(X) \quad , \quad \lambda \in \mathbb{R} \quad .$$

Conversely, suppose that $\varphi : \mathbb{R}^n \to \mathbb{R}^m$ is a map of sets which has the following two properties:

(i) $\varphi(X' + X'') = \varphi(X') + \varphi(X'')$ for all $X', X'' \in \mathbb{R}^n$;

(ii) $\varphi(\lambda X) = \lambda \varphi(X)$ for all $X \in \mathbb{R}^n$, $\lambda \in \mathbb{R}$.

Then, letting $E_n^{(1)}, \dots, E_n^{(n)}$ and $E_m^{(1)}, \dots, E_m^{(m)}$ denote the standard basis columns (see subsection 3 of §1) of the spaces $\mathbb{R}^n$ and $\mathbb{R}^m$, respectively, we apply properties (i) and (ii) to an arbitrary vector $X = [x_1, x_2, \dots, x_n] = \sum_{j=1}^{n} x_j E_n^{(j)} \in \mathbb{R}^n$:

$$\varphi(X) = \varphi\left(\sum_{j=1}^{n} x_j E_n^{(j)}\right) = \sum_{j=1}^{n} x_j \varphi(E_n^{(j)}) \quad . \tag{2}$$

The relation (2) shows that the map φ is completely determined by its values on the basis column-vectors. If we set

$$\varphi(E_n^{(j)}) = \sum_{i=1}^{m} a_{ij} E_m^{(i)} = [a_{1j}, a_{2j}, \dots, a_{mj}] = A^{(j)} \in \mathbb{R}^m , \tag{3}$$

we see that giving φ is equivalent to giving an $m \times n$ rectangular matrix $A = (a_{ij})$ with columns $A^{(1)}, \dots, A^{(n)}$. Then the expressions in (2) and in (1) coincide. Thus, we may set $\varphi = \varphi_A$.

<u>Definition.</u> A map $\varphi = \varphi_A : \mathbb{R}^n \to \mathbb{R}^m$ having properties (i) and (ii) is called a <u>linear map</u> from $\mathbb{R}^n$ to $\mathbb{R}^m$. Often (especially when $n = m$), the term <u>linear transformation</u> is also used. The matrix A is called the <u>matrix of the linear map</u> φ_A .

Let $\varphi_A, \varphi_{A'}$ be two linear maps $\mathbb{R}^n \to \mathbb{R}^m$ with matrices $A = (a_{ij})$ and $A' = (a'_{ij})$. Then $\varphi_A = \varphi_{A'}$ if and only if the values $\varphi_A(X)$ and $\varphi_{A'}(X)$ coincide for all $X \in \mathbb{R}^n$. In particular, $A'^{(j)} = \varphi_{A'}(E_n^{(j)}) = \varphi_A(E_n^{(j)}) = A^{(j)}$, $1 \le j \le n$, so that $a'_{ij} = a_{ij}$ and $A' = A$.

We summarize our results:

THEOREM 1. <u>There is a one-to-one correspondence between linear maps from</u> $\mathbb{R}^n$ <u>to</u> $\mathbb{R}^m$ <u>and</u> $m \times n$ <u>matrices.</u> □

It should be emphasized that it makes no sense to speak of a linear map $S \to T$ of arbitrary sets S and T. Conditions (i) and (ii) presuppose that S and T are subspaces of vector spaces $\mathbb{R}^n$ and $\mathbb{R}^m$.

We call attention to the special case when $m = 1$. Then a linear map $\varphi : \mathbb{R}^n \to \mathbb{R}$, which is usually called a <u>linear function</u> in n variables, is given by specifying n scalars $a_1, a_2, \cdots, a_n$:

$$\varphi(X) = \varphi(x_1, x_2, \cdots, x_n) = a_1 x_1 + a_2 x_2 + \cdots + a_n x_n \quad . \tag{4}$$

(Note that this terminology is different from that used in high school, where, in the case of one variable x, a linear function is defined as a function of the form $x \mapsto ax + b$.)

For fixed n and m, linear maps $\mathbb{R}^n \to \mathbb{R}^m$ can be added and multiplied by scalars. Namely, suppose that $\varphi_A, \varphi_B : \mathbb{R}^n \to \mathbb{R}^m$ are two linear maps. We define the map

$$\varphi = \alpha\varphi_A + \beta\varphi_B : \mathbb{R}^n \longrightarrow \mathbb{R}^m \quad , \quad \alpha, \beta \in \mathbb{R} \quad ,$$

by setting $\varphi(X)$ equal to the corresponding linear combination of the values of φ_A and φ_B:

$$\varphi(X) = \alpha\varphi_A(X) + \beta\varphi_B(X) \quad .$$

The expression on the right is an ordinary linear combination of column-vectors.

Since

$$\begin{aligned}\varphi(X' + X'') &= \alpha\varphi_A(X' + X'') + \beta\varphi_B(X' + X'') = \\ &= \alpha\{\varphi_A(X') + \varphi_A(X'')\} + \beta\{\varphi_B(X') + \varphi_B(X'')\} = \\ &= \{\alpha\varphi_A(X') + \beta\varphi_B(X')\} + \{\alpha\varphi_A(X'') + \beta\varphi_B(X'')\} = \varphi(X') + \varphi(X'') ;\end{aligned}$$

$$\begin{aligned}\varphi(\lambda X) &= \alpha\varphi_A(\lambda X) + \beta\varphi_B(\lambda X) = \alpha\lambda\varphi_A(X) + \beta\lambda\varphi_B(X) = \\ &= \lambda\{\alpha\varphi_A(X) + \beta\varphi_B(X)\} = \lambda\varphi(X)\end{aligned}$$

(here we have implicitly used the rules VS_1 - VS_8 in §1), it follows that φ is a linear map. By Theorem 1, we may speak of its matrix $C : \varphi = \varphi_C$. In order to find C, we write the j-th column, following (3):

$$[c_{1j}, c_{2j}, \cdots, c_{mj}] = C^{(j)} = \varphi_C(E_n^{(j)}) =$$
$$= \alpha\varphi_A(E_n^{(j)}) + \beta\varphi_B(E_n^{(j)}) = \alpha A^{(j)} + \beta B^{(j)} =$$
$$= [\alpha a_{1j} + \beta b_{1j}, \alpha a_{2j} + \beta b_{2j}, \cdots, \alpha a_{mj} + \beta b_{mj}] \quad .$$

It is natural to call the matrix $C = (c_{ij})$ with entries $c_{ij} = \alpha a_{ij} + \beta b_{ij}$ the linear combination of the matrices A and B with coefficients α and β. Thus, we define

$$\alpha \begin{Vmatrix} a_{11} & \cdots & a_{1n} \\ \cdot & \cdots & \cdot \\ a_{m1} & \cdots & a_{mn} \end{Vmatrix} + \beta \begin{Vmatrix} b_{11} & \cdots & b_{1n} \\ \cdot & \cdots & \cdot \\ b_{m1} & \cdots & b_{mn} \end{Vmatrix} = \begin{Vmatrix} \alpha a_{11} + \beta b_{11} & \cdots & \alpha a_{1n} + \beta b_{1n} \\ \cdot & \cdots & \cdot \\ \alpha a_{m1} + \beta b_{m1} & \cdots & \alpha a_{mn} + \beta b_{mn} \end{Vmatrix} . \quad (5)$$

Then,

$$\alpha\varphi_A + \beta\varphi_B = \varphi_{\alpha A + \beta B} \quad . \quad (6)$$

We will very frequently make use of the fact that a linear combination of linear functions is a linear function.

Finally, we note that, if the rules VS_1 - VS_8 in §1 are rewritten with the column-vectors X, Y, Z everywhere replaced by $m \times n$ matrices, then we obtain rules VSM_1 - VSM_8 (where addition and scalar multiplication of matrices are defined by (5)), which justify our speaking of the vector space of $m \times n$ matrices. We can think of $m \times n$ matrices as a compact way of writing elements of the vector space $\mathbb{R}^{mn}$ of rows of length mn (by dividing a row into segments of length n and placing these segments under one another).

2. Matrix multiplication. The relations (5) and (6) show that the operations of addition and scalar multiplication in the set of $m \times n$ matrices and the set of linear maps $\mathbb{R}^n \to \mathbb{R}^m$ agree, i.e., these operations are preserved in the one-to-one correspondence of Theorem 1. In set theory we have the important concept of the product (composition) of

two maps (see subsection 2 of §5, Ch. 1). It is reasonable to expect that the composition of two linear maps will give us an important operation on the corresponding matrices. We now see how this works.

Let $\varphi_B : \mathbb{R}^n \to \mathbb{R}^s$ and $\varphi_A : \mathbb{R}^s \to \mathbb{R}^m$ be linear maps, and let $\varphi_C = \varphi_A \circ \varphi_B$ be their product:

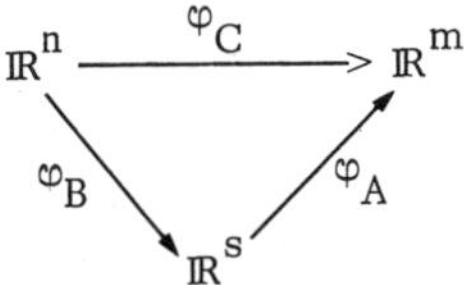

Before writing φ_C for the composition $\varphi = \varphi_A \circ \varphi_B$, we should actually verify that this product is a linear map, but this is easy:

$$(i) \quad \varphi(X' + X'') = \varphi_A(\varphi_B(X' + X'')) = \varphi_A(\varphi_B(X') + \varphi_B(X'')) =$$

$$= \varphi_A(\varphi_B(X')) + \varphi_A(\varphi_B(X'')) = \varphi(X') + \varphi(X'') ;$$

$$(ii) \quad \varphi(\lambda X) = \varphi_A(\varphi_B(\lambda X)) = \varphi_A(\lambda \varphi_B(X)) = \lambda \varphi_A(\varphi_B(X)) =$$

$$= \lambda \varphi(X) ;$$

hence, by Theorem 1, the composition φ corresponds to some matrix C.

We now write the action of the maps on columns in the chain

$$[x_1, \cdots, x_n] \overset{\varphi_B}{\longmapsto} [y_1, \cdots, y_s] \overset{\varphi_A}{\longmapsto} [z_1, \cdots, z_m]$$

explicitly using the formula (1'):

$$z_i = \sum_{k=1}^{s} a_{ik} y_k = \sum_{k=1}^{s} a_{ik} \sum_{j=1}^{n} b_{kj} x_j = \sum_{j=1}^{n} \left(\sum_{k=1}^{s} a_{ik} b_{kj} \right) x_j \quad .$$

On the other hand,

$$z_i = \sum_{j=1}^{n} c_{ij} x_j \quad , \quad i = 1, 2, \cdots, m \quad .$$

Comparing these expressions and recalling that the x_j $(j = 1, 2, \cdots, n)$ are arbitrary real numbers, we arrive at the relations

$$c_{ij} = \sum_{k=1}^{s} a_{ik} b_{kj} \quad , \quad 1 \le i \le m \quad , \quad 1 \le j \le n \quad . \tag{7}$$

We shall say that the matrix $C = (c_{ij})$ is obtained by <u>multiplying</u> the matrix A by the matrix B. It is customary to write: $C = AB$. Thus, the <u>product</u> of an $m \times s$ matrix (a_{ik}) and an $s \times n$ matrix (b_{kj}) is, by definition, the $m \times n$ matrix whose entries c_{ij} are given by (7). We have proved

THEOREM 2. <u>The product</u> $\varphi_A \varphi_B$ <u>of two linear maps with matrices</u> A <u>and</u> B <u>is the linear map with matrix</u> $C = AB$. <u>In other words,</u>

$$\varphi_A \varphi_B = \varphi_{AB} \quad . \quad \square \tag{8}$$

Relation (8) is a natural addition to relation (6).

We can forget about linear maps and find the product AB of any two matrices A and B, if, however, we keep in mind that <u>the symbol</u> AB <u>only makes sense when the number of columns in</u> A <u>equals the number of rows in</u> B. Under this condition, rule (7) says to "multiply the i-th row in A by the j-th column in B to get the ij-entry in AB":

$$c_{ij} = (a_{i1}, \dots, a_{is})[b_{1j}, \dots, b_{sj}] = A_i B^{(j)} \quad . \tag{9}$$

Note that <u>the number of rows in</u> AB <u>is equal to the number of rows in</u> A, <u>and the number of columns in</u> AB <u>is equal to the number of columns in</u> B. In particular, the product of square matrices of the same size is always defined, and the product is another square matrix of the same size.

Even in the case of square matrices, the order in which we multiply two matrices is important, since, in general $AB \neq BA$, as we see from the following example:

$$\begin{Vmatrix} 1 & 0 \\ 0 & 0 \end{Vmatrix} \begin{Vmatrix} 0 & 0 \\ 1 & 0 \end{Vmatrix} = \begin{Vmatrix} 0 & 0 \\ 0 & 0 \end{Vmatrix} \neq \begin{Vmatrix} 0 & 0 \\ 1 & 0 \end{Vmatrix} = \begin{Vmatrix} 0 & 0 \\ 1 & 0 \end{Vmatrix} \begin{Vmatrix} 1 & 0 \\ 0 & 0 \end{Vmatrix} \quad .$$

Of course, it would have been possible to define matrix multiplication in many other ways (for example, multiplying corresponding entries in matrices of the same size), but no other way can compare in importance with the above type of matrix multiplication. This is not surprising, since we arrived at it by studying the composition of linear maps, and the concept of maps and composition of maps is one of the most fundamental in mathematics.

COROLLARY. Matrix multiplication is associative:

$$A(BC) = (AB)C \quad .$$

Proof. Matrix multiplication corresponds to composition of the corresponding linear maps (Theorem 2 and relation (8)), and, by Theorem 1 of §5 Ch. 1, composition of any maps is associative. It is also possible to prove associativity of matrix multiplication by a direct computation, using (7). □

3. Square matrices. Let $M_n(\mathbb{R})$ (or simply M_n) denote the set of all square $n \times n$ matrices (a_{ij}) with real entries a_{ij}.

The identity $e_{\mathbb{R}^n} : \mathbb{R}^n \to \mathbb{R}^n$, which takes every column $X \in \mathbb{R}^n$ to itself, obviously corresponds to the matrix

$$E = \begin{Vmatrix} 1 & 0 & \dots & 0 \\ 0 & 1 & \dots & 0 \\ \dots & & \dots & \dots \\ 0 & 0 & \dots & 1 \end{Vmatrix} \quad .$$

We can write $E = (\delta_{kj})$, where

$$\delta_{kj} = \begin{cases} 1, & \text{if } k = j, \\ 0, & \text{if } k \neq j, \end{cases}$$

is the Kronecker symbol. The rule for matrix multiplication (7), with b_{kj} replaced by δ_{kj}, shows that

$$EA = A = AE \quad , \quad \forall A \in M_n(\mathbb{R}) \quad . \tag{10}$$

Of course, the matrix relations (10), which we verified by computation, follow from the relations $e\varphi = \varphi = \varphi e$ for any map φ (see subsection 2 of §5 Ch. 1), if we use Theorem 1 and the equality (8) with $\varphi_A = \varphi$, $\varphi_B = \varphi_E = e$.

We know that matrices in $M_n(\mathbb{R})$ can be multiplied by scalars: if $A = (a_{ij})$, then λA is the matrix (λa_{ij}) (see (5)). But scalar multiplication can be considered as a special case of matrix multiplication, since

$$\lambda A = \mathrm{diag}_n(\lambda) \cdot A = A\,\mathrm{diag}_n(\lambda) \quad , \tag{11}$$

where

$$\mathrm{diag}_n(\lambda) = \lambda E = \begin{Vmatrix} \lambda & 0 & \dots & 0 \\ 0 & \lambda & \dots & 0 \\ \cdot & \cdot & \cdot & \cdot \\ 0 & 0 & \dots & \lambda \end{Vmatrix}$$

is the <u>scalar matrix</u> we have encountered before (see §3 Ch. 1).

An easy verification shows that, as claimed in (11), $\mathrm{diag}_n(\lambda)$ commutes with any matrix A. The following converse is very important in applications.

THEOREM 3. <u>A matrix in</u> M_n <u>which commutes with every matrix in</u> M_n <u>must be a scalar matrix.</u>

<u>Proof.</u> Define E_{ij} to be the $n \times n$ matrix having 1 at the intersection of the i-th row and j-th column and 0 everywhere else. If $Z = (z_{ij})$ is a matrix which commutes with every matrix in M_n, then, in particular, it commutes with all of the E_{ij}:

$$ZE_{ij} = E_{ij}Z \quad , \quad i,j = 1, 2, \cdots, n \quad .$$

Multiplying the matrices on the left and the right of this equality, we obtain the matrices

$$\begin{Vmatrix} 0 & \dots & z_{1i} & \dots & 0 \\ 0 & \dots & z_{2i} & \dots & 0 \\ \cdot & \cdot & \cdot & \cdot & \cdot \\ 0 & \dots & z_{ni} & \dots & 0 \end{Vmatrix} \text{ and } \begin{Vmatrix} 0 & 0 & \dots & 0 \\ \cdot & \cdot & \cdot & \cdot \\ z_{j1} & z_{j2} & \dots & z_{jn} \\ \cdot & \cdot & \cdot & \cdot \\ 0 & 0 & \dots & 0 \end{Vmatrix} (i)$$

(j)

where the matrix on the left has all zero columns except for the j-th, and the matrix on the right has all zero rows except for the i-th. Equating these matrices immediately leads to the relations $z_{ki} = 0$ for $k \neq i$ and $z_{ii} = z_{jj}$. Letting i and j vary, we conclude that Z must be a scalar matrix. □

We further note the relations $\lambda(AB) = (\lambda A)B = A(\lambda B)$, which follow immediately from the definition of matrix multiplication and scalar multiplication of matrices, or, if we prefer, from (11) and the associativity of matrix multiplication.

Given a matrix $A \in M_n(\mathbb{R})$, we can try to find a matrix $B \in M_n(\mathbb{R})$ such that

$$AB = E = BA \quad . \tag{12}$$

If such a matrix B exists, then (12) translates into the following relation for the corresponding linear maps:

$$\varphi_A \varphi_B = e = \varphi_B \varphi_A \quad . \tag{12'}$$

In other words, $\varphi_B = \varphi_A^{-1}$ is the inverse of φ_A. According to Theorem 2 of §5 of Ch. 1, φ_A^{-1} exists if and only if φ_A is a bijective map. In that case φ_A^{-1} is uniquely determined. Since $\varphi_A(0) = 0$, bijectivity of φ_A means that, in particular,

$$X \neq 0 \ , \ X \in \mathbb{R}^n \implies \varphi_A(X) \neq 0 \quad . \tag{13}$$

Now let φ_A be a bijective linear map from $\mathbb{R}^n$ to $\mathbb{R}^n$. We already know that it has an inverse map φ_A^{-1}, but at this point we cannot say that this inverse is a linear map. To see that φ_A^{-1} is, in fact, a linear map, we introduce the column-vectors

$$X = \varphi_A^{-1}(X' + X'') - \varphi_A^{-1}(X') - \varphi_A^{-1}(X'') \ ,$$

$$Y = \varphi_A^{-1}(\lambda Y') - \lambda \varphi_A^{-1}(Y')$$

and apply φ_A to both sides of these equalities. Because φ_A is linear, we obtain

$$\varphi_A(X) = \varphi_A(\varphi_A^{-1}(X' + X'')) - \varphi_A(\varphi_A^{-1}(X')) - \varphi_A(\varphi_A^{-1}(X'')) \ ,$$

$$\varphi_A(Y) = \varphi_A(\varphi_A^{-1}(\lambda Y')) - \lambda \varphi_A(\varphi_A^{-1}(Y')) \quad .$$

Since $\varphi_A \varphi_A^{-1} = e$, we have

$$\varphi_A(X) = e(X' + X'') - e(X') - e(X'') = 0 \ ,$$

$$\varphi_A(Y) = e(\lambda Y') - \lambda e(Y') = 0 \ ,$$

and hence, because of (13), it follows that X and Y are the zero vector. Thus, properties (i) and (ii) of subsection 1 are fulfilled, i.e., φ_A^{-1} is a linear map. We then have $\varphi_A^{-1} = \varphi_B$ for some matrix B. If we rewrite (12') in the form $\varphi_{AB} = \varphi_e = \varphi_{BA}$ (see (8)) and again use Theorem 1, we arrive at the equality (12). We conclude: <u>a matrix</u> $A \in M_n(\mathbb{R})$ <u>has an inverse if and only if the map</u> $\varphi_A : \mathbb{R}^n \to \mathbb{R}^n$ <u>is bijective. In that case the map</u> φ_A^{-1} <u>is linear.</u>

Bijectivity of φ_A is equivalent to the condition that every column-vector $Y \in \mathbb{R}^n$ can be written in a unique way in the form (1)

$$Y = \varphi_A(X) = x_1 A^{(1)} + x_2 A^{(2)} + \cdots + x_n A^{(n)} \quad ,$$

where $A^{(1)}, A^{(2)}, \cdots, A^{(n)}$ are the columns of A. (Surjectivity of φ_A ensures the existence of X for which $Y = \varphi_A(X)$, and injectivity of φ_A gives uniqueness of X: if we had $Y = \varphi_A(X') = \varphi_A(X'')$, then we would have $\varphi_A(X' - X'') = \varphi_A(X') - \varphi_A(X'') = 0$, and so, by (13), $X' - X'' = 0$.) Hence, when φ_A is bijective, $\mathbb{R}^n$ coincides with the column space $V_v(A) = \langle A^{(1)}, \cdots, A^{(n)} \rangle$; thus, $\operatorname{rank} A = \dim \mathbb{R}^n = n$.

If A has an inverse, then it is unique: it is the matrix corresponding to φ_A^{-1}. It is customary to denote this inverse A^{-1}. Then we have

$$\varphi_A^{-1} = \varphi_{A^{-1}} \quad . \tag{14}$$

A square matrix A which has an inverse A^{-1} is called <u>non-singular.</u> The corresponding linear map φ_A is also called non-singular. Otherwise they are called <u>singular.</u>

We summarize our results in the following theorem.

THEOREM 4. <u>An</u> $n \times n$ <u>square matrix</u> A <u>is non-singular if and only if it has rank</u> n. <u>In that case the inverse map</u> φ_A^{-1} <u>of</u> φ_A <u>is linear and is given by</u> (14). □

COROLLARY. <u>If</u> φ_A <u>is non-singular, then so is</u> $\varphi_{A^{-1}}$, <u>and</u> $(A^{-1})^{-1} = A$. <u>If</u> $A, B, \cdots, C, D$ <u>are non-singular</u> $n \times n$ <u>matrices, then the product</u> $AB \cdots CD$ <u>is also non-singular, and</u> $(AB \cdots CD)^{-1} = D^{-1} C^{-1} \cdots B^{-1} A^{-1}$.

One can prove this either using the corollary to Theorem 2 of §5 Ch. 1, or else using the symmetry of the condition $A A^{-1} = E = A^{-1} A$. □

We shall give an explicit formula for A^{-1} in Chapter 3. Here we merely mention that the actual computation of A^{-1} (or even multiplication of two matrices), say by the method at the end of this chapter, usually requires a large number of arithmetical operations. In some applications one encounters matrices of size 100×100 and larger. If A and B are two $n \times n$ matrices, then the computation of $C = AB$ involves finding n^2 entries c_{ij} using the formula (7) (or (9)), and each of these n^2 computations requires $2n - 1$ multiplications or additions. Thus, $(2n-1)n^2$ operations are required in all, i.e., almost two million operations for $n = 100$. This is still an easy problem for a modern computer, but what if we want to find the m-th <u>power</u> A^m with $m \geq 1000$? Here, by definition, $A^m = A A^{m-1}$, and, in fact, it follows easily from associativity (see the corollary to Theorem 2) that $A^m = A^k A^{m-k}$ if $0 \leq k \leq m$. To compute A^m one can use various additional devices which come from either a general study of linear algebra or from some special knowledge of the nature of the matrix A. We shall give three examples.

<u>Example 1.</u> If

$$A = \operatorname{diag}\{\alpha_1, \cdots, \alpha_n\} = \begin{Vmatrix} \alpha_1 & \cdots & 0 \\ \cdot & \cdots & \cdot \\ 0 & & \alpha_n \end{Vmatrix},$$

then obviously

$$A^m = \operatorname{diag}\{\alpha_1^m, \cdots, \alpha_n^m\} = \begin{Vmatrix} \alpha_1^m & \cdots & 0 \\ \cdot & \cdots & \cdot \\ 0 & & \alpha_n^m \end{Vmatrix}.$$

<u>Example 2.</u> Let

$$A = \begin{Vmatrix} a & c \\ 0 & b \end{Vmatrix}.$$

Then, using induction on m, one can show that

$$A^m = \begin{Vmatrix} a^m & c\,\dfrac{a^m - b^m}{a - b} \\ 0 & b^m \end{Vmatrix},$$

where $\frac{a^m - b^m}{a - b} = a^{m-1} + a^{m-2}b + \cdots + ab^{m-2} + b^{m-1}$. In particular, if $a = b$, we have

$$\begin{Vmatrix} a & c \\ 0 & a \end{Vmatrix}^m = \begin{Vmatrix} a^m & ma^{m-1}c \\ 0 & a^m \end{Vmatrix} .$$

Example 3. Using induction on m, it is not hard to show that the m-th power of the matrix

$$A = \begin{Vmatrix} 0 & 1 \\ 1 & 1 \end{Vmatrix}$$

has thc form

$$A^m = \begin{Vmatrix} f_{m-1} & f_m \\ f_m & f_{m+1} \end{Vmatrix} , \tag{15}$$

where the integers $f_0 = 0$, $f_1 = 1$, $f_2 = 1$, $f_3 = 2, \cdots$ are defined recursively by: $f_{m+1} = f_m + f_{m-1}$. These are precisely the Fibonacci numbers (see example 2 at the end of §3 Ch. 1).

We introduce the matrix

$$B = \begin{Vmatrix} -\frac{\lambda_2}{5} & \frac{1}{5} \\ -\sqrt{5}\,\lambda_1 & \sqrt{5} \end{Vmatrix}$$

with determinant 1 (see §4 Ch. 1), where

$$\lambda_1 = \frac{1 + \sqrt{5}}{2} , \quad \lambda_2 = \frac{1 - \sqrt{5}}{2} .$$

A simple computation shows that

$$B^{-1} = \begin{Vmatrix} \sqrt{5} & -\frac{1}{5} \\ \sqrt{5}\,\lambda_1 & -\frac{\lambda_2}{5} \end{Vmatrix} \quad \text{and} \quad A = B^{-1} \cdot \begin{Vmatrix} \lambda_1 & 0 \\ 0 & \lambda_2 \end{Vmatrix} \cdot B .$$

But if three $n \times n$ matrices A, B, C, with B non-singular, are connected by the relationship $A = B^{-1}CB$, then

$$A^m = B^{-1}CB \cdot B^{-1}CB \cdot B^{-1}CB \cdots B^{-1}CB = B^{-1}C^m B$$

(the inner factors BB^{-1} equal E, and so "cancel"). In our case, using example 1 and the relation (15), we have

$$\begin{Vmatrix} f_{m-1} & f_m \\ f_m & f_{m+1} \end{Vmatrix} = A^m = B^{-1}\begin{Vmatrix} \lambda_1 & 0 \\ 0 & \lambda_2 \end{Vmatrix}^m B = B^{-1}\begin{Vmatrix} \lambda_1^m & 0 \\ 0 & \lambda_2^m \end{Vmatrix} B =$$

$$= \begin{Vmatrix} \sqrt{5} & -\frac{1}{5} \\ \sqrt{5}\,\lambda_1 & -\frac{\lambda_2}{5} \end{Vmatrix} \begin{Vmatrix} \lambda_1^m & 0 \\ 0 & \lambda_2^m \end{Vmatrix} B = \begin{Vmatrix} \sqrt{5}\,\lambda_1^m & -\frac{\lambda_2^m}{5} \\ \sqrt{5}\,\lambda_1^{m+1} & -\frac{\lambda_2^{m+1}}{5} \end{Vmatrix} \begin{Vmatrix} -\frac{\lambda_2}{5} & \frac{1}{5} \\ -\sqrt{5}\,\lambda_1 & \sqrt{5} \end{Vmatrix} =$$

$$= \begin{Vmatrix} * & \frac{1}{\sqrt{5}}(\lambda_1^m - \lambda_2^m) \\ * & * \end{Vmatrix}$$

(the stars stand for entries which we are not interested in).

Comparing the upper right entry in the first and last matrices in this sequence of equalities, we obtain the following formula for the m-th Fibonacci number:

$$f_m = \frac{\lambda_1^m - \lambda_2^m}{\sqrt{5}} = \frac{1}{\sqrt{5}}\left\{\left(\frac{1+\sqrt{5}}{2}\right)^m - \left(\frac{1-\sqrt{5}}{2}\right)^m\right\} .$$

We see that $f_m \sim \frac{1}{\sqrt{5}}\lambda_1^m$ for large m (i.e., the Fibonacci numbers start to look like a geometric progression), since $\lim_{m\to\infty}\left(\frac{1-\sqrt{5}}{2}\right)^m = 0$.

We have obtained quite a few rules for working with square matrices: VSM_1 - VSM_8 (see the remark at the end of subsection 1), associativity (the corollary to Theorem 2), the relation (10), and Theorem 4. We further note the so-called <u>distributive laws</u>:

$$(A + B)C = AC + BC \ , \quad C(A + B) = CA + CB \ , \tag{16}$$

where A, B, C are any matrices in $M_n(\mathbb{R})$. To verify this, we set $A = (a_{ij})$, $B = (b_{ij})$, $C = (c_{ij})$. Using the distributive law for $\mathbb{R}$, we see that for any $i, j = 1, \cdots, n$, we have

$$\sum_{k=1}^{n} (a_{ik} + b_{ik}) c_{kj} = \sum_{k=1}^{n} a_{ik} c_{kj} + \sum_{k=1}^{n} b_{ik} c_{kj} \quad .$$

The left side of this equality is the entry in the i-th row and j-th column of the matrix $(A + B)C$, and the two sums on the right are the (i, j)-entries in the matrices AC and BC, respectively. The second distributive law in (16) is verified in exactly the same way. (We really do have to verify both laws, since multiplication is not commutative in $M_n(\mathbb{R})$.)

The distributive laws

$$(\varphi + \psi)\xi = \varphi\xi + \psi\xi \quad , \quad \xi(\varphi + \psi) = \xi\varphi + \xi\psi \tag{16'}$$

for linear maps φ, ψ, ξ from $\mathbb{R}^n$ to $\mathbb{R}^n$ follow immediately from (16), because of the correspondence between maps and matrices. Alternately, we could have first proved (16'), using the following sequence of equalities, and then derived (16) from (16'):

$$\begin{aligned} ((\varphi + \psi)\xi)(X) &= (\varphi + \psi)(\xi X) = \varphi(\xi X) + \psi(\xi X) = \\ &= (\varphi\xi)(X) + (\psi\xi)(X) = (\varphi\xi + \psi\xi)(X) \quad , \quad X \in \mathbb{R}^n \quad . \end{aligned}$$

EXERCISES

1. Which of the following maps are linear?

   a) $[x_1, x_2, \cdots, x_n] \longmapsto [x_n, \cdots, x_2, x_1]$;

   b) $[x_1, x_2, \cdots, x_n] \longmapsto [x_1, x_2^2, \cdots, x_n^n]$;

   c) $[x_1, x_2, \cdots, x_n] \longmapsto [x_1, x_1 + x_2, \cdots, x_1 + x_2 + \cdots + x_n]$.

2. Prove that

$$A = \begin{Vmatrix} a & b \\ c & d \end{Vmatrix} \quad , \quad ad - bc \neq 0 \Longrightarrow A^{-1} = \frac{1}{ad - bc} \begin{Vmatrix} d & -b \\ -c & a \end{Vmatrix} \quad .$$

In particular, $ad - bc = 1 \Longrightarrow A^{-1} = \begin{Vmatrix} d & -b \\ -c & a \end{Vmatrix}$. Does A^{-1} exist if $ad - bc = 0$?

3. Prove that any matrix

$$A = \begin{Vmatrix} a & b \\ c & d \end{Vmatrix}$$

satisfies the relation

$$A^2 = (a + d)A - (ad - bc)E$$

(in other words, A is a "root" of the quadratic equation $x^2 - (a+d)x + (ad - bc) = 0$).

4. If $ad - bc \neq 0$, use the relation in Exercise 3 to find the inverse matrix A^{-1}.

5. Prove that

$$\begin{Vmatrix} 1 & a & c \\ 0 & 1 & b \\ 0 & 0 & 1 \end{Vmatrix}^m = \begin{Vmatrix} 1 & ma & \frac{m(m-1)}{2}ab + mc \\ 0 & 1 & mb \\ 0 & 0 & 1 \end{Vmatrix} .$$

Find the inverse of the matrix $\begin{Vmatrix} 1 & a & c \\ 0 & 1 & b \\ 0 & 0 & 1 \end{Vmatrix}$.

6. Verify that

$$\begin{Vmatrix} 0 & -1 \\ 1 & -1 \end{Vmatrix}^3 = E .$$

7. Prove that if

$$\begin{Vmatrix} a & b \\ c & d \end{Vmatrix}^m = 0 , \quad \text{then} \quad \begin{Vmatrix} a & b \\ c & d \end{Vmatrix}^2 = 0 .$$

8. Matrices of the following form, called Markov or stochastic matrices, play an important role in applications:

$$P = (p_{ij}) , \quad p_{ij} \geq 0 , \quad \sum_{i=1}^{n} p_{ij} = 1 , \quad j = 1, 2, \cdots, n .$$

The linear maps φ_P corresponding to Markov matrices are usually applied to the following special type of "probability" column-vectors:

$$X = [x_1, \cdots, x_n] \; , \; x_i \geq 0 \; , \; \sum_{i=1}^{n} x_i = 1 \; .$$

These definitions come from problems in the natural sciences. The compatibility of the definitions with one another is clear from the following assertions, which the reader should prove at least in the case $n = 2$.

(a) A matrix $P \in M_n(\mathbb{R})$ is Markov if and only if PX is a probability vector whenever X is a probability vector (where we denote $PX = \varphi_P(X)$).

(b) If P is a <u>positive</u> Markov matrix (i.e., $p_{ij} > 0$, $\forall\, i,j$), then every probability vector X corresponds to a <u>positive</u> probability vector PX (all of whose components are strictly positive).

(c) If P and Q are Markov matrices, then PQ is also Markov. In particular, any power P^k of a Markov matrix is Markov.

§4. The space of solutions

1. <u>Solving a homogeneous linear system.</u> It follows from the introductory remarks at the beginning of §§2,3 that an $m \times n$ system of equations with matrix A and column of free terms $B \in \mathbb{R}^m$ can be briefly written in the form

$$\varphi_A(X) = B \; , \tag{1}$$

or

$$AX = B \tag{1'}$$

(the left side is the product of an $m \times n$ matrix and an $n \times 1$ matrix).

Suppose for a minute that $m = n$ and the square matrix A is non-singular (see subsection 3 of §3). We then obtain a solution of (1), in fact a unique solution, if we multiply both sides of the matrix equation A^{-1} on the left: $X = EX = (A^{-1}A)X = A^{-1}(AX) = A^{-1}B$. This convenient notation for expressing the solution of the square system does not free us from having to make computations, since the matrix A^{-1} is not given to us in advance. But we should permit ourselves to take some satisfaction, at least aesthetic, in this use of the matrix apparatus developed in §3. We now make use of this apparatus to study all solutions of the system (1) in the general case. We start by considering the associated homogeneous system, obtained by setting $B = [0, 0, \cdots, 0] = 0$.

By the kernel of a linear map $\varphi_A : \mathbb{R}^n \to \mathbb{R}^m$ we mean the set

$$\operatorname{Ker} \varphi_A = \{X \in \mathbb{R}^n \mid \varphi_A(X) = 0\} \quad .$$

In other words, $\operatorname{Ker} \varphi_A$ is the set of solutions of the homogeneous system with matrix A. In fact, $\operatorname{Ker} \varphi_A$ is actually a subspace of $\mathbb{R}^n$ (called the space of solutions of the homogeneous linear system), as we already noted at the beginning of §1, and as easily follows from the linearity of the map φ_A:

$$X', X'' \in \operatorname{Ker} \varphi_A \Longrightarrow \varphi_A(\alpha X' + \beta X'') =$$

$$= \alpha\varphi_A(X') + \beta\varphi_A(X'') = 0 \Longrightarrow \alpha X' + \beta X'' \in \operatorname{Ker} \varphi_A \quad .$$

Next, we note that the image $\operatorname{Im} \varphi_A$ of the map φ_A is a subspace of $\mathbb{R}^m$: if $B' = \varphi_A(X')$ and $B'' = \varphi_A(X'')$, then we have

$$\alpha B' + \beta B'' = \alpha\varphi_A(X') + \beta\varphi_A(X'') = \varphi_A(\alpha X' + \beta X'') \in \operatorname{Im} \varphi_A \quad .$$

To say that the system (1) is compatible is equivalent to saying that $B \in \operatorname{Im} \varphi_A$. Let

$$s = \dim \operatorname{Ker} \varphi_A \quad , \quad r = \dim \operatorname{Im} \varphi_A$$

be the dimensions of the spaces $\operatorname{Ker} \varphi_A$ and $\operatorname{Im} \varphi_A$. It follows from the definition of dimension in §1 that $s \leq n$, $r \leq m$. We also have $r \leq n$, since a linearly independent set $\varphi_A(X^{(1)}), \cdots, \varphi_A(X^{(k)})$ in $\operatorname{Im} \varphi_A$ can only be obtained if the set $X^{(1)}, \cdots, X^{(k)}$ is linearly independent in $\mathbb{R}^n$. We obtain more precise information from the following theorem.

THEOREM 1. *The equality* $r + s = n$ *holds. Furthermore, the number* $r = \dim \operatorname{Im} \varphi_A$ *coincides with the rank of the matrix* A (and so is called the *rank of the linear map* φ_A).

Proof. We choose a basis $X^{(1)}, \cdots, X^{(s)}$ of the subspace $\operatorname{Ker} \varphi_A \subset \mathbb{R}^n$ and complete it to a basis $X^{(1)}, \cdots, X^{(s)}, X^{(s+1)}, \cdots, X^{(n)}$ of the entire space $\mathbb{R}^n$. This can always be done, as shown in the proof of Theorem 2 of §1 (and Exercise 5 of §1). For any vector $X = \sum_i \alpha_i X^{(i)} \in \mathbb{R}^n$ we have

$$\varphi_A(X) = \sum_{i=1}^{n} \alpha_i \varphi_A(X^{(i)}) = \alpha_{s+1} \varphi_A(X^{(s+1)}) + \cdots + \alpha_n \varphi_A(X^{(n)}) ,$$

and hence $\operatorname{Im} \varphi_A = \langle \varphi_A(X^{(s+1)}), \cdots, \varphi_A(X^{(n)}) \rangle$, so that $r \leq n - s$. The vectors $\varphi_A(X^{(s+1)}), \cdots, \varphi_A(X^{(n)})$ are linearly independent, since if we have

$0 = \sum_{k \geq s+1} \alpha_k \varphi_A(X^{(k)}) = \varphi_A \left(\sum_{k \geq s+1} \alpha_k X^{(k)} \right)$ then this means that

$\sum_{k \geq s+1} \alpha_k X^{(k)} \in \operatorname{Ker} \varphi_A$, but, by our choice of $X^{(s+1)}, \cdots, X^{(n)}$, this is only possible if $\alpha_{s+1} = \cdots = \alpha_n = 0$. Thus, $r = n - s$. Next, by the definition of φ_A, if $X = [x_1, \cdots, x_n]$, then we have

$$\varphi_A(X) = x_1 A^{(1)} + \cdots + x_n A^{(n)}, \quad \text{i.e.} \quad \operatorname{Im} \varphi_A = \langle A^{(1)}, \cdots, A^{(n)} \rangle .$$

But the dimension of the linear span $\langle A^{(1)}, \cdots, A^{(n)} \rangle$ of the columns of A is nothing other than the rank of A. □

We have already encountered a special case of Theorem 1: if A is a non-singular square matrix of order n, then A and φ_A both have the maximal possible rank n.

In order to find a basis for the space of solutions of a homogeneous linear system $AX = 0$ of rank r, we choose r basis columns in A (a practical method for doing this will be given in the next chapter). If we permute the columns (equivalently, re-index the unknowns), we may assume that the first r columns $A^{(1)}, \cdots, A^{(r)}$ are basis columns. Any set of $r + 1$ columns $A^{(1)}, \cdots, A^{(r)}, A^{(k)}$, $k > r$, is then linearly dependent, and, by Theorem 1 (v) of §1, we can write a system of relations

$$x_1^{(k)} A^{(1)} + x_2^{(k)} A^{(2)} + \cdots + x_r^{(k)} A^{(r)} + A^{(k)} = 0 \quad ,$$

$$k = r+1, r+2, \cdots, n \quad .$$

The $n-r$ column-vectors

$$\begin{aligned} X^{(1)} &= [x_1^{(r+1)}, x_2^{(r+1)}, \cdots, x_r^{(r+1)}, 1, 0, \cdots, 0] \ , \\ X^{(2)} &= [x_1^{(r+2)}, x_2^{(r+2)}, \cdots, x_r^{(r+2)}, 0, 1, \cdots, 0] \ , \\ & \cdots\cdots\cdots\cdots\cdots\cdots\cdots \\ X^{(n-r)} &= [x_1^{(n)}, \ x_2^{(n)}, \ \cdots, x_r^{(n)}, \ 0, 0, \cdots, 1] \end{aligned} \tag{2}$$

are obviously linearly independent (because of the special form of the last $n-r$ components). Since these column-vectors are solutions of the homogeneous system associated to (1), it follows by Theorem 1 that they are a basis of the space $\operatorname{Ker} \varphi_A$ of all solutions.

Any basis for the space of solutions of a homogeneous linear system $AX = 0$ is called a fundamental system of solutions. A set of vectors of the form (2) is called a normalized fundamental system. According to the corollary to Theorem 1 of §2, the rank of this system $s = \dim \operatorname{Im} \varphi_A = n - r$ is equal to the number of free variables in the linear system.

The following assertion (which we shall not need later) reveals a certain "geometrical" interpretation for systems of linear equations.

THEOREM 2. *Every subspace $V \subset \mathbb{R}^n$ of dimension s is the space of solutions for some homogeneous linear system of rank* $r = n - s$.

Proof. Let $V = \langle A^{(1)}, \cdots, A^{(s)} \rangle$. As in the proof of Theorem 1, we complete the linearly independent set $A^{(1)}, \cdots, A^{(s)}$ to a basis $A^{(1)}, \cdots, A^{(s)}, A^{(s+1)}, \cdots, A^{(n)}$ of the entire space $\mathbb{R}^n$. Any column-vector $X = [x_1, \cdots, x_n] \in \mathbb{R}^n$ can be written in a unique way in the form

$$X = \sum_{j=1}^{n} x_j' A^{(j)} = AX' \quad , \tag{3}$$

where A is the $n \times n$ square matrix made up of the columns $A^{(j)}$, and

$X' = [x'_1, x'_2, \cdots, x'_n]$. Since the columns $A^{(1)}, \cdots, A^{(n)}$ are linearly independent, it follows that $\operatorname{rank} A = n$, and, by Theorem 4 of §3, the matrix A has an inverse $A^{-1} = (\bar{a}_{ij})$. We have

$$[x'_1, \cdots, x'_n] = A^{-1}X = \left[\sum_{j=1}^{n} \bar{a}_{1j}x_j, \cdots, \sum_{j=1}^{n} \bar{a}_{nj}x_j\right] .$$

The corollary to Theorem 4 of §3 shows that $\operatorname{rank} A^{-1} = n$, and hence any r rows of A^{-1} are linearly independent. Thus, the homogeneous system

$$\sum_{j=1}^{n} \bar{a}_{kj}x_j = 0 \quad , \quad k = s+1, \cdots, n \quad ,$$

has rank $r = n - s$. But the set of solutions of this system consists precisely of those column-vectors X of the form (3) for which $x'_{s+1} = 0, \cdots, x'_n = 0$, i.e., it consists precisely of the vectors in the subspace V. □

2. <u>Linear manifolds. Solving a non-homogeneous system.</u> Let V be a subspace of $\mathbb{R}^n$, and let X^0 be a fixed vector in $\mathbb{R}^n$. The set

$$V + X^0 = \{X + X^0 \mid X \in V\} = X^0 + V$$

is called a <u>linear manifold</u> of type V and dimension $\dim V$. The geometrical picture

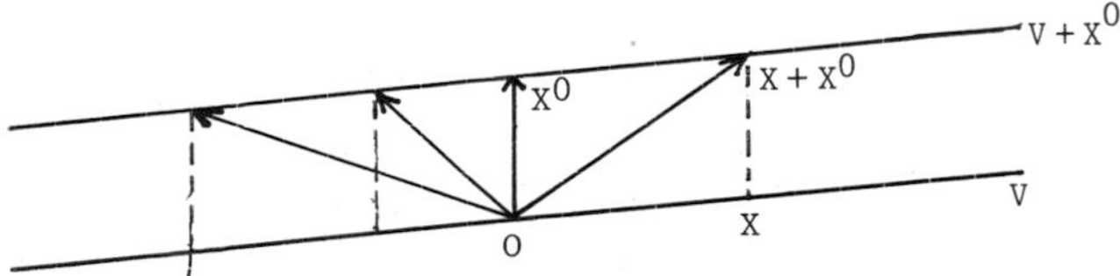

illustrates what should be an intuitively clear notion: $V + X^0$ is the space V translated (shifted) by the vector X^0. The subspace V itself is also a linear manifold, corresponding to translation by the vector $X^0 = 0$. Two linear manifolds coincide if and only if they are obtained by translation by vectors X' and X'' such that $X' - X'' \in V$ (the verification is left to the reader). In particular, if X' is any vector in the linear manifold $V + X^0$, then $V + X'$ coincides with $V + X^0$.

For example, let $V = \langle E^{(1)}, E^{(2)}, E^{(3)} \rangle \subset \mathbb{R}^5$, $X^0 = [0,0,1,1,0]$, $X' = [0,0,0,1,0]$. Then

$$V + X^0 = V + X' = \{[x,y,z,1,0] \mid x,y,z \in \mathbb{R}\} \quad .$$

We now turn to the non-homogeneous system of linear equations (1). Suppose that (1) is compatible, i.e., by Theorem 2 of §2, the rank of A and the rank of $(A|B)$ coincide. Let $X^0 = [x_1^0, \cdots, x_n^0]$ be a fixed solution of the system, so that $\varphi_A(X^0) = B$. If X' is any other solution of (1), then $\varphi_A(X'-X^0) = \varphi_A(X') - \varphi_A(X^0) = B - B = 0$. Hence, the difference $X'' = X' - X^0$ of two solutions to (1) is always a solution to the corresponding homogeneous system, and $X' = X'' + X^0$. Conversely, if $\varphi_A(X) = 0$, then

$$\varphi_A(X + X^0) = \varphi_A(X) + \varphi_A(X^0) = 0 + B = B \quad .$$

We thus have the following assertion.

THEOREM 3. <u>The solutions of a compatible non-homogeneous linear system form a linear manifold of type</u> V, <u>where</u> $V = \operatorname{Ker} \varphi_A$ <u>is the linear subspace of solutions to the corresponding homogeneous system.</u> □

3. <u>The rank of a product of matrices.</u> The action of a product $\tau \cdots \sigma\varphi\psi$ of maps can be roughly depicted by the diagram

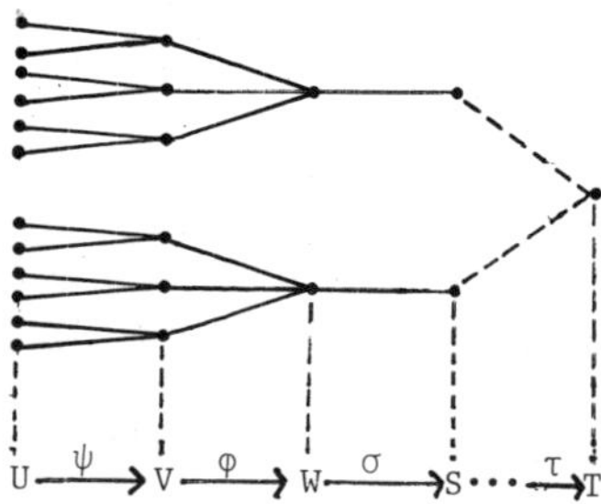

which in the case of linear maps of vector spaces is meant to convey the implication: $\varphi\psi(U) \subset \varphi(V) \Longrightarrow \operatorname{rank} \varphi\psi \le \operatorname{rank} \varphi$. Furthermore, a basis of $\psi(U)$ maps onto a set of vectors containing a basis of $\varphi\psi(U)$, and hence

$$\operatorname{rank} \varphi\psi \leq \operatorname{rank} \psi \quad .$$

Thus,

$$\operatorname{rank} \varphi\psi \leq \min\{\operatorname{rank} \varphi, \operatorname{rank} \psi\} \quad .$$

But $\operatorname{rank} \varphi_A = \operatorname{rank} A$ and $\operatorname{rank} AB = \operatorname{rank} \varphi_{AB} = \operatorname{rank} \varphi_A \varphi_B$; hence, the inequality (4) leads to the following useful fact.

THEOREM 4. <u>The rank of a product of matrices is less than or equal to the rank of each factor:</u>

$$\operatorname{rank} AB \leq \min\{\operatorname{rank} A, \operatorname{rank} B\} \quad . \quad \square \tag{4'}$$

COROLLARY 1. <u>If</u> B <u>and</u> C <u>are non-singular square matrices of order</u> m <u>and</u> n , <u>respectively, and if</u> A <u>is any</u> $m \times n$ <u>matrix, then</u>

$$\operatorname{rank} BAC = \operatorname{rank} A \quad .$$

<u>Proof.</u> By Theorem 4 , we have

$$\begin{aligned} \operatorname{rank} BAC &\leq \operatorname{rank} BA = \\ &= \operatorname{rank} BA(CC^{-1}) = \operatorname{rank} (BAC)C^{-1} \leq \operatorname{rank} BAC \quad , \end{aligned}$$

so that $\operatorname{rank} BAC = \operatorname{rank} BA$. We similarly prove that $\operatorname{rank} BA = \operatorname{rank} A$. $\square$

COROLLARY 2. <u>A square matrix which has a left or right inverse is non-singular.</u>

<u>Proof.</u> Suppose that $AB = E$, where A and B are $n \times n$ matrices. Since $\operatorname{rank} E = n$, inequality (4') can be rewritten in the form $n \leq \min\{\operatorname{rank} A, \operatorname{rank} B\}$, which implies that $\operatorname{rank} A = \operatorname{rank} B = n$. But this is equivalent to A and B being non-singular (see Theorem 4 of §3). We similarly show that A is non-singular when there exists C for which $CA = E$. $\square$

According to Corollary 2, a linear map $\varphi_A : \mathbb{R}^n \to \mathbb{R}^n$ which has either a left or right inverse must have a two-sided inverse. This is an example of a basic distinction between linear maps and general set maps (see Exercise 2 of §5 Ch. 1).

4. <u>Equivalence classes of matrices.</u> As in subsection 3 of §3 , we let E_{st}

denote the $m \times m$ matrix having a one at the intersection of the s-th row and t-th column and zeros everywhere else.

We further introduce the so-called elementary matrices in $M_m(\mathbb{R})$:

(I) $$F_{s,t} = E - E_{ss} - E_{tt} + E_{st} + E_{ts} = \left\| \begin{matrix} 1 & & & & & & \\ & \ddots & & & & & \\ & & 0 & \cdots & 1 & & \\ & & \vdots & 1 & \vdots & & \\ & & 1 & \cdots & 0 & & \\ & & & & & \ddots & \\ & & & & & & 1 \end{matrix} \right\| , \quad s \neq t \; ;$$

(II) $$F_{s,t}(\lambda) = E + \lambda E_{st} = \left\| \begin{matrix} 1 & & & \\ & \ddots & & \\ \cdots & & 1 \cdots \lambda \cdots & \\ & & \ddots & \\ & & & 1 \end{matrix} \right\| , \quad s \neq t \; ;$$

(III) $$F_s(\lambda) = E + (\lambda - 1) E_{ss} = \operatorname{diag}\{1, \cdots, 1, \lambda, 1, \cdots, 1\} , \quad \lambda \neq 0 \quad .$$

Let A be any $m \times n$ matrix. One easily checks that if F is one of the first two types of matrix, then the matrix $A' = FA$ is obtained from A by an elementary transformation of the rows of type (I) or (II), depending on whether $F = F_{s,t}$ or $F = F_{s,t}(\lambda)$, respectively. If $F = F_s(\lambda)$, we shall speak of an elementary transformation of type (III) (multiplying the s-th row A_s by λ). Similarly, the matrix $A'' = AF$ is obtained from A by an elementary transformation on the columns. From subsection 2 of §2 and Exercise 2 of §2, we know that a matrix A can be reduced to diagonal form by elementary transformations of type (I) and (II) on the rows and columns. Since

$$\begin{Vmatrix} a_1 & & & & & & \\ & a_2 & & & & 0 & \\ & & \ddots & & & & \\ & & & a_r & & & \\ & & & & 0 & & \\ 0 & & & & & \ddots & \\ & & & & & & 0 \end{Vmatrix} = F_1(a_1)F_2(a_2)\cdots F_r(a_r)\begin{Vmatrix} 1 & & & & & & \\ & 1 & & & & 0 & \\ & & \ddots & & & & \\ & & & 1 & & & \\ & & & & 0 & & \\ 0 & & & & & \ddots & \\ & & & & & & 0 \end{Vmatrix},$$

it follows that, if we allow elementary transformations of type (III), we can reduce A to a matrix of the form

$$\begin{Vmatrix} E_r & 0 \\ 0 & 0 \end{Vmatrix} \tag{5}$$

(the zeros here denote matrices of size $r \times (n-r)$, $(m-r) \times r$, and $(m-r) \times (n-r)$).

Thus,

$$P_s P_{s-1} \cdots P_1 A Q_1 Q_2 \cdots Q_t = \begin{Vmatrix} E_r & 0 \\ 0 & 0 \end{Vmatrix}, \tag{6}$$

where P_i and Q_j are elementary matrices of order m and n, respectively. We have often stated that the elementary operations are invertible. This agrees with the fact that the inverse matrices exist and are matrices of the same type:

$$(F_{s,t})^{-1} = F_{s,t},\ F_{s,t}(\lambda)^{-1} = F_{s,t}(-\lambda),\ F_s(\lambda)^{-1} = F_s(\lambda^{-1}) \quad .$$

By the corollary to Theorem 4 of §3, the matrices $P = P_s P_{s-1} \cdots P_1$ and $Q = Q_1 Q_2 \cdots Q_t$ also have inverses: $P^{-1} = P_1^{-1} \cdots P_{s-1}^{-1} P_s^{-1}$, $Q^{-1} = Q_t^{-1} \cdots Q_2^{-1} Q_1^{-1}$. Note that the P_i^{-1} and Q_j^{-1} are elementary matrices.

We call two $m \times n$ matrices A and B <u>equivalent</u> and write $A \sim B$ if there exist non-singular matrices P and Q of size $m \times m$ and $n \times n$, respectively, such that $B = PAQ$.

It is easy to see that $\sim$ is an equivalence relation: (i) $A \sim A$ (choose $P = E_m$, $Q = E_n$); (ii) $A \sim B \Rightarrow B \sim A$, since $B = PAQ \Rightarrow A = P^{-1}BQ^{-1}$; (iii) $B = P'AQ'$

and $C = P''BQ'' \Rightarrow C = PAQ$, where $P = P''P'$, $Q = Q'Q''$. As always in the case of an equivalence relation (see §6 of Ch. 1), the set of all $m \times n$ matrices splits up into a disjoint union of equivalence classes. Since the ranks of equivalent matrices are equal (see Corollary 1 of Theorem 4), the argument leading to (6) shows that the matrices of the form (5) can be taken as representatives of equivalence classes. We have obtained the following fact.

THEOREM 5. _The set of_ $m \times n$ _matrices divides up into_ $\min(m,n) + 1$ _equivalence classes. All matrices of rank_ r _lie in the same class as the matrix (5)._ □

COROLLARY. _Every non-singular_ $n \times n$ _matrix is a product of elementary matrices._

Proof. Any such matrix is equivalent to the identity matrix, since both have rank n. Then the relation (6)

$$P_s P_{s-1} \cdots P_1 A Q_1 Q_2 \cdots Q_t = E \quad ,$$

when written in the form

$$A = P_1^{-1} \cdots P_{s-1}^{-1} P_s^{-1} Q_t^{-1} \cdots Q_2^{-1} Q_1^{-1} \quad , \tag{7}$$

gives the corollary. □

The corollary does not say that the representation of A as a product of elementary matrices is unique, but even the fact that such a representation exists is very useful. For example, such a representation can be used to find the inverse matrix. Namely, (7) gives us:

$$A^{-1} = Q_1 Q_2 \cdots Q_t P_s P_{s-1} \cdots P_1 = QP \quad .$$

To find A^{-1}, we first multiply A on the left by P_i to reduce it to triangular form. As we do this, we keep track of the sequence of matrices $E, P_1, P_2 P_1, \cdots$. Once we have A reduced to the triangular matrix $P_s \cdots P_2 P_1 A$, we start multiplying on the right by the Q_j to reduce it to the identity matrix, all the time keeping track of the sequence of matrices

$E, Q_1, Q_1Q_2, \cdots$. Since each matrix P_i or Q_j merely corresponds to an elementary transformation, it follows that the chain of transformations performed simultaneously on E and A is not hard to carry out:

$$E\{A \to P_1\{P_1A \to \cdots \to P_s \cdots P_1\{P_s \cdots P_1A \to$$

$$\to P_s \cdots P_1\{P_s \cdots P_1AQ_1\{Q_1 \to \cdots$$

$$\cdots \to P_s \cdots P_1\{P_s \cdots P_1AQ_1 \cdots Q_t\{Q_1 \cdots Q_t$$

(although there may be a large number $s+t$ of steps). The wavy lines separate the products we are interested in: the result of applying elementary row transformations on the identity matrix (in the first part), and the result of applying elementary column transformations on the identity matrix (in the second part). If it turns out that $r < n$, we conclude that A is a singular matrix and so does not have an inverse. If $r = n$, we need only multiply Q and P in order to obtain A^{-1}. Note that the order of performing the row and column operations can be reversed.

We consider two examples. For the matrix

$$A = \begin{Vmatrix} 1 & 2 & 3 \\ 4 & 5 & 6 \\ 7 & 8 & 9 \end{Vmatrix}$$

we have

$$E\left\{\begin{Vmatrix} 1 & 2 & 3 \\ 4 & 5 & 6 \\ 7 & 8 & 9 \end{Vmatrix} \to F_{2,1}(-4)\left\{\begin{Vmatrix} 1 & 2 & 3 \\ 0 & -3 & -6 \\ 7 & 8 & 9 \end{Vmatrix} \to\right.\right.$$

$$\to F_{3,1}(-7)F_{2,1}(-4)\left\{\begin{Vmatrix} 1 & 2 & 3 \\ 0 & -3 & -6 \\ 0 & -6 & -12 \end{Vmatrix} \to\right.$$

$$\to F_{3,2}(-2)\cdot F_{3,1}(-7)\cdot F_{2,1}(-4)\left\{\begin{Vmatrix} 1 & 2 & 3 \\ 0 & -3 & -6 \\ 0 & 0 & 0 \end{Vmatrix}\right. .$$

Since the right side has a matrix in step form of rank 2, we must have $\operatorname{rank} A = 2$. Hence, A does not have an inverse.

Here is another example. From a sequence

$$E\left\{\begin{Vmatrix}0&0&0&1\\1&0&0&0\\0&1&0&0\\0&0&1&0\end{Vmatrix}\right. \longrightarrow F_{1,2}\left\{\begin{Vmatrix}1&0&0&0\\0&0&0&1\\0&1&0&0\\0&0&1&0\end{Vmatrix}\right. \longrightarrow F_{2,3}F_{1,2}\left\{\begin{Vmatrix}1&0&0&0\\0&1&0&0\\0&0&0&1\\0&0&1&0\end{Vmatrix}\right. \longrightarrow$$

$$\longrightarrow F_{3,4}F_{2,3}F_{1,2}\left\{\begin{Vmatrix}1&0&0&0\\0&1&0&0\\0&0&1&0\\0&0&0&1\end{Vmatrix}\right.$$

we find that

$$\begin{Vmatrix}0&0&0&1\\1&0&0&0\\0&1&0&0\\0&0&1&0\end{Vmatrix}^{-1} = F_{3,4}F_{2,3}F_{1,2} = \begin{Vmatrix}0&1&0&0\\0&0&1&0\\0&0&0&1\\1&0&0&0\end{Vmatrix} .$$

This method, sometimes called (P,Q)-reduction of a matrix to the normal form (5), is rather convenient for computing inverse matrices; however, it is premature to speak now of the advantages and disadvantages of this method, since our examples did not even use all of the types of matrices $F_{s,t}$, $F_{s,t}(\lambda)$, $F_s(\lambda)$.

EXERCISES

1. Prove the following rules for working with transpose matrices (see Exercise 1 of §2):

$${}^t(A+B) = {}^tA + {}^tB \quad ;$$

$${}^t(AB) = {}^tB \cdot {}^tA \quad .$$

2. Prove that $\operatorname{rank} AB \leq \min\{\operatorname{rank} A, \operatorname{rank} B\}$ by a direct argument with matrices.

3. Prove Sylvester's inequality:

$$\dim \operatorname{Ker} \varphi\psi \leq \dim \operatorname{Ker} \varphi + \dim \operatorname{Ker} \psi$$

for any two linear maps $\mathbb{R}^n \xrightarrow{\psi} \mathbb{R}^m \xrightarrow{\varphi} \mathbb{R}^\ell$.

4. Prove that every linear map $\varphi : \mathbb{R}^n \to \mathbb{R}^m$ of rank r can be written as a sum $\varphi = \varphi_1 + \cdots + \varphi_r$ of maps φ_i of rank 1.

5. Find the rank of the matrix

$$A = \begin{Vmatrix} x_1y_1 & x_1y_2 & \cdots & x_1y_n \\ x_2y_1 & x_2y_2 & \cdots & x_2y_n \\ \cdot & \cdot & \cdot & \cdot \\ x_ny_1 & x_ny_2 & \cdots & x_ny_n \end{Vmatrix} .$$

6. Show that the inverse of a matrix can be found by applying elementary transformations only to the rows (or only to the columns) of $E \{ A$.

Chapter 3. Determinants

The formulas (3) and (9) in §4 Ch. 1 for solving square linear systems of order $n = 2, 3$ suggest the possibility of similar formulas for any n. In essence, what we need in order to generalize these formulas is a good interpretation of the numerators and denominators in those formulas. We shall show how to regard them as the values of a "universal" function $\det : M_n(\mathbb{R}) \to \mathbb{R}$ from the set of $n \times n$ matrices to $\mathbb{R}$. The construction of the determinant function det will also answer many other questions concerning matrices which were raised in Chapter 2. In fact, the theory of determinants has much wider applications than those which we shall touch upon, and each application of this theory suggests a different possible method for constructing the determinant. One of the most natural approaches is the geometrical approach, which considers the determinant of a matrix as the volume of a multi-dimensional figure (see Exercise 3 of §4 Ch. 1). This method, based on the notion of exterior n-forms, would require us to go further into geometry, so we shall stick with an "analytic" approach. *

§1. Determinants: construction and basic properties

1. Construction by induction. We take the determinant of a 1×1 matrix (a_{11}) to be the number a_{11}. The determinants of 2×2 and 3×3 matrices are defined by formulas (2) and (8) of §4 Ch. 1. In the formula for the determinant of a 3×3 matrix, the 2×2 determinants which occur were intentionally left "un-expanded", in order to hint at the induction which we shall use to construct the determinant of an $n \times n$ matrix.

Suppose that we have already introduced determinants for matrices of order $1, 2, \cdots, n-1$. We define the determinant of an $n \times n$ matrix $A = (a_{ij})$ to be

$$D = a_{11}D_1 - a_{21}D_2 + \cdots + (-1)^{n-1}a_{n1}D_n \, , \tag{1'}$$

where D_k is the determinant of the $(n-1) \times (n-1)$ matrix

$$\left\| \begin{matrix} a_{12} & \cdots & a_{1n} \\ \cdots & \cdots & \cdots \\ a_{k-1,2} & \cdots & a_{k-1,n} \\ a_{k+1,2} & \cdots & a_{k+1,n} \\ \cdots & \cdots & \cdots \\ a_{n2} & \cdots & a_{nn} \end{matrix} \right\| ,$$

which is obtained from A by crossing out the first column and the k-th row.

It is easy to see that the expression (1') in the case $n = 2, 3$ agrees with the formulas (2) and (8) of §4 Ch. 1. The determinant of a matrix $A = (a_{ij})$ can be denoted $|A|$, $|a_{ij}|_1^n$, or $\det A$. The vertical lines are used primarily when the matrix A has been given to us explicitly.

If we cross out the i-th row and j-th column in a matrix A, we obtain a

---

* There are many analytic methods for developing the theory of determinants. In this chapter, as in §4 of Ch. 1, we shall follow Shafarevich's lecture notes (Moscow University, 1971). In the first place, the practice in the use of induction is useful for its own sake. Moreover, the most useful methods for computing determinants are developed quickly. On the other hand, approaches based on "completely expanding the determinant" (see §3 Ch. 4) are in some sense simpler.

square matrix of order $n - 1$. The determinant of this $(n - 1) \times (n - 1)$ matrix is denoted M_{ij} and is called the minor of A corresponding to the entry a_{ij}. Using this notation, (1') can be written

$$\det A = a_{11}M_{11} - a_{21}M_{21} + \cdots + (-1)^{n-1}a_{n1}M_{n1} \quad . \tag{1}$$

This formula for the determinant can be expressed in words as follows: the determinant of an $n \times n$ matrix is the sum of the products of an element in the first column by the corresponding minor, where the products are taken with alternating sign.

If we take the k-th column instead of the first, and replace the minors M_{i1} by the minors M_{ik}, then, as we shall see later, the resulting expression differs from $\det A$ at most by a sign.

In what follows, we adopt the convention of Chapter 2 that

$$A_i = (a_{i1}, a_{i2}, \cdots, a_{in}) \quad , \quad i = 1, 2, \cdots, n \quad ,$$

$$A^{(j)} = [a_{1j}, a_{2j}, \cdots, a_{nj}] \quad , \quad j = 1, 2, \cdots, n \quad ,$$

denote the i-th row and the j-th column, respectively, of $A = (a_{ij})$. The matrix A can be represented either by listing its rows (i.e., as a column of rows)

$$A = [A_1, A_2, \cdots, A_n] \quad ,$$

or by listing its columns (i.e., as a row of columns)

$$A = (A^{(1)}, A^{(2)}, \cdots, A^{(n)}) \quad .$$

We shall sometimes refer to the rows and columns of an $n \times n$ matrix A as the rows and columns of the n-th order determinant $|a_{ij}|$.

By definition, $|\ | = \det$ is a function which associates to a matrix A a number $|A| = \det A$. Our first task is to study the behavior of this function when we change the rows or columns of A, considered as elements (vectors) of the vector space $\mathbb{R}^n$. It is sometimes useful to think of $\det A$ as a function

$$\det[A_1, \cdots, A_n] \quad \text{or} \quad \det(A^{(1)}, \cdots, A^{(n)})$$

of n variables which are vectors in $\mathbb{R}^n$. Functions of n variables were introduced in subsection 2 of §5 Ch. 1; such a function can be thought of either as a function of n variables $x_i \in X$ or as a function of one variable $x \in X^n$. In our example here, $f = \det$ and $X = \mathbb{R}^n$. We now discuss some general properties which a function $\mathscr{D}$ of n variables $A_i \in \mathbb{R}^n$ may have.

We shall call a function $\mathscr{D} : [A_1, \cdots, A_n] \mapsto \mathscr{D}(A_1, \cdots, A_n)$ <u>multi linear</u> if it is linear in each argument A_i, i.e.,

$$\mathscr{D}(A_1, \cdots, \alpha A_i' + \beta A_i'', \cdots, A_n) =$$

$$= \alpha\mathscr{D}(A_1, \cdots, A_i', \cdots, A_n) + \beta\mathscr{D}(A_1, \cdots, A_i'', \cdots, A_n)$$

(compare with subsection 1 of §3 Ch. 2). Such a multilinear function $\mathscr{D}$ is called <u>skew-symmetric</u> if

$$\mathscr{D}(A_1,\cdots,A_i, A_{i+1},\cdots,A_n) = -\mathscr{D}(A_1,\cdots,A_{i+1}, A_i,\cdots,A_n) \quad , \quad 1 \leq i \leq n-1. \quad (2)$$

<u>Remark 1.</u> The definition of a linear function (see (4) of §3 Ch. 2) tells us that a function $\mathscr{D}$ is multilinear if and only if, for fixed $A_1,\cdots,A_{i-1}, A_{i+1},\cdots,A_n$ and for $A_i = X = (x_1, \cdots, x_n)$ we have

$$\mathscr{D}(A_1, \cdots, A_n) = \alpha_1 x_1 + \alpha_2 x_2 + \cdots + \alpha_n x_n \quad ,$$

where $\alpha_1, \cdots, \alpha_n$ are scalars which do not depend on $x_1, \cdots, x_n$.

<u>Remark 2.</u> A semi-linear function $\mathscr{D}$ is skew-symmetric if and only if the following equality holds for all X:

$$\mathscr{D}(A_1, \cdots, A_{i-1}, X, X, A_{i+2}, \cdots, A_n) = 0 \quad , \quad 1 \leq i \leq n-1 \quad . \quad (2')$$

To see this, in one direction, if we set $A_i = A_{i+1} = X$ in (2), we arrive at (2'). Conversely, setting $X = A_i + A_{i+1}$ in (2') and using the multilinearity of $\mathscr{D}$, we obtain

$$\mathcal{D}(\cdots, A_i, A_i, \cdots) + \mathcal{D}(\cdots, A_{i+1}, A_{i+1}, \cdots) +$$
$$+ \mathcal{D}(\cdots, A_i, A_{i+1}, \cdots) + \mathcal{D}(\cdots, A_{i+1}, A_i, \cdots) =$$
$$= \mathcal{D}(\cdots, A_i + A_{i+1}, A_i + A_{i+1}, \cdots) = 0 \quad .$$

The first two terms vanish (as we see by setting $X = A_i$ and $X = A_{i+1}$ in (2')); hence, the sum of the last two terms is zero, and this gives us (2).

The same definitions and remarks carry over to functions $\mathcal{D}(A^{(1)}, \cdots, A^{(n)})$ of column-vectors. (Note that the skew-symmetry condition (2) makes sense for a function $\mathcal{D} : M^n \to \mathbb{R}$ from a cartesian power of any set M.)

We shall later need:

LEMMA 1. <u>If any two arguments of a skew-symmetric function are interchanged, the value of the function changes sign.</u>

<u>Proof.</u> Suppose we interchange the i-th and j-th arguments, where $i < j$. We proceed by induction on the number $k = j - i - 1$ of arguments between the two we are interchanging. If $k = 0$, the lemma coincides with the definition of a skew-symmetric function. Suppose that the lemma is true whenever $j - i - 1 < k$. Then

$$\mathcal{D}(\cdots, X_i, X_{i+1}, \cdots, X_{j-1}, X_j, \cdots) =$$
$$= -\mathcal{D}(\cdots, X_{i+1}, X_i, \cdots, X_{j-1}, X_j, \cdots) =$$
$$= \mathcal{D}(\cdots, X_{i+1}, X_j, \cdots, X_{j-1}, X_i, \cdots)$$
$$= -\mathcal{D}(\cdots, X_j, X_{i+1}, \cdots, X_{j-1}, X_i, \cdots) \quad . \quad \square$$

2. <u>Basic properties of determinants.</u> The notion of determinant only becomes usable when we derive the properties which are important both from a theoretical and computational point of view.

The trivial relation $\det(a + b) = \det a + \det b$ for first order determinants might suggest the mistaken conclusion that it holds for larger n, for example, for second order determinants. But if we look at the case $n = 2$, we find that the correct relationship is as follows:

$$\begin{vmatrix} \alpha x'_1 + \beta x''_1 & \alpha x'_2 + \beta x''_2 \\ a_{21} & a_{22} \end{vmatrix} = (\alpha x'_1 + \beta x''_1) a_{22} - (\alpha x'_2 + \beta x''_2) a_{21} =$$

$$= \alpha (x'_1 a_{22} - x'_2 a_{21}) + \beta (x''_1 a_{22} - x''_2 a_{21}) = \alpha \begin{vmatrix} x'_1 & x'_2 \\ a_{21} & a_{22} \end{vmatrix} + \beta \begin{vmatrix} x''_1 & x''_2 \\ a_{21} & a_{22} \end{vmatrix} .$$

(To see that $\det(a+b) \neq \det a + \det b$, take, for example, $a = b = E$, or $a = E$, $b = -E$.) We also note that

$$\begin{vmatrix} a_{11} & a_{12} \\ a_{21} & a_{22} \end{vmatrix} = - \begin{vmatrix} a_{21} & a_{22} \\ a_{11} & a_{12} \end{vmatrix} , \quad \begin{vmatrix} 1 & 0 \\ 0 & 1 \end{vmatrix} = 1 .$$

We thus have evidence from the 2×2 case for the validity of

THEOREM 1. The function $A \mapsto \det A$ on the set $M_n(\mathbb{R})$ has the following properties:

D1. $\det A$ is a multilinear function of the rows of A, i.e., the determinant of a matrix is a linear function of the elements of any row A_1.

D2. $\det A$ is a skew-symmetric function of the rows of A, i.e., it vanishes if any two neighboring rows coincide.

D3. $\det E = 1$.

Proof. We use induction on n. Properties D1-D3 have been verified when $n = 1, 2$. Suppose that they hold for all determinants of order $< n$. We use the formula (1) to prove D1-D3 for n-th order determinants. We start with D3.

D3. If

$$A = E = \begin{Vmatrix} 1 & 0 & \dots & 0 \\ 0 & 1 & \dots & 0 \\ \dots & \dots & \dots & \dots \\ 0 & 0 & \dots & 1 \end{Vmatrix} ,$$

then in (1) we have $a_{i1} = 0$ for $i \neq 1$ and $a_{11} = 1$; hence $\det E = M_{11}$. The determinant M_{11} is the determinant of the $(n-1) \times (n-1)$ identity matrix, so, by the induction assumption, $M_{11} = 1$, and hence $\det E = 1$.

We shall prove properties D1 and D2 in a somewhat more general situation, which is described by the following lemma.

LEMMA 2. Let $\mathscr{D}_j : M_n(\mathbb{R}) \to \mathbb{R}$ be the function defined by the formula

$$\mathscr{D}_j(A) = a_{1j}M_{1j} - a_{2j}M_{2j} + \cdots + (-1)^{n-1}a_{nj}M_{nj} \quad . \tag{3}$$

Then:

$D_j1.$ $\mathscr{D}_j$ is a multilinear function of the rows of A.

$D_j2.$ $\mathscr{D}_j$ is a skew-symmetric function of the rows of A.

Proof. $D_j1.$ In order to emphasize that we are considering the elements of the i-th row to be variables, we set $x_s = a_{is}$, $s = 1, \cdots, n$:

$$A = \begin{Vmatrix} a_{11} & \cdots & a_{1j} & \cdots & a_{1n} \\ \cdot & \cdot & \cdot & \cdot & \cdot \\ a_{i-1,1} & \cdots & a_{i-1,j} & \cdots & a_{i-1,n} \\ x_1 & \cdots & x_j & \cdots & x_n \\ a_{i+1,1} & \cdots & a_{i+1,j} & \cdots & a_{i+1,n} \\ \cdot & \cdot & \cdot & \cdot & \cdot \\ a_{n1} & \cdots & a_{nj} & \cdots & a_{nn} \end{Vmatrix} .$$

The minor M_{ij} does not depend on $x_1, \cdots, x_n$, and hence $\alpha_j = (-1)^{i-1}M_{ij}$ is a constant. Any other minor M_{kj}, $k \neq i$, has $(x_1, \cdots, x_{j-1}, x_{j+1}, \cdots, x_n)$ as a row, and all of its other rows are constant. By the induction assumption, M_{kj} is a linear function of the variables $x_1, \cdots, x_{j-1}, x_{j+1}, \cdots, x_n$, i.e., by Remark 1,

$$M_{kj} = \sum_{s \neq j} \alpha_{ks} x_s \quad , \quad k \neq i \quad .$$

Now setting $\alpha_s = \sum_{k \neq i} (-1)^{k-1} \alpha_{ks} a_{kj}$, $s \neq j$, we arrive at the expression

$$\mathscr{D}_j(A) = \sum_{k=1}^{n} (-1)^{k-1} a_{kj} M_{kj} = \alpha_j x_j + \sum_{k \neq i} (-1)^{k-1} a_{kj} \sum_{s \neq j} \alpha_{ks} x_s =$$

$$= \alpha_j x_j + \sum_{s \neq j} \left(\sum_{k \neq i} (-1)^{k-1} \alpha_{ks} a_{kj} x_s \right) = \sum_{s=1}^{n} \alpha_s x_s \quad ,$$

which tells us that $\mathcal{D}_j(A)$ is a linear function of the elements $x_1, \cdots, x_n$ of the i-th row of A.

D_j2. According to Remark 2 of subsection 2, we may just as well prove that $\mathcal{D}_j(A) = 0$ for any matrix

$$A = \begin{Vmatrix} a_{11} & \cdots & a_{1j} & \cdots & a_{1n} \\ \cdot & \cdot & \cdot & \cdot & \cdot \\ x_1 & \cdots & x_j & \cdots & x_n \\ x_1 & \cdots & x_j & \cdots & x_n \\ \cdot & \cdot & \cdot & \cdot & \cdot \\ a_{n1} & \cdots & a_{nj} & \cdots & a_{nn} \end{Vmatrix}$$

with two identical rows $A_i = A_{i+1} = (x_1, \cdots, x_j, \cdots, x_n)$. The minor M_{kj}, $k \neq i$, $i+1$, also has two identical rows, namely, the row $(x_1, \cdots, x_{j-1}, x_{j+1}, \cdots, x_n)$ of length $n-1$. By the induction assumption, $M_{kj} = 0$ for $k \neq i$, $i+1$. The formula (3) can now be rewritten in the form

$$\mathcal{D}_j(A) = (-1)^{i-1} x_j M_{ij} + (-1)^i x_j M_{i+1,j} \quad .$$

But obviously $M_{i,j} = M_{i+1,j}$. Hence,

$$\mathcal{D}_j(A) = (-1)^{i-1} x_j (M_{ij} - M_{i+1,j}) = 0 \quad . \qquad \square$$

Setting $j = 1$ in (3) and comparing the resulting expression with (1), we conclude that

$$\mathcal{D}_1(A) = \det A \quad . \tag{4}$$

Hence, properties D1 and D2 of determinants are contained in the lemma. Theorem 1 is proved. $\square$

We write out property D1 in more detail:

D1'. $\det[A_1, \cdots, \lambda A_i, \cdots, A_n] = \lambda \det[A_1, \cdots, A_i, \cdots, A_n]$, i.e., <u>if a row A_i in a determinant is multiplied by λ, the determinant is also multiplied by λ. In particular, if all of the rows are multiplied by λ, one obtains the formula</u>

$$\det \lambda A = \lambda^n \det A \quad .$$

D1". *If for some* i *all of the elements in* A_i *are of the form* $a_{ij} = a'_j + a''_j$, *then* $\det A = \det A' + \det A''$, *where* $A'_j = A''_j = A_j$ *for* $j \neq i$ *and* $A'_i = (a'_1, \cdots, a'_n)$, $A''_i = (a''_1, \cdots, a''_n)$.

Theorem 1 implies several other simple assertions which we shall state in the form of properties of determinants, but shall prove for any of the functions $\mathcal{D}_j$ given by (3) . We know from (4) that the determinant is a special case of $\mathcal{D}_j$.

D4. *A determinant with a zero row is equal to zero.*

Suppose that $A_i = (0, 0, \cdots, 0)$. Then also $2A_i = (0, \cdots, 0)$. Consequently, by D_j1:

$$\mathcal{D}_j(A) = \mathcal{D}_j(A_1, \cdots, A_i, \cdots, A_n) = \mathcal{D}_j(A_1, \cdots, 2A_i, \cdots, A_n) =$$
$$= 2\mathcal{D}_j(A_1, \cdots, A_i, \cdots, A_n) = 2\mathcal{D}_j(A) \quad ,$$

and hence $\mathcal{D}_j(A) = 0$. □

D5. *If any two rows (not necessarily neighboring ones) are interchanged, the determinant changes sign.*

This property follows from D_j2 and Lemma 1 for any function $\mathcal{D}_j(A)$. □

D6. *If any two rows coincide, then the determinant is zero.*

We again take any function $\mathcal{D}_j(A)$. Suppose A_s and A_t are the rows that coincide. If we interchange A_s and A_t , the matrix remains the same. On the other hand, according to D5 , which we proved for any function $\mathcal{D}_j$, interchanging these two rows changes the sign of the determinant. Thus, $\mathcal{D}_j(A) = -\mathcal{D}_j(A)$, and hence $\mathcal{D}_j(A) = 0$. □

D7. *The determinant does not change if an elementary transformation of type (II) is performed on its rows.*

Suppose we add λ times the t-th row of A to the s-th row of A to obtain a matrix A' . Then, by properties D1 and D6 (which we proved for $\mathcal{D}_j$) , we have

$$\mathcal{D}_j(A') = \mathcal{D}_j(A_1, \cdots, A_s + \lambda A_t, \cdots, A_n) =$$

$$= \mathcal{D}_j(\cdots, A_s, \cdots) + \lambda \mathcal{D}_j(\cdots, A_t, \cdots, A_t, \cdots) =$$

$$= \mathcal{D}_j(A_1, \cdots, A_n) = \mathcal{D}_j(A) \quad . \quad \square$$

These properties make it relatively simple to compute an n-th order determinant. One method for doing this is as follows. We know that we can reduce a matrix $A = (a_{ij})$ to triangular form (see §3 of Ch. 1) by elementary row transformations. We thereby obtain a matrix

$$\overline{A} = \begin{Vmatrix} \overline{a}_{11} & \overline{a}_{12} & \cdots & \overline{a}_{1n} \\ 0 & \overline{a}_{22} & \cdots & \overline{a}_{2n} \\ \cdots & \cdots & \cdots & \cdots \\ 0 & 0 & \cdots & \overline{a}_{nn} \end{Vmatrix} \quad . \tag{5}$$

Suppose that q elementary transformations of type (I) were used in the reduction process. Since elementary transformations of type (II) do not change the determinant (property D7), while each elementary transformation of type (I) multiplies the determinant by (-1), it follows that $\det \overline{A} = (-1)^q \det A$. We shall prove that

$$\det \overline{A} = \overline{a}_{11} \overline{a}_{22} \cdots \overline{a}_{nn} \quad .$$

Once we prove this, we shall have the formula

$$\det A = (-1)^q \overline{a}_{11} \overline{a}_{22} \cdots \overline{a}_{nn} \quad . \tag{6}$$

The row reduction procedure, combined with this formula, is one method for computing $\det A$.

We prove the formula for $\det \overline{A}$ by induction on n. Since $\overline{a}_{21} = \cdots = \overline{a}_{n1} = 0$, it follows by (1) that $\det \overline{A} = \overline{a}_{11} \overline{M}_{11}$, where

$$\overline{M}_{11} = \begin{vmatrix} \overline{a}_{22} & \overline{a}_{23} & \cdots & \overline{a}_{2n} \\ 0 & \overline{a}_{33} & \cdots & \overline{a}_{3n} \\ \cdot & \cdot & \cdot & \cdot \\ 0 & 0 & \cdots & \overline{a}_{nn} \end{vmatrix}$$

is a determinant of order $n - 1$. By the induction assumption, $\overline{M}_{11} = \overline{a}_{22}\overline{a}_{33} \cdots \overline{a}_{nn}$. Hence, $\det \overline{A} = \overline{a}_{11}\overline{M}_{11} = \overline{a}_{11}\overline{a}_{22} \cdots \overline{a}_{nn}$.

Now, based on (6), we establish an important fact concerning the role of properties D1-D3 of a determinant. Namely, up to a constant factor det is the only function satisfying D1 and D2 (and that constant factor is 1 if D3 also holds).

THEOREM 2. Let $\mathcal{D} : M_n(\mathbb{R}) \to \mathbb{R}$ be any function having the following properties:

(i) $\mathcal{D}(A)$ is a linear function of the elements of each row of the matrix $A \in M_n(\mathbb{R})$;

(ii) when two neighboring rows are interchanged, $\mathcal{D}(A)$ changes sign (in other words, $\mathcal{D}(A)$ is a multilinear skew-symmetric function of the rows of the matrix).

Then there exists a constant ρ independent of A such that

$$\mathcal{D}(A) = \rho \cdot \det A \quad .$$

The number ρ is determined by the equality $\rho = \mathcal{D}(E)$, where E is the identity matrix.

Proof. By Lemma 1, $\mathcal{D}(A)$ changes sign when any two rows are interchanged, i.e., under any elementary transformation of type (I). Furthermore, an argument similar to that given in the proof of property D7 shows that $\mathcal{D}(A)$ does not change if an elementary transformation of type (II) is performed on the rows of A.

Using elementary transformations, we reduce A to the triangular form (5), where, of course, some of the $\overline{a}_{ii}$ might equal zero. Using what we now know, we have the two formulas

$$\det A = (-1)^q \det \overline{A} = (-1)^q \overline{a}_{11}\overline{a}_{22} \cdots \overline{a}_{nn} \qquad \text{(see (6))}$$

$$\mathcal{D}(A) = (-1)^q \mathcal{D}(\overline{A}) \quad ,$$

where q is the number of elementary transformations of type (I) used to go from A to $\overline{A}$. The required equality $\mathcal{D}(A) = \rho \cdot \det A$, where $\rho = \mathcal{D}(E)$, will then follow if we prove the formula

$$\mathcal{D}(\overline{A}) = \mathcal{D}(E) \cdot \overline{a}_{11} \cdots \overline{a}_{nn} \quad . \tag{7}$$

We now prove (7). By condition (i) of the theorem, we can bring $\overline{a}_{nn}$ outside the $\mathcal{D}$:

$$\mathcal{D}(\overline{A}) = \overline{a}_{nn} \mathcal{D}\left(\begin{Vmatrix} \overline{a}_{11} & \cdots & \overline{a}_{1,n-1} & \overline{a}_{1n} \\ \cdot & \cdot & \cdot & \cdot \\ 0 & \cdots & \overline{a}_{n-1,n-1} & \overline{a}_{n-1,n} \\ 0 & & 0 & 1 \end{Vmatrix}\right) .$$

We now apply an elementary transformation of type (II) to $\overline{A}$: for each i, subtract $\overline{a}_{in}$ times the last row of the matrix after the $\mathcal{D}$ on the right from the i-th row of this matrix. This changes all of the elements in the last column to zero (except for $\overline{a}_{nn} = 1$), but keeps all other elements in the matrix the same. We then move to the second-to-last row and proceed in the same manner, and so on. Each time we take the element $\overline{a}_{ii}$ outside the $\mathcal{D}$. After doing this n times, we finally obtain

$$\mathcal{D}(A) = \overline{a}_{nn} \cdots \overline{a}_{11} \cdot \mathcal{D}\left(\begin{Vmatrix} 1 & \cdots & 0 \\ \cdot & \cdot & \cdot \\ 0 & \cdots & 1 \end{Vmatrix}\right) ,$$

which is precisely (7). □

Thus, properties D1-D3 characterize the det function uniquely. For this reason, we consider those three properties to be the most fundamental properties of determinants. It would have been possible from the very beginning to define the determinant to be any function $\mathcal{D}$ having properties D1-D3, but in that case it would have been necessary to prove the existence of such a function. Our approach guaranteed the existence of det, because we constructed it by formula (1).

Because of later applications of Theorem 2, we did not insist on the normalization $\mathcal{D}(E) = 1$ for the function $\mathcal{D}$ in the theorem.

EXERCISES

1. Using formula (1) and the rule for the signs in the expansion of a third order determinant (Exercise 1 of §4 Ch. 1), write out completely all of the products occurring in the expansion of a fourth order determinant. Take note of the total number of terms in the expansion, and also try to find a rule for determining the sign of each term.

2. There are n terms on the right in formula (1). Each minor M_{i1}, in turn, is equal to a linear combination of $n-1$ minors of order $n-2$, and so on. Altogether the expansion of an n-th order determinant $\det(a_{ij})$ consists of $n(n-1)\ldots 3\cdot 2\cdot 1 = n!$ ("n factorial") products of the form $a_{i_1 1}a_{i_2 2}\cdots a_{i_n n}$, each with a $+$ or $-$. Show that

$$\det(a_{ij}) = a_{11}a_{22}\cdots a_{nn} + (-1)^{\frac{n(n-1)}{2}} a_{n1}a_{n-1,2}\cdots a_{1n} + \cdots .$$

3. Using the remarks in the preceding exercise, applied to the determinant $\det(a_{ij})$ of the matrix all of whose entries a_{ij} equal 1, prove that, in the expansion of an n-th order determinant, exactly half of the products $a_{i_1 1}a_{i_2 2}\cdots a_{i_n n}$ occur with a $+$ sign.

4. Write the following skew-symmetric function $\Delta : \mathbb{R}^3 \to \mathbb{R}$ of three variables x, y, z in the form of a third order determinant:

$$\Delta(x,y,z) = (y-x)(z-x)(z-y)$$

§2. Further properties of determinants

1. Expanding the determinant along an arbitrary column. We are now in a position to answer the question which naturally arose when we first constructed the function det : does the first column play any special role that explains its appearance in formula (1) of §1 ? The answer to this question is given by the following formula, which expresses the determinant in terms of the minors obtained from going down *any* j-th column:

$$\det A = \sum_{i=1}^{n} (-1)^{i+j} a_{ij} M_{ij} \quad . \tag{1}$$

To prove (1), we apply Theorem 2 of §1 to the function $\mathcal{D}_j$ in Lemma 2 of §1. We obtain the relation:

$$\mathcal{D}_j(A) = \mathcal{D}_j(E) \cdot \det A \quad .$$

But, by formula (3) of §1, $\mathcal{D}_j(E) = (-1)^{j-1}$. Hence, $\mathcal{D}_j(A) = (-1)^{j-1} \det A$. After multiplying both sides of this equation by $(-1)^{j-1}$, we have: $\det A = (-1)^{j-1} \mathcal{D}_j(A)$, which is merely another way of writing (1). This formula becomes more symmetrical if we introduce the so-called <u>cofactor</u> $A_{ij} = (-1)^{i+j} M_{ij}$ of the element a_{ij} in the matrix A. We state our result in the following theorem.

THEOREM 1. *The determinant of a matrix* A *is equal to the sum of the products of each element of a given column by the corresponding cofactor:*

$$\det A = \sum_{i=1}^{n} a_{ij} A_{ij} \quad . \quad \square \tag{2}$$

This theorem shows that all columns can play the same role. If $j = 1$, we obtain our original formula (1) of §1, which we used to define the determinant. We say that formulas (1) and (2) give the *expansion of a determinant along the* j-*th column*.

It is tempting to compare (2) with the analogous expression one obtains if one sums over the second index for fixed i : $\sum_{j=1}^{n} a_{ij} A_{ij}$. We shall soon see that this gives the same value $\det A$.

2. *The properties of determinants relating to columns.* As an application of Theorem 1, we obtain a whole new series of properties of determinants.

THEOREM 2. *Properties* D1-D7 *of* §1 *hold for the columns as well as the rows of a determinant.*

Proof. It is clear from §1 that the properties D4-D7 are completely formal consequences of D1-D3, and hence, in order to prove the analogous properties for the columns, it suffices to prove the first three of them for the columns. Notice that the "normalization" property D3 is of a special sort, and does not relate to the rows or columns, i.e., it is not affected if we replace rows by columns in the list of properties. Thus, we need only consider properties D1 and D2.

We first want to show that for any j, if the entries of A not in the j-th column are fixed, then $\det A$ is a linear function of the a_{ij}, $i = 1, \cdots, n$. We simply use formula (2). It shows directly that $\det A$ is a linear function of the elements of the j-th column, since the cofactor A_{ij} does not depend on these elements. This gives us D1.

We now prove D2 -- the skew-symmetry of the function $\det(A^{(1)}, \cdots, A^{(n)})$ -- by induction on n. If $n = 1$, property D2 does not assert anything. If $n = 2$, then D2 is easy to verify directly:

$$\begin{vmatrix} a & b \\ c & d \end{vmatrix} = ad - bc = - \begin{vmatrix} b & a \\ d & c \end{vmatrix} .$$

Suppose $n > 2$. Suppose we interchange the columns $A^{(k)}$ and $A^{(k+1)}$. We use formula (2) with $j \neq k, k+1$. Both of the columns $A^{(k)}, A^{(k+1)}$ occur in the minor M_{ij} (or, equivalently, in the cofactor A_{ij}), but they are shortened: they occur without the elements a_{ik}, $a_{i,k+1}$. By the induction assumption, when the two columns are interchanged each minor changes sign. Thus, we have

$$\det(\cdots, A^{(k)}, A^{(k+1)}, \cdots) = - \det(\cdots, A^{(k+1)}, A^{(k)}, \cdots) . \quad \square$$

3. <u>The transpose determinant.</u> We first recall a concept introduced in Exercise 1 of §2 Ch. 2. The $n \times m$ rectangular matrix whose i-th column, $i = 1, 2, \cdots, m$, coincides with the i-th row of the $m \times n$ matrix A is called the transpose of A. We denote the transpose of A by the symbol ${}^tA$ or else A'. Thus, if $A = (a_{ij})$ and ${}^tA = (a'_{ji})$, then $a'_{ji} = a_{ij}$. For example,

$$ {}^{t}\begin{Vmatrix} 1 & 2 & 3 & 4 \\ 5 & 6 & 7 & 8 \end{Vmatrix} = \begin{Vmatrix} 1 & 5 \\ 2 & 6 \\ 3 & 7 \\ 4 & 8 \end{Vmatrix} . $$

A column may be considered as the transpose of a row:

$$ [x_1, \cdots, x_n] = {}^{t}(x_1, \cdots, x_n) . $$

In the case of a square matrix, one sometimes refers to the determinant of the transpose

$$ \det {}^{t}A = \begin{vmatrix} a_{11} & a_{21} & \cdots & a_{n1} \\ a_{12} & a_{22} & \cdots & a_{n2} \\ \cdot & \cdot & \cdot & \cdot \\ a_{1n} & a_{2n} & \cdots & a_{nn} \end{vmatrix} $$

as the <u>transpose determinant.</u> The operation of transposing a matrix or determinant can be represented visually as rotation around the main diagonal, which consists of the elements a_{ii}.

THEOREM 3. <u>The determinant of the transpose of a matrix is equal to the determinant of the original matrix:</u>

$$ \det {}^{t}A = \det A . $$

<u>Proof.</u> We consider the function $\mathcal{D} : M_n(\mathbb{R}) \to \mathbb{R}$ which is the composition $A \mapsto {}^{t}A \mapsto \det {}^{t}A$ of the transposing function with the determinant function. The function $\mathcal{D}$ has properties (i) and (ii) in Theorem 2 of §1. Namely, by Theorem 2 of this section, the function ${}^{t}A \mapsto \det {}^{t}A$ has properties D1-D7 relative to the columns of ${}^{t}A$, i.e., relative to the rows of A. Thus, $\mathcal{D}$ is a multilinear skew-symmetric function of the rows of the matrix. By Theorem 2 of §1, we have $\mathcal{D}(A) = \mathcal{D}(E) \cdot \det A = \det {}^{t}E \cdot \det A$. But ${}^{t}E = E$, so that $\det {}^{t}E = 1$. Hence, $\mathcal{D}(A) = \det A$. □

Note how useful Theorem 2 of §1 has been in providing short, non-computational proofs of Theorems 1 and 3 of this section.

Theorem 3 tells us that the rows and columns of a determinant play equivalent roles:

the properties expressed in terms of the rows can also be expressed in terms of the columns, and vice-versa. For example, in addition to Theorem 1 on expanding a determinant along a column, we have

THEOREM 1'. The determinant of a matrix A is equal to the sum of the products of each element of any fixed row by the corresponding cofactor:

$$\det A = \sum_{j=1}^{n} a_{ij} A_{ij} \quad . \quad \square$$

We also have the following useful criterion for the vanishing of a determinant: if any row (any column) of $\det A$ is a linear combination of the other rows (resp. columns), then $\det A = 0$ (see properties D1', D1" and their analogs for columns).

The following two examples illustrate the properties of determinants.

Example 1. The so-called Vandermonde determinant

$$\Delta_n = \begin{vmatrix} 1 & 1 & \cdots & 1 \\ x_1 & x_2 & \cdots & x_n \\ x_1^2 & x_2^2 & \cdots & x_n^2 \\ \cdot & \cdot & \cdot & \cdot \\ x_1^{n-1} & x_2^{n-1} & \cdots & x_n^{n-1} \end{vmatrix} = \Delta(x_1, x_2, \cdots, x_n) \quad ,$$

is given by the formula

$$\Delta_n = \prod_{1 \leq i < j \leq n} (x_j - x_i) \quad , \tag{3}$$

or, writing this expression out fully, we have

$$\Delta_n = (x_2 - x_1)(x_3 - x_1) \cdots (x_n - x_1)(x_3 - x_2) \cdots (x_n - x_2) \cdots (x_n - x_{n-1}) \quad ,$$

(in this connection, it is useful to look back at Exercise 4 of §1). In particular, if the elements $x_1, \cdots, x_n$ are pair-wise distinct, the Vandermonde determinant is non-zero. This property is often useful. By Theorem 3, we also have

$$\Delta_n = \begin{vmatrix} 1 & x_1 & x_1^2 & \cdots & x_1^{n-1} \\ 1 & x_2 & x_2^2 & \cdots & x_2^{n-1} \\ \cdot & \cdot & \cdot & \cdot & \cdot \\ 1 & x_n & x_n^2 & \cdots & x_n^{n-1} \end{vmatrix} .$$

We use induction on n to prove (3). Suppose that Δ_m is given by (3) for $m < n$. Using property D7, we subtract x_1 times the $(i-1)$-th row from the i-th row, for each i:

$$\Delta_n = \begin{vmatrix} 1 & 1 & \cdots & 1 \\ 0 & x_2 - x_1 & \cdots & x_n - x_1 \\ 0 & x_2^2 - x_2 x_1 & \cdots & x_n^2 - x_n x_1 \\ \cdot & \cdot & \cdot & \cdot \\ 0 & x_2^{n-1} - x_2^{n-2} x_1 & \cdots & x_n^{n-1} - x_n^{n-2} x_1 \end{vmatrix} .$$

The natural next step is to expand Δ_n along the first column, and take the common factor $x_{j+1} - x_1$ in the j-th column $(j = 1, 2, \cdots, n-1)$ of the resulting $(n-1)$-th order determinant outside the determinant sign (using property D1' for columns). We arrive at the expression

$$\Delta_n = (x_n - x_1)(x_{n-1} - x_1) \cdots (x_2 - x_1) \begin{vmatrix} 1 & 1 & \cdots & 1 \\ x_2 & x_3 & \cdots & x_n \\ \cdot & \cdot & \cdot & \cdot \\ x_2^{n-2} & x_3^{n-2} & \cdots & x_n^{n-2} \end{vmatrix} =$$

$$= (x_n - x_1)(x_{n-1} - x_1) \cdots (x_2 - x_1) \cdot \Delta(x_2, x_3, \cdots, x_n) ,$$

which coincides with (3), since, by the induction assumption,

$$\Delta(x_2, \cdots, x_n) = \prod_{2 \le i < j \le n} (x_j - x_i) .$$

Example 2. A matrix $A = (a_{ij})$ of the form

$$A = \begin{Vmatrix} 0 & a_{12} & a_{13} & \cdots & a_{1n} \\ -a_{12} & 0 & a_{23} & \cdots & a_{2n} \\ -a_{13} & -a_{23} & 0 & \cdots & a_{3n} \\ \cdots & \cdots & \cdots & \cdots & \cdots \\ -a_{1n} & -a_{2n} & -a_{3n} & \cdots & 0 \end{Vmatrix}$$

is called skew-symmetric (its determinant is also called skew-symmetric). In other words, ${}^tA = -A$. By Theorem 3, we have

$$\det A = \det {}^tA = \det(-A) = (-1)^n \det A ,$$

so that $[1 + (-1)^{n-1}] \det A = 0$. For odd n we obtain $\det A = 0$, i.e., every skew-symmetric matrix of odd order has zero determinant.

4. Determinants of special matrices. The more zeros among the elements of A and the "better" their location, the easier it is to compute $\det A$. This intuitive idea in certain cases leads to an exact formula. For example, we know (see subsection 2 of §1) that the determinant of an (upper or lower) triangular matrix is equal to the product of the elements on the main diagonal. Another important special case is

THEOREM 4. *A determinant D of order $n + m$ containing zeros in the intersection of the first n columns and the last m rows is given by the formula*

$$\begin{vmatrix} a_{11} & \cdots & a_{1n} & a_{1,n+1} & \cdots & a_{1,n+m} \\ \cdots & \cdots & \cdots & \cdots & \cdots & \cdots \\ a_{n1} & \cdots & a_{nn} & a_{n,n+1} & \cdots & a_{n,n+m} \\ 0 & & 0 & b_{11} & \cdots & b_{1m} \\ \cdots & \cdots & \cdots & \cdots & \cdots & \cdots \\ 0 & & 0 & b_{m1} & \cdots & b_{mm} \end{vmatrix} = \begin{vmatrix} a_{11} & \cdots & a_{1n} \\ \cdots & \cdots & \cdots \\ a_{n1} & \cdots & a_{nn} \end{vmatrix} \cdot \begin{vmatrix} b_{11} & \cdots & b_{1m} \\ \cdots & \cdots & \cdots \\ b_{m1} & \cdots & b_{mm} \end{vmatrix}$$

(a determinant of the form on the left is sometimes called a quasi-triangular determinant or a determinant with zero corner).

Proof. First fix the $n(n + m)$ elements a_{ij} and consider D as a function of the elements $b_{k\ell}$, which make up an m-th order square matrix B. We thus consider the determinant as a function of the matrix B: $D = \mathscr{D}(B)$.

Clearly, because the determinant D is multilinear and skew-symmetric relative to the last m rows, the function $\mathscr{D}(B)$ has the same properties relative to the rows of B. Hence, we may apply Theorem 2 of §1 to $\mathscr{D}(B)$, and conclude that $\mathscr{D}(B) = \mathscr{D}(E) \cdot \det B$. By the definition of $\mathscr{D}$, we have:

$$\mathscr{D}(E) = \begin{vmatrix} a_{11} & \cdots & a_{1n} & a_{1,n+1} & \cdots & a_{1,n+m} \\ \cdot & \cdot & \cdot & \cdot & \cdot & \cdot \\ a_{n1} & \cdots & a_{nn} & a_{n,n+1} & \cdots & a_{n,n+m} \\ 0 & \cdots & 0 & 1 & \cdots & 0 \\ \cdot & \cdot & \cdot & \cdot & \cdot & \cdot \\ 0 & \cdots & 0 & 0 & \cdots & 1 \end{vmatrix} .$$

We expand $\mathscr{D}(E)$ along the last row (see (2)), obtaining a minor which we then expand along its last row, i.e., the second-to-last row of this matrix, and so on. Repeating this operation m times, we see that $\mathscr{D}(E) = \det A$, where

$$A = \begin{Vmatrix} a_{11} & \cdots & a_{1n} \\ \cdot & \cdot & \cdot \\ a_{n1} & \cdots & a_{nn} \end{Vmatrix} .$$

We finally obtain: $D = \mathscr{D}(B) = \det A \cdot \det B$. □

We can introduce a more compact notation for Theorem 4:

$$\det \begin{Vmatrix} A & C \\ 0 & B \end{Vmatrix} = \det A \cdot \det B \quad . \tag{4}$$

Here A and B are square matrices, and the 0 matrix and C are rectangular matrices. Combining Theorems 3 and 4 (or else repeating an argument completely analogous to the proof of Theorem 4), we easily see that, similarly,

$$\det \begin{Vmatrix} A & 0 \\ C & B \end{Vmatrix} = \det A \cdot \det B \quad .$$

One might try to write a similar formula for $\det \begin{Vmatrix} C & A \\ B & 0 \end{Vmatrix}$, but note that the simplest possible example $\begin{vmatrix} 0 & 1 \\ 1 & 0 \end{vmatrix} = -1$ shows that there is a problem with the sign. To obtain a correct formula, we might first permute the rows or columns, in order to transform the matrix $\begin{Vmatrix} C & A \\ B & 0 \end{Vmatrix}$ to the form $\begin{Vmatrix} B & 0 \\ C & A \end{Vmatrix}$ or $\begin{Vmatrix} A & C \\ 0 & B \end{Vmatrix}$.

A simpler approach is based on Theorem 2 of §1, which we have already used several times. Namely, using that theorem as in the proof of Theorem 4 above, we obtain

$$\det \begin{Vmatrix} C & A \\ B & 0 \end{Vmatrix} = \det \begin{Vmatrix} C & A \\ E & 0 \end{Vmatrix} \cdot \det B \quad .$$

Next, by formula (1) applied m times, we find

$$\det \begin{Vmatrix} C & A \\ E & 0 \end{Vmatrix} = \begin{vmatrix} & * & & a_{11} & \cdots & a_{1n} \\ & & & a_{n1} & \cdots & a_{nn} \\ 1 & \cdots & 0 & 0 & \cdots & 0 \\ \cdot & \cdot & \cdot & \cdot & \cdot & \cdot \\ 0 & \cdots & 1 & 0 & \cdots & 0 \end{vmatrix} =$$

$$= (-1)^{(n+2)+(n+4)+\cdots+(n+2m)} \det A = (-1)^{nm} \det A \quad .$$

We conclude that, if A and B are square matrices of order n and m, respectively, then

$$\det \begin{Vmatrix} C & A \\ B & 0 \end{Vmatrix} = (-1)^{nm} \det A \cdot \det B \quad . \tag{5}$$

Formulas (4) and (5) are both special cases of a general theorem of Laplace on expanding determinants. But this theorem is not often used, and we shall not discuss it. We are also in no hurry to derive the so-called complete determinant expansion theorem (see §3 of Chapter 4), since that complete expansion is of little use from a computational point of view.

A very important property of determinants is

THEOREM 5. *If* A *and* B *are* $n \times n$ *matrices, then*

$$\det AB = \det A \cdot \det B \quad .$$

Proof. According to formulas (7) and (9) of §3 Ch. 2, which express the entries c_{ij} in the matrix $(c_{ij}) = AB = (a_{ij})(b_{ij})$ in terms of the entries in the matrices A and B, the i-th row $(AB)_i$ is given by

$$(AB)_i = (A_i B^{(1)}, A_i B^{(2)}, \cdots, A_i B^{(n)}) ; \quad A_i B^{(j)} = \sum_{k=1}^{n} a_{ik} b_{kj} .$$

Fix a matrix B, and for any matrix A set

$$\mathscr{D}(A) = \det AB .$$

We show that the function $\mathscr{D}$ satisfies the conditions (i), (ii) of Theorem 2 §1. In fact, we know that $\det AB$ is a linear function of the elements of the i-th row $(AB)_i$:

$$\det AB = \lambda_1 A_i B^{(1)} + \lambda_2 A_i B^{(2)} + \cdots + \lambda_n A_i B^{(n)} .$$

Hence,

$$\mathscr{D}(A) = \sum_{j=1}^{n} \lambda_j \sum_{k=1}^{n} a_{ik} b_{kj} = \sum_{k=1}^{n} a_{ik} \sum_{j=1}^{n} \lambda_j b_{kj} = \sum_{k=1}^{n} \mu_k a_{ik} ,$$

where $\mu_k = \sum_{j=1}^{n} \lambda_j b_{kj}$ is a scalar which does not depend on the elements of the i-th row A_i of A. We thus see that $\mathscr{D}(A)$ depends linearly on the elements of the i-th row A_i of A.

Now suppose we interchange A_s and A_t. Since the s-th and t-th rows of AB have the form

$$(A_s B^{(1)}, \cdots, A_s B^{(n)}) ,$$

$$(A_t B^{(1)}, \cdots, A_t B^{(n)}) ,$$

it follows that they are also interchanged when we interchange A_s and A_t. Thus, by Theorem 1,

$$\mathcal{D}(\cdots, A_s, \cdots, A_t, \cdots) = \mathcal{D}(A) = \det AB =$$

$$= \det[\cdots, (AB)_s, \cdots, (AB)_t, \cdots] =$$

$$= -\det[\cdots, (AB)_t, \cdots, (AB)_s, \cdots] =$$

$$= -\mathcal{D}(\cdots, A_t, \cdots, A_s, \cdots) \quad .$$

Thus, both conditions of Theorem 2 §1 are fulfilled, and hence $\mathcal{D}(A) = \mathcal{D}(E) \cdot \det A$. But, by definition, $\mathcal{D}(E) = \det EB = \det B$. This gives us the desired formula. □

5. Building up a theory of determinants. Theorems 1 and 2 of §1 essentially give us an axiomatic description of the det function, even though we started out by defining det by an explicit construction.

We shall now give another method which can be used to construct a theory of determinants. Namely, suppose we have a function $\mathcal{D}: M_n(\mathbb{R}) \longrightarrow \mathbb{R}$ with the following properties:

(i) $\mathcal{D}(AB) = \mathcal{D}(A) \cdot \mathcal{D}(B)$ for any matrices $A, B \in M_n(\mathbb{R})$;

(ii) $\mathcal{D}(F_{s,t}) = -1$ for every elementary matrix $F_{s,t}$ (see subsection 4 of §4 Ch. 2);

(iii) $\mathcal{D}(A) = \lambda$ for any upper triangular matrix of the form

$$A = \begin{Vmatrix} \lambda & & & \\ & 1 & & * \\ & & \ddots & \\ 0 & & & 1 \end{Vmatrix}, \quad \lambda \in \mathbb{R} .$$

We claim that $\mathcal{D} = \det$. To prove this, we first apply properties (i) and (ii) to the elementary matrix

$$F_s(\lambda) = \begin{Vmatrix} 1 & & & & 0 \\ & \ddots & & & \\ & & \lambda & & \\ & & & \ddots & \\ 0 & & & & 1 \end{Vmatrix} = F_{1,s} \cdot \begin{Vmatrix} \lambda & & & 0 \\ & 1 & & \\ & & \ddots & \\ 0 & & & 1 \end{Vmatrix} \cdot F_{1,s} .$$

We find that $\mathcal{D}(F_s(\lambda)) = (-1) \cdot \lambda \cdot (-1) = \lambda$, $\lambda \neq 0$. According to (iii), $\mathcal{D}(F_{s,t}(\lambda)) = 1$ for an elementary matrix $F_{s,t}(\lambda)$ with $s < t$. Since

$$F_{s,t} \cdot F_{s,t}(\lambda) \cdot F_{s,t} = F_{t,s}(\lambda),$$

it follows that $\mathcal{D}(F_{t,s}(\lambda))$ is also 1, and so $\mathcal{D}(F_{s,t}(\lambda)) = 1$ for any indices $s \neq t$. Furthermore,

$$\begin{Vmatrix} E_r & 0 \\ 0 & 0 \end{Vmatrix} = F_{r+1}(0)\cdots F_n(0),$$

and hence

$$\mathcal{D}\left(\begin{Vmatrix} E_r & 0 \\ 0 & 0 \end{Vmatrix}\right) = \begin{cases} 0 & \text{if } r < n; \\ 1 & \text{if } r = n. \end{cases}$$

Thus, $\mathcal{D}(F_{s,t}) = -1 = \det F_{s,t}$, $\mathcal{D}(F_{s,t}(\lambda)) = 1 = \det F_{s,t}(\lambda)$, and $\mathcal{D}(F_s(\lambda)) = \lambda = \det F_s(\lambda)$. Since any matrix $A \in M_n(\mathbb{R})$ can be written in the form $A = P\cdot\begin{Vmatrix} E_r & 0 \\ 0 & 0 \end{Vmatrix}\cdot Q$, $r \leq n$, where P and Q are products of elementary matrices (see the argument before Theorem 5 of §4 Ch. 2), property (i) enables us to conclude that $\mathcal{D}(A) = \det A$.

It would be a good idea for the reader to try to suggest and justify his own version of an axiomatic description of the det function.

EXERCISES

1. The integers $1798 = 31\cdot 58$, $2139 = 31\cdot 69$, $3255 = 31\cdot 105$, $4867 = 31\cdot 157$ are divisible by 31. Without any computations, prove that the determinant of the following fourth order determinant is also divisible by 31:

$$\begin{vmatrix} 1 & 7 & 9 & 8 \\ 2 & 1 & 3 & 9 \\ 3 & 2 & 5 & 5 \\ 4 & 8 & 6 & 7 \end{vmatrix}.$$

2. Show that every fourth order skew-symmetric determinant $|a_{ij}|$ with $a_{ij} \in \mathbb{Z}$ is the square of an integer. (Note. This is true for a skew-symmetric determinant of any order.)

3. Prove that $\det AB = \det A \cdot \det B$ (Theorem 5) by performing elementary transformations of type (II) on the rows of the auxiliary matrix $C = \begin{Vmatrix} E & B \\ -A & 0 \end{Vmatrix}$ of size $2n \times 2n$ in such a way as to reduce it to the form $C' = \begin{Vmatrix} E & B \\ 0 & AB \end{Vmatrix}$.

4. Show that ${}^t(AB) = {}^tB\,{}^tA$ for any $m \times r$ matrix A and $r \times n$ matrix B.

5. Show that $\det B^{-1}AB = \det A$ for any $A \in M_n(\mathbb{R})$ and any invertible $B \in M_n(\mathbb{R})$.

6. Let

$$C_n(\lambda_1, \cdots, \lambda_n) = \begin{Vmatrix} \lambda_1 & 1 & 0 & \cdots & 0 & 0 & 0 \\ -1 & \lambda_2 & & \cdots & 0 & 0 & 0 \\ \cdot & \cdot & \cdot & \cdot & \cdot & \cdot & \cdot \\ 0 & 0 & 0 & \cdots & \lambda_{n-2} & 1 & 0 \\ 0 & 0 & 0 & \cdots & -1 & \lambda_{n-1} & 1 \\ 0 & 0 & 0 & \cdots & 0 & -1 & \lambda_n \end{Vmatrix} .$$

Show that $\det C_n = \lambda_n \det C_{n-1} + \det C_{n-2}$. If $\lambda_1 = \lambda_2 = \cdots = \lambda_n = 1$, compute the value of $\det C_n$.

7. Show that the determinant of the $n \times n$ matrix

$$A_n = \begin{Vmatrix} 2 & -1 & 0 & 0 & \cdots & 0 & 0 & 0 \\ -1 & 2 & -1 & 0 & \cdots & 0 & 0 & 0 \\ 0 & -1 & 2 & -1 & \cdots & 0 & 0 & 0 \\ \cdot & \cdot & \cdot & \cdot & \cdot & \cdot & \cdot & \cdot \\ 0 & 0 & 0 & 0 & \cdots & -1 & 2 & -1 \\ 0 & 0 & 0 & 0 & \cdots & 0 & -1 & 2 \end{Vmatrix}$$

is equal to $n + 1$.

§3. Applications of determinants

1. <u>Criterion for a matrix to be non-singular.</u> In §3 Ch. 2, we said that a square matrix A is called non-singular if it has an inverse A^{-1}. If we apply Theorem 5 of §2 to the relation $AA^{-1} = A^{-1}A = E$, we see that $\det A \cdot \det A^{-1} = 1$. Thus, <u>the determinant of a non-singular matrix is non-zero</u>, and

$$\det A^{-1} = (\det A)^{-1} .$$

Given a matrix A, we may consider its <u>classical adjoint</u> matrix

$$A^{\vee} = \begin{Vmatrix} A_{11} & \cdots & A_{n1} \\ \cdot & \cdots & \cdot \\ A_{1n} & \cdots & A_{nn} \end{Vmatrix} .$$

The matrix $A^{\vee}$ is obtained by replacing each entry a_{ij} in A by the corresponding cofactor A_{ij} and then taking the transpose matrix.

THEOREM 1. <u>A matrix</u> $A \in M_n(\mathbb{R})$ <u>is non-singular (invertible) if and only if</u> $\det A \neq 0$. <u>If</u> $\det A \neq 0$, <u>then</u> $A^{-1} = (\det A)^{-1} A^{\vee}$, <u>i.e.</u>,

$$\begin{Vmatrix} a_{11} & \cdots & a_{1n} \\ \cdot & \cdots & \cdot \\ a_{n1} & \cdots & a_{nn} \end{Vmatrix}^{-1} = \begin{Vmatrix} \frac{A_{11}}{\det A} & \cdots & \frac{A_{n1}}{\det A} \\ \cdot & \cdots & \cdot \\ \frac{A_{1n}}{\det A} & \cdots & \frac{A_{nn}}{\det A} \end{Vmatrix} .$$

To prove this theorem we first need a lemma.

LEMMA. <u>Let</u> $A \in M_n(\mathbb{R})$. <u>Then</u>:

$$a_{i1}A_{j1} + a_{i2}A_{j2} + \cdots + a_{in}A_{jn} = \delta_{ij} \det A , \tag{1}$$

$$a_{1i}A_{1j} + a_{2i}A_{2j} + \cdots + a_{ni}A_{nj} = \delta_{ij} \det A , \tag{2}$$

<u>where</u> δ_{ij} <u>is the Kronecker symbol</u> (if $i \neq j$, this can be thought of as expanding the determinant using the cofactors of the wrong row or the wrong column, respectively).

<u>Proof.</u> If $i = j$, the lemma coincides with Theorems 1 and 1' of §2. So suppose that $i \neq j$, in which case $\delta_{ij} = 0$. To prove the lemma in this case, we introduce the matrix

$$A' = [A_1, \cdots, A_i, \cdots, A_i, \cdots, A_n] = \begin{Vmatrix} a_{11} & a_{12} & \cdots & a_{1n} \\ \cdot & \cdot & \cdots & \cdot \\ a_{i1} & a_{i2} & \cdots & a_{in} \\ \cdot & \cdot & \cdots & \cdot \\ a_{i1} & a_{i2} & \cdots & a_{in} \\ \cdot & \cdot & \cdots & \cdot \\ a_{n1} & a_{n2} & \cdots & a_{nn} \end{Vmatrix} ,$$

which is obtained from $A = [\cdots, A_i, \cdots, A_j, \cdots]$ by replacing the j-th row by the i-th (keeping the i-th row in place). As always with a square matrix with two identical rows, we have $\det A' = 0$. On the other hand, the cofactor A'_{jk} $(k = 1, \cdots, n)$ is obtained by crossing out the j-th row $A'_j = A_i$ and the k-th column of the determinant; hence, $A'_{jk} = A_{jk}$. If we expand the determinant of $A' = (a'_{st})$ along the j-th row, we obtain the relation

$$0 = \det A' = \sum_{k=1}^{n} a'_{jk} A'_{jk} = \sum_{k=1}^{n} a_{ik} A_{jk} ,$$

which is precisely equation (1) in the lemma. The second equality in the lemma is obtained by the same argument, applied to the columns. □

Proceeding to the proof of the theorem, we simple note that the left side of (1) is nothing other than the entry c_{ij} in the matrix $C = A A^{\vee}$:

$$\begin{Vmatrix} c_{11} & \cdots & c_{1n} \\ \cdot & \cdots & \cdot \\ c_{n1} & \cdots & c_{nn} \end{Vmatrix} = \begin{Vmatrix} a_{11} & \cdots & a_{1n} \\ \cdot & \cdots & \cdot \\ a_{n1} & \cdots & a_{nn} \end{Vmatrix} \begin{Vmatrix} A_{11} & \cdots & A_{n1} \\ \cdot & \cdots & \cdot \\ A_{1n} & \cdots & A_{nn} \end{Vmatrix} .$$

Using (1), we have $(c_{ij}) = (\delta_{ij} \det A) = (\det A) E$. Thus,

$$A A^{\vee} = (\det A) E ,$$

so that when $\det A \neq 0$ we obtain

$$(\det A)^{-1} (A A^{\vee}) = A (\det A)^{-1} A^{\vee} = E .$$

The left side of (2) is the entry c'_{ji} in the matrix $C' = A^{\vee} A$. Since the right sides of (1) and (2) are the same, we arrive at the following equalities when $\det A \neq 0$:

$$A (\det A)^{-1} A^{\vee} = (\det A)^{-1} A^{\vee} A = E ,$$

and so we have $A^{-1} = (\det A)^{-1} A^{\vee}$. □

COROLLARY 1. <u>The determinant vanishes if and only if the rows (or columns) are linearly dependent.</u>

Proof. This criterion, part of which has already been given (see subsection 3 of §2), could have been proved much earlier, but we had no need of it. To prove the criterion, we know by Theorem 1 that $\det A = 0$ if and only if A is singular, and, by Theorem 4 of §3 Ch. 2, this is equivalent to the condition $\operatorname{rank} A < n$ (where A is an $n \times n$ matrix). Finally, by Theorem 1 of §2 Ch. 2, this condition precisely characterizes those $n \times n$ matrices with linearly dependent rows (or columns). □

Theorem 1 is of greater theoretical than practical value. From a computational point of view, especially for large matrices, it is usually more convenient to use (P, Q)-reduction (see the corollary to Theorem 5 of §4 Ch. 2) to find A^{-1}.

We now derive formulas for solving a system of n linear equations with n unknowns, which, after all, was one of our main reasons for developing the theory of determinants in the first place.

COROLLARY 2. (Cramer's rule). If a linear system

$$\begin{aligned} a_{11}x_1 + \cdots + a_{1n}x_n &= b_1 \ , \\ \cdots\cdots\cdots&\cdots\cdots \\ a_{n1}x_1 + \cdots + a_{nn}x_n &= b_n \end{aligned}$$

has non-zero determinant (i. e., $\det(a_{ij}) \neq 0$), then its unique solution is given by the formulas

$$x_k^0 = \frac{\begin{vmatrix} a_{11} & \cdots & b_1 & \cdots & a_{1n} \\ & & & & \\ a_{n1} & \cdots & b_n & \cdots & a_{nn} \end{vmatrix}}{\begin{vmatrix} a_{11} & \cdots & a_{1k} & \cdots & a_{1n} \\ \cdot & \cdot & \cdot & \cdot & \cdot \\ a_{n1} & \cdots & a_{nk} & \cdots & a_{nn} \end{vmatrix}} \ , \quad k = 1, 2, \cdots, n$$

(the numerator D_k is obtained by replacing the k-th column in $D = \det(a_{ij})$ by the column of free terms).

Proof. By Theorem 1, the matrix $A = (a_{ij})$ is invertible. Hence, if we write our system in the form $AX = B$, we obtain, as in §4 of Chapter 2,

$$X = A^{-1}B = \frac{(A_{ji})B}{\det A},$$

and hence

$$x_k^0 = \frac{\sum_j A_{jk} b_j}{\det A}.$$

It is this expression in the numerator which we obtain by expanding the determinant D_k along the k-th column (see (2)). Thus, any solution $x = (x_1^0, \cdots, x_n^0)$ must be given by the formula in the corollary.

If we go through all of these steps backwards, we see that $(D_1/\det A, \cdots, D_n/\det A)$ actually is the solution of our system. □

Note that the formulas (3) and (9) of §4 Ch. 1 coincide with Cramer's rule for $n = 2$ and 3. Although convenient for small n, for large n Cramer's rule has a largely theoretical value. For example, if we apply it to the linear system in example 2 of subsection 4 of §3 Ch. 1, we obtain the following expression for the n-th Fibonacci number (using the fact that $\det A = 1$):

$$f_n = \begin{vmatrix} 1 & 0 & 0 & \dots & 0 & 0 & 1 \\ 0 & 1 & 0 & \dots & 0 & 0 & 1 \\ -1 & -1 & 1 & \dots & 0 & 0 & 1 \\ \cdot & \cdot & \cdot & \cdot & \cdot & \cdot & \cdot \\ 0 & 0 & 0 & \dots & -1 & 1 & 0 \\ 0 & 0 & 0 & \dots & -1 & -1 & 0 \end{vmatrix}.$$

This is clearly a long way off from the nice, explicit expression for f_n which we found at the end of §3 Ch. 2.

2. Computing the rank of a matrix. In §§2 and 4 of Ch. 2, we found how to give a complete description of the set of solutions of a general rectangular system of linear equations. The notion of the rank of a matrix played an important role in this description. If we translate this notion into the language of determinants, we shall have at our disposal both another method for computing the rank and a convenient way of expressing linear independence

of a set of vectors in the vector space $\mathbb{R}^m$.

Thus, let

$$A = \begin{Vmatrix} a_{11} & \cdots & a_{1r} & \cdots & a_{1n} \\ \cdot & \cdot & \cdot & \cdot & \cdot \\ a_{r1} & \cdots & a_{rr} & \cdots & a_{rn} \\ \cdot & \cdot & \cdot & \cdot & \cdot \\ a_{m1} & \cdots & a_{mr} & \cdots & a_{mn} \end{Vmatrix}$$

be any $m \times n$ rectangular matrix with entries $a_{ij} \in \mathbb{R}$. By a k-th order minor of A we mean the determinant of any matrix obtained by taking the intersection of k rows and k columns, where $k \leq \min(m, n)$.

Suppose that A has rank r. By Theorem 1 of §2 Ch. 2, this means that r is the maximal number of linearly independent rows in A, and also the maximal number of linearly independent columns in A. If we now use Theorem 5 of §4 Ch. 2 and its corollary, we can write

$$A = B \begin{Vmatrix} E_r & 0 \\ & \\ 0 & 0 \end{Vmatrix} C ,$$

where B and C are non-singular $m \times m$ and $n \times n$ matrices, respectively, which are written as a product of elementary matrices. Since the matrix $\begin{Vmatrix} E_r & 0 \\ 0 & 0 \end{Vmatrix}$ has the non-zero r-th order minor $M = |E_r| = 1$, but does not have any non-zero minor of order $> r$, and since this property is preserved when elementary transformations are performed on the rows and columns, we have proved the following

THEOREM 2. _The rank of an_ $m \times n$ _matrix_ A _is equal to the maximal order of a non-zero minor._ □

Any non-zero minor of _maximal_ order in A is called a _basis minor._ The columns (or rows) of A which intersect a given basis minor are called basis columns (respectively, basis rows), in agreement with the terminology in Chapter 2. As before, if we interpret the rows and columns of an $m \times n$ matrix A as vectors in $\mathbb{R}^n$ and $\mathbb{R}^m$, respectively, and if we use the basic properties of a linearly independent set of vectors (the fact that it can be completed to a basis; see Exercise 5 of §1 Ch. 2), we easily see that the search for a basis minor can be much simplified if we successively look for higher order minors containing non-zero lower order ones. Namely, if we have any non-zero k-th order minor M in A, then the next step consists only in checking those $(k+1)$-th order minors which contain M, i.e., from which M is obtained by crossing out a row and column. If all of these $(k+1)$-th order minors vanish, then $\operatorname{rank} A = k$. (Why? By Theorem 2, this would mean that every column of A can be expressed as a linear combination of the k columns which intersect M.) If not all of these minors vanish, then take any which is non-zero and then proceed to check all $(k+2)$-th order minors containing it.

This method for determining the rank is rather practical, especially when we want to know not only the rank, but also a maximal linearly independent set of rows or columns of A. Of course, we have to be careful to remember that this information is lost if we perform elementary transformations on the matrix.

EXERCISES

1. Show that the following relations hold:

$$(AB)^{\vee} = B^{\vee} A^{\vee} ; \quad ({}^tA)^{\vee} = {}^t(A^{\vee}) ; \quad (\lambda A)^{\vee} = \lambda^{n-1} A^{\vee} ;$$

$$(A^{\vee})^{\vee} = (\det A)^{n-2} A \quad .$$

2. Express $\operatorname{rank} A^{\vee}$ in terms of $\operatorname{rank} A$.

3. Prove that a square system of homogeneous linear equations has non-trivial solutions if and only if the determinant of the system is zero.

4. Using the results of subsection 1 of §4 Ch. 2 and Corollary 2 of Theorem 1,

show that a homogeneous system

$$\begin{aligned} a_{11}x_1 + \cdots + a_{1n}x_n &= 0 , \\ \cdots\cdots&\cdots\cdots \\ a_{n-1,1}x_1 + \cdots + a_{n-1,n}x_n &= 0 \end{aligned}$$

of rank $r = n-1$ has as fundamental set of solutions the single column-vector

$$X^o = [D_1, -D_2, D_3, \cdots, (-1)^{n-1} D_n] ,$$

where D_i is the determinant of the matrix obtained from $A = (a_{ij})$ by crossing out the i-th column. Thus, every solution has the form $X = \lambda X^o$.

5. Suppose that $A = (a_{ij}) \in M_n(\mathbb{R})$ and $(n-1)|a_{ij}| < |a_{ii}|$ for all $i \neq j$. Prove that $\det A \neq 0$.

6. Prove the following fact. Let $A = (a_{ij})$ and $B = (b_{k\ell})$ be matrices of size $n \times m$ and $m \times n$, respectively, and let $C = AB$. Then

$$\det C = \sum_{1 \le j_1 < \cdots < j_n \le m} \begin{vmatrix} a_{1j_1} & a_{2j_1} & \cdots & a_{nj_1} \\ a_{1j_2} & a_{2j_2} & \cdots & a_{nj_2} \\ \cdots & \cdots & \cdots & \cdots \\ a_{1j_n} & a_{2j_n} & \cdots & a_{nj_n} \end{vmatrix} \begin{vmatrix} b_{j_1 1} & b_{j_1 2} & \cdots & b_{j_1 n} \\ b_{j_2 1} & b_{j_2 2} & \cdots & b_{j_2 n} \\ \cdots & \cdots & \cdots & \cdots \\ b_{j_n 1} & b_{j_n 2} & \cdots & b_{j_n n} \end{vmatrix} .$$

The summation on the right is over all $\binom{m}{n}$ combinations of n numbers $\{j_1, j_2, \cdots, j_n\}$ from $1, 2, \cdots, m$. In particular, $\det C = \det A \cdot \det B$ when $m = n$, and $\det C = 0$ when $n > m$.

7. Using the preceding exercise, show that, if A is an $m \times n$ matrix, $m \geq n$, then

$$\det {}^tAA = \sum_M M^2 ,$$

where M runs through all $\binom{m}{n}$ n-th order minors of A.

Chapter 4. Algebraic Structures—Groups, Rings, Fields

The preceding chapters have provided us with a great deal of concrete material, which it is now time to consider from a more general point of view. For this purpose, we introduce and study (still on an elementary level) the concepts of groups, rings, and fields, which play a fundamental role in all of algebra.

§1. Sets with algebraic operations

1. Binary operations. Let X be any set. An algebraic binary operation (or composition law) on X is any fixed map $\tau : X \times X \to X$ of the cartesian square $X^2 = X \times X$ to X. Thus, to every ordered pair (a,b) of elements $a, b \in X$ there corresponds a unique element $\tau(a,b)$ of the same set X. We sometimes write $a\tau b$ instead of $\tau(a,b)$, and in fact usually we introduce a special symbol $*, \circ, \cdot, +$, etc. to designate a binary operation on X. In our examples we shall follow this practice, and shall call $a \cdot b$ (or simply ab with no symbol between a and b) the product and $a + b$ the sum of the elements $a, b \in X$. In most cases one of these conventions will be convenient.

It is possible for the same set X to have more than one binary operation defined on it. When we choose one such operation, and want to think of X in conjunction with that particular binary operation $*$, we write $(X, *)$ and say that $*$ determines an algebraic structure on X or that $(X, *)$ is an algebraic system. For example, in addition to the usual operations $+$ (addition) and $\cdot$ (multiplication) on the set $\mathbb{Z}$ of integers, it is easy to give new operations made up from $+$ (or $-$) and $\cdot$: $n \circ m = n + m - nm$, $n * m = -n - m$, etc. We thereby obtain different algebraic structures $(\mathbb{Z}, +)$, $(\mathbb{Z}, \cdot)$, $(\mathbb{Z}, \circ)$, $(\mathbb{Z}, *)$.

Clearly, the imagination can find a boundless expanse in constructing all sorts of binary operations on a set X. But the problem of studying arbitrary algebraic structures is too general to lead to conclusions that have any concrete value. For this reason the problem is studied under certain restrictions to various special types of algebraic structures.

2. Semigroups and monoids. A binary operation $*$ on a set X is called associative if $(a * b) * c = a * (b * c)$ for all $a, b, c \in X$; it is called commutative if $a * b = b * a$ for all $a, b \in X$. We use the same terms to apply to the algebraic structure $(X, *)$. The conditions of associativity and commutativity are independent, i.e., neither implies the other. For example, the operation $*$ on $\mathbb{Z}$ given by $n * m = -n - m$ is obviously commutative, but it is not associative, since, for example, $(1 * 2) * 3 = (-1 - 2) * 3 = -(-1 - 2) - 3 = 0$, while $1 * (2 * 3) = 4$. On the other hand, the set $M_n(\mathbb{R})$ of all $n \times n$ square matrices is associative but not commutative under multiplication (if $n > 1$), as shown in subsection 2 of §3 Ch. 2.

An element $e \in X$ is called a unit element (or a neutral element) relative to a given binary operation $*$ if we have: $e * x = x * e = x$ for all $x \in X$. If e' is another unit element, then it follows immediately from the definition that $e' = e' * e = e$. Thus, an algebraic structure $(X, *)$ can have at most one unit element.

A set X with an associative binary operation is called a semigroup. A semigroup having a unit element is called a monoid (or simply a semigroup with unit).

As with any set, the cardinality of a monoid $M = (M, *)$ is denoted Card M or $|M|$. If it has finitely many elements, we call M a finite monoid of order $|M|$.

We now give some examples of semigroups and monoids.

1) Let Ω be any set, and let $M(\Omega)$ be the set of all transformations of Ω (maps from Ω to itself). It follows from the properties of sets and maps in §5 Ch. 1 that $M(\Omega)$ is a monoid. Of course, we have in mind as our binary operation the composition of maps $\circ$. $M(\Omega)$ has the unit element e_Ω which is the identity map.

Consider the special case when Ω is a finite set of $n = |\Omega|$ elements, which we simply denote by the integers $1, 2, \cdots, n$. Every map $f : \Omega \to \Omega$ is determined by giving an ordered sequence $f(1), f(2), \cdots, f(n)$, where each $f(i)$ is an element of Ω. We allow the possibility that $f(i) = f(j)$ for $i \neq j$. There are precisely n^n possible sequences, i.e., n^n transformations. Thus, $|M(\Omega)| = \text{Card } M(\Omega) = n^n$. For example, take $n = 2$. The four elements e, f, g, h of the monoid $M(\{1,2\})$ and their products (compositions) are completely given by the tables

	1	2
e	1	2
f	2	1
g	1	1
h	2	2

•	e	f	g	h
e	e	f	g	h
f	f	e	h	g
g	g	g	g	g
h	h	h	h	h

It is clear from the table to the right that $M(\{1,2\})$ is a non-commutative monoid.

2) Again let Ω be an arbitrary set, and let $\mathcal{P}(\Omega)$ be the set of all subsets (see Exercise 4 of §5 Ch. 1). Since $(A \cap B) \cap C = A \cap (B \cap C)$ and $(A \cup B) \cup C = A \cup (B \cup C)$, it follows that there are two natural associative binary operations defined on $\mathcal{P}(\Omega)$. Obviously, $\emptyset \cup A = A$ and $A \cap \Omega = A$. So we have two commutative monoids $(\mathcal{P}(\Omega), \cup, \emptyset)$ and $(\mathcal{P}(\Omega), \cap, \Omega)$, where we denote a monoid as a triple (set, binary operation, unit element). We know that $|\mathcal{P}(\Omega)| = 2^n$ if $|\Omega| = n$.

3) $(M_n(\mathbb{R}), +, 0)$ is a commutative monoid whose neutral element is the zero matrix, and $(M_n(\mathbb{R}), \cdot, E)$ is a non-commutative monoid whose neutral element is the identity matrix. This follows immediately from the properties of matrix addition and multiplication which we encountered in Chapter 2.

4) Let $n\mathbb{Z} = \{mn \mid m \in \mathbb{Z}\}$ be the set of integers divisible by n. It is clear that $(n\mathbb{Z}, +, 0)$ is a commutative monoid, and $(n\mathbb{Z}, \cdot)$ is a commutative semigroup without unit (if $n > 1$).

5) The set $P_n(\mathbb{R})$ of stochastic matrices of order n (see Exercise 7 of §3 Ch. 2) is a monoid under the usual matrix multiplication.

A subset S' of a semigroup S with binary operation $*$ is called a <u>subsemigroup</u> if $x * y \in S'$ whenever $x, y \in S'$. In this case we say that <u>the subset</u> $S' \subset S$ <u>is closed under</u> $*$. If $(M, *)$ is a monoid, and the subset $M' \subset M$ is not only closed under $*$, but also contains the unit element, then we say that M' is a <u>submonoid</u> of M. For example, $(n\mathbb{Z}, \cdot)$ is a subsemigroup of $(\mathbb{Z}, \cdot)$, and $(n\mathbb{Z}, +, 0)$ is a submonoid of $(\mathbb{Z}, +, 0)$. Any submonoid of the monoid $M(\Omega)$ of maps from a set to itself is called a <u>monoid of transformations</u> (of the set Ω).

3. <u>Generalized associativity; powers.</u> Let $(X, \cdot)$ be any algebraic structure. For simplicity, we shall omit the $\cdot$ and write xy instead of $x \cdot y$. Let $x_1, \cdots, x_n$ be an ordered sequence of elements in X. Without changing the order, there are many different ways we can form the product of n elements. Let ℓ_n be the number of ways:

$$\ell_2 = 1 : x_1 x_2 ;$$

$$\ell_3 = 2 : (x_1 x_2) x_3 , \; x_1 (x_2 x_3) ;$$

$$\ell_4 = 5 : ((x_1 x_2) x_3) x_4 , \; (x_1 (x_2 x_3)) x_4 , \; x_1 ((x_2 x_3) x_4) , \; x_1 (x_2 (x_3 x_4)) ,$$

$(x_1 x_2)(x_3 x_4)$; and so on.

It is clear that we can obtain all ℓ_n possibilities if, for each k, $1 \le k \le n-1$, we run through all possible products of $x_1, \dots, x_k$ and all possible products of $x_{k+1}, \dots, x_n$, and then take the product of these two products. It is of great importance that the location of the parentheses does not matter if $(X, \cdot)$ is a semigroup.

THEOREM 1. If the binary operation on X is associative, then the result of applying it successively to n elements of X does not depend on the location of the parentheses.

Proof. If $n = 1$ or 2, there is nothing to prove. If $n = 3$, the theorem coincides with the associative law. We now proceed by induction on n. Suppose that $n > 3$ and that the theorem holds when the number of elements is $< n$. We need only show that

$$(x_1 \dots x_k) \cdot (x_{k+1} \dots x_n) = (x_1 \dots x_\ell) \cdot (x_{\ell+1} \dots x_n) \tag{1}$$

for any k and ℓ, $1 \leq k, \ell \leq n-1$. We have only written out the outside parentheses, since, by the induction assumption, the location of the inner paretheses does not matter. In particular, we can set $x_1 x_2 \dots x_k = (\dots((x_1 x_2)x_3)\dots x_{k-1})x_k$, which is called the left-normalized product. We distinguish between two cases:

a) $k = n-1$. Then $(x_1 \dots x_{n-1})x_n = (\dots(x_1 x_2)\dots x_{n-1})x_n$ is a left-normalized product.

b) $k < n-1$. By associativity, we have

$$(x_1 \dots x_k)(x_{k+1} \dots x_n) = (x_1 \dots x_k)((x_{k+1} \dots x_{n-1})x_n) =$$

$$= ((x_1 \dots x_k)(x_{k+1} \dots x_{n-1}))x_n =$$

$$= (\dots((\dots(x_1 x_2)\dots x_k)x_{k+1})\dots x_{n-1})x_n ,$$

i.e., we again obtain a left-normalized product. The right side of the desired equality (1) can be reduced to the same form. □

In §2 Ch. 2 we introduced the summation sign Σx_i. It can be used in any additive commutative monoid. The analogous symbol in a multiplicative monoid is the product sign:

$$\prod_{i=1}^{2} x_i = x_1 x_2 \quad , \quad \prod_{i=1}^{3} x_i = (x_1 x_2) x_3 \quad , \quad \prod_{i=1}^{n} x_i = \left(\prod_{i=1}^{n-1} x_i \right) x_n \quad .$$

By Theorem 1, the parentheses are not needed when writing (or computing) a product $x_1 x_2 \dots x_n$ of elements in a monoid. The only care that must be exercised is in the order of the elements, and even that is unnecessary if the elements commute with one another. In particular, if $x_1 = x_2 = \dots = x_n = x$, then, as with ordinary numbers, we let x^n denote the product $xx\dots x$, which we call the n-th <u>power</u> of the element x. As a consequence of Theorem 1, we have the relations

$$x^m x^n = x^{m+n} \quad , \quad (x^m)^n = x^{mn} \quad , \quad m, n \in \mathbb{N} \quad . \tag{2}$$

If $(M, \cdot, e)$ is a monoid, we further set $x^0 = e$ for every $x \in M$.

In an additive monoid, i.e., in a commutative monoid in which the operation is denoted by $+$, the "power" x^n is written $nx = x + x + \dots + x$ and is called a <u>multiple</u> of x. Then (2) becomes the following rules for multiples:

$$mx + nx = (m + n)x \quad , \quad n(mx) = (nm)x \quad . \tag{2'}$$

We note another useful fact. If $xy = yx$ in a monoid M, then

$$(xy)^n = x^n y^n \quad , \quad n = 0, 1, 2, \dots \quad . \tag{3}$$

In particular, this is always the case in a commutative monoid. (3) is proved by induction on n:

$$(xy)^n = (xy)^{n-1}(xy) = (x^{n-1} y^{n-1})(xy) = (x^{n-1} y^{n-1} x) y =$$
$$= (x^{n-1} x y^{n-1}) y = (x^{n-1} x)(y^{n-1} y) = x^n y^n \quad .$$

More generally, if we use (3) and induction on m, we obtain

$$x_i x_j = x_j x_i \ , \ 1 \le i, j \le m \implies (x_1 \dots x_m)^n = x_1^n \dots x_m^n \quad . \tag{4}$$

The analogous rules for multiples are:

$$n(x+y) = nx + ny \,, \qquad n = 0,1,2,\cdots \,, \tag{3'}$$

$$n(x_1 + \cdots + x_m) = nx_1 + \cdots + nx_m \,, \qquad n = 0,1,2,\cdots \,. \tag{4'}$$

Normally, a monoid which is written $(M, \cdot, e)$ is called a <u>multiplicative</u> monoid, and one which is written $(M, +, 0)$ is called an <u>additive</u> monoid. The additive notation is usually only used for commutative monoids.

4. <u>Invertible elements.</u> An element a in a monoid $(M, \cdot, e)$ is called <u>invertible</u> if there exists an element $b \in M$ such that $ab = e = ba$. (Clearly, in that case b is also invertible.) If we also had $ab' = e = b'a$, then it would follow that $b' = eb' = (ba)b' = b(ab') = be = b$. Thus, we can speak of <u>the</u> inverse element a^{-1} when a is invertible: $a^{-1}a = e = aa^{-1}$.

It is clear that $(a^{-1})^{-1} = a$. The notion of an invertible element in a monoid is a natural generalization of the notion of an invertible matrix in the multiplicative monoid $(M_n(\mathbb{R}), \cdot, E)$.

Since $(xy)(y^{-1}x^{-1}) = x(yy^{-1})x^{-1} = xex^{-1} = xx^{-1} = e$, and similarly $(y^{-1}x^{-1})(xy) = e$, we have: $(xy)^{-1} = y^{-1}x^{-1}$. Hence, <u>the set of all invertible elements in a monoid</u> $(M, \cdot, e)$ <u>is closed under</u> $\cdot$ <u>and is a submonoid of</u> M.

EXERCISES

1. Subsection 2 contains the example of the commutative but non-associative operation $* : n * m = -n - m$ on $\mathbb{Z}$. The algebraic system $(\mathbb{Z}, *)$ has the following identities: $(n * m) * m = n$, $m * (m * n) = n$. Now suppose that we are given an arbitrary algebraic system $(X, *)$ in which $(x * y) * y = x$ and $y * (y * x) = x$ for all $x, y \in X$. Prove that $x * y = y * x$, i.e., $*$ is commutative.

2. Show that

$$M_n^0(\mathbb{R}) = \left\{ A = (a_{ij}) \in M_n(\mathbb{R}) \;\middle|\; \sum_{j=1}^{n} a_{ij} = 0 \ , \ i = 1, 2, \cdots, n \right\}$$

is a semigroup under the usual operation of matrix multiplication. Is $(M_n^0(\mathbb{R}), \cdot)$ a monoid?

3. In a multiplicative monoid M we choose an arbitrary element t and introduce the new operation $* : x * y = xty$. Show that $(M, *)$ is a semigroup, and that $(M, *)$ is a monoid if and only if t is invertible, in which case the neutral element is t^{-1}.

4. Show that the set $\mathbb{Z}$ with the operation $\circ : n \circ m = n + m + nm = (1+n)(1+m) - 1$ is a commutative monoid. What is the neutral element in $(\mathbb{Z}, \circ)$? Find all invertible elements in $(\mathbb{Z}, \circ)$.

§2. Groups

1. Definition and examples. Consider the set $GL(n, \mathbb{R})$ of all square $n \times n$ matrices with non-zero determinant. By Theorem 5 of §2 Ch. 3, if $\det A \neq 0$ and $\det B \neq 0$, then $\det AB \neq 0$. Thus, $A, B \in GL(n, \mathbb{R}) \Rightarrow AB \in GL(n, \mathbb{R})$. In addition, $(AB)C = A(BC)$, and there is a special matrix E such that $AE = EA = A$ for all $A \in GL(n, \mathbb{R})$. Finally, every matrix $A \in GL(n, \mathbb{R})$ has an "opposite", i.e., an inverse A^{-1} such that $AA^{-1} = A^{-1}A = E$.

The set $GL(n, \mathbb{R})$ considered with the composition law (binary operation) $(A, B) \mapsto AB$ is called the general linear group of order n over $\mathbb{R}$. Following the terminology in §1, we could define $GL(n, \mathbb{R})$ as simply the submonoid of all invertible elements in the monoid $(M_n(\mathbb{R}), \cdot, E)$. But this submonoid is extremely important in its own right, and is a key example of the following abstract definition.

Definition. A monoid G all of whose elements are invertible is called a group. In other words, the following axioms must hold:

(G1) a binary operation $(x,y) \mapsto xy$ is defined on the set G ;

(G2) this operation is associative: $(xy)z = x(yz)$ for all $x,y,z \in G$;

(G3) G has a neutral (unit) element $e : xe = ex = x$ for all $x \in G$;

(G4) every element $x \in G$ has an inverse $x^{-1} : xx^{-1} = x^{-1}x = e$.

Surprisingly, one of the oldest and richest areas of algebra, playing a fundamental role in geometry and in applications of mathematics to the natural sciences, is based on such a simple set of axioms.

A group whose binary operation is commutative is called a commutative group or else an abelian group (in honor of the Norwegian mathematician Abel). The term "group" itself was introduced by the French mathematician Galois, the founder of group theory. The ideas of group theory were "in the air" (as often happens with fundamental mathematical ideas) long before Galois; some of the theorems of group theory were actually proved, although in a more naive form, by Lagrange. The brilliant work of Galois was at first poorly understood, and its importance became fully recognized only after the appearance of Jordan's book "A course in the theory of permutations and algebraic equations" (1870). It was only toward the end of the nineteenth century that group theory "completely left the realm of fantasy. Instead, a logical skeleton was carefully prepared". (F. Klein, "Lectures on the development of mathematics in the nineteenth century").

The symbols Card G , $|G|$, and $(G:e)$ are all used to denote the number of elements in a group (its cardinality). All of the facts about monoids in §1 apply, of course, to groups. However, some new words are introduced. A subset $H \subset G$ is called a subgroup of G if $e \in H$; $h_1, h_2 \in H \Rightarrow h_1h_2 \in H$; and $h \in H \Rightarrow h^{-1} \in H$. A subgroup $H \subset G$ is called proper if $H \neq G$.

We now give some examples of groups.

1) In the general linear group $GL(n, \mathbb{R})$, consider the subset $SL(n, \mathbb{R})$ of matrices with determinant 1 :

$$SL(n, \mathbb{R}) = \{A \in GL(n, \mathbb{R}) \mid \det A = 1\} \quad .$$

Obviously, $E \in SL(n, \mathbb{R})$. By the results in Ch. 3 about determinants, $\det A = 1$, $\det B = 1 \Rightarrow \det AB = 1$, and $\det A^{-1} = (\det A)^{-1} = 1$. Thus, $SL(n, \mathbb{R})$ is a subgroup of $GL(n, \mathbb{R})$; it is called the <u>special linear group of order</u> n over $\mathbb{R}$. It is also called the <u>unimodular</u> group, although this name is sometimes used for the group of matrices with determinant ± 1.

The group $GL(n, \mathbb{R})$, which contains many interesting groups, has been for mathematicians of several generations a seemingly inexhaustible source of new ideas and unsolved problems.

2) If we replace the real numbers by the rational numbers, we obtain the general linear group $GL(n, \mathbb{Q})$ of order n over $\mathbb{Q}$ and the subgroup $SL(n, \mathbb{Q})$. The group $SL(n, \mathbb{Q})$ contains the interesting subgroup $SL(n, \mathbb{Z})$ of matrices with integer entries and determinant 1. Theorem 1 of §3 Ch. 3, which gives an explicit formula for the entries in an inverse matrix, shows that $SL(n, \mathbb{Z})$ actually is a group. The groups $SL(n, \mathbb{Q})$ and $SL(n, \mathbb{Z})$ occupy an important place in number theory. Figure 11 depicts the partially order set (see subsection 3 of §6 Ch. 1) of these subgroups of $GL(n, \mathbb{R})$.

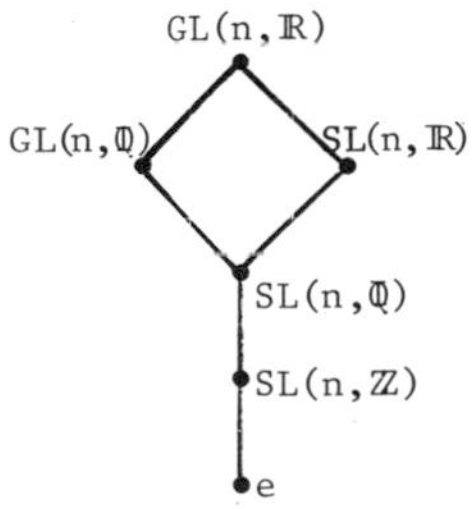

GL(n, ℝ)
GL(n, ℚ)
SL(n, ℝ)
SL(n, ℚ)
SL(n, ℤ)
e

Fig. 11

3) If we set $n = 1$ in examples 1) and 2), we obtain, first of all, the multiplicative groups $\mathbb{R}^* = \mathbb{R}\setminus\{0\} = GL(1, \mathbb{R})$ and $\mathbb{Q}^* = \mathbb{Q}\setminus\{0\} = GL(1, \mathbb{Q})$ of real and rational numbers, respectively. These are obviously infinite groups. Since the only invertible elements in $(\mathbb{Z}, \cdot, 1)$ are 1 and -1, we have $GL(1, \mathbb{Z}) = \{\pm 1\}$. Furthermore, $SL(1, \mathbb{R}) = SL(1, \mathbb{Q}) = SL(1, \mathbb{Z}) = 1$. But if $n = 2$, even the group $SL(2, \mathbb{Z})$ is infinite: it contains, for example, all of the matrices

$$\begin{pmatrix} 1 & m \\ 0 & 1 \end{pmatrix}, \begin{pmatrix} 1 & 0 \\ m & 1 \end{pmatrix}, \begin{pmatrix} m & m-1 \\ 1 & 0 \end{pmatrix}, \quad m \in \mathbb{Z} \quad .$$

We further note the infinite additive groups:

$$(\mathbb{R}, +, 0) \;,\; (\mathbb{Q}, +, 0) \;,\; (\mathbb{Z}, +, 0) \quad .$$

4) Let Ω be any set, and let $S(\Omega)$ be the set of all bijective (one-to-one) transformations $f : \Omega \to \Omega$. Using the results of §5 Ch. 1 on set maps (Theorems 1 and 2 and the corollary to Theorem 2), we immediately conclude that $S(\Omega)$ is a group under the natural binary operation of composing maps. $S(\Omega)$ is the submonoid of all invertible elements in the monoid $M(\Omega)$ in example 1) of §1. The group $S(\Omega)$ and its various subgroups are called transformation groups. Such groups are the basic type of groups that arise in applications of group theory. In 1872, F. Klein announced his "Erlangen program", which sought to classify different geometries using the notion of transformation groups. If we take $\Omega = \mathbb{R}^n$, we obtain a very large group $S(\mathbb{R}^n)$, which is difficult to imagine in its entirety. But $S(\mathbb{R}^n)$ contains the subgroup of invertible (bijective) linear transformations $\varphi_A : \mathbb{R}^n \to \mathbb{R}^n$, which we found to be in one-to-one correspondence with the $n \times n$ non-singular matrices A (see §3 Ch. 2). Thus, we have an imbedding of $GL(n, \mathbb{R})$ in $S(\mathbb{R}^n)$. The significance of this imbedding will become clearer after we introduce the important concept of an isomorphism of groups.

2. Systems of generators. Given a subset S of a group G, we try to find a subgroup $H \subset G$ containing S such that every subgroup $K \subset G$ containing S must also contain H. There cannot be two distinct subgroups H, H' which play the role of a minimal subgroup containing S:

$$S \subset H \,,\, S \subset H' \implies H \subset H' \subset H \implies H' = H \quad .$$

Thus, the minimal subgroup H containing S must coincide with the intersection of all subgroups containing S, if we show that this intersection must be a subgroup in G. But we have the following simple result.

THEOREM 1. The intersection $\bigcap_{i \in I} H_i$ of any family $\{H_i \mid i \in I\}$ of subgroups of a group G is a subgroup.

Proof. Let e be the unit element of G. The properties $e \in \cap H_i$, $x, y \in \cap H_i \Rightarrow xy \in \cap H_i$; $x \in \cap H_i \Rightarrow x^{-1} \in \cap H_i$, which characterize a subgroup, must hold in $\cap H_i$, because they hold in each subgroup H_i separately. □

Now take for $\{H_i \mid i \in I\}$ the family of all subgroups containing the given subset $S \subset G$. Then, by Theorem 1 and the remarks before the theorem, the intersection

$$\langle S \rangle = \bigcap_{S \subseteq H} H$$

is precisely the minimal subgroup containing S. We call $\langle S \rangle$ the subgroup generated by S in G, and we call S a set of generators for the subgroup $\langle S \rangle$. At first glance, it seems that $\langle S \rangle$ is defined ineffectively, since we have to find all subgroups containing S. But there is an easier way to determine $\langle S \rangle$, as we see from the following corollary of Theorem 1.

COROLLARY. *The subgroup $\langle S \rangle$ coincides with the set T consisting of the unit element e and all possible products*

$$t_1 t_2 \cdots t_n, \quad n = 1, 2, 3, \cdots,$$

where either $t_i \in S$ or $t_i^{-1} \in S$, $1 \le i \le n$.

Proof. Since $t_1 \cdots t_n \in T$, $t'_1 \cdots t'_m \in T \Rightarrow t''_1 \cdots t''_{n+m} = t_1 \cdots t_n t'_1 \cdots t'_m \in T$ and $t_1 \cdots t_n \in T \Rightarrow (t_1 \cdots t_n)^{-1} = t_n^{-1} \cdots t_1^{-1} \in T$, it follows that the set T is a subgroup of G. On the other hand, every subgroup H containing all $x_i \in S$ must contain all of the inverses x_i^{-1}, and, hence, must contain all products of the form $t_1 t_2 \cdots t_n$. Hence, $H \supset T$, and T coincides with the intersection of all such subgroups. □

It should be noted that by no means are all of the products $t_1 t_2 \cdots t_n$ distinct elements of $\langle S \rangle$, even if one agrees (as one usually does) to cancel any pair of successive t_i of the form $a a^{-1}$ or $a^{-1} a$. In general, when $|S| > 1$, the question of when two products of the form $t_1 t_2 \cdots t_n$ are equal is a difficult one, and we shall only briefly discuss it in Chapter 7.

Every group G has some set of generators S: for example, we can take S to be the whole group G. For simplicity, we consider a group G which is generated by a finite set of elements (such groups are said to be finitely generated). If we remove from S all "extra" elements, i.e., those which can be written as products of the other elements (and their inverses), we obtain a minimal set M of generators of G. To say that M is minimal means that $\langle M\rangle = G$, but $\langle M'\rangle \neq G$ if M' is obtained from M by removing an element. Let $M = \{g_1, \cdots, g_d\}$. Then we also write $G = \langle g_1, g_2, \cdots, g_d\rangle$ as well as $G = \langle M\rangle$. If $d = 1$, we call G a cyclic group.

3. Cyclic groups. If G is any group and g is an element in G, then, by definition, $\langle g\rangle$ is a cyclic subgroup of G.

Because of Theorem 1 and the properties of the powers of an element in a monoid, we might expect that any cyclic group $\langle a\rangle$ with generator a is an abelian group of the form $\langle a\rangle = \{a^n \mid n \in \mathbb{Z}\}$, or $\langle a\rangle = \{na \mid n \in \mathbb{Z}\}$ if the group operation is written additively (this notation is not meant to imply that all of the elements a^n or na are distinct). This is in fact the case, once we agree to denote $(a^{-1})^k = a^{-k}$ and prove the following fact.

THEOREM 2. For any $m, n \in \mathbb{Z}$,

$$a^m a^n = a^{m+n}, \quad (a^m)^n = a^{mn}$$

(or, in additive notation, $ma + na = (m+n)a$, $n(ma) = (nm)a$).

Proof. If m and n are non-negative, see relations (2) and (2') of subsection 4 §1. If $m < 0$ and $n < 0$, then $m' = -m > 0$, $n' = -n > 0$, and

$$a^m a^n = (a^{-1})^{m'} (a^{-1})^{n'} = (a^{-1})^{m'+n'} = a^{-(m'+n')} = a^{m+n} .$$

If $m' = -m > 0$ and $n > 0$, we have

$$a^m a^n = (a^{-1})^{m'} a^n = (\underbrace{a^{-1} \ldots a^{-1}}_{m'})(\underbrace{a \ldots a}_{n}) = a^{n-m'} \text{ (or } (a^{-1})^{m'-n}\text{, if } m' \geq n) = a^{m+n} .$$

We similarly treat the case when $m > 0$ and $n < 0$. The equality $(a^m)^n = a^{mn}$ is easy to prove using the first equality, just proved, and the definition of the powers of an element. $\square$

The simplest example of a cyclic group is the additive group of integers $(\mathbb{Z}, +, 0)$, which is generated by 1 or by -1. Also, it is easy to see that the matrix $\left\|\begin{matrix} 1 & 1 \\ 0 & 1 \end{matrix}\right\|$ generates an infinite cyclic subgroup of $SL(2, \mathbb{Z})$. The set $\{1, -1\}$ under multiplication is a cyclic group of order 2.

We can construct an example of a cyclic group of order n by considering all rotations of the plane around a point O which take a regular n-gon P_n with center O to itself. These rotations clearly form a group; the group operation is successively performing the rotations. Our group C_n contains the rotations $\varphi_0, \varphi_1, \cdots, \varphi_{n-1}$ through the angles 0, $2\pi/n, \cdots, (n-1)2\pi/n$, counterclockwise. Here $\varphi_s = \varphi_1^s$, and it is geometrically obvious that $\varphi_s^{-1} = \varphi_1^{n-s}$ and $\varphi_1^n = \varphi_0$, the identity transformation. Thus, $|C_n| = n$, and $C_n = \langle \varphi_1 \rangle$. Note that the cyclic group C_n is a proper subgroup of the group D_n of all symmetries of the n-gon P_n (i.e., rigid transformations of P_n).

Again suppose that G is any group and a is an element of G. There are two possibilities: 1) All powers of a are distinct, i.e., $m \neq n \Rightarrow a^m \neq a^n$. In this case we say that a has <u>infinite order.</u> 2) We have $a^m = a^n$ for $m \neq n$. If, say, $m > n$, then $a^{m-n} = e$, i.e., there is a positive power of a which is the identity element. Let q be the least positive exponent for which $a^q = e$. We then say that a is an element of <u>finite order</u> q. Of course, if G has finite order (i.e., $\operatorname{Card} G < \infty$), then all of the elements have finite order.

<u>Warning.</u> The word "order" has many meanings in mathematics. Before we spoke of square matrices of order n (i.e., $n \times n$ matrices), but a non-singular matrix A,

considered as an element of the group $GL(n, \mathbb{R})$, also has an order (perhaps infinite) in the sense just defined. But it will always be clear from the context what meaning of the word "order" we have in mind.

If we think of our example C_n of a cyclic group of order n, the following theorem becomes almost obvious.

THEOREM 3. The order of an element $a \in G$ (where G is any group) is equal to $\text{Card}\langle a \rangle$. If a is an element of finite order q, then

$$\langle a \rangle = \{e, a, \cdots, a^{q-1}\} \quad \text{and} \quad a^k = e \iff k = \ell q, \ \ell \in \mathbb{Z} \quad .$$

Proof. If a has infinite order, there is nothing left to prove. If a has order q, then, by definition, all of the elements $a, a, a^2, \cdots, a^{q-1}$ are distinct. We claim that any power a^k must coincide with one of these elements, i. e., $\langle a \rangle = \{e, a, \cdots, a^{q-1}\}$. To see this, we use the division algorithm in $\mathbb{Z}$ (see subsection 3 §8 Ch. 1) to write k in the form

$$k = \ell q + r \quad , \quad 0 \leq r \leq q-1 \quad .$$

Then, using the rules in Theorem 2, we obtain

$$a^k = (a^q)^\ell a^r = ea^r = a^r \quad .$$

In particular, $a^k = e \Rightarrow r = 0 \Rightarrow k = \ell q$. □

The property of a group being cyclic is very useful. Given a group, we are not always told in advance whether it is cyclic; sometimes this must be proved. An example is the following

PROPOSITION. Let G be any group, and let a and b be two elements that commute with one another. Suppose that a and b have finite orders s and t, and that s and t are relatively prime. Then a and b generate a cyclic subgroup of order st, and

$$\langle a, b \rangle = \langle ab \rangle \quad .$$

Proof. First of all, $D = \langle a \rangle \cap \langle b \rangle = e$, since if $d \in D$ has order q, then, by Theorem 3,

$$d = a^i = b^j \Longrightarrow d^s = (a^s)^i = e \ , \ d^t = (b^t)^j = e \Longrightarrow q|s \ , \ q|t \ ,$$

and, since s and t are relatively prime, this means that $q = 1$. Next, if $n = |\langle ab \rangle|$, we have (see relation (3) in §1)

$$a^n b^n = (ab)^n = e \Rightarrow a^n = b^{-n} \in D = e \Rightarrow a^n = e \ , \ b^n = e \Rightarrow s|n \ , \ t|n \Rightarrow \text{l.c.m.}(s,t)|n \Rightarrow st|n \ ,$$

since $st = \text{l.c.m.}(s,t) \cdot \text{g.c.d.}(s,t) = \text{l.c.m.}(s,t)$. But $(ab)^{st} = (a^s)^t (b^t)^s = e$ (using Theorem 2), so that $n|st$, and hence $n = st$. It remains to note that

$$\langle a,b \rangle = \{a^i b^j | 0 \leq i \leq s-1 \ , \ 0 \leq j \leq t-1\} \Longrightarrow \text{Card} \langle a,b \rangle \leq st \ ,$$

and, since $\langle ab \rangle \subset \langle a,b \rangle$ and $\text{Card} \langle ab \rangle = st$, it follows that $\langle a,b \rangle = \langle ab \rangle$. □

We shall return to cyclic groups, but now we examine a richer special type of group, namely, transformation groups, which will be used to illustrate the various group theoretic concepts we introduce.

4. The symmetric group and the alternating group. Let Ω be a finite set with n elements. Since we shall not be concerned with the nature of those elements, we may as well assume that $\Omega = \{1, 2, \cdots, n\}$. The group $S(\Omega)$ (see Example 4 above) of all one-to-one correspondences $\Omega \to \Omega$ is called the n-th symmetric group (or the symmetric group on n elements), and is usually denoted S_n. The elements of S_n, which are usually denoted by small Greek letters, are call permutations.

Written out visually, a permutation $\pi : i \mapsto \pi(i)$, $i = 1, 2, \cdots, n$, is represented as follows:

$$\pi = \begin{pmatrix} 1 & 2 & \dots & n \\ i_1 & i_2 & \dots & i_n \end{pmatrix} ,$$

where all of the images are given explicitly:

$$\pi : \begin{array}{cccc} 1 & 2 & \cdots & n \\ \downarrow & \downarrow & & \downarrow \\ i_1 & i_2 & \cdots & i_n \end{array} ;$$

here $i_k = \pi(k)$, $k = 1, \cdots, n$, are the permuted elements $1, 2, \cdots, n$. As usual, e denotes the identity permutation (even though it is a Latin and not a Greek letter): $e(i) = i$, $\forall i$.

Two permutations $\sigma, \tau \in S_n$ are multiplied by the usual rule for composing maps: $(\sigma\tau)(i) = \sigma(\tau(i))$. For example, if

$$\sigma = \begin{pmatrix} 1 & 2 & 3 & 4 \\ 2 & 3 & 4 & 1 \end{pmatrix} , \quad \tau = \begin{pmatrix} 1 & 2 & 3 & 4 \\ 4 & 3 & 2 & 1 \end{pmatrix}$$

then we have

$$\sigma\tau = \begin{pmatrix} 1 & 2 & 3 & 4 \\ 2 & 3 & 4 & 1 \end{pmatrix} \begin{pmatrix} 1 & 2 & 3 & 4 \\ 4 & 3 & 2 & 1 \end{pmatrix} = \begin{array}{cccc} 1 & 2 & 3 & 4 \\ \downarrow & \downarrow & \downarrow & \downarrow \\ 4 & 3 & 2 & 1 \\ \downarrow & \downarrow & \downarrow & \downarrow \\ 1 & 4 & 3 & 2 \end{array} = \begin{pmatrix} 1 & 2 & 3 & 4 \\ 1 & 4 & 3 & 2 \end{pmatrix} .$$

Notice that

$$\tau\sigma = \begin{pmatrix} 1 & 2 & 3 & 4 \\ 4 & 3 & 2 & 1 \end{pmatrix} \begin{pmatrix} 1 & 2 & 3 & 4 \\ 2 & 3 & 4 & 1 \end{pmatrix} = \begin{pmatrix} 1 & 2 & 3 & 4 \\ 3 & 2 & 1 & 4 \end{pmatrix} ,$$

so that $\sigma\tau \neq \tau\sigma$.

We now find the order of the group S_n. A permutation σ can take the element 1 into any of n possible elements. Once $\sigma(1)$ is fixed, we can choose $\sigma(2)$ to be any of the remaining $n-1$ elements; thus, there are $n(n-1)$ possible choices for the pair $\sigma(1)$, $\sigma(2)$. Then we can choose $\sigma(3)$ to be any of the $n-2$ numbers which have not already been taken to be $\sigma(1)$ or $\sigma(2)$. Continuing in this manner, we see that the number of possible choices for the $\sigma(1), \sigma(2), \cdots, \sigma(n)$ is $n(n-1)(n-2) \cdots 3 \cdot 2 \cdot 1 = n!$ ("n factorial"). Thus,

$$\text{Card } S_n = |S_n| = (S_n : e) = n! \quad .$$

The permutations in S_n can be decomposed into products of simpler permutations. We illustrate by drawing diagrams for the two examples $\sigma, \tau \in S_4$ given above:

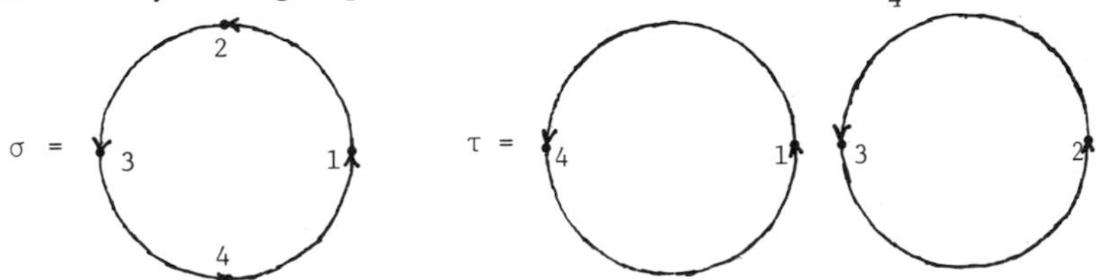

We abbreviate the permutation σ, which is a type of permutation called a "cycle" of length 4 (meaning that it "rotates" 4 elements), by writing $\sigma = (1234)$, or, equivalently, $\sigma = (2341) = (3412) = (4123)$. The permutation τ is the product of two "disjoint" cycles of length 2, namely the cycles (14) and (23), and so can be written $\tau = (14)(23)$. Note that $\sigma^2 = (13)(24)$, $\sigma^4 = e$, $\tau^2 = e$.

Returning to the general case, we call two elements $i, j \in \Omega$ <u>equivalent</u> under a cyclic subgroup $\langle \pi \rangle$ of S_n (or simply π-<u>equivalent</u>) if there is an integer s for which $j = \pi^s(i) = \pi(\ldots\pi(i)\ldots)$. Since S_n is a finite group, all of its subgroups are also finite. By Theorem 3, if $\text{Card}\,\langle \pi \rangle = q$, we may take $0 \leq s < q$. This relation is in fact an equivalence relation, i.e., it is reflexive, symmetric, and transitive (see subsection 2 of §6 Ch. 1), since $i = \pi^0 i = e(i)$; $j = \pi^k(i) \Rightarrow i = \pi^{-k}(j)$ and $j = \pi^s(i)$, $k = \pi^t(j) \Rightarrow k = \pi^{s+t}(i)$. By the usual property of equivalence relations, we obtain a partition

$$\Omega = \Omega_1 \cup \ldots \cup \Omega_p \tag{1}$$

of the set Ω into pair-wise disjoint subsets (equivalence classes) $\Omega_1, \cdots, \Omega_p$, which are called π-<u>orbits</u>. This name makes sense: every $i \in \Omega$ belongs to precisely one orbit, and if Ω_k is the orbit to which i belongs, then Ω_k consists of the images of i under the action of the powers of π : $i, \pi(i), \pi^2(i), \cdots, \pi^{\ell_k - 1}(i)$, where $\ell_k = |\Omega_k|$ is the length of the π-orbit Ω_k. It is obvious that $\ell_k \leq q = \text{Card}\,\langle \pi \rangle$; $\pi^{\ell_k}(i) = i$; and ℓ_k is the least positive integer such that $\pi^{\ell_k}(i) = i$. If we set

$$\pi_k = (i\pi(i)\dots\pi^{l_k-1}(i)) = \begin{pmatrix} i & \pi(i) & \dots & \pi^{l_k-2}(i) \\ \pi(i) & \pi^2(i) & \dots & \pi^{l_k-1}(i) \end{pmatrix},$$

we obtain a permutation which is a cycle of length ℓ_k.

Sometimes for clarity we write the cycle $(123\dots\ell)$ as $(1,2,3,\dots,\ell)$, separating the numbers by commas. The cycle π_k is the permutation which leaves all elements in the set $\Omega\setminus\Omega_k$ fixed and takes j to $\pi(j)$ for all $j \in \Omega_k$. We therefore call π_s and π_t <u>independent</u> or <u>disjoint</u> cycles when $s \neq t$, since they effect disjoint sets of elements. Note that $\pi_k^{\ell_k} = e$.

Thus, associated with the partition (1) we have a corresponding decomposition of the permutation π into a product of disjoint cycles:

$$\pi = \pi_1\pi_2\dots\pi_p, \tag{2}$$

where all of the cycles commute with one another: $\pi = \pi_1\pi_2\dots\pi_p = \pi_{i_1}\pi_{i_2}\dots\pi_{i_p}$.

For example, we may arrange the cycles so that $\ell_1 \geq \ell_2 \geq \dots \geq \ell_m > \ell_{m+1} = \dots = \ell_p = 1$ (i.e., the last $p-m$ cycles correspond to orbits consisting of one element, so that as permutations $\pi_{m+1} = \dots = \pi_p = e$). Since a cycle of length one acts as the identity, it is natural to omit such cycles in (2), and write

$$\pi = \pi_1\pi_2\dots\pi_m; \ l_k > 1, \ 1 \leq k \leq m. \tag{3}$$

For example, we write the permutation

$$\pi = \begin{pmatrix} 1 & 2 & 3 & 4 & 5 & 6 & 7 & 8 \\ 2 & 3 & 4 & 5 & 1 & 7 & 6 & 8 \end{pmatrix} \in S_8$$

in the form

$$\pi = (1\ 2\ 3\ 4\ 5)(6\ 7)(8) = (1\ 2\ 3\ 4\ 5)(6\ 7). \tag{4}$$

It may seem a little unpleasant that, for example, $(12345)(67)$ can be considered as a permutation in S_n for any $n \geq 7$, since the total number of elements in the set is not indicated in the notation; but when n is fixed in a given context, there is no ambiguity.

Moreover, we claim that the decomposition of a permutation into a product of cycles is unique. Suppose that we have another decomposition $\pi = \alpha_1 \alpha_2 \dots \alpha_r$ into a product of disjoint cycles, and let i be an element which is not fixed by π. Then $\pi_s(i) \neq i$ and $\alpha_t(i) \neq i$ for one (and only one) of the $\pi_1, \cdots, \pi_m$ and one and only one of the $\alpha_1, \cdots, \alpha_r$. We have

$$\pi_s^k(i) = \pi^k(i) = \alpha_t^k(i) \quad , \quad k = 0, 1, 2, \dots \quad .$$

But a cycle is uniquely determined by the action of all of its powers on any one element which it does not leave fixed. Thus, $\pi_s - \alpha_t$. We continue in this way, using induction on m (or r) to show that the cycles occurring in the two decompositions are the same. We have proved:

THEOREM 4. *Every permutation* $\pi \neq e$ *in* S_n *is a product of disjoint cycles of length* ≥ 2. *This decomposition is uniquely determined except for the order of the cycles.* □

This decomposition (3) is convenient for many reasons. For example, it makes it easy to find the order of a permutation.

COROLLARY 1. *The order of a permutation* $\pi \in S_n$ *(i.e., the order of the cyclic subgroup* $\langle \pi \rangle$*) is equal to the least common multiple of the lengths of the disjoint cycles in the decomposition of* π.

Proof. As noted before, the disjoint cycles in the decomposition of π commute with one another. Hence, by relation (4) in §1,

$$\pi^s = \pi_1^s \dots \pi_m^s \quad , \quad s = 0, 1, 2, \dots \quad .$$

Since the cycles $\pi_1, \cdots, \pi_m$ are independent (they act on disjoint sets $\Omega_1, \cdots, \Omega_m$), it follows that $\pi^q = e \Leftrightarrow \pi_k^q = e$ for $k = 1, \cdots, m$. Thus, q is a multiple of all of the orders of the cycles π_k, which, we have seen, coincide with their lengths ℓ_k. If q is the least natural number for which $\pi^q = e$, then $q = \mathrm{Card}\,\langle \pi \rangle$ and

$q = \text{l.c.m.}(\ell_1, \cdots, \ell_m)$ is the integer defined in subsection 2 of §8 Ch. 1 (see also the proposition at the end of subsection 3). □

As an example, we can immediately say that the permutation (4) has order 10. As another example, suppose we wanted to know the maximum possible order of an element in S_8? If we go through all possible ways that 8 can be written as a sum of positive integers (in non-increasing order), we see that the following numbers occur as orders of elements $\neq e$ in S_8: 2, 3, 4, 5, 6, 7, 8, 10, 12, 15. An example of a permutation of maximal order 15 is (12345)(678).

<u>Definition.</u> A cycle of length 2 is called a <u>transposition.</u>

A transposition has the form $\tau = (ij)$; it leaves fixed all elements besides i and j. Theorem 4 implies the following

COROLLARY 2. <u>Every permutation</u> $\pi \in S_n$ <u>is a product of transpositions.</u>

<u>Proof.</u> By Theorem 4, it suffices to write each cycle as a product of transpositions. But this can be done, for example, by taking

$$(1\,2 \ldots \ell{-}1\ \ell) = (1\,\ell)(1\ \ell-1) \ldots (13)(12) \quad . \quad \square$$

Corollary 2 can be expressed using the notion of a set of generators of a group (see subsection 2):

$$S_n = \langle (12), \ldots, (1n), (23), \ldots, (2n), \ldots, (n{-}1\ n) \rangle \quad .$$

Of course, this set of generators is not a minimal set. For example,

$$S_3 = \langle (12), (13), (23) \rangle = \langle (12), (13) \rangle \quad .$$

Notice that we cannot hope for any uniqueness assertion about expressing a permutation as a product of transpositions: in general, transpositions do not commute with one another, and the number of transpositions which appear when we write a permutation as a product of transpositions is not fixed. For example, in S_4 we have

$$(123) = (13)(12) = (23)(13) = (13)(24)(12)(14) \quad .$$

In fact, the non-uniqueness of the transposition decomposition is immediately clear if we note than $\sigma\tau^2 = \sigma$ for any transpositions σ and τ. Nevertheless, there is one thing about the transposition decomposition which is fixed. In order to discover it in the most natural possible way, we consider the action of S_n on functions.

Definition. Suppose that $\pi \in S_n$ and $f(X_1, \cdots, X_n)$ is a function of n variables. Set

$$(\pi \circ f)(X_1, \cdots, X_n) = f\left(X_{\pi^{-1}(1)}, \cdots, X_{\pi^{-1}(n)}\right) . \tag{5}$$

We say that the function $g = \pi \circ f$ is obtained by letting π act on f.

For example, if $\pi = (123)$ and $f(X_1, X_2, X_3) = X_1 + 2X_2^2 + 3X_3^3$, then $(\pi \circ f)(X_1, X_2, X_3) = X_3 + 2X_1^2 + 3X_2^3$.

As in §1 of Ch. 3, a function f is called skew-symmetric if $\tau \circ f = -f$ for every transposition τ, i.e.,

$$f(\ldots, X_j, \ldots, X_i, \ldots) = -f(\ldots, X_i, \ldots, X_j, \ldots) .$$

LEMMA. Let α and β be permutations in S_n. Then

$$(\alpha\beta) \circ f = \alpha \circ (\beta \circ f) .$$

Proof. Using the definition (5), we have

$$((\alpha\beta) \circ f)(X_1, \ldots, X_n) = f\left(X_{(\alpha\beta)^{-1}(1)}, \ldots, X_{(\alpha\beta)^{-1}(n)}\right) =$$

$$= f\left(X_{(\beta^{-1}\alpha^{-1})(1)}, \ldots, X_{(\beta^{-1}\alpha^{-1})(n)}\right) =$$

$$= f\left(X_{\beta^{-1}(\alpha^{-1}1)}, \ldots, X_{\beta^{-1}(\alpha^{-1}n)}\right) =$$

$$= (\beta \circ f)\left(X_{\alpha^{-1}(1)}, \ldots, X_{\alpha^{-1}(n)}\right) = (\alpha \circ (\beta \circ f))(X_1, \ldots, X_n) . \qquad \square$$

THEOREM 5. Let π be a permutation in S_n, and let

$$\pi = \tau_1 \tau_2 \cdots \tau_k \tag{6}$$

be any decomposition of π into a product of transpositions. Then the number

$$\varepsilon_\pi = (-1)^k \quad , \tag{7}$$

which is called the "parity" (or "signature" or "sign") of π is completely determined by π, and does not depend on which decomposition (6) is used, i.e., the parity of k is always the same for a given π. In addition,

$$\varepsilon_{\alpha\beta} = \varepsilon_\alpha \varepsilon_\beta \tag{8}$$

for all $\alpha, \beta \in S_n$.

Proof. Take any skew-symmetric function f in n variables $X_1, \cdots, X_n$. By the lemma, the action of π on f reduces to the composition of the actions of the transpositions $\tau_k, \tau_{k-1}, \cdots, \tau_1$, i.e., to multiplication by -1 k times:

$$\pi \circ f = (\tau_1 \cdots \tau_{k-1}) \circ (\tau_k \circ f) = -(\tau_1 \cdots \tau_{k-1}) \circ f = \cdots = (-1)^k f = \varepsilon_\pi f \quad .$$

Since the left side of this equality depends only on π (not on what choice of decomposition (6) is taken), it follows that the map $\varepsilon : \pi \mapsto \varepsilon_\pi$ defined by (7) must be completely determined by π, provided, of course, that f is not the identically zero function. But we know that there exist skew-symmetric functions which are not identically zero; for example, the Vandermonde determinant $\Delta_n(X_1, \cdots, X_n)$ is such a function.

Next, if we apply the permutation $\alpha\beta$ to f and use the rule in the lemma, we obtain

$$\varepsilon_{\alpha\beta} f = (\alpha\beta) \circ f = \alpha \circ (\beta \circ f) = \alpha \circ (\varepsilon_\beta f) = \varepsilon_\beta(\alpha \circ f) = \varepsilon_\beta(\varepsilon_\alpha f) = (\varepsilon_\alpha \varepsilon_\beta) f \quad ,$$

which gives us (8). □

Definition. A permutation $\pi \in S_n$ is called even if $\varepsilon_\pi = 1$ and odd if $\varepsilon_\pi = -1$. Thus, every transposition is an odd permutation.

COROLLARY 1. <u>The even permutations in</u> S_n <u>form a subgroup</u> A_n <u>of order</u> $|A_n| = n!/2$ (called the <u>alternating group</u> on n elements).

<u>Proof.</u> By (8), we see that $\epsilon_{\alpha\beta} = 1$ if $\epsilon_\alpha = \epsilon_\beta = 1$, and $\epsilon_{\pi^{-1}} = \epsilon_\pi$, since $\epsilon_e = 1$. It is then easy to see that all of the group axioms are fulfilled in A_n.

We write S_n as the disjoint union $S_n = A_n \cup \overline{A}_n$, where $\overline{A}_n$ is the set of all odd permutations in S_n. The map of S_n to itself defined by the rule

$$\rho_{(12)} : \pi \mapsto (12)\pi$$

is bijective. (Namely, it is injective, since $(12)\alpha = (12)\beta \Rightarrow \alpha = \beta$; then apply Theorem 3 of §5 Ch. 1. Or else, simply observe that $(\rho_{(12)})^2$ is the identity map, from which bijectivity follows.) Since $\epsilon_{(12)\pi} = \epsilon_{(12)}\epsilon_\pi = -\epsilon_\pi$, we have $\rho_{(12)} A_n = \overline{A}_n$, $\rho_{(12)} \overline{A}_n = A_n$. Thus, the number of even permutations in S_n coincides with the number of odd permutations; hence, $|A_n| = \frac{1}{2}|S_n| = \frac{n!}{2}$. □

COROLLARY 2. <u>Suppose that we write</u> $\pi \in S_n$ <u>as a product of disjoint cycles of length</u> $\ell_1, \ell_2, \cdots, \ell_m$. <u>Then</u>

$$\epsilon_\pi = (-1)^{\sum_{k=1}^{m} (\ell_k - 1)} .$$

<u>Proof.</u> By Theorem 5, we have $\epsilon_\pi = \epsilon_{\pi_1 \cdots \pi_m} = \epsilon_{\pi_1} \cdots \epsilon_{\pi_m}$. We also have $\epsilon_{\pi_k} = (-1)^{\ell_k - 1}$, since π_k can be written as a product of $\ell_k - 1$ transpositions (see the proof of Corollary 2 of Theorem 4). We conclude:

$$\epsilon_\pi = (-1)^{\ell_1 - 1} \cdots (-1)^{\ell_m - 1} = (-1)^{\sum_{k=1}^{m} (\ell_k - 1)} . \qquad \square$$

We end this section by taking a break from serious things and considering the game "fifteen". Fifteen numbered flat square markers, all of the same size, are placed on a square board so as to occupy all but one of 16 squares of the same size. The free square can be used

i_1	i_2	i_3	i_4
i_5	i_6	i_7	i_8
i_9	i_{10}	i_{11}	i_{12}
i_{13}	i_{14}	i_{15}	

(a)

1	2	3	4
5	6	7	8
9	10	11	12
13	14	15	

(b)

Fig. 12

to move the markers horizontally or vertically (without lifting them from the board). Given an arbitrary set-up for the markers (see Fig. 12(a); we may suppose that we start with the free square in the lower right corner), we are required to move them to the set-up in Fig. 12(b). When is this possible? Elementary group theory killed this game when it was at the "height of fashion". Namely, we associate a permutation $\pi \in S_{15}$ to the diagrams 12(a) and (b). It is not hard to see (and it is worthwhile to really convince yourself of this) that it is possible to move the markers to the desired position if and only if the parity ϵ_π of the permutation π is 1, i.e., if and only if $\pi \in A_{15}$.

EXERCISES

1. Show that if $M = \langle S \rangle$ is the monoid generated by a set S, and if every element $s \in S$ has an inverse in M, then M is a group.

2. Prove that a group is a monoid G in which all equations of the form $ax = b$ or $ya = b$ have a unique solution, if $a, b \in G$.

3. Show that the set $A_1(\mathbb{R})$ of so-called affine transformations $\varphi_{a,b} : x \mapsto ax + b$ $(a, b \in \mathbb{R};\ a \neq 0)$ of the real line $\mathbb{R}$ is a group with multiplication law $\varphi_{a,b}\varphi_{c,d} = \varphi_{ac,ad+b}$. The group $A_1(\mathbb{R})$ contains the subgroup $GL(1, \mathbb{R})$ of affine transformations which leave the point $x = 0$ fixed, and the subgroup of "translations" $x \mapsto x + b$.

4. The group $SL(2, \mathbb{Z})$ contains the elements $A = \left\|\begin{matrix} 0 & 1 \\ -1 & 0 \end{matrix}\right\|$ and $B = \left\|\begin{matrix} 0 & 1 \\ -1 & -1 \end{matrix}\right\|$ of orders 4 and 3, respectively. Show that $\langle AB \rangle$ is an infinite cyclic subgroup of $SL(2, \mathbb{Z})$. This shows that the product of two elements of finite order in a group is not necessarily of finite order. What happens in an abelian group?

5. Prove that a group G of even order $|G| = 2n$ must contain an element $g \neq e$ of order 2.

6. Prove that $S_n = \langle (12), (13), \cdots, (1n) \rangle$.

7. Prove that $S_n = \langle (12), (1234 \ldots n) \rangle$.

8. Prove that the alternating group A_n, $n \geq 3$, is generated by cycles of length 3; more precisely, prove that

$$A_n = \langle (123), (124), \cdots, (12n) \rangle .$$

9. Find the sign of the permutation

$$\pi = \begin{pmatrix} 1 & 2 & 3 & \ldots & n-1 & n \\ n & n-1 & n-2 & \ldots & 2 & 1 \end{pmatrix} .$$

10. Let $\Omega = \{1, 2, \cdots, n\}$, and let $\Omega \times \Omega$ be the cartesian square. Call a pair $(i, j) \in \Omega \times \Omega$ an <u>inversion relative to the permutation</u> $\sigma \in S_n$ (or simply a σ-<u>inversion</u>) if $i < j$ but $\sigma(i) > \sigma(j)$. Set

$$\operatorname{sgn} \sigma = \prod_{1 \leq i < j \leq n} \frac{\sigma(j) - \sigma(i)}{j - i} .$$

Since $(\sigma(j) - \sigma(i))/(j - i)$ is a non-zero rational number which is negative if and only if (i, j) is a σ-inversion, and since $\sigma : \Omega \to \Omega$ is a bijective map, it follows that $\operatorname{sgn} \sigma = (-1)^k$, where k is the total number of σ-inversions. If $\tau = (ij)$ is a transposition, then $\operatorname{sgn} \tau = -1$. It is easy to see that

$$(\sigma(j)\,\sigma(i))\sigma = \begin{pmatrix} \ldots & \sigma(j) & \ldots & \sigma(i) & \ldots \\ \ldots & \sigma(i) & \ldots & \sigma(j) & \ldots \end{pmatrix} \begin{pmatrix} \ldots & i & \ldots & j & \ldots \\ \ldots & \sigma(i) & \ldots & \sigma(j) & \ldots \end{pmatrix} = \begin{pmatrix} \ldots & i & \ldots & j & \ldots \\ \ldots & \sigma(j) & \ldots & \sigma(i) & \ldots \end{pmatrix} ,$$

so that a σ-inversion (ij) is not an inversion relative to the permutation $\tau\sigma$, where τ is the transposition $(\sigma(j)\ \sigma(i))$. Show that it is possible to find k transpositions $\tau_1, \cdots, \tau_k$ such that $\tau_k \tau_{k-1} \cdots \tau_1 \sigma = e$ is the identity. Then $\sigma = \tau_1 \cdots \tau_{k-1} \tau_k$, and $\operatorname{sgn} \sigma = (-1)^k = \varepsilon_\sigma$, so we have two designations for the same thing: sgn and ε. This gives us another useful method for determining the sign of a permutation. Suppose that the set of inversions relative to a permutation π consists of five pairs $(1,5)$, $(2,5)$, $(3,5)$, $(4,5)$, $(6,7)$; then $\operatorname{sgn} \pi = -1$. In practice, this method amounts to counting the number of j in the lower row of a permutation which are greater than i but come before i, as $i = 1, 2, \cdots, n-1$.

11. Prove that a non-empty subset H of a finite (multiplicative) group G is a subgroup if H is closed under multiplication. That is, in that case the requirement that H has an identity element and an inverse h^{-1} for each $h \in H$ is superfluous.

12. Give a possible set of generators for the multiplicative group of positive rational numbers.

13. Prove that the k-th power π^k of the cycle $\pi = (12 \ldots n) \in S_n$ is the product of $d = \text{g.c.d.}(n,k)$ independent cycles, each of length $q = \text{l.c.m.}(n,k) = n/d$.

14. Suppose that $A, B \in M_n(\mathbb{R})$ and $(AB)^m = E$ for some integer m. Is it necessarily true that $(BA)^m = E$?

§3. Morphisms of groups

1. Isomorphisms. As already noted, the three rotations $\varphi_0, \varphi_1, \varphi_2$ counter-clockwise through 0°, 120°, 240° take the equilateral triangle P_3 to itself. But there are also three reflections about an axis of symmetry, which we denote ψ_1, ψ_2, ψ_3; in Fig. 13, the axes of symmetry are 1--1', 2--2', and 3--3'. For each of the six transformations of P_3 there is a corresponding permutation of the set of vertices of the triangle. We have the correspondence

$$\varphi_0 \sim e\,, \qquad \varphi_1 \sim (123)\,, \quad \varphi_2 \sim (132)$$

$$\psi_1 \sim (23)\,, \quad \psi_2 \sim (13)\,, \quad \psi_3 \sim (12)$$

Since this exhausts all of S_3, we can say that the group D_3 of all symmetry transformations of an equilateral triangle is very much like the symmetric group S_3.

In the same sense, the two groups C_n (the cyclic group of order n, see the example in subsection 3 of §2) and $\langle(12\ldots n)\rangle \subset S_n$ are very similar to one another. These examples, along with some general thought about the nature of groups, inevitably lead to a very natural question about the most essential properties of groups. At first glance, complete information is contained in the multiplication table for a group G, sometimes called the Cayley table:

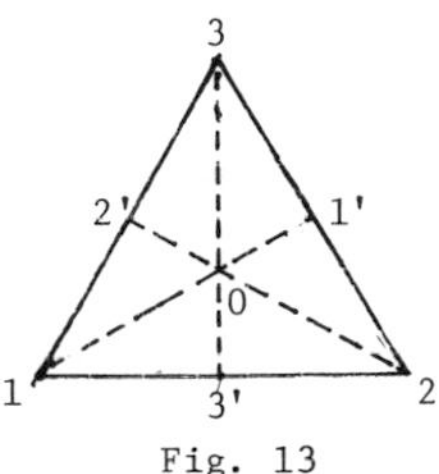

Fig. 13

	g_1	g_2	$\cdots$	g_n	$\cdots$
g_1	g_1g_1	g_1g_2	$\cdots$	g_1g_n	$\cdots$
g_2	g_2g_1	g_2g_2	$\cdots$	g_2g_n	
$\vdots$	. . .	. . .	. . .	. . .	. . .
g_n	g_ng_1	g_ng_2	$\cdots$	g_ng_n	$\cdots$
$\vdots$	. . .	. . .	. . .	. . .	. . .

And, in fact, many properties of a group can be seen from the Cayley table, i.e., from the $n \times n$-matrix $M = (m_{ij})$, where $n = (G:e)$, with entries $m_{ij} = g_ig_j \in G$. We note, for example, that in each row or column of M every element of G occurs exactly once (see the proof of Theorem 2 below). A group G is abelian if and only if M is symmetric, i.e., $m_{ij} = m_{ji}$. There are other such properties, but we soon see that it is rather hard to compare two multiplication tables, say for groups G and G', because the appearance of M depends on how the elements of the group are indexed; moreover, the situation becomes more complicated when the groups are infinite.

The best and most profound approach to comparing two groups is based on the notion of an isomorphism.

Definition. Two groups G and G' with operations $*$ and $\circ$ are called isomorphic if there exists a map $f : G \to G'$ such that:

(i) $f(a * b) = f(a) \circ f(b)$ for all $a, b \in G$.

(ii) f is bijective.

Such a map f is called an isomorphism between G and G'. We use the notation $G \cong G'$ when G and G' are isomorphic.

We give the simplest properties of an isomorphism.

1) The identity goes to the identity. To see this, let e be the identity of G. Since $e * a = a * e = a$, we have $f(e) \circ f(a) = f(a) \circ f(e) = f(a)$ for all $a \in G$; hence, $f(e) = e'$ is the identity in G'. Note that, in addition to property (i), we have used property (ii), since surjectivity of f is needed to be able to write any element of G' in the form $f(a)$. □

2) $f(a^{-1}) = f(a)^{-1}$. Namely, by 1), we have $f(a) \circ f(a^{-1}) = f(a * a^{-1}) = f(e) = e'$, where e' is the identity in G'. Thus,

$$f(a)^{-1} = f(a)^{-1} \circ e' = f(a)^{-1} \circ (f(a) \circ f(a^{-1})) =$$
$$= (f(a)^{-1} \circ f(a)) \circ f(a^{-1}) = e' \circ f(a^{-1}) = f(a^{-1}) . \quad \square$$

3) The inverse map $f^{-1} : G' \to G$ (which exists by property (ii)) is also an isomorphism.

By the corollary of Theorem 2 of §5 Ch. 1, we need only verify property (i) for f^{-1}. Let $a', b' \in G'$. Then, since f is bijective, we can write $a' = f(a)$, $b' = f(b)$ for some $a, b \in G$. Because f is an isomorphism, we have $a' \circ b' = f(a) \circ f(b) = f(a * b)$. This gives $a * b = f^{-1}(a' \circ b')$, and, since $a = f^{-1}(a')$ and $b = f^{-1}(b')$, it follows that $f^{-1}(a' \circ b') = f^{-1}(a') * f^{-1}(b')$. □

A simple verification shows that our correspondence $\sim$ between the groups D_3 and S_3 in the example above is actually an isomorphism.

The function $f = \ln$ is an isomorphism of the multiplicative group of positive real numbers with the additive group of all real numbers. The basic property of the logarithm $\ln ab = \ln a + \ln b$ is precisely property (i) in the definition of an isomorphism. The inverse mapping is $x \mapsto e^x$.

We now prove two general theorems which illustrate the role of isomorphisms in group theory.

THEOREM 1. Any two cyclic groups of the same order (in particular, any two infinite cyclic groups) are isomorphic.

Proof. First, if $\langle g \rangle$ is an infinite cyclic group, then all powers g^n are distinct, and we obtain an isomorphism $f : \langle g \rangle \to (\mathbb{Z}, +)$ by setting $g^n \mapsto n$. f is obviously bijective, and the property $f(g^m g^n) = f(g^m) + f(g^n)$ follows from Theorem 2 of §2.

Now suppose that $G = \{e, g, \cdots, g^{q-1}\}$ and $G' = \{e', g', \cdots, g'^{q-1}\}$ are two cyclic groups of order q (where we are using multiplicative notation for the group operation in both groups). We define a bijective map by setting

$$f : g^k \longmapsto (g')^k, \qquad k = 0, 1, \cdots, q-1 \quad .$$

Setting $n + m = \ell q + r$, $0 \leq r \leq q - 1$, for any $n, m = 0, 1, \cdots, q - 1$ and reasoning as in the proof of Theorem 3 of §2, we have

$$f(g^{n+m}) = f(g^r) = (g')^r = (g')^{n+m} = (g')^n (g')^m = f(g^n)\, f(g^m) \quad . \quad \square$$

THEOREM 2 (Cayley). Any finite group of order n is isomorphic to some subgroup of the symmetric group S_n.

Proof. Let G be a group with $n = |G|$. We may take S_n to be the group of all bijective maps of the set G to itself, since the nature of the elements permuted by the elements of S_n is immaterial.

For any $a \in G$, consider the map $L_a : G \to G$ given by the formula

$$L_a(g) = ag \quad .$$

If $e = g_1, g_2, \cdots, g_n$ are all elements of G, then $a, ag_2, \cdots, ag_n$ are the same elements but in a different order (recall the Cayley table!). (To see why these elements are distinct, we have:

$$ag_i = ag_j \Rightarrow a^{-1}(ag_i) = a^{-1}(ag_j) \Rightarrow (a^{-1}a)g_i = (a^{-1}a)g_j \Rightarrow g_i = g_j \quad .)$$

Hence, L_a is a bijective map (permutation), whose inverse will be $L_a^{-1} = L_{a^{-1}}$. Of course, L_e is the identity permutation. If we use the associativity of the group operation, we obtain: $L_{ab}(g) = (ab)g = a(bg) = L_a(L_b g)$, i.e., $L_{ab} = L_a \circ L_b$.

Thus, the set $L_e, L_{g_2}, \cdots, L_{g_n}$ forms a subgroup -- call it H -- in the group of all bijective maps of G to itself, i.e., in S_n. The group G is isomorphic to the subgroup H using the correspondence $a \mapsto L_a$, which, by what was said in the last paragraph, has all of the properties of an isomorphism. □

Despite its simplicity, Cayley's theorem has an important meaning for group theory. It shows the existence of a sort of "universal object" -- the family $\{S_n | n = 1, 2, \ldots\}$ of symmetric groups -- in which all finite groups (considered up to isomorphism) live. The phrase "up to isomorphism" is so typical, not only of group theory, but of all mathematics, which tends to consider at once all objects having common properties; without such abstraction and generalization, the whole point of the subject would be lost.

If $G' = G$ in the definition of an isomorphism, we have the concept of an isomorphism $\varphi : G \to G$ of a group G to itself. Such an isomorphism is called an <u>automorphism</u> of G. For example, the identity map $e_G : g \mapsto g$ (henceforth denoted simply 1) is an automorphism. But, in general, a group G also has non-trivial automorphisms. Property 3) of isomorphisms shows that the inverse of an automorphism is also an automorphism. Furthermore, if φ and ψ are automorphisms of G, then

$(\varphi \circ \psi)(ab) = \varphi(\psi(ab)) = \varphi(\psi(a)\,\psi(b)) = (\varphi \circ \psi)(a) \cdot (\varphi \circ \psi)(b)$ for any $a, b \in G$. Hence, the set $\mathrm{Aut}(G)$ of all automorphisms of a group G forms a group, in fact, a subgroup of the group $S(G)$ of all bijective maps $G \to G$.

2. Homomorphisms. The group of automorphisms $\mathrm{Aut}(G)$ of a group G contains a very special subgroup, which is denoted $\mathrm{Inn}(G)$ and is called the group of inner automorphisms. Its elements are the maps

$$I_a : g \mapsto aga^{-1} \quad .$$

A simple exercise shows that I_a really has all of the properties required of an automorphism, and that $I_a^{-1} = I_{a^{-1}}$, $I_e = 1$ is the identity automorphism, and $I_a \circ I_b = I_{ab}$ (because

$$(I_a \circ I_b)(g) = I_a(I_b(g)) = I_a(bgb^{-1}) = abgb^{-1}a^{-1} = abg(ab)^{-1} = I_{ab}(g) \ .)$$

This last fact about I_a shows that the map

$$f : G \to \mathrm{Inn}(G), \quad f(a) = I_a \quad \text{for} \quad a \in G \quad ,$$

from the group G to the group $\mathrm{Inn}(G)$ of inner automorphisms of G satisfies property (i) in the definition of an isomorphism: $f(a) \circ f(b) = f(ab)$. However, property (ii) is not necessarily satisfied. For example, if G is an abelian group, then $aga^{-1} = g$ for all $a, g \in G$, i.e., $I_a = I_e$ for all $a \in G$, and $\mathrm{Inn}(G)$ only consists of the identity I_e. The example of this map f makes it natural to introduce the following

Definition. A map $f : G \to G'$ from the group $(G, *)$ to the group $(G', \circ)$ is called a homomorphism if

$$f(a * b) = f(a) \circ f(b) , \qquad \forall\, a, b \in G$$

(in other words, property (ii) in the definition of an isomorphism is omitted).

By the kernel of the homomorphism f, we mean the set

$$\mathrm{Ker}\, f = \{g \in G \,|\, f(g) = e' , \quad \text{where } e' \text{ is the identity of } G'\} \quad .$$

A homomorphism from a group to itself is called an endomorphism.

In the definition of a homomorphism, f need be neither injective nor surjective. We can make f into a surjective map by replacing G by $\operatorname{Im} f \subset G'$, which is obviously a subgroup of G'. So the "main" difference between a homomorphism and an isomorphism is the presence of a non-trivial kernel $\operatorname{Ker} f$, which is, one might say, a measure of the non-injectivity of f. If $\operatorname{Ker} f = \{e\}$, then $f : G \to \operatorname{Im} f$ is an isomorphism.

Note that

$$f(a) = e', \; f(b) = e' \Rightarrow f(a * b) = f(a) \circ f(b) = e' \circ e' = e'$$

and

$$f(a^{-1}) = f(a)^{-1} = (e')^{-1} = e' \quad .$$

Hence $\operatorname{Ker} f$ is a subgroup of G. Let $H = \operatorname{Ker} f \subset G$. Then (we are now omitting the $*$ and $\circ$):

$$f(ghg^{-1}) = f(g)\, f(h)\, f(g)^{-1} = f(g)\, e'\, f(g)^{-1} = e', \qquad \forall\, h \in H,\ g \in G \quad ,$$

i.e., $ghg^{-1} \in H$; hence, $gHg^{-1} \subset H$. If we replace g by g^{-1} here, we obtain $g^{-1}Hg \subset H$, so that $H \subset gHg^{-1}$. Thus,

$$gHg^{-1} = H, \qquad \forall\, g \in G \quad .$$

A subgroup which has this property is called a normal subgroup (sometimes an invariant subgroup or a normal divisor). We have thereby proved

THEOREM 3. The kernel of a homomorphism is always a normal subgroup. □

We shall see the importance of this fact somewhat later. For now, we note that far from every subgroup is normal. For example, in S_3 the cyclic subgroup $\langle(123)\rangle = A_3$ is normal, but $\langle(12)\rangle = \{e, (12)\}$ is not normal.

3. Glossary. Examples. The terms "surjective map" (map "onto"), "injective map" (imbedding), "bijective map" (one-to-one correspondence), which can be used for maps of any sets (with or without operations), are often replaced by other terms when used for groups

(the same happens for other algebraic systems). We use the terms epimorphism (homomorphism "onto"), monomorphism (homomorphism whose kernel is the identity element), and isomorphism (a one-to-one homomorphism, i.e., a homomorphism which is both an epimorphism and a monomorphism). There is a tendency to replace the term homomorphism by the word morphism. These words are useful to know when reading mathematical literature, but the reader can, if he chooses, get along using only the terms isomorphism and homomorphism with the prepositions "into" and "onto".

We now give some further examples of group homomorphisms.

1) The additive group of integers $\mathbb{Z}$ maps homomorphically onto the finite cyclic group $\langle g \rangle$ of order q if we set $f : n \mapsto g^n$ (see Theorem 2 of §2). Here $\operatorname{Ker} f = \{\ell q \mid \ell \in \mathbb{Z}\}$. Namely, it is clear that $\{\ell q\} \subset \operatorname{Ker} f$, and the reverse inclusion follows from Theorem 3 of §2.

2) The map $f : \mathbb{R} \to T = SO(2)$ of the additive group of real numbers onto the group T of rotations of the plane about the origin, which is given by $f(\lambda) = \Phi_\lambda$ (Φ_λ is counterclockwise rotation through an angle of $2\pi\lambda$) is a homomorphism. Since rotation through a multiple of 2π coincides with rotation through an angle of zero, i.e., the identity map, we have: $\operatorname{Ker} f = \mathbb{Z}$. We also say that f gives a homomorphism of $\mathbb{R}$ onto the unit circle S^1, since there is a one-to-one correspondence between $SO(2)$ and S^1, namely, Φ_λ corresponds to the point with polar coordinates $(1, 2\pi\lambda)$.

3) The general linear group $GL(n)$, consisting of matrices A with entries in $\mathbb{R}$ and non-zero determinant, maps homomorphically onto the multiplicative group $\mathbb{R}^*$ of non-zero real numbers, if we set $f = \det$. The condition that f is a homomorphism: $f(AB) = f(A)\,f(B)$, is merely another way of stating Theorem 5 of §2 Ch. 3. By definition, $SL(n) = \operatorname{Ker} f$.

4) Consider the cyclic group $C_2 = \langle -1 \rangle = \{1, -1\}$ of order 2. If we want, we can give this group by writing its Cayley table:

$\cdot$	1	-1
C_2 : 1	1	-1
-1	-1	1

The map $S_n \to C_2$ given by our function $\varepsilon = \mathrm{sgn} : \pi \mapsto \varepsilon_\pi$ is a homomorphism. Here $\mathrm{Ker}\,\varepsilon = A_n$, by the definition of the alternating group.

5) An infinite group can be isomorphic to a proper subgroup. For example, the additive group $(\mathbb{Z}, +)$ contains the proper subgroup $n\mathbb{Z} = \{nk \mid k \in \mathbb{Z}\}$, where $n > 1$ is a fixed natural number. It is easy to check that the map $g_n : \mathbb{Z} \to n\mathbb{Z}$ given by $g_n(k) = nk$ is an isomorphism. Incidentally, note that $\mathbb{Z}$ and $n\mathbb{Z}$ are infinite cyclic groups with generator 1 or -1 and n or $-n$, respectively; hence g_n and the map $k \mapsto -kn$ are all possible isomorphisms $\mathbb{Z} \to n\mathbb{Z}$.

6) The group $\mathrm{Aut}(G)$, and even a single element $\varphi \in \mathrm{Aut}(G)$ which is not the identity, can be a source of important information about the group G. Here is an example of this. Let G be a finite group having an automorphism φ of order 2 (i.e., $\varphi^2 = 1$) which has no fixed points:

$$a \neq e \Longrightarrow \varphi(a) \neq a \quad .$$

Suppose that $\varphi(a)\,a^{-1} = \varphi(b)\,b^{-1}$ for any $a, b \in G$. If we multiply this equality on the left by $\varphi(b)^{-1}$ and on the right by a, we see that $\varphi(b)^{-1}\varphi(a) = b^{-1}a$, i.e., $\varphi(b^{-1}a) = b^{-1}a$, which implies that $b^{-1}a = e$, and so $b = a$. Thus, as a runs through the elements of G, so does $\varphi(a)a^{-1}$; equivalently, any element $g \in G$ can be written in the form $g = \varphi(a)a^{-1}$. But then $\varphi(g) = \varphi(\varphi(a))\,\varphi(a^{-1}) = \varphi^2(a)\,\varphi(a^{-1}) = a\varphi(a)^{-1} = (\varphi(a)a^{-1})^{-1} = g^{-1}$. Thus, φ is precisely the map $g \mapsto g^{-1}$. With this in mind, we obtain $ab = \varphi(a^{-1})\,\varphi(b^{-1}) = \varphi(a^{-1}b^{-1}) = (a^{-1}b^{-1})^{-1} = ba$, i.e., the group G turns out to be abelian! In addition, $(G:e)$ is an odd number, since G consists of e and disjoint pairs of elements $g_i, g_i^{-1} = \varphi(g_i)$.

7) The following example shows how much one can alter the group operation without

changing the group itself, i.e., only changing it to an isomorphic group (see also Exercise 3 of §1). Let G be any group, and let t be a fixed element of G. Introduce a new operation on the set G:

$$(g, h) \longmapsto g * h = gth \quad .$$

We immediately verify that $(g_1 * g_2) * g_3 = g_1 * (g_2 * g_3)$, i.e., the operation $*$ is associative. In addition, $g * t^{-1} = t^{-1} * g = g$, and $g * (t^{-1}g^{-1}t^{-1}) = (t^{-1}g^{-1}t^{-1}) * g = t^{-1}$; this means that $(G, *)$ is a group with identity element $e_* = t^{-1}$. The inverse of an element g in $(G, *)$ is $g_*^{-1} = t^{-1}g^{-1}t^{-1}$. The map $f : g \mapsto gt^{-1}$ gives an isomorphism between $(G, \cdot)$ and $(G, *)$.

All of these examples illustrate the general principle that studying the morphisms of a group G gives much information about G itself.

4. Cosets of a subgroup. It is clear from the definition of a homomorphism $f : G \to G'$ that all of the elements of the set

$$a \operatorname{Ker} f = \{ab \mid b \in \operatorname{Ker} f\}, \qquad a \in G \quad ,$$

are mapped to the same element f(a) in $G' : f(ab) = f(a)f(b) = f(a)e' = f(a)$ if $b \in \operatorname{Ker} f$. Conversely, if $f(g) = f(a)$, then $f(a^{-1}g) = f(a^{-1})f(g) = f(a)^{-1}f(g) = e'$, so that $a^{-1}g = b \in \operatorname{Ker} f$ and $g = ab \in a \operatorname{Ker} f$. This fact shows the usefulness of partitioning G into subsets of the form $a \operatorname{Ker} f$. We now forget about homomorphisms for a moment, and study such partitions in their own right.

Definition. Let H be a subgroup of G. A left coset of H in G is a set of the form gH, consisting of all elements of the form gh, where g is a fixed element of G and h runs through all elements of the subgroup H. The element g is called a coset representative for gH.

We similarly define a right coset Hg. (Sometimes the terminology "right" and "left" is reversed, i.e., used to refer to the position of H rather than of g; the important thing is to be consistent, whichever convention is adopted.) If $H = \operatorname{Ker} f$ is the kernel of a homomorphism, then $gH = Hg$, because H is normal in G (see subsection 2). Note that the subgroup H itself is a coset: $H = He = eH$. But none of the other cosets can be a subgroup, because, if gH were a subgroup, then we would have $e \in gH$, so that $e = gh$, $g = h^{-1}$, and $gH = h^{-1}H = H$.

THEOREM 4. <u>Two left cosets of</u> H <u>in</u> G <u>must either coincide or be disjoint. The partition of</u> G <u>into left cosets of</u> H <u>gives an equivalence relation on</u> G.

<u>Proof.</u> Suppose that the cosets g_1H and g_2H have an element in common: $a = g_1h_1 = g_2h_2$. Then $g_2 = g_1h_1h_2^{-1}$, and any element g_2h of the coset g_2H has the form $g_1h_1h_2^{-1}h = g_1h'$, where $h' = h_1h_2^{-1}h \in H$. Thus, $g_2H \subset g_1H$. We similarly prove that every element in g_1H is contained in g_2H. Hence, $g_1H = g_2H$.

Since each element $g \in G$ is contained in the coset gH, we may conclude that G is a union of disjoint left cosets of H:

$$G = \cup\, g_iH \quad .$$

By the general principle in §6 of Chapter 1, this partition induces an equivalence relation on G, which is defined in the obvious way:

$$a \sim b \iff a^{-1}b \in H \quad .$$

If we want, we can verify reflexivity, symmetry, and transitivity of this relation directly: $a \sim a$, because $a^{-1}a = e \in H$; $a \sim b \Leftrightarrow a^{-1}b = h \Leftrightarrow b^{-1}a = h^{-1} \in H \Leftrightarrow b \sim a$; $a \sim b$, $b \sim c \Rightarrow b^{-1}a = h_1$, $c^{-1}b = h_2 \Rightarrow c^{-1}a = c^{-1}bh_1 = h_2h_1 \in H \Rightarrow a \sim c$. □

The analogous theorem holds for right cosets.

The partition into cosets arises naturally in permutation groups. Let $G = S_n$ be the symmetric group, acting on the set $\Omega = \{1, 2, \cdots, n\}$. If we consider the set H of elements $\pi \in S_n$ for which $\pi(n) = n$, then we easily see that H is a subgroup of S_n, which can be identified with S_{n-1}. Let $\tau_0 = e$, and let $\tau_i = (in)$ be the transposition taking n to i $(i = 1, 2, \cdots, n-1)$. It is clear that

$$S_n = \bigcup_{k=0}^{n-1} \tau_k S_{n-1} \quad .$$

Here is the partition of S_3 into left and right cosets of the subgroup $\langle(12)\rangle = S_2$:

$$S_3 = \{e, (12)\} \cup \{(13), (123)\} \cup \{(23), (132)\} \quad ;$$

$$S_3 = \{e, (12)\} \cup \{(13), (132)\} \cup \{(23), (123)\} \quad .$$

We see that the set of left cosets gS_2 is not the same as the set of right cosets S_2g'. Nevertheless, the sets $\{gH\}$ and $\{Hg'\}$ are always in bijective correspondence, as follows:

$$x = gh \in gH \longleftrightarrow x^{-1} = h^{-1}g^{-1} \in Hg^{-1} \quad .$$

In fact, if, say, $h_1g_1^{-1} = h_2g_2^{-1}$, then $g_1 = g_2h_2^{-1}h_1$, and $g_1H = g_2H$. In particular, <u>if $\{e, x, y, z, \cdots\}$ is a set of left (respectively, right) coset representatives, then $\{e, x^{-1}, y^{-1}, z^{-1}, \cdots\}$ is a set of right (respectively, left) coset representatives. Both sets have the same cardinality.</u>

We denote the set of all left cosets of H in G by the symbol G/H (or $(G/H)_\ell$ if we have to consider simultaneously the set $(G/H)_r$ of right cosets). We call the cardinality of this set Card G/H the "index of the subgroup H in G", which we denote (G:H). (This agrees with the notation (G:e) introduced before for the order $|G|$ of G, i.e., the number of cosets of the trivial subgroup $\{e\}$.) Since we have the one-to-one correspondence $a \mapsto ga$ between H and gH (see the proof of Cayley's theorem), it follows that Card gH = (H:e). We thus have the following simple formula:

$$(G:e) = (G:H)\,(H:e) \quad ,$$

which implies the classical

THEOREM 5 (Lagrange). <u>The order of a finite group is divisible by the order of any subgroup.</u> □

COROLLARY. <u>The order of any element divides the order of the group. A group whose order is a prime</u> p <u>is always cyclic; such a group is unique up to isomorphism.</u>

<u>Proof.</u> The order of an element $g \in G$ is the same as the order of the cyclic subgroup generated by g (Theorem 3 of §2), and so divides $|G|$ by Theorem 5. Next, if $|G| = p$ is a prime, and if H is a non-trivial subgroup (i.e., $H \neq \{e\}$), then Theorem 5 implies that $|H| = p$, and hence $H = G$. Thus, G coincides with the cyclic subgroup generated by any element $g \neq e$. Since all cyclic groups of a given order are isomorphic (Theorem 1), we have the uniqueness assertion in the corollary. $\square$

Lagrange's theorem leads one to want to find a subgroup of order m for every m dividing $n = |G|$. But this cannot always be done. For example, the reader can verify that the alternating group A_4, which has order 12, has no subgroup of order 6.

But in some groups the "converse Lagrange theorem" does hold. For example:

THEOREM 6. <u>Every subgroup of a cyclic group is cyclic. The subgroups of the infinite cyclic group</u> $(\mathbb{Z}, +)$ <u>are precisely the (infinite cyclic) groups</u> $(m\mathbb{Z}, +)$, $m \in \mathbb{N}$; <u>and the subgroups of a cyclic group of order</u> q <u>are in one-to-one correspondence with the (positive) divisors</u> d <u>of</u> q.

<u>Proof.</u> Let $A = \langle a \rangle$ be a cyclic group. For variety, let's use additive notation. Thus, every element has the form ka, where $k \in \mathbb{Z}$ or else $k = 0, 1, \cdots, q-1$ if A is a finite group of order q (see Theorem 3 of §2). Let B be a non-zero subgroup of A. If $ka \in B$ for some $k \neq 0$, then we also have $-ka \in B$. Among all elements $ka \in B$ with positive k, let ma be the element for which m is minimal. If we write any $k > 0$ in the form $k = \ell m + r$, $0 \leq r < m$, we see that $ka \in B$ implies $ra = ka - \ell(ma) \in B$; hence, $r = 0$. Thus, $B = \langle ma \rangle$ is a cyclic group.

All infinite cyclic groups are isomorphic (by Theorem 1). So without loss of generality we may take $(\mathbb{Z}, +)$ to be the model of an infinite cyclic group. It has generator 1 or -1, so that, by the last paragraph, any subgroup of $(\mathbb{Z}, +)$ is determined by a natural number m and has the from

$$m\mathbb{Z} = \langle m \cdot 1 \rangle = \{0, \pm m, \pm 2m, \cdots\} \quad .$$

It is obvious that all of these subgroups are infinite.

Now suppose that $\langle a \rangle = \{0, a, \cdots, (q-1)a\}$, $qa = 0$. We found that $B = \{0, ma, 2ma, \cdots\}$, where $m \in \mathbb{N}$ and $sa \in B$ for $s \in \mathbb{N}$ only if s is a multiple of m. We claim that m divides q. In fact, let $q = dm + r$, $0 \leq r < m$. Then

$$0 = qa = d(ma) + ra \quad ,$$

so that $ra = -d(ma) \in B$. Since m is minimal, we have $r = 0$, and so $q = dm$. Thus,

$$B = \{0, ma, 2ma, \cdots, (d-1)ma\} = mA$$

is a subgroup of A of order d. As m runs through all positive divisors of q, so does d, and we obtain exactly one subgroup for each order d dividing q. □

COROLLARY. _In a cyclic group_ $\langle a \rangle$ _of order_ q, _the subgroup of order_ $d|p$ _is precisely the set of elements_ $b \in \langle a \rangle$ _such that_ $db = 0$.

Proof. If $dm = q$, then $b \in B = mA$, and $db = 0$. Conversely, suppose that $b = \ell a \in \langle a \rangle$ and $db = 0$. The condition $d\ell a = 0$ implies that $d\ell = qk = dmk$, so that $\ell = mk$, and $b = \ell a = k(ma) \in mA$. □

5. _The monomorphism_ $S_n \to GL(n)$. Recall that a monomorphism of groups $G \to G'$ is an injective homomorphism from G to G'.

For example, here is a monomorphism $f : S_3 \to GL(3)$:

$$e \longmapsto \begin{Vmatrix} 1 & 0 & 0 \\ 0 & 1 & 0 \\ 0 & 0 & 1 \end{Vmatrix}, \quad (12) \longmapsto \begin{Vmatrix} 0 & 1 & 0 \\ 1 & 0 & 0 \\ 0 & 0 & 1 \end{Vmatrix}, \quad (13) \longmapsto \begin{Vmatrix} 0 & 0 & 1 \\ 0 & 1 & 0 \\ 1 & 0 & 0 \end{Vmatrix},$$

$$(23) \longmapsto \begin{Vmatrix} 1 & 0 & 0 \\ 0 & 0 & 1 \\ 0 & 1 & 0 \end{Vmatrix}, \quad (123) \longmapsto \begin{Vmatrix} 0 & 0 & 1 \\ 1 & 0 & 0 \\ 0 & 1 & 0 \end{Vmatrix}, \quad (132) \longmapsto \begin{Vmatrix} 0 & 1 & 0 \\ 0 & 0 & 1 \\ 1 & 0 & 0 \end{Vmatrix}.$$

The reader can check that f is really a monomorphism, and that for each $\pi \in S_3$ the determinant of $f(\pi)$ is ± 1, depending on the signature of the permutation π.

THEOREM 7. There exists a monomorphism $f : S_n \to GL(n)$ such that the matrix $f(\pi)$, $\pi \in S_n$, has determinant $|f(\pi)| = \varepsilon_\pi$.

Proof. We shall write an $n \times n$ matrix (a_{ij}) as a sequence of columns: $(a_{ij}) = (A^{(1)}, A^{(2)}, \ldots, A^{(n)})$. In particular, let

$$E^{(1)} = \begin{Vmatrix} 1 \\ 0 \\ \vdots \\ 0 \end{Vmatrix}, \quad E^{(2)} = \begin{Vmatrix} 0 \\ 1 \\ \vdots \\ 0 \end{Vmatrix}, \ldots, \quad E^{(n)} = \begin{Vmatrix} 0 \\ 0 \\ \vdots \\ 1 \end{Vmatrix}$$

be the columns of the identity matrix E. We define the map $f : S_n \to GL(n)$ by setting

$$\pi \longmapsto f(\pi) = (E^{(\pi(1))}, E^{(\pi(2))}, \ldots, E^{(\pi(n))}) \quad . \tag{1}$$

Thus, $f(\pi)$ is an $n \times n$ matrix in which each row and each column has one 1 and the other entries zero. It is easy to see that $f(\pi) \in GL(n)$.

Let σ and τ be any permutations, and let $\pi = \sigma\tau$ be their product. By definition, the non-zero entries in the i-th row of $f(\sigma) = (a_{is})$ and in the j-th column of $f(\tau) = (b_{k\ell})$ are $a_{i,\sigma^{-1}(i)} = 1$ and $b_{\tau(j),j} = 1$, respectively. Thus, in the matrix $f(\sigma) f(\tau) = (c_{ij})$, the condition $c_{ij} \neq 0$ is equivalent to $\sigma^{-1}(i) = \tau(j)$, i.e., $i = \sigma\tau(j) = \pi(j)$; but this means that $f(\sigma) f(\tau) = f(\sigma\tau)$. Consequently, f is a homomorphism.

The property $\operatorname{Ker} f = \{e\}$ is obvious, since it is clear from (1) that $f(\pi) = E \Rightarrow \pi = e$. Thus, f is a monomorphism.

Finally, since f, $\det$, and ε all preserve products, i.e.,

$$|f(\sigma\tau)| = |f(\sigma)\ f(\tau)| = |f(\sigma)||f(\tau)|, \qquad \varepsilon_{\sigma\tau} = \varepsilon_{\sigma}\,\varepsilon_{\tau},$$

and since any π can be decomposed into a product of transpositions, it suffices to prove the equality $|f(\pi)| = \varepsilon_{\pi}$ when $\pi = (ij)$ is a transposition. But then $f(\pi)$ is obtained from E by interchanging the i-th and j-th columns; hence, $|f(\pi)| = -|E| = -1 = \varepsilon_{(ij)}$. □

Matrices of the form $f(\pi)$, $\pi \in S_n$, are called <u>permutation matrices.</u> The restriction of the monomorphism f to A_n is a monomorphism into $SL(n, \mathbb{R})$. Given any finite group G, the composition $f \circ L$ of the map $L : G \to S_n$ (see Theorem 2) and the map $f : S_n \to GL(n)$ gives a monomorphism $G \to GL(n)$.

Using Theorem 7, we can easily prove the so-called <u>theorem on the complete expansion of the determinant:</u>

THEOREM 8. <u>The determinant</u>

$$\det A = \begin{vmatrix} a_{11} & a_{12} & \cdots & a_{1n} \\ a_{21} & a_{22} & \cdots & a_{2n} \\ \cdot & \cdot & \cdot & \cdot \\ a_{n1} & a_{n2} & \cdots & a_{nn} \end{vmatrix}$$

<u>can be written as a sum of</u> $n!$ <u>terms, each of which is a product of</u> n <u>entries in</u> A:

$$\det A = \sum_{\pi \in S_n} \varepsilon_{\pi}\, a_{\pi(1),1}\, a_{\pi(2),2} \cdots a_{\pi(n),n} \quad . \tag{2}$$

<u>Proof.</u> Let $|A^{(1)}, \dots, A^{(j-1)}, E^{(i)}, A^{(j+1)}, \dots, A^{(n)}|$ denote the determinant obtained from $|A|$ by replacing the column $A^{(j)}$ with the i-th column $E^{(i)}$ of the identity matrix. Using the formula for expanding a determinant along the j-th column, we see that the cofactor A_{ij} of the element a_{ij} in $|A|$ can be written as the following n-th order determinant:

$$A_{ij} = |A^{(1)}, \cdots, A^{(j-1)}, E^{(i)}, A^{(j+1)}, \cdots, A^{(n)}| ,$$

and, so, expanding $|A|$ along the j-th column, we have:

$$\det A = \sum_i a_{ij} A_{ij} = \sum_i a_{ij} |A^{(1)}, \cdots, A^{(j-1)}, E^{(i)}, A^{(j+1)}, \cdots, A^{(n)}| .$$

If we first do this for $j = 1$, and then evaluate each of the n determinants in the sum by repeating the method for $j = 2$, and so on, we end up with an expression for $\det A$ containing first n, then $n^2, \cdots,$ finally n^n determinants:

$$\det A = \sum_{i_1} a_{i_1,1} |E^{(i_1)}, A^{(2)}, \cdots, A^{(n)}| =$$

$$= \sum_{i_1} a_{i_1,1} \sum_{i_2} a_{i_2,2} |E^{(i_1)}, E^{(i_2)}, A^{(3)}, \cdots, A^{(n)}| =$$

$$= \sum_{i_1,i_2} a_{i_1,1} a_{i_2,2} |E^{(i_1)}, E^{(i_2)}, \cdots, A^{(n)}| =$$

$$= \cdots = \sum_{i_1,i_2,\cdots,i_n} a_{i_1,1} a_{i_2,2}, \cdots, a_{i_n,n} |E^{(i_1)}, E^{(i_2)}, \cdots, E^{(i_n)}| .$$

Here $i_1, i_2, \cdots, i_n$ run through all sets of n numbers in $\{1, 2, \cdots, n\}$ (with repetitions allowed). In other words, taking all n^n maps $\pi : \{1, 2, \cdots, n\} \to \{1, 2, \cdots, n\}$ (see Example 1 in subsection 2 of §1), where $\pi(1) = i_1, \cdots, \pi(n) = i_n$, we write $\det A$ in the form

$$\det A = \sum_{\pi} a_{\pi(1),1} a_{\pi(2),2} \cdots a_{\pi(n),n} |E^{(\pi(1))}, E^{(\pi(2))}, \cdots, E^{(\pi(n))}| .$$

It remains for us to note that, if $\pi(i) = \pi(j)$ for any $i \neq j$, then the determinant $|E^{(\pi(1))}, \cdots, E^{(\pi(n))}|$ has two identical columns, and hence vanishes. Consequently, the terms in the above summation are only non-zero when π is bijective, i.e., when it is a permutation. But in that case, by Theorem 7, we have $|E^{(\pi(1))}, \cdots, E^{(\pi(n))}| = |f(\pi)| = \varepsilon_\pi$. $\square$

Remark. Of course, Theorem 8 can be proved directly by induction on n, but the appearance of the sign ε_π becomes easier to understand when one adopts the point of view of group theory. Theorem 7 is also of independent interest, in addition to its use in proving Theorem 8.

Theorem 8 can be taken as the starting point for the theory of determinants (as is often done). Namely, after defining det A by (2), we would then prove all of the properties, including the formula for expanding det A along the elements of the first (or j-th) column, which was our point of departure in Chapter 3.

EXERCISES

1. Prove that, up to isomorphism, there are only a finite number $\rho(n)$ of groups of given order n.

2. Using Exercise 7 of §2, show that every finite group can be imbedded in a finite group with two generators (i. e., there exists a monomorphism into such a group).

3. Prove that a subgroup of index 2 must be normal.

4. Using Exercise 3, try to prove that, up to isomorphism, S_3 is the only non-abelian group of order 6.

5. Try to show that all of the subgroups of the alternating group A_4 are depicted in Fig. 14. Here we let V_4 denote the so-called <u>Klein four-group</u>:

$V_4 = \{e, (12)(34), (13)(24), (14)(23)\}$;

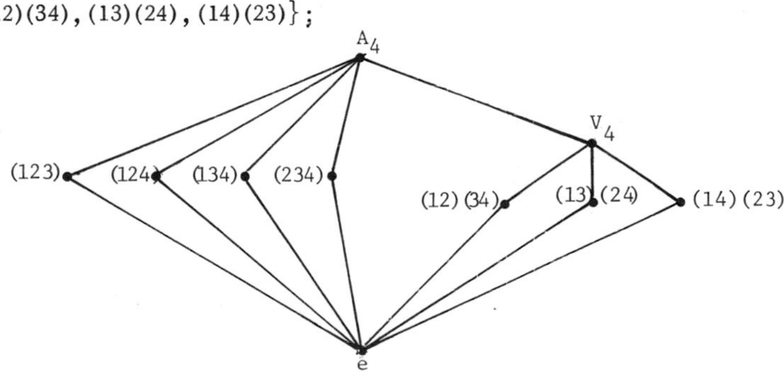

Fig. 14

the other vertices correspond to cyclic groups whose generators are given in the diagram.

6. Show that any group of order 4 is abelian, and is isomorphic to one of the permutation groups: $U = \langle(1234)\rangle$ or the Klein four-group V_4. Alternately, it is isomorphic to one of the matrix groups

$$L_1 = \left\{ \begin{Vmatrix} 1 & 0 \\ 0 & 1 \end{Vmatrix}, \begin{Vmatrix} 0 & 1 \\ -1 & 0 \end{Vmatrix}, \begin{Vmatrix} -1 & 0 \\ 0 & -1 \end{Vmatrix}, \begin{Vmatrix} 0 & -1 \\ 1 & 0 \end{Vmatrix} \right\} \subset GL(2, \mathbb{R}) ,$$

$$L_2 = \left\{ \begin{Vmatrix} 1 & 0 \\ 0 & 1 \end{Vmatrix}, \begin{Vmatrix} 1 & 0 \\ 0 & -1 \end{Vmatrix}, \begin{Vmatrix} -1 & 0 \\ 0 & 1 \end{Vmatrix}, \begin{Vmatrix} -1 & 0 \\ 0 & -1 \end{Vmatrix} \right\} \subset GL(2, \mathbb{R}) .$$

Write out explicitly isomorphisms $U \to L_1$, $V_4 \to L_2$.

§4. Rings and fields

1. <u>The definition and general properties of rings.</u> The algebraic structures $(\mathbb{Z}, +)$ and $(\mathbb{Z}, \cdot)$ were our first examples of monoids, and then $(\mathbb{Z}, +)$ turned out to be an additive abelian group (actually, a cyclic group). But usually one thinks of these two structures together, as a single structure known as a ring. There are important relations of arithmetic which involve combining the additive and multiplicative structures of $\mathbb{Z}$, the most basic of which is the distributive law $(a + b)c = ac + bc$. This law seems trivial to us only because we are so accustomed to using it. If we try, for example, to combine the algebraic structures $(\mathbb{Z}, +)$ and $(\mathbb{Z}, \circ)$ where $n \circ m = n + m + nm$, we find that it is not so easy to find a basic relationship between the two binary operations. Before proceeding to more examples, we give the precise definition of a ring.

<u>Definition.</u> Let R be a non-empty set with two binary operations $+$ (addition) and $\cdot$ (multiplication) which satisfy the following conditions:

(R1) $(R, +)$ is an abelian group;

(R2) $(R, \cdot)$ is a semi-group;

(R3) The operations of addition and multiplication are related by the left and right distributive laws (in other words: multiplication is distributive with respect to addition):

$$(a + b)c = ac + bc \quad , \qquad c(a + b) = ca + cb$$

for all $a, b, c \in R$.

Then $(R, +, \cdot)$ is called a ring.

We call $(R, +)$ the additive group of the ring, and we call $(R, \cdot)$ its multiplicative semi-group. If $(R, \cdot)$ is a monoid, then we say that $(R, +, \cdot)$ is a ring with unit.

It is customary to let 1 denote the unit element of a ring with unit. Sometimes, the existence of a unit element is stipulated in the definition of a ring, but we shall not do this, and shall maintain the distinction between a ring and a ring with unit.

Sometimes in applications and in general ring theory (which now exists in a very developed form), one considers algebraic systems in which the axiom (R2) is either completely dispensed with or else replaced with a weaker axiom, depending on the concrete problem. In such cases one speaks of non-associative rings. But here we shall only deal with the usual (associative) rings. Thus, we shall freely use Theorem 1 of §1, which allows us to disregard parentheses in a product $a_1 a_2 \cdots a_k$ of any number k of elements of a ring.

A subset L of a ring R is called a subring if

$$x, y \in L \Rightarrow x - y \in L \quad \text{and} \quad xy \in L \quad ,$$

i.e., if L is a subgroup of the additive group and a subsemigroup of the multiplicative semigroup of the ring.

It is clear that the intersection of any family of subrings of R is a subring (the argument is the same as in the case of groups), and so it makes sense to speak of the subring $\langle T \rangle \subset R$ generated by the subset $T \subset K$. By definition, $\langle T \rangle$ is the intersection of all subrings of R which contain T. If T itself is a subring, then

$\langle T \rangle = T$.

A ring is said to be commutative if $xy = yx$ for all $x, y \in K$ (unlike for groups, the word "abelian" is not usually used for commutative rings!).

This notion of a ring is very broad. Even the class of commutative rings, which at first glance seems to be rather specialized, has been the object of intensive study for many decades, and the theory of commutative rings is now interwoven with algebraic geometry -- a beautiful mathematical discipline which overlaps with algebra, geometry and topology.

Examples. 1) $(\mathbb{Z}, +, \cdot)$, the ring of integers with the usual operations of addition and multiplication. The set $m\mathbb{Z}$ of integers divisible by m is a subring of $\mathbb{Z}$ (which does not have a unit element if $m > 1$). Similarly, $\mathbb{Q}$ and $\mathbb{R}$ are rings with unit, and the natural inclusions $\mathbb{Z} \subset \mathbb{Q} \subset \mathbb{R}$ give a chain of subrings of the ring $\mathbb{R}$.

2) The properties of addition and multiplication in $M_n(\mathbb{R})$, which were introduced and studied in detail in Chapter 2, show that $M_n(\mathbb{R})$ is a ring with unit element $1 = E$. It is called the full matrix ring over $\mathbb{R}$, or else the ring of $n \times n$ matrices over $\mathbb{R}$. This is one of the most important examples of a ring. Since matrices do not generally commute with one another if $n > 1$, $M_n(\mathbb{R})$ is a non-commutative ring for $n > 1$. The rings $M_n(\mathbb{Q})$ and $M_n(\mathbb{Z})$ of $n \times n$ matrices with entries in $\mathbb{Q}$ and $\mathbb{Z}$, respectively, are contained as subrings in $M_n(\mathbb{R})$. Actually, $M_n(\mathbb{R})$ is full of a wide variety of subrings. Many of them will arise in various contexts later in the book. We further note that it is possible to consider the ring $M_n(R)$ of $n \times n$ matrices with entries in any commutative ring R, since the sum or product of two matrices in $M_n(R)$ will again have entries in R, and the distributive laws for $M_n(R)$ follow from the distributive laws for R. All of this is a direct consequence of the rules for matrix operations, which were summarized at the end of subsections 1 and 3 of §3 Ch. 2.

3) In many areas of mathematics the concept of a ring of functions plays a vital role. Let X be any set, and let R be any ring. Let $R^X = \{X \to R\}$ denote the set of all functions (i.e., all maps) $f : X \to R$, considered along with the two binary operations

<u>pointwise sum</u> $f+g$ and <u>pointwise product</u> fg, which are defined as follows:

$$(f+g)(x) = f(x) \oplus g(x) \quad ,$$

$$(fg)(x) = f(x) \odot g(x)$$

($\oplus$ and $\odot$ are the addition and multiplication operations in R). Here multiplication is obviously not the composition of functions which in the case of linear maps led us to the ring M_n. Rather, pointwise multiplication reflects the point of view in calculus, where $X = \mathbb{R}$, $R = \mathbb{R}$, and, for example, the product of tan and sin is $\tan \cdot \sin : x \mapsto \tan x \sin x$, and <u>not</u> $\tan \circ \sin : x \mapsto \tan(\sin x)$.

It is easy to check that R^X satisfies all of the ring axioms. For example, the distributive law in R gives

$$[f(x) \oplus g(x)] \odot h(x) = f(x) \odot h(x) \oplus g(x) \odot h(x)$$

for any three functions $f, g, h \in R^X$ and any $x \in X$; by the definition of the operations in R^X, this gives $(f+g)h = fh + gh$. If 0 and 1 denote the zero element and the unit element in R, then the zero element and the unit element in R^X are the <u>constant</u> functions

$$0_X : x \longmapsto 0 \quad , \quad 1_X : x \longmapsto 1 \quad .$$

If R is commutative, then so is R^X.

The ring R^X contains many subrings, which can be defined by various special properties of functions. For example, let $X = [0,1]$ be the closed interval in $\mathbb{R}$, and let $R = \mathbb{R}$. Then the ring $\mathbb{R}^{[0,1]}$ of all real-valued functions defined on $[0,1]$ contains the subring $\mathbb{R}^{[0,1]}_{bd}$ of all bounded functions, the subring $\mathbb{R}^{[0,1]}_{cont}$ of all continuous functions, the ring $\mathbb{R}^{[0,1]}_{diff}$ of all differentiable functions, and so on, since all of these properties are preserved under addition (and subtraction) and multiplication of functions.

To every element $a \in R$ there corresponds the <u>constant</u> function a_X defined by

$a_X(x) = a$ for all $x \in X$, and the imbedding which takes each $a \in R$ to $a_X \in R^X$ allows us to consider R as a subring of R^X.

4) Every additive abelian group $(A, +)$ has the structure of a ring with zero multiplication if we set $xy = 0$ for all $x, y \in A$.

Many of the properties of rings are simply reformulations of properties of groups or, more generally, of sets with an associative operation. For example, $a^m a^n = a^{m+n}$ and $(a^m)^n = a^{mn}$ for all non-negative integers m and n and for all $a \in R$ (compare with relation (2) in §1). Other properties, which are more specific properties of rings resulting from the ring axioms, are suggested by the properties of $\mathbb{Z}$. We note a few of them. First of all,

$$a \cdot 0 = 0 \cdot a = 0 \quad \text{for all} \quad a \in R \quad . \tag{1}$$

Namely, $a + 0 = a \Rightarrow a(a + 0) = aa \Rightarrow a^2 + a \cdot 0 = a^2 \Rightarrow a^2 + a \cdot 0 = a^2 + 0 \Rightarrow a \cdot 0 = 0$ and similarly $0 \cdot a = 0$).

Now suppose for a moment that $0 = 1$. We obtain: $a = a \cdot 1 = a \cdot 0 = 0$ for all $a \in R$, i.e., R only contains the element 0. Thus, $0 \neq 1$ in a non-trivial ring R. Next, we have

$$(-a) \cdot b = a \cdot (-b) = -(a \cdot b) \quad , \tag{2}$$

since, for example, (1) and the distributive laws imply

$$0 = a \cdot 0 = a(b - b) = ab + a(-b) \Longrightarrow a(-b) = -(ab) \quad . \tag{3}$$

Since $-(-a) = a$, it follows from (2) that $(-a)(-b) = ab$ (in particular, $(-1)(-1) = 1$) and $-a = (-1) \cdot a$.

The distributive laws imply the following general distributive law:

$$(a_1 + \cdots + a_n)(b_1 + \cdots + b_m) = \sum_{i=1}^{n} \sum_{j=1}^{m} a_i b_j \quad , \tag{4}$$

which one easily derives by induction, first (with $m = 1$) using induction on n, and then using induction on m. Now using (1), (2) and (3), we obtain

$$n(ab) = (na)b = a(nb)$$

for all $n \in \mathbb{Z}$ and all $a, b \in R$.

Finally, we note the binomial formula of Newton

$$(a+b)^n = \sum_{i=0}^{n} \binom{n}{i} a^i b^{n-i} , \tag{5}$$

which holds for all $a, b \in R$, but only if R is a commutative ring. To prove (5), one uses (4) and proceeds just as in §7 Ch. 1, where we considered the special case $R = \mathbb{Z}$.

2. Congruences. The ring of residue classes. According to Theorem 6 of §2, the only non-zero subgroups of the group $(\mathbb{Z}, +)$ are the groups $m\mathbb{Z}$, where m runs through the set $\mathbb{N}$ of natural numbers. But the set $m\mathbb{Z}$ is obviously closed under multiplication as well as addition, and all of the ring axioms are satisfied. Thus, we have the following assertion: every non-zero subring of $\mathbb{Z}$ has the form $m\mathbb{Z}$, where $m \in \mathbb{N}$.

We now try to use the subring $m\mathbb{Z} \subset \mathbb{Z}$ to construct a non-zero ring having only finitely many elements. To do this we introduce the

Definition. Two integers n and n' are said to be congruent mod m (in words: modulo m) if they have the same remainder when divided by m, i.e., if $n - n'$ is divisible by m. In that case we write $n \equiv n' \pmod{m}$ or simply $n \equiv n'\ (m)$.

In this way $\mathbb{Z}$ is partitioned into equivalence classes of numbers congruent to one another mod m; these classes are called residue classes mod m. Each residue class has the form

$$\{r\}_m = r + m\mathbb{Z} = \{r + mk \mid k \in \mathbb{Z}\} ,$$

so that we may write

$$\mathbb{Z} = \{0\}_m \cup \{1\}_m \cup \ldots \cup \{m-1\}_m . \tag{6}$$

Note that the residue classes are the cosets of the subgroup $m\mathbb{Z}$ in the additive group $\mathbb{Z}$, and the partition (6) is the one given in Theorem 4 of §2.

By definition, $n \equiv n'\ (m) \Leftrightarrow n - n'$ is divisible by m. But the notation $n \equiv n'\ (m)$ is more convenient than $m \mid (n - n')$, because it is possible to operate with congruences in the same way as with equations. Namely, if $k \equiv k'\ (m)$ and $\ell \equiv \ell'\ (m)$, then $k \pm \ell \equiv k' \pm \ell'\ (m)$ and $k\ell \equiv k'\ell'\ (m)$. In particular, $k \equiv k'\ (m) \Rightarrow ks \equiv k's\ (m)$ for any $s \in \mathbb{Z}$.

Thus, given two residue classes $\{k\}_m$ and $\{\ell\}_m$, we can define their sum, difference or product independently of the choice of coset representative. In other words, we have operations $\oplus$ (addition) and $\odot$ (multiplication) defined on the set $\mathbb{Z}_m = \mathbb{Z}/m\mathbb{Z}$ of residue classes mod m :

$$\begin{aligned} \{k\}_m \oplus \{\ell\}_m &= \{k + \ell\}_m , \\ \{k\}_m \odot \{\ell\}_m &= \{k\ell\}_m . \end{aligned} \tag{7}$$

Since these operations are defined using the usual operations on integers, it follows that $(\mathbb{Z}_m, \oplus, \odot)$, like $\mathbb{Z}$, is a commutative ring with unit $\{1\}_m = 1 + m\mathbb{Z}$. This ring is called the ring of residue classes mod m. If one is dealing with a fixed m, one often writes $\bar{k}$ instead of $\{k\}_m$, so that

$$\begin{aligned} \bar{k} \oplus \bar{\ell} &= \overline{k + \ell} , \\ \bar{k} \odot \bar{\ell} &= \overline{k\ell} . \end{aligned}$$

It is often especially convenient to forget about bars and curly brackets entirely, and simply choose the so-called reduced system of residues mod m, namely, the set $\{0, 1, \ldots, m-1\}$ and work with this fixed set of representatives. For example, if we use this convention, we have: $-k = m - k$, $2(m-1) = -2 = m - 2$.

Thus, there is such a thing as finite rings. Here are three examples of the rings $\mathbb{Z}_m$, with the addition and multiplication tables given separately:

$\mathbb{Z}_2$:

+	0	1
0	0	1
1	1	0

•	0	1
0	0	0
1	0	1

$\mathbb{Z}_3$:

+	0	1	2
0	0	1	2
1	1	2	0
2	2	0	1

•	0	1	2
0	0	0	0
1	0	1	2
2	0	2	1

$\mathbb{Z}_4$:

+	0	1	2	3
0	0	1	2	3
1	1	2	3	0
2	2	3	0	1
3	3	0	1	2

•	0	1	2	3
0	0	0	0	0
1	0	1	2	3
2	0	2	0	2
3	0	3	2	1

The residue rings $\mathbb{Z}_m$ have been of interest to number theorists for a long time, and in algebra they served as a point of departure for various important generalizations.

3. Ring homomorphisms and ideals. By (7), the map $f : n \mapsto \{n\}_m$ has the following properties: $f(k+\ell) = f(k) \oplus f(\ell)$, $f(k\ell) = f(k) \odot f(\ell)$. This suggests that we should call f a homomorphism of the rings $\mathbb{Z}$ and $\mathbb{Z}_m$ and make the following general definition.

Definition. Let $(R, +, \cdot)$ and $(R', \oplus, \odot)$ be two rings. A map $f : R \to R'$ is called a homomorphism if it preserves both operations, i.e., if

$$f(a+b) = f(a) \oplus f(b) ,$$

$$f(ab) = f(a) \odot f(b) .$$

Of course, in this case $f(0) = 0'$ and $f(na) = nf(a)$ for $n \in \mathbb{Z}$.

The kernel of a homomorphism f is the set

$$\operatorname{Ker} f = \{a \in R \mid f(a) = 0'\} .$$

It is clear that $\operatorname{Ker} f$ is a subring of R. But $\operatorname{Ker} f$ is much more than just a subring. Namely, for all $x \in R$ we have $x \cdot \operatorname{Ker} f \subset \operatorname{Ker} f$ (since for all $k \in \operatorname{Ker} f$ we have $f(xk) = f(x) \odot f(k) = f(x) \odot 0' = 0'$) and similarly $(\operatorname{Ker} f) \cdot x \subset \operatorname{Ker} f$. Thus, if we denote $L = \operatorname{Ker} f$, we have $RL \subset L$ and $LR \subset L$. A subring L having these two properties is called a (two-sided) ideal of the ring R. Thus, the kernel of a homomorphism is always an ideal.

As in the case of groups (see the glossary in subsection 3 of §3), a homomorphism $f : R \to R'$ is called a monomorphism if $\operatorname{Ker} f = 0$, an epimorphism if its image (the set of all $a' \in R'$ of the form $f(a)$ for $a \in R$) is all of R', and an isomorphism if it is both a monomorphism and an epimorphism. If R and R' are isomorphic, we write $R \cong R'$.

The map $f : n \mapsto \{n\}_m$ is clearly an epimorphism $\mathbb{Z} \to \mathbb{Z}_m$ with kernel $\operatorname{Ker} f = m\mathbb{Z}$. When we constructed $\mathbb{Z}_m$ we implicitly used the fact that $m\mathbb{Z}$ is an ideal of the ring $\mathbb{Z}$.

It so happens that every non-zero subring in the ring $\mathbb{Z}$ is an ideal, but that is unusual. For example, in the matrix ring $M_2(\mathbb{Z})$ the set

$$\left\{ \begin{pmatrix} \alpha & \beta \\ 0 & \delta \end{pmatrix} \middle| \alpha, \beta, \delta \in \mathbb{Z} \right\}$$

is a subring but not an ideal.

The example of $m\mathbb{Z}$ suggests a method for constructing ideals (not necessarily all ideals) in an arbitrary commutative ring R: if a is an arbitrary fixed element of R, then the set aR is always an ideal in R. This is because

$$ax + ay = a(x+y) \quad , \quad (ax)y = a(xy) \quad .$$

We say that aR is the principal ideal generated by the element $a \in R$.

Note that if we insist on defining rings to be rings with unit, then ideals are not generally subrings. In that case, we define an ideal to be a subgroup of the additive group of the ring which is taken to itself under left or right multiplication by any element of the ring. If we are using this definition of a ring, then we also insist that $f(1) = 1'$ in the definition of a homomorphism. Of course, even with our broader definition of a homomorphism, this condition $f(1) = 1'$ holds whenever f is an epimorphism.

Isomorphic rings are identical in their algebraic properties, and only properties which are preserved under isomorphisms are of real mathematical interest. It is for this reason, for example, that we allow ourselves to think of the ring $\mathbb{Z}_m$ either as a set of residue classes or as an arbitrary chosen set of representative numbers from the residue classes.

4. <u>The concept of quotient group and quotient ring.</u> Normal subgroups of a group and ideals in a ring have a common origin -- they are kernels of homomorphisms. This common element is reflected in the notion of forming a quotient, which we shall briefly discuss. We shall return to this theme in the second part of the book.

Let us start with groups. The equivalence relation $\sim$ on a group G which is obtained from the partition of G into cosets of a normal subgroup H has a remarkable property. Namely, if a and b are arbitrary elements of G, and if $a \sim c$ and $b \sim d$, then, by definition (see the proof of Theorem 4 in §3), we have $a^{-1}c = h_1 \in H$, $b^{-1}d = h_2 \in H$, and so

$$(ab)^{-1}cd = b^{-1}a^{-1}cd = b^{-1}(a^{-1}c)d = b^{-1}h_1b(b^{-1}d) = h_1'h_2 \in H$$

which means that $ab \sim cd$. Here we used the fact that H is normal in G: $b^{-1}h_1b = h_1' \in H$. Thus,

$$a \sim c, b \sim d \Longrightarrow ab \sim cd \quad .$$

This says that the multiplicative operation on G induces a multiplication of the quotient set $G/\sim$ (see subsection 3 of §6 Ch. 1), which we agreed to denote G/H.

It makes sense to speak of the product of any two subsets A and B of a group G, where we mean the set AB of all products ab with $a \in A$ and $b \in B$. Because G is associative, we have

$$(AB)C = \{(ab)c\} = \{a(bc)\} = A(BC) \quad .$$

A subset $H \subset G$ is a subgroup if and only if $H^2 = H$ and $H^{-1} = \{h^{-1} | h \in H\} \subset H$.

From this point of view, the coset aH is the product of the one-element set $\{a\}$ and the subgroup H. The product of two cosets aH and bH is the set $aH \cdot bH$, which does not necessarily have to be a coset of H. For example, the partition of S_3 into cosets of $H = \{e, (12)\}$, which we considered in subsection 4 of §3, shows that

$$H \cdot (13)H = (13)H \cup (23)H \quad .$$

However, if H is a normal subgroup of G, then the product of two cosets aH, bH will turn out to be a coset of H. Namely, since $gH = Hg$ for all $g \in G$, it follows that

$$aH \cdot bH = a(Hb)H = a(bH)H = abH^2 = abH ,$$

and the same reasoning as used above shows that the coset abH does not depend on the representatives a and b of the cosets aH and bH. The properties

$$aH \cdot bH = abH ,$$

$$H \cdot aH = aH \cdot H = aH ,$$

$$a^{-1}H \cdot aH = aH \cdot a^{-1}H = eH = H$$

show that we have the following

THEOREM 1. *If* H *is a normal subgroup of* G, *then the operation* $aH \cdot bH = abH$ *gives the quotient set* G/H *the structure of a group, which is called the quotient group of* G *by* H. *The coset* H *is the identity element in* G/H, *and* $a^{-1}H$ *is the inverse of* aH. □

If G is a finite group, then the order of the quotient group G/H is given by the formula

$$|G/H| = \frac{|G|}{|H|} = (G:H) ,$$

which should come as no surprise in view of Lagrange's theorem (subsection 4 of §3).

If G is an abelian group whose binary operation is written additively, then the operation induced on the quotient group G/H is written

$$(a + H) + (b + H) = (a + b) + H .$$

In this case, G/H is often called the group G modulo H. If we apply this to the pair $G = \mathbb{Z}$ and $H = m\mathbb{Z}$, we also use the expression "the group $\mathbb{Z}$ modulo m".

We now proceed to the idea of a *quotient ring* R/L, where R is a ring and L

is an ideal. We base the construction on the additive group of the ring R. Thus, the elements of R/L are the cosets $a + L$, which we call the residue classes of R modulo the ideal L. They are added by the usual rule:

$$(a + L) \oplus (b + L) = (a + b) + L \quad , \qquad \ominus (a + L) = -a + L \quad . \tag{8}$$

We take the product of these residue classes to be

$$(a + L) \odot (b + L) = ab + L \quad . \tag{9}$$

We have to be careful that this multiplication is correctly defined, i.e., that it does not depend on the choice of coset representatives. Suppose that $a' = a + x$, $b' = b + y$, where $x, y \in L$. Then

$$a'b' = ab + ay+xb+xy = ab + z \quad ,$$

where $z = ay + xb + xy$ is an element of L, because L is a two-sided ideal. Hence, $a'b'$ is in the same coset as ab, and this means that the product (9) is correctly defined. For brevity we write $\bar{a} = a + L$, so that

$$\bar{a} \oplus \bar{b} = \overline{a + b} \quad , \quad \bar{a} \odot \bar{b} = \overline{ab} \quad .$$

In particular, $\bar{0} = L$ and $\bar{1} = 1 + L$ (if R has a 1). We should also check that the set $\bar{R} = R/L = \{\bar{a} \mid a \in R\}$ with the operations $\oplus$ and $\odot$ satisfy all the ring axioms, but this is fairly obvious, because the operations on $\bar{R}$ are defined in terms of the operations on elements of R. For example, the distributive law is verified as follows:

$$(\bar{a} \oplus \bar{b}) \odot \bar{c} = (\overline{a + b}) \odot \bar{c} = \overline{(a + b)c} = \overline{ac} + \overline{bc} = \overline{ac} \oplus \overline{bc} = \bar{a} \odot \bar{c} \oplus \bar{b} \odot \bar{c} \quad .$$

All of this shows that the map

$$\pi : a \longmapsto \bar{a}$$

is an epimorphism of rings $R \to R'$ with kernel $\operatorname{Ker} \pi = L$. Starting with the special case of the quotient ring $\mathbb{Z}_m = \mathbb{Z}/m\mathbb{Z}$ and the epimorphism $\mathbb{Z} \to \mathbb{Z}_m$, we have found an analogous situation occurring in arbitrary rings.

It is worth mentioning, although it goes beyond our immediate goal (which is to explain the construction of $\mathbb{Z}_m$ from a general algebraic point of view), that the quotient rings of R by its ideals essentially exhaust all possible images of R under homomorphisms. Namely, if $f : R \to R'$ is a homomorphism and $f(R)$ is the image of R under f, then, if we consider $f(R) \subset R'$ in place of R', we obtain an epimorphism. In order to simplify notation, suppose that f was an epimorphism from the very start, i.e., that $f(R) = R'$. According to the general principle in subsection 3 of §6 Ch. 1, f determines an equivalence relation O_f on R; in our present situation, O_f is the partition of R into cosets $a + \operatorname{Ker} f = C_a$. The map f leads to a bijective correspondence f' between the elements a' of R' and the classes C_a, namely, $f' : C_a \mapsto a'$ if $a' = f(a)$. Under this correspondence f' we have

$$f'(C_a + C_b) = f'(C_{a+b}) = f(a+b) = f(a) + f(b) = f'(C_a) + f'(C_b) ,$$

$$f'(C_a \cdot C_b) = f'(C_{ab}) = f(ab) = f(a) \cdot f(b) = f'(C_a) \cdot f'(C_b) ,$$

so that the bijective map f' is an isomorphism (for simplicity, we have denoted addition and multiplication in the rings R, $R/\operatorname{Ker} f$, and R' in the same way).

We have proved the following

THEOREM 2 (the fundamental theorem on ring homomorphisms). *Any ideal* L *of a ring* R *determines a ring structure on the quotient set* R/L *(using the formulas* (8) *and* (9)), *and* R/L *is a homomorphic image of* R *under a map having kernel* L. *Conversely, any homomorphic image* $R' = f(R)$ *of the ring* R *is isomorphic to the quotient ring* $R/\operatorname{Ker} f$. □

Remark. The right side of (9) is not, in general, the same thing as the product of the residue classes $a + L$ and $b + L$ in the set-theoretic sense. For example, if $R = \mathbb{Z}$ and $L = 8\mathbb{Z}$, then $24 \in 16 + 8\mathbb{Z}$ cannot be written in the form $(4 + 8s)(4 + 8t)$, since the latter is always divisible by 16.

5. Types of rings. Fields. In our familiar rings $\mathbb{Z}$, $\mathbb{Q}$ and $\mathbb{R}$, whenever $ab = 0$ we must have either $a = 0$ or $b = 0$. But the ring of square matrices M_n does not have this property. For example, using the notation E_{ij} (see the proof of Theorem 3 of §3 Ch. 2), we have $E_{ij} E_{k\ell} = 0$ if $j \neq k$, although, of course, $E_{ij} \neq 0$ and $E_{k\ell} \neq 0$. One might think that this is because of the unpleasant phenomenon of non-commutativity in M_n, but that has nothing to do with it. As we saw in subsection 2, in the ring $\mathbb{Z}_4$ we have $2 \odot 2 = 0$ (despite the well-known platitude "twice two is four"!).

Here are two more examples.

Example 1. Pairs of numbers (a,b) (where we may take a, b in $\mathbb{Z}$, $\mathbb{Q}$ or $\mathbb{R}$) with addition and multiplication defined by the formulas

$$(a_1, b_1) + (a_2, b_2) = (a_1 + a_2, b_1 + b_2) ,$$

$$(a_1, b_1) \cdot (a_2, b_2) = (a_1 a_2, b_1 b_2) ,$$

obviously make up a commutative ring with unit $(1,1)$. Here we encounter the same phenomenon: $(1,0) \cdot (0,1) = (0,0) = 0$.

Example 2. In the ring $\mathbb{R}^{\mathbb{R}}$ of real-valued functions (see Example 3 in subsection 1), the functions $f : x \mapsto |x| + x$ (which is 0 when $x \leq 0$) and $g : x \mapsto |x| - x$ (which is 0 when $x \geq 0$) have the property that their product is the zero function, even though $f \neq 0$ and $g \neq 0$.

Definition. If $ab = 0$ for $a \neq 0$ and $b \neq 0$ in the ring R, then a is called a left zero divisor and b is called a right zero divisor (if R is commutative, then there is no distinction, and we speak simply of zero divisors). The zero element in R can be considered the trivial zero divisor. If there are no zero divisors (except 0), then R is called a ring without zero divisors. If R is a commutative ring with unit $1 \neq 0$ not having zero divisors, then R is called an integral domain.

THEOREM 3. A non-trivial commutative ring R with unit is an integral domain if and only if the law of cancellation holds:

$$ab = ac \quad , \quad a \neq 0 \Longrightarrow b = c$$

for all $a, b, c \in R$.

Proof. If R has the cancellation law, then whenever $ab = 0 = a \cdot 0$, we then have either $a = 0$ or else $a \neq 0$, in which case $b = 0$. Conversely, if R is an integral domain, then $ab = ac$, $a \neq 0 \Rightarrow a(b-c) = 0 \Rightarrow b - c = 0 \Rightarrow b = c$. □

If R is a ring with unit, it is natural to consider the set of invertible elements: an element a is called invertible (or a unit, not to be confused with the alternative use of this word for the element 1) if there exists an element a^{-1} such that $aa^{-1} = 1 = a^{-1}a$. More precisely, one might want to speak of right invertible or left invertible elements (if $ab = 1$ or $ba = 1$ can be solved for b, respectively). But if R is commutative or is without zero divisors, then right (or left) invertibility implies invertibility. Namely, suppose, for example, that $ab = 1$ in a ring without zero divisors. Then $aba = a$, so that $a(ba - 1) = 0$. Since $a \neq 0$, we must have $ba - 1 = 0$, i.e., $ba = 1$.

As an example, we already know that in the ring $M_n(\mathbb{R})$ or $M_n(\mathbb{Q})$ the invertible elements are precisely the matrices with non-zero determinant.

An invertible element a cannot be a zero divisor: $ab = 0 \Rightarrow a^{-1}(ab) = 0 \Rightarrow$ $\Rightarrow (a^{-1}a)b = 0 \Rightarrow 1 \cdot b = 0 \Rightarrow b = 0$ (similarly $ba = 0 \Rightarrow b = 0$).

THEOREM 4. The set U(R) of all invertible elements in a ring R with unit is a group under multiplication.

Proof. Since the set U(R) contains the identity element, and multiplication is associative in R, it remains for us to see that U(R) is closed under multiplication, i.e., we must check that the product ab of two invertible elements is invertible. But this is obvious, since $b^{-1}a^{-1}$ can be seen to be the inverse of ab (for example, $(ab)(b^{-1}a^{-1}) = a(bb^{-1})a^{-1} = a \cdot 1 \cdot a^{-1} = aa^{-1} = 1$). □

As an example, it is easy to see that $U(\mathbb{Z}) = \{\pm 1\}$ is a cyclic group of order 2.

If we replace the ring axiom (R 2) by the much stronger axiom (R 2') : the set R^* of non-zero elements in R forms a group under multiplication, then we obtain a very interesting class of rings, called <u>division rings</u> or <u>skew fields.</u> Thus, a division ring must be a ring without zero divisors in which every non-zero element is invertible. A commutative division ring -- in which multiplication has essentially all of the properties of addition (i.e., the non-zero elements form an abelian group) -- is called a <u>field.</u> Thus, to repeat:

<u>Definition.</u> <u>A field K is a commutative ring with unit $1 \neq 0$ in which every non-zero element is invertible. The group $K^* = U(K)$ is called the multiplicative group of the field.</u>

A field is sort of a hybrid of two abelian groups -- the additive and the multiplicative -- which are connected by the distributive law. A product of the form ab^{-1} is usually written in the form of a <u>fraction</u> a/b. Thus, the fraction a/b, which only makes sense when $b \neq 0$, is the unique solution of the equation $bx = a$. Operations with fractions are subject to the rules:

$$\begin{aligned}
&\frac{a}{b} = \frac{c}{d} \iff ad = bc, && b, d \neq 0, \\
&\frac{a}{b} + \frac{c}{d} = \frac{ad + bc}{bd}, && b, d \neq 0, \\
&-\frac{a}{b} = \frac{-a}{b} = \frac{a}{-b}, && b \neq 0, \\
&\frac{a}{b} \cdot \frac{c}{d} = \frac{ac}{bd}, && b, d \neq 0, \\
&\left(\frac{a}{b}\right)^{-1} = \frac{b}{a}, && a, b \neq 0.
\end{aligned} \tag{10}$$

These are the usual rules of "grammar school arithmetic", but, rather than taking them on faith and memorizing them, the reader should derive them from the axioms of a field; but this is not hard. For example, here is a derivation of the second rule in (10). Let $x = a/b$ and $y = c/d$ be the solutions of the equations $bx = a$ and $dy = c$. It follows from these equations that $dbx = da$, $bdy = bc \Rightarrow bd(x+y) = da + bc \Rightarrow t = x + y = (da+bc)/bd$ is the unique solution of the equation $bdt = da + bc$.

A <u>subfield</u> of a field is a subring which itself is a field. For example, the field $\mathbb{Q}$ of rational numbers is a subfield of the field $\mathbb{R}$ of real numbers. If F is a subfield of a field K , then we also say that K is an <u>extension field</u> of F . It follows from the definition of a subfield that the zero and identity elements in K will also be contained in F , and will be the zero and identity elements of F . If in the field K we take the intersection F_1 of all subfields which contain F and also contain a given element $a \in K$ not in F , then F_1 will be the minimal field containing the set $\{F, a\}$ (the reasoning is the same as for groups, see subsection 2 of §2). In this case we say that F_1 is the extension of F obtained by <u>adjoining</u> a to F , and we write $F_1 = F(a)$. Similarly, we speak of the subfield $F_1 = F(a_1, \cdots, a_n)$ of K obtained by adjoining to F the n elements $a_1, \cdots, a_n \in K$.

For example, if we adjoin $\sqrt{2}$ to the subfield $\mathbb{Q} \subset \mathbb{R}$, the resulting field $\mathbb{Q}(\sqrt{2})$ is easily seen to be the set of numbers of the form $a + b\sqrt{2}$, $a, b \in \mathbb{Q}$, since $(\sqrt{2})^2 = 2$ and $1/(a+b\sqrt{2}) = (a/(a^2-2b^2)) - (b/(a^2-2b^2))\sqrt{2}$ if $a + b\sqrt{2} \neq 0$. We have a similar situation for $\mathbb{Q}(\sqrt{3})$, $\mathbb{Q}(\sqrt{5})$, and so on.

Two fields are said to be <u>isomorphic</u> if they are isomorphic considered as rings. By definition, if f is an isomorphism, then $f(0) = 0$ and $f(1) = 1$ (where we are using the same notation for the zero and identity elements in both rings). There is no such thing as a non-trivial homomorphism of fields with non-zero kernel, since $\operatorname{Ker} f \neq 0 \Rightarrow f(a) = 0$, $a \neq 0 \Rightarrow f(1) = f(aa^{-1}) = f(a)\, f(a^{-1}) = 0 \cdot f(a^{-1}) = 0 \Rightarrow f(b) = f(1 \cdot b) = f(1)\, f(b) = 0 \cdot f(b) = 0$, $\forall b \Rightarrow \operatorname{Ker} f = K$.

The <u>automorphisms</u> of a field K , i.e., the isomorphisms from K to itself, are connected with the deepest properties of fields, and are a powerful instrument for studying these properties. This subject is called <u>Galois theory</u>.

The notion of a field extension is a reflection of the tendency of mankind, from time immemorial, to try to increase the supply of number systems to work with. This lengthy historical process can be roughly summarized by the diagram:

$$\{\text{one}\} \rightsquigarrow \{\text{one plus one is two}\} \rightsquigarrow \mathbb{N} \rightsquigarrow \{\mathbb{N}, 0\} \rightsquigarrow \mathbb{Z} \rightsquigarrow \mathbb{Q} \rightsquigarrow \mathbb{Q}(\sqrt{2}) \rightsquigarrow \mathbb{R}.$$

It has continued right up to modern times, and has led to the study of a very extensive network of fields, which range far from the usual number systems of everyday use. Not all of the steps in constructing field extensions are purely algebraic. For example, we go from the rational numbers to the real numbers using the concepts of continuity and completeness (limits of Cauchy sequences), and this is normally done in an axiomatic calculus course. It turns out that there is a completely analogous construction of the so-called p-adic number fields, which we shall not deal with here. The resulting theory of p-adic analysis is the worthy offspring of three areas of mathematics -- number theory, algebra, and analysis.

6. <u>The characteristic of a field.</u> In subsection 2 we constructed the finite ring $\mathbb{Z}_m$ of residue classes

$$\overline{0}, \overline{1}, \overline{2}, \ldots, \overline{m-1}$$

with addition and multiplication operations $\overline{k} + \overline{\ell} = \overline{k+\ell}$, $\overline{k} \cdot \overline{\ell} = \overline{k\ell}$ (we are no longer using the special symbols $\oplus$ and $\odot$). If $m = st$, $s > 1$, $t > 1$, then $\overline{s} \cdot \overline{t} = \overline{m} = \overline{0}$, i.e., $\overline{s}$ and $\overline{t}$ are zero divisors in $\mathbb{Z}_m$.

Now suppose that $m = p$ is a prime number. We claim that $\mathbb{Z}_p$ is a field (having p elements). In the cases $p = 2, 3$, this is immediately clear from the multiplication tables in subsection 2. In the general case, it is sufficient to show that, for every $\overline{s} \in \mathbb{Z}_p^*$ there exists an inverse element $\overline{s'}$ (where we only allow numbers s and s' which are not divisible by p). To see this, we consider the elements

$$\overline{s}, \overline{2s}, \ldots, \overline{(p-1)s} \quad . \tag{11}$$

They are all non-zero, since $s \not\equiv 0 \pmod p \Rightarrow ks \not\equiv 0 \pmod p$ for $k = 1, 2, \cdots, p-1$; this is because p is a prime number. For the same reason, all of the elements in (11) are distinct: $\overline{ks} = \overline{\ell s}$ for $k < \ell$ would imply that $\overline{(\ell - k)s} = \overline{0}$, which is false. Thus, the sequence of elements (11) is the same, except for the order in which they are written, as the sequence

$$\bar{1}, \bar{2}, \cdots, \overline{p-1} \quad .$$

In particular, there exists s', $1 \leq s' \leq p-1$, for which $\overline{s's} = 1$. But this means that $\overline{s'} \cdot \bar{s} = \bar{1}$, i.e., $\overline{s'}$ is an inverse for $\bar{s}$. We have proved

THEOREM 5. <u>The ring of residue classes</u> $\mathbb{Z}_m$ <u>is a field if and only if</u> m <u>is a prime number.</u> □

COROLLARY (Fermat's Little Theorem). <u>For any integer</u> a <u>not divisible by the prime</u> p,

$$a^{p-1} \equiv 1 \pmod{p} \quad .$$

<u>Proof.</u> The multiplicative group $\mathbb{Z}_p^*$ has order $p-1$. By Lagrange's theorem (§3), $p-1$ is divisible by the order of any element in $\mathbb{Z}_p^*$, in particular, by the order of $\bar{a}$. Thus, $\bar{1} = (\bar{a})^{p-1} = \overline{a^{p-1}}$, i.e., $\overline{a^{p-1}-1} = \bar{0}$. □

An alternative proof of this corollary can be obtained by replacing s by a in (11) and multiplying together all of the elements in the sequence; that product must be congruent to the product of $1, 2, \ldots, p-1$.

Even though the fields $\mathbb{Z}_2, \mathbb{Z}_3, \mathbb{Z}_5, \ldots$ seem so different from the familiar field $\mathbb{Q}$, they have one basic property in common with $\mathbb{Q}$: they have no smaller subfields.

Let K be a field. As we already noticed, the intersection of any family of subfields of K is itself a subfield of K.

<u>Definition.</u> A field which does not contain any proper subfield is called a <u>prime field.</u>

THEOREM 6. <u>Every field</u> K <u>contains precisely one prime field</u> K_0. <u>This prime field is isomorphic either to</u> $\mathbb{Q}$ <u>or to</u> $\mathbb{Z}_p$ <u>for some prime</u> p.

<u>Proof.</u> If K contained two fields K' and K'', their intersection would be a

field (non-empty, since both K' and K'' contain 0 and 1) which is distinct from K' and K''. But this is impossible if K' and K'' are prime fields. Thus, the prime field K_0, which is the intersection of all fields contained in K, is unique.

Since K_0 contains 1, it must contain all multiplies $n \cdot 1 = 1 + \cdots + 1$. It follows from the general properties of addition and multiplication in rings (see the end of subsection 1) that

$$s \cdot 1 + t \cdot 1 = (s+t) \cdot 1 , \quad (s \cdot 1)(t \cdot 1) = (st) \cdot 1 ; \quad s, t \in \mathbb{Z} \quad . \tag{12}$$

Hence, the map f of the ring $\mathbb{Z}$ to K which is defined by $f(n) = n \cdot 1$ is a homomorphism. Its kernel is an ideal in $\mathbb{Z}$: $\operatorname{Ker} f = m\mathbb{Z}$. If $m = 0$, then f is an isomorphism, in which case the fractions $(s \cdot 1)/(t \cdot 1)$, $s, t \in \mathbb{Z}$, which make sense because K is a field, form a field K_0 which is isomorphic to $\mathbb{Q}$. This field is clearly the prime field in K.

If, on the other hand, $m > 0$, then the map f^* defined by setting

$$f^* : \bar{k} = \{k\}_m \longmapsto f(k) \quad ,$$

is obviously an imbedding $\mathbb{Z}_m \to K$. By Theorem 5, this is only possible if $m = p$ is a prime number. Thus, $f^*(\mathbb{Z}_p)$ is the prime subfield of K. □

<u>Definition.</u> We say that a field K has <u>characteristic zero</u> if its prime field K_0 is isomorphic to $\mathbb{Q}$. We say that K has <u>characteristic</u> p if $K_0 \cong \mathbb{Z}_p$. We write $\operatorname{char} K = 0$ or $\operatorname{char} K = p > 0$, respectively.

Instead of $\mathbb{Z}_p$ the notation $\mathbb{F}_p$ or $GF(p)$ (GF for Galois field) is often used to denote the "abstract" field of p elements. It turns out that there exists a finite field $GF(q)$ with $q = p^n$ elements for any prime p and positive integer n. We shall return to this interesting question in Chapter 9, but for now we merely give the example of the field of four elements $\{0, 1, \alpha, \beta\}$:

GF(4):

+	0	1	α	β
0	0	1	α	β
1	1	0	β	α
α	α	β	0	1
β	β	α	1	0

$\cdot$	0	1	α	β
0	0	0	0	0
1	0	1	α	β
α	0	α	β	1
β	0	β	1	α

(At this point we are not concerned with what α and β are.) The reader should check, for example, that the distributive law holds.

A field K of characteristic zero has the property that 1 has infinite order in the additive group of K. If K has characteristic p, then any non-zero element has order exactly p in the additive group:

$$px = x + \ldots + x = p(1\cdot x) = 1\cdot x + \ldots + 1\cdot x = (1 + \ldots + 1)x = (p\cdot 1)x = 0 \quad .$$

7. <u>A remark on linear systems.</u> The time has come to cast a thoughtful glance at the theory of systems of linear equations and determinants developed in the earlier chapters. In those chapters the coefficients in the equations and the corresponding matrix entries were numbers (rational or real), but the exact nature of rational and real numbers was never used. There is nothing to stop us now from allowing the coefficients and matrix entries to be elements of any given field K. Of course, then the results must be formulated in terms of the field K: the components of a solution to the linear system and the values of the det function will lie in K. Gauss' method for solving systems of linear equations, the theory of determinants, Cramer's rule, and so on, all remain valid for an arbitrary field K.

<u>Example 1.</u> Suppose we are given a homogeneous system of linear equations $AX = 0$ with square matrix

$$A = (a_{ij}) = \begin{Vmatrix} 1 & 2 & 3 & 4 \\ -10 & 13 & 14 & 15 \\ 12 & -9 & 14 & 15 \\ 12 & 13 & -8 & 15 \end{Vmatrix}$$

and column of unknowns $X = [x_1, x_2, x_3, x_4]$. A direct computation shows that $\det A = 2^3 \cdot 11^3$. Hence, if we consider a_{ij}, x_k to be in K, where K is any field of characteristic zero or of characteristic $p \neq 2, 11$ (i.e., the integers $1, 2, 3, 4, -10, \cdots, 15$ are replaced by their residue classes in characteristic p), then the system is determined, and so only has the trivial solution $X = 0$.

If $\operatorname{char} K = 2$ (for example, if $K = \mathbb{Z}_2$), then we conclude from the congruence

$$\begin{Vmatrix} 1 & 2 & 3 & 4 \\ -10 & 13 & 14 & 15 \\ 12 & -9 & 14 & 15 \\ 12 & 13 & -8 & 15 \end{Vmatrix} \equiv \begin{Vmatrix} 1 & 0 & 1 & 0 \\ 0 & 1 & 0 & 1 \\ 0 & 1 & 0 & 1 \\ 0 & 1 & 0 & 1 \end{Vmatrix} \pmod 2$$

that the rank of the system is equal to two, and so the system has two independent solutions $X_1 = [1, 0, 1, 0]$ and $X_2 = [0, 1, 0, 1]$. To avoid confusion we should write $X_1 = [\bar{1}, \bar{0}, \bar{1}, \bar{0}]$, $X_2 = [\bar{0}, \bar{1}, \bar{0}, \bar{1}]$, but we now have had enough experience to adopt the simpler notation, ignoring the bars denoting residue class.

If $\operatorname{char} K = 11$, then it follows from the congruence

$$\begin{Vmatrix} 1 & 2 & 3 & 4 \\ -10 & 13 & 14 & 15 \\ 12 & -9 & 14 & 15 \\ 12 & 13 & -8 & 15 \end{Vmatrix} \equiv \begin{Vmatrix} 1 & 2 & 3 & 4 \\ 1 & 2 & 3 & 4 \\ 1 & 2 & 3 & 4 \\ 1 & 2 & 3 & 4 \end{Vmatrix} \pmod{11}$$

that the system has three independent solutions

$$X_1 = [9, 1, 0, 0], \quad X_2 = [8, 0, 1, 0], \quad X_3 = [7, 0, 0, 1] \quad .$$

As we see, the answer to our questions about the system depends on our field K, although the procedure is the same in all cases. Thus, one of the advantages of generalizing beyond $\mathbb{R}$ and $\mathbb{Q}$ to an arbitrary field is the avoidance of duplication of the same arguments. But there are even more important advantages.

When we spoke about the general linear group, we have so far meant the group of all non-singular matrices with coefficients in $\mathbb{Q}$ or $\mathbb{R}$. We now let $M_n(K)$ denote the ring of $n \times n$ matrices with entries in an arbitrary field K, and we define the general linear group $GL(n, K)$ to be the subset of all non-singular matrices $A \in M_n(K)$ (i.e., matrices with $\det A \neq 0$). If we vary K, for example, if we take $K = \mathbb{F}_p$, many important groups arise in a natural way (see Ch. 7).

Fields of the type $\mathbb{R}$, $\mathbb{Q}$, $\mathbb{Q}(\sqrt{2})$, etc. are usually called number fields. The field $\mathbb{F}_p$ is an example of a field which is not a number field. We do not think of $\mathbb{F}_p$ as a "number field", even though its elements can be identified with the elements of the set $\{0, 1, \ldots, p-1\}$.

In §2 Ch. 1 we posed the problem (Problem 3) of using finite fields in coding theory. Here is a small example related to this theme.

Example 2. In order to transmit the word PEACE, in principle it is sufficient to use four elementary message units $P = (0,0)$, $E = (1,0)$, $A = (0,1)$, $C = (1,1)$, which we interpret as row-vectors in the vector space $\mathbb{F}_2^2$ over the field $\mathbb{F}_2 \cong \mathbb{Z}_2 = \{0,1\}$ containing two elements. But suppose some static arises during the transmission, some interference which occasionally switches a 0 and 1. This could cause the receiver to pick up, for example, the message APACE. According to a fundamental theorem of Shannon, the static can always be overcome at the cost of increasing the length of the elementary message units (i.e., lengthening the time of transmission). Suppose, for example, that we know that there can be no more than one distortion in every elementary message unit of length five. Then take a subset S_0 of so-called code vectors in the vector space $S = \mathbb{F}_2^5$, say: $S_0 = \{P = (0,0,1,1,0),\ E = (1,0,0,1,1),\ A = (0,1,1,0,1),\ C = (1,1,0,0,0)\}$. From the table

	P	E	A	C
Code Vectors	00110	10011	01101	11000
Possible vectors obtained from the code vectors as a result of the distortions	00010	00011	00101	01000
	00100	10001	01001	10000
	00111	10010	01100	11100
	01110	10111	01111	11001
	10110	11011	11101	11010

it is clear that the sets of distorted vectors in the different columns do not overlap, and, hence, accurate decoding is possible, i.e., the true message can be re-established.

We have obtained a code S_0 which can correct one error. If we use $\mathbb{F}_2^n$ for n sufficiently large, we can construct a similar code which can be used to transmit the Latin alphabet, and hence any text, accurately. If we want to avoid unnecessarily long and cumbersome decoding, we should choose S_0 carefully. There are many techniques for doing this, including some purely algebraic approaches based on using the finite field $\mathbb{F}_q$.

EXERCISES

1. Developing the idea in Example 2) of §1, show that the set $\mathbb{P}(\Omega)$ with the operations

$$A + B = (A \cup B) \setminus (A \cap B), \quad AB = A \cap B; \quad A, B \in \Omega \quad ,$$

is a ring with unit in which all of the elements of the additive group have order two.

2. Prove that any ring in which $x^2 = x$ for every element x must be commutative. Is this true if $x^3 = x$ for every x?

3. Are the fields $\mathbb{Q}(\sqrt{2})$ and $\mathbb{Q}(\sqrt{5})$ isomorphic?

4. Do the non-invertible elements of the following rings form an ideal: 1) $\mathbb{Z}_{16}$; 2) $\mathbb{Z}_{24}$?

5. Show that the image of a commutative ring under an epimorphism is a commutative ring.

6. Show that, if R is a ring with unit and L is an ideal, then the quotient ring R/L also has a unit element.

7. Show that any finite integral domain is a field.

8. Let p be a prime number, and let R be a commutative ring with unit such that $px = 0$ for all $x \in R$. Show that then

$$(x + y)^{p^m} = x^{p^m} + y^{p^m} \quad , \quad m = 1, 2, \ldots \quad .$$

9. Prove that a ring consisting of five elements is either a ring with zero multiplication or is isomorphic to $\mathbb{Z}_5$.

10. The set $T = \{ \begin{Vmatrix} a & c \\ 0 & b \end{Vmatrix} \mid a, b \in \mathbb{Z} \}$ of upper triangular matrices is a subring in $M_2(\mathbb{Z})$. Prove this, and give a description of all of the ideals of the ring T.

11. A non-zero element x in a ring R is called <u>nilpotent</u> if $x^n = 0$ for some $n \in \mathbb{N}$. Show that:

(i) if R is a ring with unit and x is nilpotent, then the element $1 - x$ is invertible;

(ii) the ring $\mathbb{Z}_m = \mathbb{Z}/m\mathbb{Z}$ contains nilpotent elements if and only if m is divisible by the square of a natural number greater than 1.

12. Prove that, in a commutative ring with unit R of infinite cardinality, there cannot be a finite number $n \geq 1$ of non-invertible non-zero elements.

13. Let R be any associative ring with unit 1, and let $a, b \in R$. Show that

$$(1 - ab)c = 1 = c(1 - ab) \implies (1 - ba)d = 1 = d(1 - ba) \quad ,$$

where $d = 1 + bca$, i.e., if $1 - ab$ is invertible in R, then so is $1 - ba$. What does the element $1 + adb$ equal?

14. Show that the matrices $\begin{Vmatrix} a & b \\ -b & a \end{Vmatrix}$ with $a, b \in \mathbb{Z}_3$ form a field of 9 elements, and that the multiplicative group of this field is cyclic of order 8.

15. Can the code S_0 in Example 2 at the end of the section correct two errors?

Chapter 5. Complex Numbers and Polynomials

In this chapter we study some very concrete algebraic systems, which are somewhat familiar from elementary mathematics, but which deserve more detailed examination. The point of view developed in the preceding chapter allows us to take a fresh look at what was the traditional domain of algebra in earlier times. Meanwhile, such concepts as ring extensions and unique factorization in integral domains become more tangible and understandable by looking at the example of polynomial rings.

§1. The field of complex numbers

The history of mathematics saw a stubborn and protracted struggle between the supporters and detractors of the "imaginary" numbers that arise from the algebraic equation

$$x^2 + 1 = 0 \quad . \tag{1}$$

One could simply agree to write the solutions of (1) in the form $\pm\sqrt{-1}$ as a formal notational convention. But this is not a satisfactory answer to the problem; we would like to attach some meaning to the notation. We shall discuss this on several levels. First we give some heuristic considerations.

1. An auxiliary construction. We would like to extend the field of real numbers $\mathbb{R}$ in such a way that equation (1) has a solution in the new field. As a model (example) of such a field extension, let us take the set K of all square matrices

$$\begin{Vmatrix} a & b \\ -b & a \end{Vmatrix} \in M_2(\mathbb{R}) \quad . \tag{2}$$

We claim that K is a field (compare with Exercise 14 in §4 Ch. 4). First of all, K contains the zero 0 and the identity E of the ring $M_2(\mathbb{R})$. Next, the relations

$$\begin{aligned} \begin{Vmatrix} a & b \\ -b & a \end{Vmatrix} + \begin{Vmatrix} c & d \\ -d & c \end{Vmatrix} &= \begin{Vmatrix} a+c & b+d \\ -(b+d) & a+c \end{Vmatrix} , \\ - \begin{Vmatrix} a & b \\ -b & a \end{Vmatrix} &= \begin{Vmatrix} -a & -b \\ -(-b) & -a \end{Vmatrix} , \\ \begin{Vmatrix} a & b \\ -b & a \end{Vmatrix} \begin{Vmatrix} c & d \\ -d & c \end{Vmatrix} &= \begin{Vmatrix} ac-bd & ad+bc \\ -(ad+bc) & ac-bd \end{Vmatrix} \end{aligned} \tag{3}$$

imply that K is closed under addition and multiplication. The associativity of these operations follows from the associativity of addition and multiplication in the ring $M_2(\mathbb{R})$. The same goes for the distributive laws. Thus, K is a subring of $M_2(\mathbb{R})$. K is commutative by (3). It remains to show that any matrix (2) with a and b not both zero has an inverse in K. (Note that the determinant of the matrix is $a^2 + b^2 \neq 0$). Either by using the formula for the entries in the inverse matrix (see Theorem 1 of §3 Ch. 3) or by solving the linear system which comes from the condition

$$\begin{Vmatrix} a & b \\ -b & a \end{Vmatrix} \begin{Vmatrix} x & y \\ -y & x \end{Vmatrix} = \begin{Vmatrix} 1 & 0 \\ 0 & 1 \end{Vmatrix} ,$$

we find that

$$\begin{Vmatrix} a & b \\ -b & a \end{Vmatrix}^{-1} = \begin{Vmatrix} c & d \\ -d & c \end{Vmatrix} , \quad \text{where } c = \frac{a}{a^2+b^2} , \quad d = \frac{-b}{a^2+b^2} \quad . \tag{4}$$

Thus, K is a field.

Using the rule (5) in §3 Ch. 2 for multiplying matrices by scalars, we can write any element of K in the form

$$\begin{Vmatrix} a & b \\ -b & a \end{Vmatrix} = aE + bJ, \quad \text{where} \quad a, b \in \mathbb{R}, J = \begin{Vmatrix} 0 & 1 \\ -1 & 0 \end{Vmatrix} \quad . \tag{5}$$

The field K contains the subfield $\{aE \mid a \in \mathbb{R}\} \cong \mathbb{R}$, and the relation

$$J^2 + E = 0$$

shows that "up to isomorphism" the element $J \in K$ is a solution of equation (1). So there is no need for any mysticism about J being an "imaginary quantity".

But it is not the field K which is called the field of complex numbers, but rather a certain field isomorphic to it whose elements are identified with the points of a plane. It is natural to want to think of K in terms of a geometrical realization; after all, the real number field $\mathbb{R}$ is inseparable in our minds from the number line, i.e., with a line having a point denoted 0 and a fixed scale of distance to the point 1.

2. <u>The complex plane.</u> Thus, we would like to construct a field $\mathbb{C}$ whose elements are the points of the plane $\mathbb{R}^2$, having addition and multiplication operations which satisfy the field axioms, and solving our problem of being able to solve equation (1). On the cartesian plane, take the usual rectangular coordinate system with x- and y-axes. We write (a,b) for the point with x-coordinate a and y-coordinate b. We define the sum and product of two points (a,b) and (c,d) according to the rules

$$\begin{aligned} (a,b) + (c,d) &= (a+c\,,\; b+d)\,, \\ (a,b)\,(c,d) &= (ac - bd\,,\; ad + bc) \end{aligned} \tag{6}$$

(we use the same symbols $+$ and $\cdot$ as in the field $\mathbb{R}$, but this should not cause confusion). We could verify by a straightforward but tedious series of calculations that these operations satisfy all of the field axioms, i.e., the operations (6) make the set of pairs (points in the plane) into a field. But happily, there is no need for this verification, because if we let each point of the plane $\mathbb{C}$ correspond to an element of our earlier field K as follows

$$(a,b) \longmapsto \begin{Vmatrix} a & b \\ -b & a \end{Vmatrix},$$

by comparing (3) and (6) we see that the set $\mathbb{C}$ is a field isomorphic to K. $\mathbb{C}$ is called the field of complex numbers. Because of the geometrical realization of $\mathbb{C}$ it is also often called the complex plane.

The points on the x-axis, i.e., the set of points of the form $(a,0)$ has the same properties as the real number line, so we set $(a,0) = a$. Then the zero $(0,0)$ and the identity $(1,0)$ remain the same as on the real number line. The point $(0,1)$ on the y-axis is traditionally denoted i (the "imaginary" unit); it is a root of equation (1), since $i^2 = (0,1)(0,1) = (-1,0) = -1$. An arbitrary complex number $z = (x,y)$ can now be written in the customary form

$$z = x + iy, \quad x,y \in \mathbb{R}, \tag{7}$$

which is very close to the form (5) for the elements of the field K. Note that $\mathbb{Q} \subset \mathbb{R} \subset \mathbb{C}$. Hence, $\mathbb{C}$ is a field of characteristic zero (see subsection 6 of §4 Ch. 4).

3. Geometrical interpretation of operations with complex numbers. The x-axis in the complex plane is usually called the real axis, and the y-axis is called the imaginary axis. A number iy on the imaginary axis is called a purely imaginary number, although the word "imaginary" has lost its original meaning, which implied that such numbers are less legitimate than real numbers. If z is a complex number written in the form (7), x is called its real part and y is called its imaginary part. The map that associates to every complex number $z = x + iy$ the complex number $\bar{z} = x - iy$ (called the complex conjugate of z) is called complex conjugation. Geometrically, it amounts to reflection of the complex plane about the real axis (Fig. 15). A very remarkable and important fact is

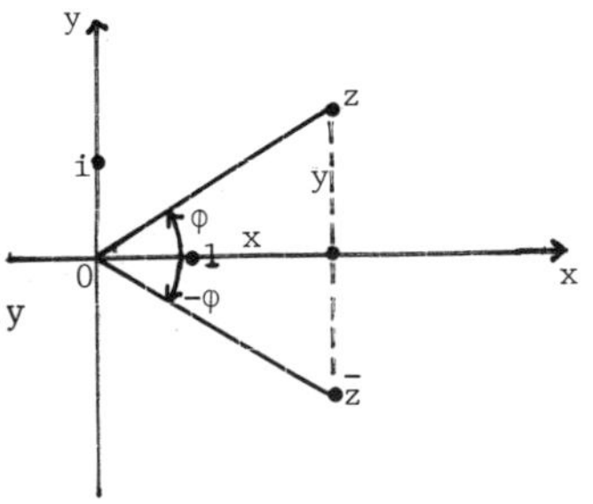

Fig. 15

THEOREM 1. _The map_ $z \mapsto \bar{z}$ _is an automorphism of the field_ $\mathbb{C}$ _having order 2 which keeps all real numbers fixed. The sum and the product of a number with its conjugate are real numbers._

Proof. It follows immediately from the definition of complex conjugation that $\bar{x} = x$ for $x \in \mathbb{R}$. In particular, $\bar{0} = 0$ and $\bar{1} = 1$. The claim that complex conjugation has order 2 (i.e., performing it twice gives the identity map) is also obvious: $\bar{\bar{z}} = z$. We must now verify that

$$\overline{z_1 + z_2} = \bar{z}_1 + \bar{z}_2, \quad \overline{z_1 z_2} = \bar{z}_1 \cdot \bar{z}_2, \tag{8}$$

but this follows immediately from the formulas (6), which should be rewritten in the form

$$\begin{aligned} (x_1 + iy_1) + (x_2 + iy_2) &= (x_1 + x_2) + i(y_1 + y_2) \\ (x_1 + iy_2) \cdot (x_2 + iy_2) &= (x_1 x_2 - y_1 y_2) + i(x_1 y_2 + x_2 y_1) \end{aligned} \tag{9}$$

Finally, as a special case of the formulas in (9), we find the sum and product of $z = x + iy$ and $\bar{z} = x - iy$ to be: $z + \bar{z} = 2x$, $z\bar{z} = x^2 + y^2$. □

Remark. Among all of the automorphisms of the field $\mathbb{C}$ (of which there are many), complex conjugation is the only continuous automorphism other than the identity (i.e., it takes nearby points in the plane $\mathbb{C}$ to nearby points). We shall not give a precise definition or proof of this assertion.

The _modulus_ (or _absolute value_) of a complex number $z = x + iy$ is the non-negative real number $|z| = \sqrt{z\bar{z}} = \sqrt{x^2 + y^2}$. The position of a point z on the plane is completely determined by giving its distance $r = |z|$ from the origin $(0,0)$ and the angle φ measured counterclockwise from the positive real axis to the line from $(0,0)$ to z (see Fig. 15); these are the well-known polar coordinates. The angle φ is called the _argument_ of z and is denoted $\arg z$. Although $\arg z$ can take any positive or negative value, for given r the angles which differ by integer multiples of 2π correspond to one and the same point. The argument of 0 is not defined, but its modulus is $|0| = 0$. Note that the relationship "greater than" or "less than" makes no sense for complex numbers, i.e., they cannot be related to one another by an inequality sign: unlike in the case of real numbers, whose arguments can only take the value 0 (for positive numbers) or π (for negative numbers), complex numbers are not ordered.

The polar coordinates r and φ determine x and y by the well-known formulas

$$x = r\cos\varphi, \quad y = r\sin\varphi, \quad z = r(\cos\varphi + i\sin\varphi) \quad . \tag{10}$$

This is the so-called trigonometric form for z.

The operation of adding two complex numbers z and z' is expressed easily in cartesian coordinates, namely, by the parallelogram rule, which tells us how to add directed line segments (vectors) from the origin (see Fig. 16). We obtain an important inequality if we look at this picture and compare the sides of the triangle with vertices $0, z,$ and $z + z'$ (and note that the lengths of the sides are given by absolute values of the corresponding complex numbers):

$$|z + z'| \leq |z| + |z'| \quad . \tag{11}$$

Note that the inequality (11), which can also be written in more general form

$$|z| - |z'| \leq |z \pm z'| \leq |z| + |z'| \quad ,$$

is completely analogous to a similar inequality for real numbers.

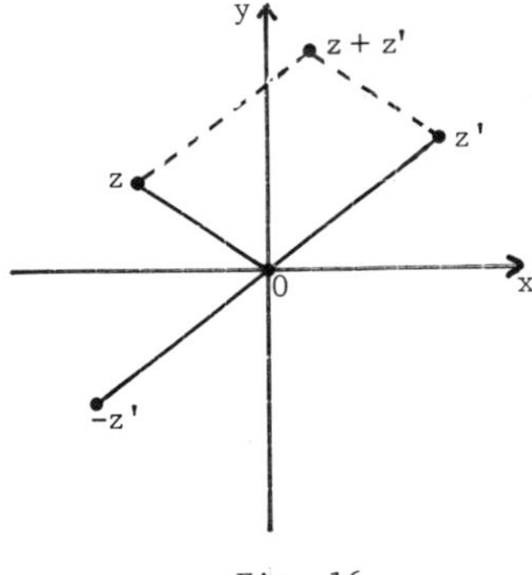

Fig. 16

The operation of multiplying two complex numbers can be conveniently expressed in polar coordinates.

THEOREM 2. The modulus of the product of complex numbers is equal to the product of the moduli, and the argument of the product is equal to the sum of the arguments:

$$|zz'| = |z| \cdot |z'|, \quad \arg zz' = \arg z + \arg z' \quad . \tag{12}$$

Similarly,

$$|z/z'| = |z| / |z'| , \quad \arg z/z' = \arg z - \arg z' \quad .$$

Proof. Suppose that the triginometric form (10) for z and z' are

$$z = r(\cos\varphi + i\sin\varphi) , \quad z' = r'(\cos\varphi' + i\sin\varphi') \quad .$$

Using (9) or simply multiplying directly, we obtain

$$zz' = rr' [(\cos\varphi \cos\varphi' - \sin\varphi \sin\varphi') + i(\cos\varphi \sin\varphi' + \sin\varphi \cos\varphi')] \quad ,$$

and, using the well-known formulas from trigonometry, we can re-write this as the trigonometric form for zz' :

$$zz' = |z| \cdot |z'| \cdot [\cos(\varphi + \varphi') + i \sin(\varphi + \varphi')] \quad .$$

Next, if $z'' = z/z'$, then $z = z'z''$, and, using (12) for the product $z'z''$, we obtain the desired formula for the fraction z/z' . □

In particular, $z^{-1} = |z|^{-1} [\cos(-\varphi) + i \sin(-\varphi)]$. To find z^{-1} in the complex plane (see Fig. 17), first find the point z' which is obtained from z by "inversion" with respect to the unit circle, and then reflect z' about the real axis. ("Inversion" means that the distance from z' to 0 is the reciprocal of the distance from z to 0.)

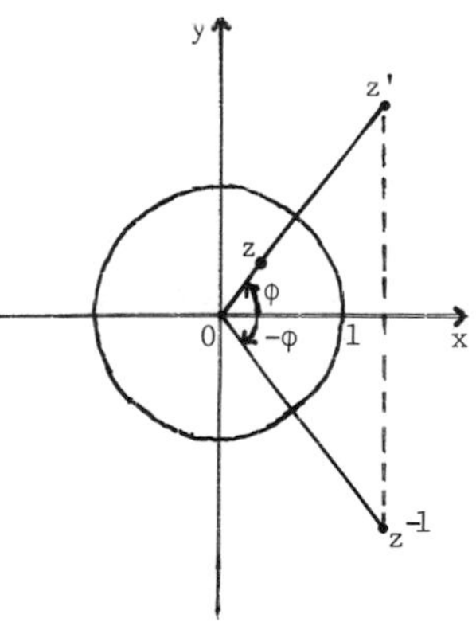

Fig. 17

Actually, the assertions about the modulus of a product and the modulus of a sum can also be easily derived from Theorem 1. In the first place, we have

$$|zz'|^2 = zz'\overline{zz'} = zz'\bar{z}\bar{z}' = z\bar{z} \cdot z'\bar{z}' = |z|^2 |z'|^2 \quad ,$$

and hence $|zz'| = |z| \cdot |z'|$. Next, since $|z| = \sqrt{x^2 + y^2} \geq \sqrt{x^2} = |x|$, we obtain

$$|1 + z|^2 = (1 + z)(1 + \bar{z}) = 1 + (z + \bar{z}) + z\bar{z} = 1 + 2x + |z|^2 \leq 1 + 2|z| + |z|^2 = (1 + |z|)^2,$$

and hence $|1+z| \leq 1+|z|$. Now if $z \neq 0$ and $z' \neq 0$, it follows that

$$|z+z'| = |z(1+z^{-1}z')| = |z| \cdot |1+z^{-1}z'| \leq$$

$$\leq |z| \cdot (1+|z^{-1}z'|) = |z|(1+|z|^{-1}|z'|) = |z|+|z'| .$$

The above results lead us to the general conclusion: the usual form (7) for complex numbers is convenient for expressing additive properties, and the trigonometric form (10) is best suited for multiplicative properties. If we violate this principle, we end up with very complicated formulas which obscure what's going on.

4. Raising to powers and extracting roots. The formula (12) for multiplying complex numbers given in trigonometric form implies the following de Moivre's formula:

$$[r(\cos\varphi + i\sin\varphi)]^n = r^n(\cos n\varphi + i\sin n\varphi) , \tag{13}$$

which holds for all $n \in \mathbb{Z}$ (another way of writing it: $|z^n| = |z|^n$, $\arg z^n = n \cdot \arg z$). If we use the special case of (13) when $r = 1$, together with the binomial formula (1) of §7 Ch. 1 and the relations

$$i^2 = -1 , \quad i^3 = -i , \quad i^4 = 1 , \quad i^{4k+\ell} = i^{\ell}$$

we obtain formulas for the cosines and sines of multiples of an angle:

$$\begin{aligned} \cos n\varphi &= \sum_{k\geq 0} (-1)^k \binom{n}{2k} \cos^{n-2k}\varphi \cdot \sin^{2k}\varphi , \\ \sin n\varphi &= \sum_{k\geq 0} (-1)^k \binom{n}{2k+1} \cos^{n-1-2k}\varphi \cdot \sin^{2k+1}\varphi . \end{aligned} \tag{14}$$

Remark. Let $e^{\alpha} = \lim_{n\to\infty} (1+\frac{\alpha}{n})^n$. In a first course on functions of a complex variable it is proved (by power series expansion) that e^{α} for complex α can be evaluated using Euler's formula

$$e^{i\varphi} = \cos\varphi + i\sin\varphi , \tag{15}$$

which, in turn, can be used to derive all of the results above. One need merely note that

$$e^{i\varphi} e^{i\varphi'} = e^{i(\varphi+\varphi')}, \quad (e^{i\varphi})^n = e^{in\varphi} \quad .$$

Thus, the trigonometric form for a complex number z amounts to writing: $z = |z| \cdot e^{i\varphi}$.

We would next like to learn how to extract roots of complex numbers, and the first question that arises is whether or not it is always possible to extract an arbitrary root of any complex number. It turns out that it is always possible, and de Moivre's formula essentially gives a complete answer to the problem. Suppose we are given a complex number $z = r(\cos\varphi + i\sin\varphi)$, and we would like to find a number $z' = r'(\cos\varphi' + i\sin\varphi')$ such that $(z')^n = z$. Using de Moivre's formula to express $(z')^n$, and then equating the moduli and arguments in both sides of the equation $(z')^n = z$, we find that $(r')^n = r$ and $n\varphi' = \varphi + 2\pi k$ (where we have to add the term $2\pi k$ because φ is only determined up to a multiple of 2π). Thus,

$$r' = \sqrt[n]{r}, \quad \varphi' = \frac{\varphi + 2\pi k}{n}$$

(by $\sqrt[n]{r}$ we mean the positive root of the positive real number r). Thus, the root $\sqrt[n]{z}$ exists but is not uniquely determined. For each $k = 0, 1, \cdots, n-1$ we obtain a different value for z'; but these n values are all possible roots, since if we write $k = nq + r$, $0 \leq r \leq n-1$, then we have

$$\varphi' = \frac{\varphi + 2\pi r}{n} + 2\pi q \quad .$$

We have proved

THEOREM 3. <u>It is always possible to extract an n-th root of a complex number $z = |z|(\cos\varphi + i\sin\varphi)$. There are n n-th roots of z, which are located at the vertices of a regular n-gon inscribed in the circle centered at the origin with radius</u> $\sqrt[n]{|z|}$:

$$\sqrt[n]{z} = \sqrt[n]{|z|}\left(\cos\frac{\varphi + 2\pi k}{n} + i \sin\frac{\varphi + 2\pi k}{n}\right) , \tag{16}$$

$k = 0, 1, \cdots, n-1$. □

COROLLARY. The n-th roots of 1 are given by the formula

$$\sqrt[n]{1} = \varepsilon_k = \cos\frac{2\pi k}{n} + i \sin\frac{2\pi k}{n}, \qquad k = 0, 1, \cdots, n-1 . \tag{17}$$

They are located at the vertices of a regular n-gon inscribed in the unit circle. □

It is immediately clear from (16) and (17) that there will be either zero, one, or two real roots $\sqrt[n]{z}$ and either one or two real roots $\sqrt[n]{1}$.

An n-th root of 1 is called primitive if it is not a root of 1 of lower degree. For example, the following are primitive n-th roots of 1:

$$\varepsilon = \varepsilon_1 = \cos\frac{2\pi}{n} + i \sin\frac{2\pi}{n} \quad \text{and} \quad \varepsilon_{n-1} .$$

Any n-th root of 1 is a power of the primitive one: $\varepsilon_k = \varepsilon_1^k$, as we can again see from de Moivre's formula. Moreover, $\varepsilon_k \varepsilon_\ell = \varepsilon_{k+\ell}$, if we consider $k+\ell$ modulo n. In particular, $\varepsilon_k^{-1} = \varepsilon_{n-k}$ and $\varepsilon_0 = 1$. Having studied group theory, we now conclude that the n-th roots of 1 form a cyclic group $\langle \varepsilon \rangle$ of order n.

This gives us another model for the cyclic group of order n. By Theorem 6 of §3 Ch. 4, the subgroups of this cyclic group are in one-to-one correspondence with the positive divisors d of n. For each $d|n$ there is exactly one subgroup of $\langle \varepsilon \rangle$ having order d, and this is the subgroup $\langle \varepsilon^{n/d} \rangle$. A root ε_m is primitive if and only if $\langle \varepsilon_m \rangle = \langle \varepsilon \rangle$, i.e., if and only if $\text{Card}\langle \varepsilon^m \rangle = n$, and this holds if and only if m and n are relatively prime. For example, if $n = 12$, then the primitive roots are $\varepsilon, \varepsilon^5, \varepsilon^7$, and ε^{11}. If $n = p$ is a prime, then all n-th roots of 1 except for 1 are primitive. From the algebraic point of view (disregarding their location on the complex plane), all of the primitive roots of a given degree n are equivalent to one another.

Returning now to the problem of extracting the n-th roots of an arbitrary non-zero complex number z, we note that, if z' is any fixed root (for example, $z' = \sqrt[n]{|z|}\left(\cos\frac{\varphi}{n} + i\sin\frac{\varphi}{n}\right)$), then all of the other roots have the form $z'\varepsilon_k$, $k = 0, 1, \cdots, n-1$. This agrees with formula (16).

5. Uniqueness theorem. We are not yet ready to appreciate fully the advantage $\mathbb{C}$ has over $\mathbb{R}$, but already the fact that $\mathbb{C}$ contains all roots of 1 justifies our taking a special interest in the complex numbers. A natural question which arises is whether there are other fields having the same properties as $\mathbb{C}$. It turns out that we have the following uniqueness theorem for complex numbers.

THEOREM 4. *Let F be a field isomorphic to $\mathbb{R}$ (for example, $F = \mathbb{R}$), and let K be an extension obtained from F by adjoining a root j of the equation $x^2 + 1 = 0$. Then K is isomorphic to $\mathbb{C}$.*

Proof. By the definition in subsection 5 of §4 Ch. 4, $K = F(j)$ is the minimal subfield of some field L containing F and j. Since we were given the field L, we can consider elements of the form $a + jb$, $a, b \in F$, where the sum and product are taken in the sense of the operations in L. Distinct pairs $a, b \in F$ correspond to distinct elements $a + jb$, since otherwise we would have an element $a' + jb' = 0$ with $a' \neq 0$ or $b' \neq 0$. If $b' = 0$, then clearly $a' = 0$. But if $b' \neq 0$, then we would obtain $j = -a'/b' \in F$, which is absurd, since $F \cong \mathbb{R}$ and $\mathbb{R}$ does not contain a solution to the equation $x^2 + 1 = 0$; thus, $j \notin F$. Next, we use the equality $j^2 = -1$ and the operations in L to compute

$$\begin{aligned}(a_1 + jb_1) + (a_2 + jb_2) &= (a_1 + a_2) + j(b_1 + b_2) \quad, \\ (a_1 + jb_1) \cdot (a_2 + jb_2) &= (a_1a_2 - b_1b_2) + j(a_1b_2 + a_2b_1) \quad.\end{aligned} \tag{18}$$

In addition,

$$(a + jb)^{-1} = \frac{a}{a^2 + b^2} + j\,\frac{-b}{a^2 + b^2} \qquad \text{if} \qquad a^2 + b^2 \neq 0 \quad .$$

This shows that the set $\{a + jb \mid a, b \in F\}$ in K is closed under all of the operations in L and so forms a subfield. Since K is the minimal such field, we must have

$$K = \{a + jb \mid a, b \in F\} \quad .$$

In addition, the formulas (18) coincide with (9).

If $f : F \to \mathbb{R}$ is our given isomorphism, then the map

$$f^* : a + jb \longmapsto (f(a), f(b)) \quad ,$$

which takes each element of K to the point of the complex plane $\mathbb{C}$ with coordinates $(f(a), f(b))$ is an isomorphism between K and $\mathbb{C}$, because of the above formulas. □

In Subsection 1 we constructed such a field K in $M_2(\mathbb{R})$. But there are many constructions for a field obtained by solving the equation $x^2 + 1 = 0$ -- we shall give one more such construction in the next section. By Theorem 4, all such fields are isomorphic. Note that in the statement of the theorem we really should have written $x^2 + \tilde{1} = \tilde{0}$, where $\tilde{1}$ and $\tilde{0}$ are the identity and zero elements in F. For example, in the case of $K \subset M_2(\mathbb{R})$ we have $J^2 + \tilde{1} = \tilde{0}$, where $\tilde{1} = E$ and $\tilde{0}$ is the zero matrix.

There are many other subfields of $\mathbb{C}$ besides $\mathbb{Q}$ and $\mathbb{R}$. Especially interesting examples are the extensions of $\mathbb{Q}$ obtained by adjoining some element of $\mathbb{C}$ not in $\mathbb{Q}$.

Example 1 (quadratic fields). Let d be a non-zero integer (positive or negative) such that $\sqrt{d} \notin \mathbb{Q}$. The field $\mathbb{Q}(\sqrt{d}) \subset \mathbb{C}$ is called a real quadratic field if $d > 0$ and an imaginary quadratic field if $d < 0$. The field $\mathbb{Q}(\sqrt{2})$ was discussed briefly in §4 of Chapter 4. If we use the same argument as in the proof of Theorem 4, with j replaced by $\sqrt{d}$ and the relation $j^2 = -1$ replaced by $(\sqrt{d})^2 = d$, we find that

$$\mathbb{Q}(\sqrt{d}) = \{a + b\sqrt{d} \mid a, b \in \mathbb{Q}\} \quad .$$

In particular, in place of (18) we have

$$
\begin{aligned}
(a_1 + b_1\sqrt{d}) + (a_2 + b_2\sqrt{d}) &= (a_1 + a_2) + (b_1 + b_2)\sqrt{d} \quad , \\
(a_1 + b_1\sqrt{d})(a_2 + b_2\sqrt{d}) &= (a_1 a_2 + b_1 b_2 d) + (a_1 b_2 + a_2 b_1)\sqrt{d} \quad .
\end{aligned} \tag{19}
$$

In addition,

$$
(a + b\sqrt{d})^{-1} = \frac{a}{a^2 - db^2} + \frac{-b}{a^2 - db^2}\sqrt{d}
$$

if $a + b\sqrt{d} \neq 0$ (i.e., if a and b are not both zero).

Using (19), we easily verify that the map

$$
f : a + b\sqrt{d} \longmapsto a - b\sqrt{d}
$$

is an automorphism of the field $\mathbb{Q}(\sqrt{d})$ (the analog of complex conjugation). By the norm of a number $\alpha = a + b\sqrt{d}$ we mean the real number

$$
N(\alpha) = a^2 - db^2 = \alpha f(\alpha) \quad .
$$

It is obvious that $N(\alpha) = 0 \Leftrightarrow \alpha = 0$. Furthermore, since f is an automorphism, we have

$$
N(\alpha\beta) = \alpha\beta f(\alpha\beta) = \alpha\beta f(\alpha) f(\beta) = \alpha f(\alpha) \cdot \beta f(\beta) = N(\alpha) \cdot N(\beta) \quad .
$$

In particular, $N(\alpha) \cdot N(\alpha^{-1}) = N(\alpha\alpha^{-1}) = N(1) = 1$. Hence, the norm has basically the same properties as the square of the modulus in $\mathbb{C}$.

Example 2 (Constructive number fields). We suppose the points (0,0) and (1,0) to be given to us on the cartesian plane $\mathbb{R}^2$. All subsequent constructions must be realized using only a compass and straight-edge. Once we construct two points P and Q, we naturally consider the segment PQ also to have been constructed. If we have constructed a point P and a segment r, then we can also construct the circle with center P and radius r. Points of intersection of any two lines or circles which have been constructed will similarly be considered to be constructible.

A complex number $a + ib \in \mathbb{C}$ is called constructive if it is possible to construct the point $P = (a,b)$ in finitely many steps (as described in the last paragraph), starting from (0,0) and (1,0). It is not hard to see that $a + ib$ is constructive if and only if both $|a|$ and $|b|$ are constructive. We let

CS denote the set of points in the plane which can be constructed in this way by straight-edge and compass, i.e., the set of all constructive complex numbers.

Theorem 5. <u>The set</u> CS <u>is a subfield of</u> $\mathbb{C}$.

<u>Proof</u>. It follows immediately from the definition of constructivity that CS is closed under addition and taking negatives (going from $z = a + ib$ to $-z = -a - ib$).

Now suppose we have constructed line segments of length α and β. The diagrams to the right show how, by constructing similar triangles (the dotted lines), one can construct the product $\gamma = \alpha\beta$ and the ratio $\delta = \alpha/\beta$. But, in the final analysis, construction of $zz' = (a+ib)(a'+ib') = (aa'-bb') + i(ab'+a'b)$ and $1/z = a/(a^2+b^2) - ib/(a^2+b^2)$ reduces to construction of segments of the type γ and δ. Hence, the product and the reciprocal of constructive numbers are constructive. We have thereby proved that the set CS is closed under all of the field operations in $\mathbb{C}$. □

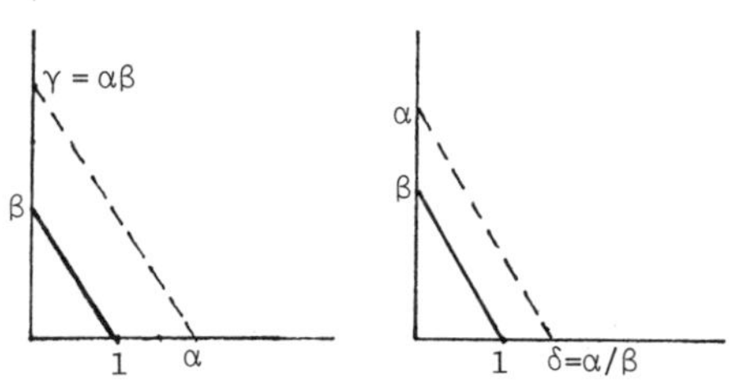

It is customary to call any subfield $K \subset CS$ a <u>constructive number field</u>. Obviously, $\mathbb{Q} \subseteq K$, and K is a field of characteristic zero.

EXERCISES

1. Find all complex numbers z of modulus 1 for which $z^2 + (1+i)z$ is purely imaginary. Draw the locus of points z with this property on the complex plane.

2. What can be said about the field $\mathbb{R}(\delta)$ which is obtained from $\mathbb{R}$ by adjoining a complex number δ which satisfies the equation $\delta^4 = -1$?

3. Let $A, B \in M_n(\mathbb{R})$. Using Theorem 1, prove that $\overline{\det(A+iB)} = \det(A-iB)$

(the bar denotes complex conjugation).

4. Let $A, B \in M_n(\mathbb{R})$, and let

$$C = \begin{Vmatrix} A & -B \\ B & A \end{Vmatrix} \in M_{2n}(\mathbb{R}) .$$

Applying elementary transformations of type I and II over the field of complex numbers $\mathbb{C}$ to the real matrix C, show that $\det C = |\det(A + iB)|^2$.

5. (Pólya and Szegö). Using Exercises 3 and 4, explain the following "strange" fact. The square homogeneous linear system

$$\begin{array}{c} d_{11} z_1 + \cdots + d_{1n} z_n = 0 \\ \cdots\cdots\cdots\cdots\cdots \\ d_{n1} z_1 + \cdots + d_{nn} z_n = 0 \end{array} \qquad (*)$$

with complex coefficients $d_{k\ell} = a_{k\ell} + i b_{k\ell}$ and unknowns $z_\ell = x_\ell + i y_\ell$ has a non-trivial solution $(z_1, \cdots, z_n)$ if and only if $\det(d_{k\ell}) = a + ib = 0$ (see the general remarks on this in Subsection 7 of §4 Ch. 4). This condition leads to the two equations $a = 0$ and $b = 0$ involving $2n^2$ real numbers $a_{k\ell}, b_{k\ell}$. On the other hand, the system $(*)$ can be written as a system of $2n$ linear homogeneous equations with $2n$ real unknowns x_ℓ, y_ℓ. In that case the condition for there to exist a non-trivial solution is the vanishing of a single real $2n \times 2n$ determinant, i.e., a single equation involving the $a_{k\ell}, b_{k\ell}$. Explain why these two conditions for a non-trivial solution are compatible.

6. Keeping in mind that an automorphism of the quadratic field $\mathbb{Q}(\sqrt{d})$ must leave the rational numbers fixed, find the automorphisms of this field.

7. Find the sum of all n-th roots of 1 $(n > 1)$. Find the sum of the primitive 12-th roots of 1 and the sum of the primitive 15-th roots of 1.

8. Find and give a geometrical picture of the kernel and image of the map $(\mathbb{R}, +) \to (\mathbb{C}^*, \cdot)$ (where $\mathbb{C}^* = \mathbb{C} \setminus \{0\}$) given by $t \mapsto e^{2\pi i t}$ (see formula (15)).

9. Show that $\xi = \frac{2+i}{2-i}$ is not a root of 1, even though $|\xi| = 1$.

§2. Rings of polynomials

Along with linear systems, which we studied in Chapters 2 and 3, the other traditional branch of algebra is the study of polynomials. A wide variety of mathematical problems can be stated or solved in the language of polynomials. There are many reasons for this, and one of them is the property of "universality" of polynomial rings, which we shall discuss briefly in Subsections 1 and 2.

Let R be a commutative (and, as usual, associative) ring with unit 1, and let A be a subring containing 1. If $t \in R$, then the smallest subring of R containing A and t is obviously the set of elements of the form

$$a(t) = a_0 + a_1 t + a_2 t^2 + \cdots + a_n t^n \quad , \qquad (*)$$

where $a_s \in A$, $n \in \mathbb{Z}$, $n \geq 0$. We denote the ring $A[t]$ and call it the ring obtained by adjoining t to A. The expression $(*)$ is called a polynomial in t with coefficients in A. It is clear from looking at a few simple examples how one computes the sum and product of polynomials:

$$a(t) + b(t) = (a_0 + a_1 t + a_2 t^2) + (b_0 + b_1 t + b_2 t^2) = (a_0 + b_0) + (a_1 + b_1)t + (a_2 + b_2)t^2 \quad ,$$

$$a(t) \cdot b(t) = a_0 b_0 + (a_0 b_1 + a_1 b_0)t + (a_0 b_2 + a_1 b_1 + a_2 b_0)t^2 + (a_1 b_2 + a_2 b_1)t^3 + a_2 b_2 t^4 \quad .$$

Clearly, we have been able to write the terms in this way because of the commutativity of all of the elements a_i, b_j, t^k.

But now we should recall that t was just an element of R chosen at random, and so it is possible for two expressions $(*)$ which look different really to be equal. For example, if $A = \mathbb{Q}$, $t = \sqrt{2}$, then $t^2 = 2$ and $t^3 = 2t$ are relations which in no way follow from the formal rules for working with polynomial expressions. In order to arrive at the customary notion of a polynomial, we must get rid of such extraneous special properties t might have if it can be any element of R. We do this by taking t to be an arbitrary symbol, not necessarily denoting an element of R. The choice of symbol is not important, but what is important is the rules for computing $a(t) + b(t)$ and $a(t)\,b(t)$. Keeping in mind these preliminary remarks, we now give the precise definition of the algebraic object which is called a polynomial and the set of such objects -- the polynomial ring.

1. Polynomials in one variable. Let A be any commutative ring with unit. We construct a new ring B whose elements are infinite ordered sequences

$$f = (f_0, f_1, f_2, \ldots), \qquad f_i \in A , \tag{1}$$

such that all of the f_i except for finitely many are equal to zero. We define the operations of addition and multiplication on B by setting

$$f + g = (f_0, f_1, f_2, \ldots) + (g_0, g_1, g_2, \ldots) = (f_0 + g_0, f_1 + g_1, f_2 + g_2, \ldots) ,$$

$$f \cdot g = h = (h_0, h_1, h_2, \ldots) ,$$

where

$$h_k = \sum_{i+j=k} f_i g_j , \qquad k = 0, 1, 2, \ldots .$$

It is clear that, after adding or multiplying, we again obtain a sequence of the form (1) with only finitely many non-zero terms, i.e., an element of B. The verification of the ring axioms (see §4 Ch. 4) is completely obvious, except perhaps for the associative law. Namely, since addition of two elements of B reduces to addition of a finite number of elements of A, it follows that $(B, +)$ is a commutative group with zero element $(0, 0, 0, \ldots)$, and any element $f = (f_0, f_1, f_2, \ldots)$ has additive inverse $-f = (-f_0, -f_1, -f_2, \ldots)$. Next, multiplication is commutative because the expression for h_k in terms of the f_i and g_j is symmetric, i.e., gives the same thing when f and g are interchanged. The expression for h_k also shows that the distributive law $(f + g)h = fh + gh$ holds in B. We now show associativity of multiplication. Let

$$f = (f_0, f_1, f_2, \ldots), \quad g = (g_0, g_1, g_2, \ldots), \quad h = (h_0, h_1, h_2, \ldots)$$

be any three elements of B. Then $fg = d = (d_0, d_1, d_2, \ldots)$, where

$d_\ell = \sum_{i+j=\ell} f_i g_j$, $\ell = 0, 1, 2, \ldots$, and $(fg)h = dh = e = (e_0, e_1, e_2, \ldots)$, where

$e_s = \sum_{\ell+k=s} d_\ell h_k = \sum_{\ell+k=s} \left(\sum_{i+j=\ell} f_i g_j\right) h_k = \sum_{i+j+k=s} f_i g_j h_k$. If we compute

$f(gh)$, we get the same result. Thus, B is a commutative associative ring with unit $(1, 0, 0, 0, \ldots)$.

Sequences of the form $(a, 0, 0, 0, \ldots)$ are added and multiplied just like the elements of A. This allows us to identify such sequences with the corresponding elements of A, i.e., to set $a = (a, 0, 0, 0, \ldots)$ for all $a \in A$. This makes A into a subring of B. Next, let X denote the sequence $(0, 1, 0, 0, \ldots)$. We call X a variable (or unknown) over A. Using the multiplication operation that was introduced in B, we find that

$$\begin{aligned} X &= (0, 1, 0, 0, \ldots) \;, \\ X^2 &= (0, 0, 1, 0, \ldots) \;, \\ &\cdots\cdots\cdots\cdots \\ X^n &= (0, 0, \ldots, 0, 1, 0, \ldots) \;. \end{aligned} \tag{2}$$

In addition, using (2) and the inclusion $A \subset B$, we have

$$(0, 0, \ldots, 0, a, 0, \ldots) = aX^n = X^n a \;.$$

Thus, if f_n is the last non-zero term in the sequence $f = (f_0, f_1, \ldots, f_n, 0, 0, \ldots)$, then in our new notation

$$\begin{aligned} f &= (f_0, \ldots, f_{n-1}, 0, 0, \ldots) + f_n X^n = (f_0, \ldots, f_{n-2}, 0, 0, \ldots) + \\ &+ f_{n-1} X^{n-1} + f_n X^n = f_0 + f_1 X + f_2 X^2 + \cdots + f_n X^n \;. \end{aligned} \tag{3}$$

This representation for f is unique, since the expression on the right is zero if and only if $f_0 = f_1 = \cdots = f_n = 0$, i.e., if and only if $f = 0$.

Definition. The ring B introduced above is denoted $A[X]$ and is called the polynomial ring over A in one variable X. Its elements are called polynomials.

Of course, the choice of a particular letter X as the name of the variable is not a major terminological advance, but is traditional and avoids misunderstanding. We intentionally chose capital X so as to be able to distinguish between the polynomial $f = X$ and the function theoretic small x used to denote a variable which runs through some set of values. (This distinction is only temporary, and we shall not adhere to it rigorously later on.) A function f can be written either in the form (3) or in decreasing powers of X:

$$f(X) = a_0 X^n + a_1 X^{n-1} + \cdots + a_n \quad .$$

We shall use both ways of writing f, depending on which seems more convenient in the given situation. The elements f_i or a_i are called the coefficients of the polynomial f. We call f the zero polynomial if all of its coefficients are zero. The coefficient f_0 of X to the zero power is called the constant term. If $f_n \neq 0$, then f_n is called the leading coefficient, and n is called the degree of the polynomial; we write $n = \deg f$. We adopt the convention of calling the degree of the zero polynomial $-\infty$ ($-\infty + (-\infty) = -\infty$, $-\infty + n = -\infty$, $-\infty < n$ for every $n \in \mathbb{N}$). Polynomials of degree $1, 2, 3, \ldots$ are called, respectively, linear, quadratic, cubic, etc.

The unit element in the ring $A[X]$ is the element 1 of the ring A, considered as a zero degree polynomial. It follows directly from the definition of addition and multiplication in $A[X]$ that for any two polynomials

$$f = f_0 + f_1 X + \cdots + f_n X^n, \qquad g = g_0 + g_1 X + \cdots + g_m X^m \tag{4}$$

of degree n and m, respectively, we have

$$\deg(f + g) \leq \max(\deg f, \deg g), \qquad \deg(fg) \leq \deg f + \deg g \quad . \tag{5}$$

The second inequality in (5) is actually an equality

$$\deg(fg) = \deg f + \deg g$$

whenever the product $f_n g_m$ of the leading coefficients in f and g is non-zero, because

$$fg = f_0 g_0 + (f_0 g_1 + f_1 g_0)X + \cdots + (f_n g_m)X^{n+m} \quad . \tag{6}$$

This last fact gives us

THEOREM 1. <u>If</u> A <u>is an integral domain, then so is the ring</u> $A[X]$. □

The importance of polynomial rings in the class of commutative rings is partly due to the following

THEOREM 2. <u>Suppose that</u> R <u>is a commutative ring containing the subring</u> A. <u>For every element</u> $t \in R$ <u>there exists a unique ring homomorphism</u> $\Pi_t : A[X] \to R$ <u>such that</u>

$$\Pi_t(a) = a \, , \quad \forall a \in A \, , \quad \Pi_t(X) = t \quad . \tag{7}$$

<u>Proof.</u> First suppose that such a homomorphism Π_t exists. Since $\Pi_t(f_i) = f_i$ for each coefficient of f written in the form (3), and $\Pi_t(X^k) = (\Pi_t(X))^k = t^k$ (by (7) and the property of being a homomorphism), it follows that

$$\Pi_t(f) = \Pi_t(f_0 + f_1 X + \cdots + f_n X^n) = f_0 + f_1 t + \cdots + f_n t^n \quad , \tag{8}$$

i.e., $\Pi_t(f)$ is uniquely determined, and is given by (8). Conversely, if we define a map Π_t by the formula (8), we obviously satisfy the condition (7), and also obtain a ring homomorphism. The map is clearly a homomorphism of the additive groups of the rings, and to see that Π_t is a homomorphism with respect to multiplication it suffices to apply Π_t to the product (6) and then use the general distributive law:

$$\Pi_t(fg) = f_0 g_0 + (f_0 g_1 + f_1 g_0)t + \cdots + (f_n g_m)t^{n+m} = \left(\sum_{i=0}^{n} f_i t^i\right)\left(\sum_{j=0}^{m} g_j t^j\right) = \Pi_t(f) \cdot \Pi_t(g) \; .$$

□

Applying the map Π_t defined by (8) to a polynomial $f = f(X)$ is called substituting t in place of X in f, or else simply finding the value of f at $X = t$; so we write $\Pi_t(f) = f(t)$. Knowing $\Pi_t(f)$ means to be able to compute the value of f at $X = t$. The homomorphisms Π_x for $x \in A$ are the connecting link between the function theoretic and algebraic points of view on polynomials. By definition, the linear polynomial $X - c$ (i.e., the sequence $(-c, 1, 0, 0, \ldots)$) is not zero, but the associated function $x \mapsto x - c$ takes the value zero when $x = c$. Here is another example: the non-zero polynomial $X^2 + X$ with coefficients in the field $\mathbb{F}_2$ (in which $1 + 1 = 0$) gives us the zero function $\tilde{f} : \mathbb{F}_2 \to \mathbb{F}_2$, since $0^2 + 0 = 0$ and $1^2 + 1 = 0$.

An element $t \in R$ is called _algebraic over_ A if $\Pi_t(f) = 0$ for some $f \in A[X]$. But if $\Pi_t : A[X] \to R$ is a monomorphism, then t is called a _transcendental_ **element** over A. In the case $A = \mathbb{Q}$ and $R = \mathbb{C}$ we use the terms _algebraic_ and _transcendental numbers._ The numbers e and π are transcendental, and examples of algebraic numbers are $\sqrt{2}$, $\sqrt{3}$, $\sqrt{2} + \sqrt{3}$.

In order to measure by how much the ring $A[t] \subset R$ which was introduced at the beginning of the section differs from the polynomial ring $A[X]$, we consider the kernel $J_t = \operatorname{Ker} \Pi_t$ of the homomorphism Π_t in Theorem 2. By (7), Π_t acts as the identity map on A; hence, $A \cap J_t = 0$. Also, $J_t = 0$ if t is transcendental over A. By the theorem on ring homomorphisms (Theorem 2 of Subsection 4, §4 Ch. 4), we have:

$$A[t] \cong A[X]/J_t \quad . \tag{9}$$

The isomorphism (9) can be said to express a _universal property_ of the polynomial ring $A[X]$. The universality of the ring of polynomials comes out more clearly in the following generalization of Theorem 2.

THEOREM 3. _Let_ A _and_ R _be arbitrary commutative rings, let_ t _be an element in_ R, _and let_ $\varphi : A \to R$ _be a homomorphism. Then there exists a unique extension of_ φ _to a homomorphism_ $\varphi_t : A[X] \to R$ _which takes the variable_ X _to_ t.

Theorem 3 is proved in essentially the same way as Theorem 2; we leave the proof to the reader as an exercise. □

2. Polynomials in several variables. In the situation $A \subset R$, suppose we take n elements $t_1, \ldots, t_n \in R$ and consider the intersection of all subrings of R containing A and $t_1, \ldots, t_n$. We then obtain a ring $A[t_1, \ldots, t_n]$. As in the case $n = 1$, it is natural to introduce polynomial rings in n variables. This is simple to do. Recall that the construction of $B = A[X]$ started with an arbitrary commutative ring with unit, the ring A. Hence, we can replace A by B in our construction and obtain the ring $C = B[Y]$, where Y is a new independent variable, which plays the same role for B as X did for A. The elements of C can be uniquely written in the form $\Sigma\, b_j Y^j$, $b_j \in B$, and B is identified with a subring of C, namely, the set of elements $bY^0 = b \cdot 1$. Since any element $b_j \in B$ can, in turn, be written uniquely in the form $b_j = \Sigma\, a_{ij} X^i$, it follows that any element of C has the form

$$\sum_{i=0}^{k} \sum_{j=0}^{\ell} a_{ij} X^i Y^j , \qquad a_{ij} \in A ,$$

where, by construction, X and Y commute with each other and with the a_{ij}. C is called the ring of polynomials over A in the two variables X and Y.

If we repeat this construction, we can obtain the ring $A[X_1, \ldots, X_n]$ of polynomials over A in the n variables $X_1, \ldots, X_n$. We agree to denote an n-tuple $(i_1, \ldots, i_n)$ of non-negative integers by the symbol i. Then any element $f \in A[X_1, \ldots, X_n]$ can be written in the form

$$f = \sum_{(i)} a_{(i)} X^{(i)} , \qquad a_{(i)} \in A , \tag{10}$$

where $X^{(i)} = X_1^{i_1} \ldots X_n^{i_n}$ is a monomial. Thus, f is a linear combination of monomials with coefficients in A. It follows from the definition of a polynomial that all of the coefficients $a_{(i)}$ in (10) except for finitely many are equal to zero.

The expression (10) for the polynomial f is unique, as follows from the claim:

a polynomial f is equal to zero if and only if all of its coefficients $a_{(i)}$ are equal to zero. We have already seen this in the case $n = 1$. To prove it for $n > 1$, we use induction on n. Namely, we write

$$f = \sum a_{i_1 \ldots i_n} X_1^{i_n} \ldots X_n^{i_n} = \sum_{i_n} b_{i_n} X_n^{i_n} ,$$

where

$$b_{i_n} = \sum_{i_1, \ldots, i_{n-1}} a_{i_1 \ldots i_{n-1} i_n} X_1^{i_1} \ldots X_{n-1}^{i_{n-1}}$$

are polynomials in a smaller number of variables. The truth of the claim for $n = 1$, along with the induction assumption, show that

$$f = 0 \iff b_{i_n} = 0, \quad \forall i_n \iff a_{i_1 \ldots i_{n-1} i_n} = 0, \quad \forall (i_1, \ldots, i_n) \quad .$$

Thus, two polynomials are equal if and only if the coefficients of each monomial term are equal.

By the degree of f in X_k, denoted $\deg_k f$, we mean the greatest integer which occurs in the exponent of X_k in a term $a_{(i)} X^{(i)}$ for which $a_{(i)}$ is non-zero. For example, the polynomial $1 + X + XY^3 + X^2Y^2$ has degree 2 in X and degree 3 in Y. The integer $i_1 + \cdots + i_n$ is called the total degree of the monomial $X_1^{i_1} \ldots X_n^{i_n}$. By the total degree of a polynomial f, denoted $\deg f$, we mean the maximum of the total degrees of its non-zero monomial terms. We set $\deg 0 = -\infty$. It makes no sense to speak of the leading term of a polynomial in several variables, since there can be many monomial terms having the same maximum total degree.

Many of the results in Subsection 1 for $A[X]$ carry over to the ring $A[X_1, \ldots, X_n]$. For example, using Theorem 1 and induction on n, we immediately obtain

THEOREM 1'. *If A is an integral domain, then so is $A[X_1, \ldots, X_n]$. In* particular, the ring of polynomials in n variables over a field is an integral domain. □

Next, let A be a subring of a commutative ring R, and let $t_1, \ldots, t_n$ be elements of R. Then the map

$$\Pi_{t_1,\ldots,t_n} : f(X_1, \ldots, X_n) \longmapsto f(t_1, \ldots, t_n), \quad \forall f \in A[X_1, \ldots, X_n],$$

gives a homomorphism $A[X_1, \ldots, X_n] \to R$ (compare with Theorem 2). We say that $t_1, \ldots, t_n$ are substituted in place of $X_1, \ldots, X_n$ in f, or that f is evaluated at $X_1 = t_1, \ldots, X_n = t_n$. If $\operatorname{Ker} \Pi_{t_1,\ldots,t_n} = 0$, then $t_1, \ldots, t_n$ are called algebraically independent over A. If the elements $t_1, \ldots, t_n$ are algebraically dependent, then there exists a non-zero polynomial $f \in A[X_1, \ldots, X_n]$ for which $f(t_1, \ldots, t_n) = 0$.

Finally, the generalization of Theorem 3 is

THEOREM 3' (universality of polynomial rings). Suppose that A and R are commutative rings, $t_1, \ldots, t_n$ are elements of R, and $\varphi : A \to R$ is a ring homomorphism. Then there exists a unique extension of φ to a homomorphism $\varphi_{t_1,\ldots,t_n} : A[X_1, \ldots, X_n] \to R$ which takes X_i to t_i, $1 \le i \le n$.

The proof, like the construction of the ring $A[X_1, \ldots, X_n]$, proceeds by induction. Theorem 3 gives the assertion when $n = 1$. Suppose that we have a homomorphism $\varphi_{t_1,\ldots,t_{n-1}} : A[X_1, \ldots, X_{n-1}] \to R$ which extends φ and takes X_i to t_i for $1 \le i \le n-1$. Replacing A by the ring $A[X_1, \ldots, X_{n-1}]$ and φ by $\varphi_{t_1,\ldots,t_{n-1}}$ in Theorem 3, and using the fact that $A[X_1, \ldots, X_n] = A[X_1, \ldots, X_{n-1}][X_n]$, we find the desired homomorphism $\varphi_{t_1,\ldots,t_n} = (\varphi_{t_1,\ldots,t_{n-1}})_{t_n}$. It is clear that this homomorphism is unique, since a homomorphism from $A[X_1, \dots, X_n]$ is completely determined by its action on A and on $X_1, \dots X_n$. □

COROLLARY. <u>To every permutation</u> $\pi \in S_n$ <u>of the set</u> $\{1, 2, \dots, n\}$ <u>there corresponds a unique automorphism</u> $\tilde{\pi} : f \mapsto f$ <u>of the ring</u> $A[X_1, \dots, X_n]$ <u>which is the identity on</u> A <u>and satisfies</u>

$$(\pi f)(X_1, \dots, X_n) = f\left(X_{\pi^{-1}(1)}, \dots, X_{\pi^{-1}(n)}\right) .$$

Proof. In Theorem 3' set $R = A[X_1, \dots, X_n]$, $t_1 = X_{\pi^{-1}(1)}, \dots, t_n = X_{\pi^{-1}(n)}$, and take φ to be the inclusion of A into $A[X_1, \dots, X_n]$. Then Theorem 3' gives us a homomorphism $\tilde{\pi} = \varphi_{t_1, \dots, t_n}$ from $A[X_1, \dots, X_n]$ to itself (i.e., an endomorphism). Since $\tilde{\pi}\widetilde{\pi^{-1}} = \widetilde{\pi^{-1}}\tilde{\pi} = 1$, $\tilde{1} = 1$, and $\widetilde{\pi\rho} = \tilde{\pi}\tilde{\rho}$ (this is verified in the lemma in §2 Ch. 4), it follows that $\tilde{\pi}$ is an automorphism. □

A useful refinement of Theorem 1' is

THEOREM 4. <u>Let</u> f <u>and</u> g <u>be any two polynomials in</u> n <u>variables over an integral domain</u> A. <u>Then</u>

$$\deg(fg) = \deg f + \deg g .$$

Proof. A polynomial $h(X_1, \dots, X_n)$ all of whose monomial terms have the same total degree m is called a <u>homogeneous polynomial</u> or a <u>homogeneous form</u> of degree m. Forms of degree 1, 2, 3 are called <u>linear</u>, <u>quadratic</u>, and <u>cubic</u> forms, respectively. If we combine all monomial terms of a given total degree which occur in f (i.e., with non-zero coefficient), we can uniquely write f as the sum of forms f_m of different degrees:

$$f = f_0 + f_1 + \cdots + f_k, \qquad k = \deg f .$$

Now if

$$g = g_0 + g_1 + \cdots + g_\ell , \qquad \ell = \deg g ,$$

then obviously

$$fg = f_0 g_0 + (f_0 g_1 + f_1 g_0) + \cdots + f_k g_\ell$$

(this resembles (6), except that the f_i and g_j have a different meaning here). Hence, $\deg(fg) \leq k + \ell$. Since $f_k \neq 0$ and $g_\ell \neq 0$, Theorem 1' implies that $f_k g_\ell \neq 0$, so that $\deg(fg) = \deg(f_k g_\ell) = k + \ell = \deg f + \deg g$. □

3. <u>The division algorithm.</u> In addition to the general properties of polynomial rings studied in Subsection 2, these rings have some important special properties. We immediately discover one such property if we try to describe the ideals in a polynomial ring. We say (see Subsection 3 of §4 Ch. 4) that every ideal in the ring $\mathbb{Z}$ is principal, i.e., of the form $m\mathbb{Z}$. The proof of this fact was based on a mechanism called the <u>division algorithm</u> and described for the ring $\mathbb{Z}$ in Subsection 3 of §8 Ch. 1. It turns out that a completely analogous algorithm holds in $A[X]$ whenever A is an integral domain. In the case $A = \mathbb{R}$ this algorithm amounts to the usual long division of polynomials in high school algebra.

THEOREM 5. <u>Let</u> A <u>be an integral domain, and let</u> g <u>be a polynomial in</u> $A[X]$ <u>whose leading coefficient is an invertible element of</u> A. <u>Then for every polynomial</u> $f \in A[X]$ <u>there exists a unique pair of polynomials</u> $q, r \in A[X]$ <u>for which</u>

$$f = qg + r, \qquad \deg r < \deg g . \tag{11}$$

<u>Proof.</u> Let

$$f = a_0 X^n + a_1 X^{n-1} + \cdots + a_n ,$$

$$g = b_0 X^m + b_1 X^{m-1} + \cdots + b_m ,$$

where $a_0 b_0 \neq 0$ and $b_0 \mid 1$. We use induction on n. If $n = 0$ and $m = \deg g > \deg f = 0$, then set $q = 0$ and $r = f$; if $n = m = 0$, then set $r = 0$

and $q = a_0 b_0^{-1}$. Now suppose that the theorem is true for all polynomials of degree $< n$ (where $n > 0$). Without loss of generality we may assume that $m \leq n$, since otherwise simply take $q = 0$ and $r = f$. With this assumption, we write

$$f = a_0 b_0^{-1} X^{n-m} \cdot g + \overline{f} ,$$

where $\deg \overline{f} < n$. By the induction assumption, we find $\overline{q}$ and r for which $\overline{f} = \overline{q} g + r$, where $\deg r < m$. If we set

$$q = a_0 b_0^{-1} X^{n-m} + \overline{q} ,$$

we obtain a pair of polynomials with the required properties. q is called the quotient and r is called the remainder. It remains to prove that q and r are unique.

To see this, suppose that

$$qg + r = f = q'g + r' .$$

Then $(q' - q)g = r - r'$. By Theorem 1, we have: $\deg(r - r') = \deg(q' - q) + \deg g$, which in our situation can only happen if $r' = r$ and $q' = q$ (recall that $\deg 0 = -\infty$ and $-\infty + m = -\infty$).

Note that the coefficients of q and r belong to the same ring A as the coefficients of f and g , i.e., $f, g \in A[X] \Rightarrow q, r \in A[X]$. □

Remark. The above process of dividing f by g , which is known as the Euclidean algorithm, becomes simpler if g is a monic polynomial, which means that its leading coefficient is 1 .

Notice that in Theorem 5 the polynomial f is divisible by g if and only if $r = 0$.

COROLLARY. *If K is a field, then all ideals in the ring $K[X]$ are principal.*

Proof. Let T be a non-zero ideal in $K[X]$. Choose a polynomial $t = t(X)$ of minimal degree contained in T . Since K is a field, the leading coefficient of t must be invertible in K . If f is any polynomial in T , then the division algorithm gives us

$f = qt + r$, $\deg r < \deg t$. This equality implies that $r \in T$, since f, t, and qt are elements of T. By our choice of t this means that $r = 0$. Thus, $f(X)$ is divisible by $t(X)$, and so $T = (t) = tK[X]$, i.e., T consists precisely of all polynomials divisible by $t(X)$. □

This corollary is false for polynomial rings in several variables over a field; for example, not all ideals in $\mathbb{R}[X,Y]$ are principal.

Example. The set $T = \{Xf + Yg \mid f, g \in \mathbb{R}[X,Y]\}$, which consists of all polynomials $h(X,Y)$ such that $h(0,0) = 0$, is obviously an ideal in $\mathbb{R}[X,Y]$. If we had $T = t(X,Y)\,\mathbb{R}[X,Y]$, then, since $1 \in \mathbb{R}[X,Y]$, we would have $t \in T$, and so $t(0,0) = 0$; thus, $\deg t \geq 1$. Now apply Theorem 4 to the equalities: $X = tu$, $Y = tv$. We find that $\deg u = \deg v = 0$, i.e., $u, v \in \mathbb{R}$ and so $Y = u^{-1}vX$. This is absurd; hence, T is not a principal ideal.

The corollary to Theorem 5 is convenient for giving an explicit description of the isomorphism (9). As an example, we prove a fact which really goes with Theorem 4 of §1.

THEOREM 6. The field of complex numbers $\mathbb{C}$ is isomorphic to the quotient ring $\mathbb{R}[X]/(X^2 + 1)\,\mathbb{R}[X]$.

Proof. According to (9), $\mathbb{C} = \mathbb{R}[i] \cong \mathbb{R}[X]/J$, where $J = \{f \in \mathbb{R}[X] \mid f(i) = 0\}$. Using the fact that $a + bX \notin J$ (because $a + ib \neq 0$ if a and b are not both 0) and $X^2 + 1 \in J$ (since $i^2 + 1 = 0$), it is not hard to show that $J = (X^2 + 1)\,\mathbb{R}[X]$ (use the same argument as in the proof of the corollary to Theorem 5).

The elements of the quotient ring $\mathbb{R}[X]/J$ are the cosets $(a + bX) + J$, $a, b \in \mathbb{R}$; the map $a + ib \mapsto (a + bX) + J$ gives an isomorphism between $\mathbb{C}$ and $\mathbb{R}[X]/J$. □

EXERCISES

1. The polynomials $f(X) = X^5 + 3X^4 + X^3 + 4X^2 - 3X - 1$ and $g(X) = X^2 + X + 1$ can be considered as elements either in $\mathbb{Z}[X]$ or, for example, $\mathbb{Z}_5[X]$. Apply the division algorithm and show that in the first case $f(X)$ is not divisible by $g(X)$, but in the second case it is. Would it be possible to have an example where it was the other way around?

2. Using Theorem 3, prove that, if F is a field, then the group of all automorphisms of the ring $F[X]$ is isomorphic to the group of transformations $X \mapsto aX + b$, where $a, b \in F$ and $a \neq 0$.

3. Show that a polynomial $f \in F[X_1, \dots, X_n]$ is a form of degree m (see the proof of Theorem 4) if and only if $f(tX_1, \dots, tX_n) = t^m f(X_1, \dots, X_n)$, where t is a new variable.

4. Show that the number of different monomials in n variables of total degree m is equal to $\binom{m+n-1}{m}$.

5. Consider the set $A[[X]]$ of so-called formal power series $f(X) = \sum_{i \geq 0} a_i X^i$ in the variable X, which are defined as sequences $(a_0, a_1, a_2, \dots)$ as in Subsection 1 but with any (perhaps infinitely many) of the a_i allowed to be non-zero. The operations with formal power series follow the same rules as in the case of polynomials:

$$\left(\sum a_i X^i\right) + \left(\sum b_i X^i\right) = \sum (a_i + b_i) X^i ,$$

$$\left(\sum a_i X^i\right) \cdot \left(\sum b_j X^j\right) = \sum c_k X^k , \qquad c_k = \sum_{i+j=k} a_i b_j .$$

Show that the set $A[[X]]$ with these operations is an associative and commutative ring with unit $1 = (1, 0, 0, 0, \dots)$.

It no longer makes sense to speak of the degree of f, since a formal power series may have arbitrarily large powers of X. Instead we define the order $\omega(f)$ to be the least integer n for which $a_n \neq 0$ (we agree to set $\omega(0) = +\infty$). Show that

$$\text{(i)} \quad \omega(f-g) \geq \min\{\omega(f), \omega(g)\}; \qquad \text{(ii)} \quad \omega(fg) \geq \omega(f) + \omega(g) \quad .$$

If A is an integral domain, prove that $\omega(fg) = \omega(f) + \omega(g)$. In particular, if A is an integral domain, then so is $A[[X]]$.

Further show that $A[X]$ is a subring of $A[[X]]$.

6. Polynomials and power series are often used as generating functions for different types of numbers. We illustrate with two simple examples.

a) Prove the relation

$$\sum_{i=0}^{k} \binom{m}{i}\binom{n}{k-i} = \binom{m+n}{k} ,$$

by using the binomial formula $\Sigma \binom{n}{i} X^i = (1+X)^n$ in $\mathbb{Z}[X]$ and the obvious factoring $(1+X)^m (1+X)^n = (1+X)^{m+n}$.

b) Find the number ℓ_n of possible ways one can arrange the parentheses in a product of n elements of a set with one binary operation. To do this it is useful to introduce the formal power series ("generating function" of the ℓ_n)

$$\ell(X) = \sum_{n \geq 1} \ell_n X^n = X + X^2 + 2X^3 + \cdots ,$$

the first few of whose coefficients were computed in Subsection 3 of §1 Ch. 4. The obvious relation

$$\ell_n = \sum_{k=1}^{n-1} \ell_k \ell_{n-k}$$

implies that $\ell(X)^2 = \ell(X) - X$. Solve this quadratic equation to find

$$\ell(X) = \frac{1 - \sqrt{1-4X}}{2}$$

(the sign in front of the radical is determined by the condition that $\ell_n > 0$). But the binomial expansion applied to $(1-4x)^{\frac{1}{2}}$ and a simple computation finally gives

$$\ell_n = \frac{1}{n}\binom{2n-2}{n-1} \quad .$$

The reader should fill in the various steps.

7. The ring $A[[X,Y]]$ of formal power series in two independent variables X and Y (which commute with one another) consists of expressions $\sum_{i\geq 0, j\geq 0} a_{ij} X^i Y^j$.

Show that $B[[Y]] = A[[X,Y]] = C[[X]]$, where $B = A[[X]]$ and $C = A[[Y]]$ (repeat the construction of the ring of polynomials in several variables). Show that if A is an integral domain, then so is $A[[X]]$.

§3. Factoring in polynomial rings

1. Elementary divisibility properties. In various places, starting in Chapter 1, we have touched upon questions of divisibility in the ring of integers $\mathbb{Z}$, but we have not yet proved the so-called Fundamental Theorem of Arithmetic. It is now time not only to fill in this gap but to prove this result for a wider class of rings, in particular, for the ring $K[X]$ of polynomials over a field K.

Suppose we have an arbitrary integral domain R. The invertible elements of R can also be called the divisors of 1. For example, it is clear that a polynomial $f \in A[X]$ is invertible if and only if $\deg f = 0$ and $f = f_0$ is an invertible element in the ring A; this is because if we had $fg = 1$, we would have to have $\deg f + \deg g = \deg 1 = 0$.

We say that an element $b \in R$ is divisible by $a \in R$ (or that b is a multiple of a) if there exists $c \in R$ such that $b = ac$; this is denoted $a|b$. If both $a|b$ and $b|a$, then a and b are called associated elements. In that case $b = ua$, where $u|1$. By the remark in the last paragraph, two polynomials $f, g \in A[X]$ are associated if and only if they differ by a multiplicative factor which is an invertible element of A.

A non-zero element $p \in R$ is called prime (or irreducible) if p is not invertible and cannot be written in the form $p = ab$, where both a and b are non-invertible. A field has no prime elements, because every non-zero element is invertible. A prime element of the ring $A[X]$ is called an irreducible polynomial.

We note the following basic properties of divisibility in an integral domain R:

1) If $a|b$ and $b|c$, then $a|c$. Namely, we have $b = ab'$ and $c = bc'$, where $b', c' \in R$. Hence $c = (ab')c' = a(b'c')$.

2) If $c|a$ and $c|b$, then $c|(a \pm b)$. Namely, since $a = ca'$ and $b = cb'$ for some $a', b' \in R$, it follows by the distributive law that $a \pm b = c(a' \pm b')$.

3) If $a|b$, then $a|bc$. Clearly, if $b = ab'$, then $bc = (ab')c = a(b'c)$.

Combining 2) and 3), we obtain

4) If each element $b_1, b_2, \ldots, b_m \in R$ is divisible by $a \in R$, then so is any linear combination $b_1c_1 + b_2c_2 + \cdots + b_mc_m$, where $c_1, c_2, \ldots, c_m$ are any elements in R.

Definition. We say that an integral domain R is a unique factorization domain (or a factorial ring) if any non-zero element $a \in R$ can be represented in the form

$$a = up_1p_2 \cdots p_r, \tag{1}$$

where u is an invertible element and $p_1, p_2, \ldots, p_r$ are prime elements (not necessarily distinct from one another), and if, given another such decomposition $a = vq_1q_2 \cdots q_s$, we must have $r = s$ and

$$q_1 = u_1p_1, \ldots, q_r = u_rp_r,$$

for a suitable choice of indexing of the p's and q's and for suitable invertible elements $u_1, \ldots, u_r$.

If we allow r to equal 0 in (1), we are adopting the convention that invertible elements of R also have a decomposition into prime factors, although a trivial one. It is clear that, if p is a prime and u is an invertible element, then the element up is also a prime. For example, in $\mathbb{Z}$, which has invertible elements 1 and -1, we can agree to choose the positive prime in each pair of associated primes $\{p, -p\}$. In a polynomial ring $K[X]$ over a field K it is convenient to choose monic (i.e., leading coefficient 1) irreducible polynomials.

We have the following general

THEOREM 1. *Let R be any integral domain in which every element has a factorization* (1). *That factorization is unique* (*i.e., R is a unique factorization domain*) *if and only if any prime* $p \in R$ *dividing a product* ab, $a, b \in R$, *must divide* a *or* b.

Proof. First suppose R is a unique factorization domain. Let $ab = pc$. If

$$a = \prod a_i, \qquad b = \prod b_j, \qquad c = \prod c_k$$

are decompositions of a, b, c into prime factors, then the equation $\prod a_i \prod b_j = p \prod c_k$ implies that p must be associated with one of the primes a_i or b_j, i.e., p divides a or b.

Conversely, suppose whenever $p|ab$ we must have $p|a$ or $p|b$. We must show that R is a unique factorization domain. We use induction on the minimal number n of prime factors in a decomposition (1) of $a \in R$. We prove that any element a which is a product of n primes factors uniquely. If $n = 1$, i.e., if a itself is a prime, this assertion is easy to check, and we leave this to the reader. So suppose that the assertion holds for any element which is a product of $\leq n$ primes, and let $a \neq 0$ be a product of $n + 1$ prime factors. Let

$$a = \prod_{i=1}^{n+1} p_i = \prod_{j=1}^{m+1} r_j \tag{2}$$

be two decompositions of a with $m \geq n$. Applying the hypothesis of the theorem to $p = p_{n+1}$, we find that p_{n+1} must divide one of the elements $r_1, \ldots, r_{m+1}$. Without loss of generality (renumbering the r's if necessary), we may suppose that

$p_{n+1} | r_{m+1}$. But r_{m+1} is a prime, so that $r_{m+1} = u p_{n+1}$, where u is an invertible element. Using cancellation in R (Theorem 3 of §4 Ch. 4), we see that (2) gives us $\prod_{i=1}^{n} p_i = u \prod_{j=1}^{m} r_j$. But the left side of this equality is a product of n primes. By the induction assumption, $m = n$ and the two decompositions can only differ in the order of the primes and the possible presence of invertible elements u as ratios of corresponding p's and r's. This completes the induction step. □

In an arbitrary integral domain R it is not necessarily true that every non-zero element a has a prime decomposition (1). But what is even more interesting is the integral domains in which all elements have prime factorizations, but they are not unique. Thus, the condition in Theorem 1, which at first glance seems trivial, is not always satisfied.

Example. Consider the imaginary quadratic field $\mathbb{Q}(\sqrt{-5})$ (see the example in Subsection 5 of §1), and in this field consider the integral domain $R = \{a + b\sqrt{-5} \mid a, b \in \mathbb{Z}\}$. The norm $N(a + b\sqrt{-5}) = a^2 + 5b^2$ of any non-zero element in R is a positive integer. If α is invertible in R, then $N(\alpha)^{-1} = N(\alpha^{-1}) \in \mathbb{Z}$, and hence $N(\alpha) = 1$. This is only possible if $b = 0$ and $a = \pm 1$. Thus, in R, like in $\mathbb{Z}$, the only invertible elements are ± 1. If $\alpha = \varepsilon\alpha_1\alpha_2 \ldots \alpha_r \neq 0$, $\varepsilon = \pm 1$, then $N(\alpha) = N(\alpha_1) \ldots N(\alpha_r)$. Since $1 < N(\alpha_i) \in \mathbb{N}$, it follows that for a given α the number of factors is bounded. This implies that any element $\alpha \in R$ has a prime factorization.

But the number 9 (and plenty of others) have two essentially different prime factorizations:

$$9 = 3 \cdot 3 = (2 + \sqrt{-5})(2 - \sqrt{-5}) \quad .$$

It is obvious that the elements 3 and $2 \pm \sqrt{-5}$ are not associated. Furthermore, none of these elements of R factors further, because they each have norm 9 and if $\alpha = \alpha_1\alpha_2$ for $N(\alpha) = 9$ and neither α_1 nor α_2 equals ± 1, then it follows that

$N(\alpha_1) = N(\alpha_2) = 3$. But this is impossible, since the equation $x^2 + 5y^2 = 3$, $x, y \in \mathbb{Z}$, cannot be solved. Thus, we really do have two distinct prime factorizations of 9.

This example suggests a wide circle of questions, some of which have not yet been answered, concerning the quadratic fields $\mathbb{Q}(\sqrt{d})$. Such questions are studied in <u>algebraic number theory</u>.

Before using Theorem 1 to prove that various rings have unique factorization, we introduce some important auxiliary concepts, which also are of independent interest.

2. G. c. d. <u>and</u> l. c. m. <u>in rings.</u> Let R be an integral domain. By a <u>greatest common divisor</u> of two elements $a, b \in R$, denoted g. c. d. (a,b), we mean any element $d \in K$ having the two properties

(i) $d \mid a$, $d \mid b$;

(ii) $c \mid a$ and $c \mid b \Rightarrow c \mid d$.

Clearly, if d has properties (i) and (ii), then so does any associated element ud. Conversely, if c and d are two greatest common divisors of a and b, then we have $c \mid d$ and $d \mid c$, so that c and d are associated. The notation g. c. d. (a,b) is used for any greatest common divisor, i. e., we do not distinguish here between associated elements. If we agree for the time being not to distinguish between associated elements, we have the following further properties of the g. c. d. :

(iii) g. c. d. $(a,b) = a \Leftrightarrow a \mid b$;

(iv) g. c. d. $(a,0) = a$;

(v) g. c. d. $(ta, tb) = t$ g. c. d. (a,b);

(vi) g. c. d. (g. c. d. (a,b), c) = g. c. d. (a, g. c. d. (b,c)).

These properties are easily checked, and we leave that to the reader. Property (vi) allows us to extend the notion of g. c. d. to an arbitrary finite set of elements: if we define g. c. d. $(a_1, \ldots, a_n)$ = g. c. d. (g. c. d. (... g. c. d. $(a_1, a_2)\ldots)$, a_n), then this does not depend on the order of the a's.

Along with g.c.d. we have the companion concept of the <u>least common multiple</u> $m = \text{l.c.m.}(a,b)$, which is defined (to within multiplication by a unit u, i.e., we again do not distinguish between associated elements) by the two properties:

(i') $a \mid m$, $b \mid m$;

(ii') $a \mid c$ and $b \mid c \Rightarrow m \mid c$.

In particular, setting $c = ab$, we obtain: $m \mid ab$.

THEOREM 2. <u>Suppose that the</u> g.c.d. <u>and the</u> l.c.m. <u>exist for two elements</u> a, b <u>in an integral domain</u> R. <u>Then:</u>

(a) $\text{l.c.m.}(a,b) = 0 \Leftrightarrow a = 0$ <u>or</u> $b = 0$.

(b) <u>If</u> $a, b \neq 0$, $m = \text{l.c.m.}(a,b)$, <u>and</u> $ab = dm$, <u>then</u> $d = \text{g.c.d.}(a,b)$.

<u>Proof.</u> Assertion (a) follows immediately from the definition of $\text{l.c.m.}(a,b)$. To prove (b) we must show that the element d defined by the equation $ab = dm$ has properties (i) and (ii). First, (i') implies that $m = a'a$, $m = b'b$. Hence $ab = dm = da'a$, so that, after cancelling a, as we are allowed to do in an integral domain, we obtain $b = da'$, i.e., $d \mid b$. We similarly show that $d \mid a$, and so d satisfies (i).

Next, suppose that $a = fa''$, $b = fb''$. We must show that $f \mid d$. Set $c = fa''b''$. Then $c = ab'' = ba''$ is a common multiple of a and b. By property (ii'), we have $c = c'm$ for some $c' \in R$, and hence $fc'm = fc = f^2a''b'' = ab = dm$, i.e., $d = fc'$ and so $f \mid d$. □

Note that Theorem 2 merely gives a relationship between the g.c.d. and the l.c.m. when they are known to exist. It does not give us a method for actually computing the g.c.d. and l.c.m., nor does it guarantee that the g.c.d. or the l.c.m. of two elements will always exist.

Now suppose for the time being that R is a unique factorization domain. Let $\mathcal{P}$

denote a set of prime elements in R such that every prime in R is associated with one and only one element of $\mathcal{P}$. When we consider the prime factorizations of two elements $a, b \in R$, it is convenient to assume that the primes that appear in each are the same elements of $\mathcal{P}$, but possibly with zero exponents, i.e., we write

$$a = up_1^{k_1} \cdots p_r^{k_r}, \qquad b = vp_1^{\ell_1} \cdots p_r^{\ell_r},$$
$$u|1, \quad v|1; \quad k_i \geq 0; \quad \ell_i \geq 0; \quad p_i \in \mathcal{P}; \quad 1 \leq i \leq r \quad . \tag{3}$$

Using Theorem 1, we readily obtain the following easily remembered divisibility criterion.

Divisibility criterion. *Let* a *and* b *be two elements of a unique factorization domain* R *written in the form* (3). *Then:*

1) $a|b$ *if and only if* $k_i \leq \ell_i$, $i = 1, 2, \ldots, r$;

2) g.c.d. $(a,b) = p_1^{s_1} \cdots p_r^{s_r}$, *where* $s_i = \min\{k_i, \ell_i\}$, $i = 1, 2, \ldots, r$;

3) l.c.m. $(a,b) = p_1^{t_1} \cdots p_r^{t_r}$, *where* $t_i = \max\{k_i, \ell_i\}$, $i = 1, 2, \ldots, r$.

□

Thus, we take s_i to be the smaller and t_i to be the greater of the two exponents k_i, ℓ_i. In particular, two elements a and b are relatively prime, i.e., g.c.d. $(a,b) = 1$, if and only if the prime factors which occur with positive exponent in the factorization of one of them do not appear with positive exponent in the factorization of the other. The one trouble with this divisibility criterion is that in practice it is often very hard to obtain a decomposition of the form (3). Even in the case $R = \mathbb{Z}$, one has to be satisfied with some variation of the method of going through all primes less than some given n. For this reason, it is all the more gratifying that there is a more effective method for computing the g.c.d. and l.c.m. in a fairly large class of rings, as explained in the next subsection.

3. *Unique factorization in Euclidean rings.* The division algorithm in $\mathbb{Z}$ and

$K[X]$ (see Subsection 3 of §8 Ch. 1 and Subsection 3 of §2 in this chapter) makes it natural to consider the class of integral domains R in which to every non-zero a there is associated a non-negative integer $\delta(a)$ (i.e., δ is a map from $R^* = R\setminus\{0\}$ to $\mathbb{N} \cup \{0\}$) such that:

(E1) $\delta(ab) \geq \delta(a)$ for all $a, b \in R^*$;

(E2) given any $a, b \in R$, $b \neq 0$, there exist $q, r \in R$ (q is called the "quotient" and r is called the "remainder") such that

$$a = qb + r; \qquad \delta(r) < \delta(b) \ \underline{\text{or}} \ r = 0 \quad . \tag{4}$$

An integral domain R for which such a function δ exists is called a Euclidean ring. If we define $\delta(a) = |a|$ for $a \in \mathbb{Z}$ and define $\delta(a) = \deg a$ for $a = a(X) \in K[X]$, we see that $\mathbb{Z}$ and $K[X]$ are Euclidean rings.

In Euclidean rings there is a special algorithm, called the Euclidean algorithm, for finding the g.c.d. of two elements a and b. Let a and b be two non-zero elements of a Euclidean ring R. Applying the procedure in axiom (E2) a large enough (but finite) number of times, we obtain a sequence of equations of type (4) whose last equation has zero remainder:

$$\begin{array}{rlrl}
a &= q_1 b + r_1, & \delta(r_1) &< \delta(b) \\
b &= q_2 r_1 + r_2, & \delta(r_2) &< \delta(r_1) , \\
r_1 &= q_3 r_2 + r_3, & \delta(r_3) &< \delta(r_2) , \\
\cdots & \cdots\cdots\cdots & \cdots & \cdots\cdots \\
r_{k-2} &= q_k r_{k-1} + r_k, & \delta(r_k) &< \delta(r_{k-1}) , \\
r_{k-1} &= q_{k+1} r_k, & r_{k+1} &= 0 \quad .
\end{array} \tag{5}$$

This is the case because the strictly decreasing sequence of positive integers $\delta(b) > \delta(r_1) > \delta(r_2) > \cdots$ must terminate, and this is only possible when a remainder vanishes.

We claim that the last non-zero remainder r_k is precisely the greatest common divisor of a and b, as defined in Subsection 2. To see this, note that, by assumption,

$r_k \mid r_{k-1}$. Going from the bottom up in (5) and using property (4) of the divisibility relation (see Subsection 1), we obtain the chain of divisibilities: $r_k \mid r_{k-1}$, $r_k \mid r_{k-2}, \ldots, r_k \mid r_2$, $r_k \mid r_1$, and finally $r_k \mid b$ and $r_k \mid a$. Hence, r_k is a common divisor of a and b. Now suppose that c is any other common divisor of a and b. Then $c \mid r_1$, and, now going from the top down in (5), we obtain the chain of divisibility relations: $c \mid r_2, c \mid r_3, \ldots, c \mid r_k$. Thus, the g.c.d. of a and b exists, and is equal to r_k:

$$r_k = \text{g.c.d.}\,(a,b) \quad . \tag{6}$$

Next, notice that each remainder r_i in (5) is a linear combination of the two preceding remainders r_{i-1} and r_{i-2} with coefficients in R. This is when $i \geq 3$; as for r_2, it is a linear combination of b and r_1; and r_1 is a linear combination of a and b. By successively substituting expressions for r_{i-1} and r_{i-2} in terms of a and b, we eventually obtain an expression for r_k of the form

$$r_k = au + bv \tag{7}$$

for some $u, v \in R$.

Comparing (6) and (7) and taking into account Theorem 2(b), we obtain:

THEOREM 3. *In a Euclidean ring* R *any two elements* a *and* b *have a greatest common divisor and a least common multiple. Using the Euclidean algorithm it is possible to find* $u, v \in R$ *such that* $\text{g.c.d.}\,(a,b) = au + bv$. *In particular, two elements* $a, b \in R$ *are relatively prime if and only if there exist elements* $u, v \in R$ *such that*

$$au + bv = 1 \quad . \qquad \square$$

COROLLARY. *Let* a, b, c *be elements of a Euclidean ring* R.

(i) *If* $\text{g.c.d.}\,(a,b) = 1$ *and* $\text{g.c.d.}\,(a,c) = 1$, *then* $\text{g.c.d.}\,(a,bc) = 1$.

(ii) *If* $a \mid bc$ *and* $\text{g.c.d.}\,(a,b) = 1$, *then* $a \mid c$.

(iii) <u>If</u> $b \mid a$ <u>and</u> $c \mid a$ <u>and</u> g. c. d. $(b, c) = 1$, <u>then</u> $bc \mid a$.

<u>Proof.</u> (i) By Theorem 3, we have $au_1 + bv_1 = 1$ and $au_2 + cv_2 = 1$. Multiplying together the left and right sides of the two equations, we obtain: $a(au_1u_2 + bu_2v_1 + cu_1v_2) + bc(v_1v_2) = 1$, which gives the desired conclusion.

(ii) Since $au + bv = 1$, we have $ac \cdot u + (bc)v = c$. But $bc = aw$, and hence $c = a(cu + wv)$, i. e., $a \mid c$.

(iii) By property (ii') of the l. c. m.,

$$b \mid a,\ c \mid a \implies \text{l. c. m.}(b, c) \mid a \implies bc \mid a \quad ,$$

since $bc = \text{g. c. d.}(b, c) \cdot \text{l. c. m.}(b, c)$, and g. c. d. $(b, c) = 1$ by assumption. □

We leave it to the reader to generalize Theorem 3 to the case of the g. c. d. of a finite set of elements in a Euclidean ring.

The next lemma is an important step in proving that Euclidean rings have unique factorization.

LEMMA. <u>Every Euclidean ring</u> R <u>has factorization (i. e., any non-zero element</u> $a \in R$ <u>can be written in the form (1))</u>.

<u>Proof.</u> Suppose that the element $a \in R$ has a proper divisor b, i. e., $a = bc$, where b and c are non-invertible elements (in other words, a and b are not associated). We prove that $\delta(b) < \delta(a)$. To see this, first use (E1) to obtain: $\delta(b) \leq \delta(bc) = \delta(a)$. If we had $\delta(b) = \delta(a)$, by (E2) we could find q and r with $b = qa + r$, where $\delta(r) < \delta(a)$ or else $r = 0$. The possibility $r = 0$ would contradict the fact that a and b are not associated, i. e., $a \nmid b$. Furthermore, since c is not invertible, we have: $1 - qc \neq 0$. But then, by (E1),

$$\delta(a) = \delta(b) \leq \delta(b(1 - qc)) = \delta(b - qa) = \delta(r) < \delta(a) \quad ,$$

a contradiction. Thus, $\delta(b) < \delta(a)$.

Now if $a = a_1 a_2 \cdots a_n$, where all of the a_i are non-invertible, then for each m the element $a_{m+1} a_{m+2} \cdots a_n$ is a proper divisor of $a_m a_{m+1} \cdots a_n$, and so

$$\delta(a) = \delta(a_1 a_2 \cdots a_n) > \delta(a_2 \cdots a_n) > \cdots > \delta(a_n) \geqslant \delta(1) .$$

This strictly decreasing sequence of non-negative integers has length $n \leq \delta(a)$. Hence, there is a maximum value n such that the decomposition $a = a_1 a_2 \cdots a_n$ can split up no further, i.e., all of the a_i are prime elements. □

THEOREM 4. Every Euclidean ring R is a unique factorization domain.

Proof. Using the above lemma and the criterion in Theorem 1, we see that it suffices to prove that, if p is a prime element of R dividing the product bc of two elements $b, c \in R$, then p must divide either b or c.

First, if $b = 0$ or $c = 0$, there is nothing left to prove. But if $bc \neq 0$ and $d = \text{g.c.d.}(b, p)$, then d, since it divides p, must be either p or 1 (more precisely, an element associated to one of these two elements). If $d = 1$, then b and p are relatively prime, and part (ii) of the corollary to Theorem 3 allows us to conclude that $p \mid c$. If $d = p$, then $p \mid b$. □

COROLLARY. The rings $\mathbb{Z}$ and $K[X]$ (where K is any field) are unique factorization domains. □

The polynomial rings $K[X_1, \ldots, X_n]$ where $n > 1$, which are not Euclidean rings, are nonetheless unique factorization domains, as we shall prove in Chapter 9. We shall also give further examples there of unique factorization domains.

4. Irreducible polynomials. Recall that a special case of the above definition of a prime element of a ring is: a polynomial $f \in K[X]$ of degree greater than zero is called irreducible (over the field K) if it is not divisible by any polynomial $g \in K[X]$ for which $0 < \deg g < \deg f$. In particular, every linear (first degree) polynomial is irreducible. It

is easy to see that the question of whether or not a polynomial of degree > 1 is irreducible and the problem of decomposing it into a product of irreducible factors are intimately connected with the "ground field" K. For example, the polynomial $X^2 + 1$ is irreducible in $\mathbb{R}[X]$ but decomposes as $X^2 + 1 = (X + i)(X - i)$. The polynomial $X^4 + 4$ can be factored even over $\mathbb{Q}$, although this fact is not easy to guess at first glance:

$$X^4 + 4 = (X^2 - 2X + 2)(X^2 + 2X + 2) \quad .$$

Both of the factors on the right are irreducible not only over $\mathbb{Q}$, but even over $\mathbb{R}$; however, they are reducible over $\mathbb{C}$.

The set that plays the role of the positive prime numbers in $\mathbb{Z}$ is the set of monic (i.e., leading coefficient = 1) irreducible polynomials. As in the case of the primes in $\mathbb{Z}$ (see §8 of Chapter 1), <u>the set of monic irreducible polynomials over any field K is infinite.</u> This is obvious in the case of an infinite field K; just take the irreducible polynomials $X - c$, $c \in K$. If K is a finite field, we use an argument of Euclid. Namely, suppose we already have n irreducible polynomials $p_1, \ldots, p_n$. The polynomial $f = p_1 p_2 \ldots p_n + 1$ has at least one monic irreducible polynomial divisor, since $\deg f \geq n$. Let us denote it p_{n+1}. It is different from $p_1, \ldots, p_n$, since if we had $p_{n+1} = p_s$ for some $s \leq n$, it would follow that $p_s \mid (f - p_1 \ldots p_n) = 1$. Thus, $K[X]$ has infinitely many monic irreducible polynomials.

Since there are only finitely many polynomials of a given degree over a finite field, we have the following useful corollary of the result of the preceding paragraph: <u>There exist irreducible polynomials of arbitrarily high degree over any finite field.</u> We will obtain a more precise version of this qualitative assertion in Chapter 9.

Irreducible polynomials over $\mathbb{Q}$ play a vital role in the theory of algebraic number fields. Since we can always multiply a polynomial in $\mathbb{Q}[X]$ by a suitable natural number to obtain a polynomial in $\mathbb{Z}[X]$, we might first want to clarify the relationship between irreducibility over $\mathbb{Q}$ and over $\mathbb{Z}$. Since we will also be interested in other applications, we prove a general assertion about polynomials over a unique factorization domain R. By the content $d = d(f)$ of a polynomial $f = a_0 + a_1 X + \cdots + a_n X^n \in R[X]$ we mean the greatest common divisor of all of the coefficients. Up till now we have only discussed the g.c.d. of two elements, but the properties (i)-(vi) of the g.c.d. allow us to extend this notion to any finite set of elements of an integral domain. If $d(f)$ is an invertible element of R, then f is called a primitive polynomial.

GAUSS'S LEMMA. *Let* R *be a unique factorization domain, and let* $f, g \in R[X]$. *Then*

$$d(fg) = d(f) \cdot d(g) \quad ,$$

where equality is understood to mean to within an invertible element of R, *i.e., we do not distinguish between associated elements. In particular, the product of two primitive polynomials is a primitive polynomial.*

Proof. We first prove that the product of two primitive polynomials is primitive. Let

$$f = a_0 + a_1 X + \cdots + a_n X^n , \quad g = b_0 + b_1 X + \cdots + b_m X^m$$

be two primitive polynomials in $R[X]$ whose product fg is not primitive. Thus, there exists a prime element $p \in R$ which divides $d(fg)$. Choose the least indices s and t such that $p \nmid a_s$ and $p \nmid b_t$, as we can do, since f and g are primitive. The coefficient of X^{s+t} in fg is

$$c_{s+t} = a_s b_t + (a_{s+1} b_{t-1} + a_{s+2} b_{t-2} + \cdots) + (a_{s-1} b_{t+1} + a_{s-2} b_{t+2} + \cdots) \quad .$$

Since we have assumed that a_{s-i} and b_{t-i} are divisible by p for $i > 0$, and since $p \mid c_{s+t}$ because $p \mid d(fg)$, it follows that for suitable u and v

$$pu = a_s b_t + pv \quad ,$$

so that $p \mid a_s b_t$. Since R is a unique factorization domian, this means that $p \mid a_s$ or $p \mid b_t$, a contradiction. This proves our claim that fg is primitive.

Proceeding to the general case, we can write any two polynomials $f, g \in R[X]$ in the form

$$f = d(f) f_0 , \qquad g = d(g) g_0 ,$$

where f_0 and g_0 are primitive polynomials. Since $fg = d(f) d(g) f_0 g_0$, and we have proved that $d(f_0 g_0) = 1$, it follows that $d(fg) = d(f) d(g)$. □

COROLLARY. A polynomial $f \in \mathbb{Z}[X]$ which is irreducible over $\mathbb{Z}$ is also irreducible over $\mathbb{Q}$.

Proof. By the corollary to Theorem 4, $\mathbb{Z}$ is a unique factorization domain, so that Gauss's Lemma applies to $\mathbb{Z}[X]$. Suppose that $f = gh$, where $f \in \mathbb{Z}[X]$ and $g, h \in \mathbb{Q}[X]$. If we multiply both sides of this equation by the least common multiple of the denominators of the coefficients in g and h, we obtain $af = b g_0 h_0$, where $a, b \in \mathbb{Z}$ and g_0, h_0 are primitive polynomials in $\mathbb{Z}[X]$. By Gauss's Lemma, $a d(f) = b$, so that we can divide by a to obtain: $f = d(f) g_0 h_0$, which is a factorization of f over $\mathbb{Z}$. □

Eisenstein Irreducibility Criterion. Let

$$f(X) = X^n + a_1 X^{n-1} + \cdots + a_{n-1} X + a_n$$

be a monic polynomial over $\mathbb{Z}$ whose coefficients $a_1, \ldots, a_n$ are all divisible by some prime p, but whose coefficient a_n is not divisible by p^2. Then $f(X)$ is irreducible in $\mathbb{Q}[X]$.

Proof. If we suppose the contrary and use the corollary to Gauss's Lemma, we can write f as a product of two monic polynomials over $\mathbb{Z}$:

$$f(X) = (X^s + b_1 X^{s-1} + \cdots + b_s)(X^t + c_1 X^{t-1} + \cdots + c_t), \quad s, t > 0 .$$

This factorization is preserved when we pass to the quotient ring $\mathbb{Z}[X]/(p) \cong \mathbb{Z}_p[X]$, whose elements are obtained from polynomials with integral coefficients by reducing modulo p. By assumption, $\bar{a}_i = \bar{0}$, where $\bar{a}_i$ is the residue class mod p of a_i.

But $\mathbb{Z}_p[X]$ is a unique factorization domain, by the corollary to Theorem 4. Comparing the two decompositions

$$X^s X^t = (X^s + \bar{b}_1 X^{s-1} + \cdots)(X^t + \bar{c}_1 X^{t-1} + \cdots), \qquad s + t = n ,$$

we come to the conclusion that $\bar{b}_i = \bar{0} = \bar{c}_j$, i.e., all of the coefficients b_i and c_j are divisible by p. In that case $a_n = b_s c_t$ is divisible by p^2, a contradiction. This proves the Eisenstein Irreducibility Criterion. □

Remark. The above criterion also applies when the leading coefficient a_0 is not 1 but is still prime to p.

Example. The polynomial $f(X) = X^{p-1} + X^{p-2} + \cdots + X + 1$ is irreducible over $\mathbb{Q}$ for any prime p.

To see this, if suffices to notice that irreducibility of f is equivalent to irreducibility of the polynomial

$$f(X+1) = \frac{(X+1)^p - 1}{(X+1) - 1} = X^{p-1} + \binom{p}{1}X^{p-2} + \cdots + \binom{p}{p-2}X + \binom{p}{p-1} ,$$

to which the Eisenstein Irreducibility Criterion applies, since all of the coefficients after the leading one are divisible by p to the first power (see Exercise 8 of §4 Ch. 4 for this property of binomial coefficients).

EXERCISES

1. Show that

$$n\mathbb{Z} + m\mathbb{Z} = \mathbb{Z} \cdot \text{g.c.d.}(n, m) ,$$

$$n\mathbb{Z} \cap m\mathbb{Z} = \mathbb{Z} \cdot \text{l.c.m.}(n, m) .$$

2. Let f and g be monic polynomials in $\mathbb{Z}[X]$. Show that in the equation g.c.d. $(f, g) = fu + gv$ with $u, v \in \mathbb{Z}[X]$ we may take u and v so that

$\deg u < \deg g$ and $\deg v < \deg f$.

3. Are the rings $\mathbb{Z}[\sqrt{-3}]$ and $\mathbb{Z}_8[X]$ unique factorization domains?

4. Factor $X^n - 1$ into irreducible factors in $\mathbb{Z}[X]$ for $5 \leq n \leq 12$.

5. Prove that the irreducible factors in the factorization of a homogeneous polynomial

$$f(X,Y) = a_0 X^n + a_1 X^{n-1} Y + \cdots + a_{n-1} XY^{n-1} + a_n Y^n \in \mathbb{Q}[X, Y]$$

are homogeneous, and that $f(X, Y)$ is irreducible if and only if the polynomial $f(X, 1) = = a_0 X^n + a_1 X^{n-1} + \cdots + a_{n-1} X + a_n \in \mathbb{Q}[X]$ is irreducible.

6. Let K be a field, and let $f(X) = \sum_{i \geq 0} a_i X^i$ be a formal power series in $K[[X]]$ (see Exercise 5 in §2). The condition $a_0 \neq 0$, or equivalently $\omega(f) = 0$, is necessary and sufficient for there to exist a power series $g(X) \in K[[X]]$ such that $fg = 1$. For example, $(1 - X)^{-1} = \sum_{i \geq 0} X^i$. The element X is the only prime in $K[[X]]$ (except for associated elements). $K[[X]]$ is a unique factorization domain. Prove these assertions.

7. Prove that $\det(x_{ij}) = \sum_{\pi \in S_n} \epsilon_\pi x_{\pi(1),1} \cdots x_{\pi(n),n}$ is an irreducible homogeneous polynomial of degree n in the n^2 independent variables x_{ij}.

§4. The field of fractions

1. <u>Construction of the field of fractions of an integral domain.</u> The last two sections established many properties that $\mathbb{Z}$ and $K[X]$ have in common. Our next goal is to imbed $K[X]$ in a field (just as $\mathbb{Z}$ is imbedded in $\mathbb{Q}$). But it is actually no harder to do the same for any integral domain A.

Consider the set $A \times A^*$ $(A^* = A\setminus\{0\})$ of all pairs (a,b) of elements $a,b \in A$ with $b \neq 0$. We partition this set into classes by considering two pairs (a,b) and (c,d) to belong to the same class if $ad = bc$; in that case we write $(a,b) \sim (c,d)$. We obviously always have $(a,b) \sim (a,b)$, and also $(a,b) \sim (c,d) \Rightarrow (c,d) \sim (a,b)$. Finally, $(a,b) \sim (c,d)$ and $(c,d) \sim (e,f) \Rightarrow (a,b) \sim (e,f)$, since $ad = bc$ and $cf = de$ imply that $adf = bcf = bde$, i.e., $d(af - be) = 0$. But $d \neq 0$, and, since A is an integral domain, we obtain $af = be$, which means that $(a,b) \sim (e,f)$. Thus, the relation $\sim$ is reflexive, symmetric, and transitive, i.e. (see §6 of Ch. 1), it is an equivalence relation on the set $A \times A^*$ and so gives a partition of $A \times A^*$ into disjoint classes.

Let $Q(A)$ be the set of all equivalence classes, i.e., the quotient set $A \times A^*/\sim$. We shall let $[a,b]$ denote the class in which (a,b) lies. By definition,

$$[a,b] = [c,d] \Longleftrightarrow ad = bc \quad . \tag{1}$$

If we define addition and multiplication operations as follows on the set $A \times A^*$:

$$(a,b) + (c,d) = (ad + bc, bd); \qquad (a,b)(c,d) = (ac, bd)$$

(this makes sense, since $bd \neq 0$ whenever $b \neq 0$ and $d \neq 0$), then these binary operations can be carried over to $Q(A)$. To see this, we must show that

$$(a',b') \sim (a,b) \Longrightarrow \begin{cases} (a,b) + (c,d) \sim (a',b') + (c,d) \\ (a,b) \cdot (c,d) \sim (a',b') \cdot (c,d) \end{cases} .$$

The equivalences on the right will hold if

$$(ad + bc)b'd = (a'd + b'c)bd \quad ,$$

$$ac \cdot b'd = a'c \cdot bd \quad ,$$

and these equations follow immediately from the condition $a'b = ab'$. We have a similar result if (c,d) is replaced by (c',d'), where $cd' = c'd$. Thus, the sum and product of two elements in $Q(A)$ do not depend on the choice of representatives of the equivalence classes, and we have:

$$[a,b] + [c,d] = [ad + bc, bd]; \qquad [a,b][c,d] = [ac, bd] \quad . \tag{2}$$

Here we really should write $[a,b] \oplus [c,d]$ and $[a,b] \odot [c,d]$, but there is no loss of clarity if we use the same symbols $+$ and $\cdot$ as before for addition and multiplication in $Q(A)$.

We now show that $Q(A)$ is a field under the operations (2). First, the relations

$$[a,b] + ([c,d] + [e,f]) = [a,b] + [cf + de, df] = [adf + bcf + bde, bdf] \ ,$$

$$([a,b] + [c,d]) + [e,f] = [ad + bc, bd] + [e,f] = [adf + bcf + bde, bdf]$$

imply the associative law for addition. Associativity of multiplication is obvious. Next, the relations

$$([a,b] + [c,d]) \cdot [e,f] = [ade + bce, dbf] \ ,$$

$$[a,b][e,f] + [c,d][e,f] = [adef + bcef, bfdf] = [(ade + bce)f, (bdf)f]$$

and the condition (1) for two equivalence classes to be equal show that the distributive law holds. It is just as easy to check that addition and multiplication are commutative. The zero for addition is $[0,1]$ (since $[0,1] + [a,b] = [a,b]$) and the identity for multiplication is $[1,1]$. Next, $-[a,b] = [-a,b]$, since $[a,b] + [-a,b] = [0,b^2] = [0,1]$. So far we have shown that $Q(A)$ is a commutative ring with unit. If $[a,b] \neq [0,1]$, then $a \neq 0$ in A, so that $[b,a] \in Q(A)$ and $[a,b][b,a] = [1,1]$; thus, $[b,a]$ is the multiplicative inverse of $[a,b] \neq [0,1]$. We have thus proved that $Q(A)$ is a field.

The map $a \mapsto [a,1]$ is an injective map $f : A \to Q(A)$ which is a ring homomorphism. (It is easy to check that $f(a + b) = f(a) + f(b)$, $f(ab) = f(a)f(b)$, and $a \neq b \Rightarrow f(a) \neq f(b)$.) For any element $x = [a,b] \in Q(A)$ we have

$$[b,1]x = [a,1] \ ,$$

so that x is the "ratio" $f(a)/f(b)$ of elements in $f(A)$. For this reason, $Q(A)$ is called the <u>field of fractions</u> of A.

It is convenient to identify every element $a \in A$ with its image $f(a) = [a, 1] \in Q(A)$, i.e., to identify A with $f(A)$. We then call the element $[a,b]$ a fraction and write it in the usual form

$$[a,b] = \frac{a}{b} \quad .$$

By now it should be clear that the above rules for operating with the equivalence classes $[a,b]$ repeat the rules for operating with fractions in a field (see (10) in Subsection 5 of §4 Ch. 4). We have proved

THEOREM 1. *For every integral domain* A *there exists a field of fractions* $Q(A)$ *whose elements have the form* a/b, $a \in A$, $0 \neq b \in A$. *The operations in* $Q(A)$ *are given by* (1) *and* (2), *where* $[a,b]$ *is replaced by* a/b. □

The construction of the field of fractions of an integral domain is used fairly often in mathematics. The simplest example $Q(\mathbb{Z}) = \mathbb{Q}$ shows that it is a very natural idea. It is easy to see (the reader should check this!) that $Q(A) \cong A$ if A is a field.

Remark. It can be proved that, if the integral domain A is a subring of a field K in which every element x can be written as a ratio of two elements of A, then $K \cong Q(A)$. For example, $\mathbb{Q}(\sqrt{d}) = Q(\mathbb{Z}[\sqrt{d}])$.

2. *The field of rational functions.* Let K be a field, and let $K[X]$ be the ring of polynomials over K. The field of fractions $Q(K[X])$ of $K[X]$ is denoted $K(X)$ (i.e., square brackets are replaced by parentheses) and is called the *field of rational functions* of the variable X with coefficients in K.

Note that the field of rational functions $K(X)$ always contains an infinite number of elements. The characteristic of this field is the same as the characteristic of K. Thus, $\mathbb{F}_p(X)$ is an example of an infinite field of characteristic $p > 0$.

Every rational function in $K(X)$ can be written (in fact, in many ways) in the form f/g, where f and g are polynomials in $K[X]$ and $g \neq 0$. By definition

$f/g = f_1/g_1 \Leftrightarrow fg_1 = f_1 g$. We call f the <u>numerator</u> and g the <u>denominator</u> of the rational function f/g. The rational function does not change if both the numerator and denominator are multiplied by the same non-zero polynomial or if a non-zero common factor is canceled. In particular the (positive, negative, or zero) integer $\deg f - \deg g$ depends only on the rational function, and not on the particular way it is written in the form f/g. This number is called the <u>degree</u> of the rational function. A rational function is said to be <u>in lowest terms</u> if its numerator is relatively prime to its denominator. Up to multiplication by an element of K in the numerator and denominator, any rational function can be uniquely written as the ratio of polynomials in lowest terms. Namely, if we have the rational function in the form f/g, we can divide f and g by g.c.d. (f, g) to obtain a fraction in lowest terms; and if f/g and f_1/g_1 are two rational functions in lowest terms which are equal, we have $fg_1 = f_1 g$, from which it follows that $f = cf_1$, $c \in K$, and $g = cg_1$ (use the corollary to Theorem 4 of §3).

If $\deg(f/g) = \deg f - \deg g < 0$ and f/g is in lowest terms, we call f/g a <u>proper</u> fraction. (The zero polynomial is also considered to be a proper fraction, since by convention $\deg 0 = -\infty$.)

THEOREM 2. <u>Every rational function in</u> $K(X)$ <u>can be uniquely written as the sum of a polynomial and a proper fraction.</u>

<u>Proof.</u> If we apply the division algorithm to the numerator and denominator of f/g, we obtain $f = qg + r$, where $\deg r < \deg g$. Then $f/g = q + r/g$ is written in the desired form. If we also have $f/g = \overline{q} + \overline{r}/\overline{g}$ $(\overline{q}, \overline{r}, \overline{g} \in K[X]$, $\deg \overline{r} < \deg \overline{g})$, then we obtain

$$q - \overline{q} = \overline{r}/\overline{g} - r/g = (\overline{r}g - r\overline{g})/g\overline{g} \quad .$$

Since $q - \overline{q} \in K[X]$ and

$$\deg((\overline{r}g - r\overline{g})/g\overline{g}) = \deg(\overline{r}g - r\overline{g}) - \deg g - \deg \overline{g} < 0 \quad ,$$

this can only occur if $q - \overline{q} = 0$ and $r/g = \overline{r}/\overline{g}$. □

3. Primary rational functions. A proper rational function $f/g \in K(X)$ is called primary if $g = p^n$, $n \geq 1$, where $p = p(X)$ is an irreducible polynomial and $\deg f < \deg p$.

The fundamental theorem on rational functions is

THEOREM 3. Every proper rational function can be uniquely represented as a sum of primary rational functions.

The proof of Theorem 3 is divided into two parts -- existence and uniqueness of the desired representation.

I. Let $f/g \in K(X)$ be the given proper rational function. Without loss of generality we may assume that g is monic. Suppose that $g = g_1 g_2$ is the product of two relatively prime monic polynomials. According to the results of §3, we have the relation

$$1 = u_1 g_1 + u_2 g_2$$

for some $u_1, u_2 \in K[X]$. Multiplying both sides by f, we obtain

$$f = fu_1 g_1 + fu_2 g_2 \quad .$$

If $fu_1 = qg_2 + v_2$ with $\deg v_2 < \deg g_2$, then

$$f = v_1 g_2 + v_2 g_1 \quad , \tag{3}$$

where $v_1 = qg_1 + fu_2$. Since $\deg v_2 < \deg g_2$ and $\deg f < \deg g$ (because the rational function f/g is proper), it follows that (3) can only hold if $\deg v_1 < \deg g_1$.

Dividing both sides of (3) by $g_1 g_2$, we obtain an expression for f/g as a sum of proper rational functions:

$$f/g = v_1/g_1 + v_2/g_2 \quad .$$

If either g_1 or g_2 can be written as a product of two relatively prime polynomials, then we repeat the procedure. We finally arrive at the expression

$$\frac{f}{g} = \sum_{i=1}^{m} \frac{a_i}{p_i^{n_i}} , \tag{4}$$

where g.c.d. $(a_i, p_i) = 1$ and $\deg a_i < n_i \deg p_i$ for each i. Here the denominators are the powers $p_i^{n_i}$ of the monic irreducible polynomials in the factorization of g:

$$g = p_1^{n_1} p_2^{n_2} \cdots p_m^{n_m} \tag{5}$$

$(p_i \neq p_j$ for $i \neq j)$.

We now further decompose each proper fraction a/p^n. Since $\deg a < n \deg p$, the division algorithm gives us the sequence of equations

$$\begin{aligned} a &= q_1 p^{n-1} + r_1 , \\ r_1 &= q_2 p^{n-2} + r_2 , \\ &\cdots\cdots\cdots\cdots \\ r_{n-2} &= q_{n-1} p + r_{n-1} , \\ r_{n-1} &= q_n , \end{aligned}$$

where $\deg q_i < \deg p$ for all i. Since these equations give

$$a = q_1 p^{n-1} + q_2 p^{n-2} + \cdots + q_{n-1} p + q_n ,$$

we have

$$\frac{a}{p^n} = \frac{q_1}{p} + \frac{q_2}{p^2} + \cdots + \frac{q_{n-1}}{p^{n-1}} + \frac{q_n}{p^n} .$$

Since $\deg q_i < \deg p$, the rational functions q_i / p^i are primary, so we have the desired expression for f/g if we write each term in (4) in this form.

II. We now prove uniqueness. Suppose that, in addition to the expression derived above

$$\frac{f}{g} = \sum_{i=1}^{m} \left(\sum_{j=1}^{n_i} \frac{a_{ij}}{p_i^j} \right), \qquad \deg a_{ij} < \deg p_i \tag{6}$$

for the proper rational function f/g as a sum of primary rational functions, we had another such expression

$$\frac{f}{g} = \sum_{k=1}^{\mu} \left(\sum_{\ell=1}^{\nu_k} \frac{b_{k\ell}}{q_k^{\ell}} \right)$$

perhaps involving terms $b_{k\ell}/q_k^{\ell}$ whose denominator q_k^{ℓ} does not occur in (6). But if we add terms with zero values for a_{ij} and $b_{k\ell}$ to the two expressions for f/g, we may assume that the denominators are the same in the two expressions. Then, subtracting them, we obtain

$$\sum_{i=1}^{M} \left(\sum_{j=1}^{N_i} \frac{a_{ij} - b_{ij}}{p_i^j} \right) = 0 \quad . \tag{7}$$

Here we have M instead of m because of the possible addition of terms in which some of the q_k are taken for the p_i (namely, when $i > m$). The N_i are chosen so that

$$a_{i,N_i} - b_{i,N_i} \neq 0 \quad . \tag{8}$$

Multiplying (7) by $\prod_{i=1}^{M} p_i^{N_i}$, we obtain the polynomial identity

$$\left(a_{1,N_1} - b_{1,N_1}\right) \prod_{i=2}^{M} p_i^{N_i} + p_1 u = 0 \quad .$$

The exact nature of the polynomial u does not concern us. The important thing is that this equality implies that p_1 divides $\left(a_{1,N_1} - b_{1,N_1}\right) \prod_{i=2}^{M} p_i^{N_i}$. But g.c.d. $\left(\prod_{i=2}^{M} p_i^{N_i}, p_i \right) = 1$, so that we must have $p_1 \,\|\, \left(a_{1,N_1} - b_{1,N_1}\right)$. However, $\deg\left(a_{1,N_1} - b_{1,N_1}\right) \leq \max\left\{\deg a_{1,N_1}, \deg b_{1,N_1}\right\} < \deg p_1$. Hence

$a_{1,N_1} - b_{1,N_1} = 0$, which contradicts (8). □

The proof of Theorem 3 is completely constructive once we know the factorization (4), and can be used to actually write a proper rational function as a sum of primary rational functions.

Note that if $g = (X-c)^n h(X)$ and $h(c) \neq 0$, then for any b_1

$$\frac{f}{g} = \frac{b_1}{(X-c)^n} + \frac{f - b_1 h}{(X-c)^n h} .$$

Setting $b_1 = f(c)/h(c)$, we obtain $f(c) - b_1 h(c) = 0$, and hence $f - b_1 h = (X-c) f_1$ for some f_1 (see §1 and also Chapter 6). Thus,

$$\frac{f - b_1 h}{(X-c)^n h} = \frac{f_1}{(X-c)^{n-1} h} .$$

Applying the same procedure to this rational function, we again lower the power of $X - c$ in the denominator by one, and so on. After n steps we arrive at the expansion

$$\frac{f}{g} = \frac{f}{(X-c)^n h} = \frac{f_0}{h} + \sum_{i=1}^{n} \frac{b_{1+n-i}}{(X-c)^i} , \qquad b_k \in K .$$

If h (and so also $g = (X-c)^n h$) decomposes completely into linear factors, then, if we go through this procedure for each factor $(X-c_i)^k$ in g, we eventually obtain the expansion for f/g in the theorem.

If we know that all of the irreducible factors p_i in (4) are linear or quadratic (this will be the case if $K = \mathbb{R}$), then the primary rational functions will have the form

$$\frac{d}{(X-c)^n} \quad \text{or} \quad \frac{dX + e}{(X^2 + aX + b)^n} ; \qquad a, b, c, d, e \in K , \tag{9}$$

and it is also convenient to use the so-called <u>method of undetermined coefficients.</u> One first writes f/g as a sum of fractions of the form (9), then multiplies both sides by g, gathers together terms with the same power of X, and equates coefficients of each power of X on both sides of the equation. In the next chapter we shall see that this procedure always works if K is the field of real or complex numbers. And it is in $\mathbb{R}(X)$ and $\mathbb{C}(X)$ that we often want to decompose into primary rational functions, because this is a necessary step whenever we integrate rational functions. This technique is also called "partial fraction decomposition".

EXERCISES

1. Construct the field of fractions $\mathbb{R}((X))$ of the ring $\mathbb{R}[[X]]$ of formal power series in X with real coefficients. Using Exercise 6 of §3, show that every element of the field $\mathbb{R}((X))$ has the form of a so-called meromorphic power series

$$\varphi(X) = a_{-m}X^{-m} + a_{-m+1}X^{-m+1} + \cdots + a_{-1}X^{-1} + a_0 + a_1X + a_2X^2 + \cdots, \quad a_i \in \mathbb{R},$$

in which we allow a finite number of terms with negative exponents. In other words, we can write $\varphi(X) = X^{-m} f(X)$, where $f(X)$ is an ordinary power series in $\mathbb{R}[[X]]$.

2. Let $\mathbb{R}(X,Y)$ (respectively, $\mathbb{R}((X,Y))$) denote the field of fractions of the polynomial ring $\mathbb{R}[X,Y]$ (respectively, of the integral domain $\mathbb{R}[[X,Y]]$; see Exercise 7 of §2). Show that

$$\mathbb{R}(X,Y) = (\mathbb{R}(X))(Y) = (\mathbb{R}[X])(Y) \quad .$$

Are the fields $\mathbb{R}((X,Y))$ and $(\mathbb{R}((X)))((Y))$ isomorphic?

3. Suppose that the infinite sequence of real numbers $a_0, a_1, a_2, \ldots$ is periodic from some point on. Show that the power series $f(X) = a_0 + a_1X + a_2X^2 + \cdots$ can be written as a rational function in $\mathbb{R}(X)$.

4. Let R be a commutative ring with unit 1 (not necessarily an integral domain), let M be a submonoid of the multiplicative monoid of R, and let $S = R \times M$. Prove that the following binary relation $\Gamma \subset S^2$ is an equivalence relation on S: $\Gamma = \{((a,b),(c,d)) \in S^2 \mid (ad-bc)u = 0 \text{ for some } u \in M\}$.

5. Copying the proof of Theorem 1, show that the quotient set S/Γ, where Γ is the equivalence relation in Problem 4, has the structure of a commutative ring

with unit. This ring $Q_M(R)$ is called the <u>ring of fractions of</u> R <u>relative to</u> M. If R is an integral domain and $M = R^*$, then we obtain the usual field of fractions $Q(R)$.

Chapter 6. Roots of Polynomials

We now take up what used to be the raison d'être of algebra, the roots of polynomials. This subject has ceased to dominate algebra, but its importance is indisputable. After all, many problems in mathematics ultimately boil down to the determination of the roots of certain specific polynomials or some information about the set of roots. We shall only be able to discuss the simplest properties of roots, but those properties will be enough to convey the full importance of the special place occupied by the field $\mathbb{C}$ of complex numbers.

§1. General properties of roots

1. Roots and linear factors. Let A be a commutative ring with unit that is contained in an integral domain R.

Definition. An element $c \in R$ is called a root (or a zero) of the polynomial $f \in A[X]$ if $f(c) = 0$. We also say that c is a root of the equation $f(x) = 0$.

It is clear why we have to consider roots lying in a larger ring R than the ring A in which we found all of the coefficients of f. Namely, remember the polynomial

$f(X) = X^2 + 1 \in \mathbb{R}[X]$, which does not have any roots in $\mathbb{R}$ but has the two roots i and $-i$ in $\mathbb{C}$.

But we first consider the case $R = A$.

THEOREM 1 (Bezout's Theorem). An element $c \in A$ is a root of $f \in A[X]$ if and only if $X - c$ divides f in the ring $A[X]$.

Proof. This theorem follows from a more general assertion that we could have proved much earlier. Namely, the division algorithm (Theorem 5 of §2 Ch. 5) say that $f(X) = (X-c)\,q(X) + r(X)$, where $\deg r(X) < \deg(X-c) = 1$. Hence, $r(X)$ is a constant. Substituting c in place of X (i.e., applying the map Π_c in Theorem 2 of §2 Ch. 5) gives $f(c) = r$, so that we always have

$$f(X) = (X - c)\,q(X) + f(c) \quad . \tag{1}$$

In particular, $f(c) = 0 \Leftrightarrow f(X) = (X-c)\,q(X)$. □

Dividing a polynomial $f(X)$ with coefficients in an integral domain A by a linear polynomial $X - c$ is best done by the so-called Horner method, (also known as "synthetic division") which is simpler than the division algorithm. Namely, let

$$f(X) = a_0 X^n + a_1 X^{n-1} + \cdots + a_n , \qquad a_i \in A \quad .$$

According to formula (1),

$$q(X) = b_0 X^{n-1} + b_1 X^{n-2} + \cdots + b_{n-1} , \qquad b_j \in A \quad .$$

Comparing the coefficients of each power of X on both sides of formula (1) (beginning with the highest powers), after a little rearranging we obtain

$b_0 = a_0$	...	$b_k = b_{k-1}c + a_k$	...	$b_{n-1} = b_{n-2}c + a_{n-1}$	$f(c) = b_{n-1}c + a_n$	(2)

so that we have also computed the value of f at $X = c$. These formulas are convenient to use in computations.

In view of Theorem 1 it is natural to introduce a more general

Definition. An element $c \in A$ is called a k-fold root (or k-fold zero) of $f \in A[X]$ if f is divisible by $(X-c)^k$ but is not divisible by $(X-c)^{k+1}$. k is called the multiplicity of the root. A 1-fold root is called a simple root, and 2- and 3-fold roots are called double and triple roots, respectively.

Thus, $c \in A$ is a root of multiplicity k of $f \in A[X]$ if and only if $f(X) = (X-c)^k g(X)$, where g. c. d. $(X-c, g(X)) = 1$. By formula (1), the latter condition can also be expressed as: $g(c) \neq 0$. Note that, by Theorem 1 of §2 Ch. 5 we have $\deg f = k + \deg g$, so that $k \leq \deg f$. We now prove the important

THEOREM 2. Let A be an integral domain, let $f \neq 0$ be a polynomial in $A[X]$, and let $c_1, \ldots, c_r \in A$ be roots of f of multiplicities $k_1, \ldots, k_r$, respectively. Then

$$f(X) = (X-c_1)^{k_1} \cdots (X-c_r)^{k_r} g(X) \quad ,$$

$g(X) \in A[X]$, $g(c_i) \neq 0$, $i = 1, \ldots, r$.

In particular, the number of roots of a polynomial $f \in A[X]$ counting multiplicity does not exceed the degree of the polynomial, i. e.,

$$k_1 + k_2 + \cdots + k_r \leq \deg f \quad . \tag{3}$$

Proof. Theorem 2 can be proved by considering f as a polynomial over the field of fractions $Q(A)$ and then using the uniqueness of the factorization of f in the ring $Q(A)[X]$, i. e., the uniqueness of the prime factors $X-c_1, \ldots, X-c_r$. But there is no real need to use all these general facts (the results of §3 and 4 of Ch. 5); we shall give a direct argument.

We shall prove that f is divisible by $(X-c_1)^{k_1} \cdots (X-c_r)^{k_r}$ by induction on r; after that, the inequality (3) will follow because $\deg f = k_1 + \cdots + k_r + \deg g$. If

$r = 1$, the assertion follows directly from the definition of a multiple root. Suppose that we already know that

$$f(X) = (X - c_1)^{k_1} \ldots (X - c_{r-1})^{k_{r-1}} h(X) \quad .$$

Since we have $c_r - c_1 \neq 0, \ldots, c_r - c_{r-1} \neq 0$, and A is an integral domain, it follows that c_r is not a root of the polynomial $(X - c_1)^{k_1} \ldots (X - c_{r-1})^{k_{r-1}}$. But c_r is a k_r-fold root of f, i.e., $f(X) = (X - c_r)^{k_r} u(X)$. Hence $h(c_r) = 0$. Let c_r be an s-fold root of h, so that $h(X) = (X - c_r)^s v(X)$, $s \leq k_r$. We have

$$(X - c_r)^{k_r} u(X) = f(X) = (X - c_1)^{k_1} \ldots (X - c_{r-1})^{k_{r-1}} (X - c_r)^s v(X) \quad .$$

Using the cancellation law in the integral domain $A[X]$, we come to the conclusion that $s = k_r$, which completes the induction step. This proves Theorem 2. □

If we don't assume that A is an integral domain, Theorem 2 is no longer true. For example, let $A = \mathbb{Z}_8$, and let $f(X) = X^3$. Then $f(0) = f(2) = f(4) = f(6) = 0$, so the cubic polynomial f has four roots. In addition, the factorization of f in $\mathbb{Z}_8[X]$ is not unique: $f(X) = X^3 = X(X-4)^2 = (X-2)(X^2 + 2X + 4) = (X-6)(X^2 - 2X + 4)$.

Theorem 2 implies the following

COROLLARY. Let A be an integral domain, and let $f, g \in A[X]$ be two polynomials of degree $\leq n$. If f and g take the same value when $n + 1$ different elements of A are substituted in place of X, then $f = g$.

Proof. Set $h = f - g$, so that $\deg h \leq n$. But assumption, $h(c_1) = \cdots = h(c_{n+1}) = 0$ for distinct elements $c_1, \ldots, c_{n+1} \in A$, i.e., h has at least $n + 1$ roots. But, since $\deg h \leq n$, by (3) this can only happen if $h = 0$, i.e., if $f = g$. □

2. Polynomial functions. The corollary to Theorem 2 allows us to answer the question mentioned before (see Subsection 1 of §2 Ch. 5) of the relationship between the

function theoretic and algebraic points of view on polynomials. Every polynomial $f \in A[X]$ corresponds to a function

$$\tilde{f} : a \longmapsto f(a), \qquad \forall a \in A \quad .$$

The set of all such functions is a ring A_{pol}, called the ring of <u>polynomial functions.</u> It is a subring of the ring $A^A = \{A \to A\}$ of all functions from A to A with point-wise addition and multiplication (see Example 3 in Subsection 1 of §4 Ch. 4 and also Theorem 2 of §2 Ch. 5). Polynomial functions in several variables are defined in a completely analogous manner.

We have already mentioned, by way of example, that the polynomial $X^2 + X \in \mathbb{F}_2[X]$ gives the zero function. In general, if $f(X) \in \mathbb{F}_p[X]$ is a polynomial having the form $(X^p - X)\,g(X)$, then the corresponding function $\tilde{f}$ is the zero function, since $x^p - x = x(x^{p-1} - 1) = 0$ for all x in the field of p elements. It is only when $\deg f \leq p - 1$ that we can say that a polynomial $f \in \mathbb{F}_p[X]$ is determined once the function $\tilde{f}$ is known. An arbitrary polynomial $f \in \mathbb{F}_p[X]$ can be replaced by a uniquely determined <u>reduced polynomial</u> f^* of degree $\leq p - 1$ by taking f^* to be the remainder when f is divided by $X^p - X$. Then, obviously, $\tilde{f} = \tilde{f}^*$.

But the situation is much simpler for infinite fields.

THEOREM 3. <u>If A is an integral domain with an infinite number of elements, then the map $f \mapsto \tilde{f}$ from $A[X]$ to A_{pol} is an isomorphism of rings.</u>

<u>Proof.</u> This follows immediately from the corollary to Theorem 2, since we need only check that $\tilde{f} \neq 0$ if $f \neq 0$; if $\deg f = n$, $f \neq 0$, then the corollary to Theorem 2 says that f can have at most n zeros in A, so that it is impossible to have $f(a) = 0$ for all $a \in A$. □

Because of Theorem 3, whenever we have an infinite field K we identify polynomials $f(X)$ over K with the corresponding polynomial functions $f(x)$ (where we use a small x to emphasize that we are thinking of f as a function). The question remains of how, in practice, one reconstructs a polynomial f if one knows several values of $f(x)$.

This so-called "interpolation" problem is stated more precisely as follows. Let $b_0, b_1, \dots, b_n$ be $n+1$ elements of the field K, and let $c_0, c_1, \dots, c_n$ be $n+1$ distinct elements of K. We want to find a polynomial $f \in K[X]$ of degree $\le n$ such that $f(c_i) = b_i$ for $i = 0, 1, \dots, n$. According to the corollary to Theorem 2, if a solution to this problem exists, then it is unique. But there is always a solution to the problem, given by the Lagrange interpolation formula

$$f(X) = \sum_{i=0}^{n} b_i \frac{(X - c_0) \dots (X - c_{i-1})(X - c_{i+1}) \dots (X - c_n)}{(c_i - c_0) \dots (c_i - c_{i-1})(c_i - c_{i+1}) \dots (c_i - c_n)} . \qquad (4)$$

By the way, the existence and uniqueness of the required $f(X) = a_0 X^n + a_1 X^{n-1} + \dots + a_n$ can also be seen by considering the linear system

$$\begin{aligned} a_0 c_0^n + a_1 c_0^{n-1} + \dots + a_n &= b_0 , \\ \dots\dots\dots\dots\dots\dots\dots\dots \\ a_0 c_n^n + a_1 c_n^{n-1} + \dots + a_n &= b_n \end{aligned}$$

for the unknown coefficients $a_0, \dots, a_n$. The determinant of this system is the Vandermonde determinant, which is non-zero, so that the solution can be found by Cramer's rule. But the formula (4) is more convenient, because it is simple and easily remembered. In some situations it is more useful to use the Newton interpolation formula

$$f(X) = u_0 + u_1(X - c_0) + \dots + u_n(X - c_0)(X - c_1) \dots (X - c_{n-1}) , \qquad (5)$$

where the coefficients $u_0, u_1, \dots, u_n$ are found by successively substituting the values $X = c_0$, $X = c_1, \dots, X = c_n$. The interpolation formulas (4) and (5) are used in practical applications when computing and graphing a function $\varphi : \mathbb{R} \to \mathbb{R}$ based on a table or experimental data. If one somehow knows that the function φ on an interval I of the real number line behaves like a very "smooth" function, perhaps a polynomial function, one can find a polynomial function that agrees with φ for the known values. The so-called "interpolation points" are the (c_i, b_i) for c_i which fall in the interval I; since $\varphi(c_i) = b_i$, the desired polynomial must pass through these points. Of course, we have just presented the simplest type of interpolation. Entire fields of mathematics are devoted to the delicate questions which arise in selecting the interpolation points and applying the various methods of interpolation. It should also be mentioned that interpolation methods have played an important role in the theory of transcendental numbers (for the definition of algebraic and transcendental numbers, see §2 Ch. 5), so we can say that function theorists, number theorists, and algebraists all have an interest in this subject.

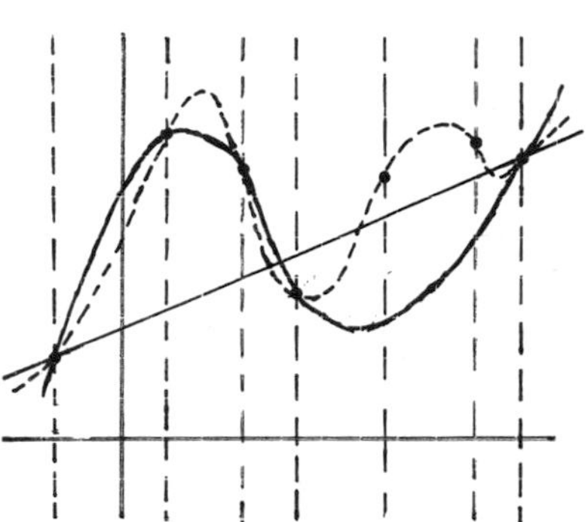

Finally, we note that whenever we are given a rational function $f/g \in K(X)$ in lowest terms (see §4 of Ch. 5) and an extension $F \supset K$ having infinitely many elements, there is a corresponding function whose domain is the set $F_{(f/g)}$ obtained from F by deleting all of the (finitely many) zeros of g in F and whose range is F. It can be shown that distinct f/g always give distinct functions $F_{(f/g)} \to F$. But we shall not need this assertion. One should be careful not to confuse the two notions of an element of $K(X)$ and the corresponding function. For example, the rational function $x \mapsto 1/x$ is not defined at the point $x = 0$, but it makes no sense to ask whether the element $1/X \in K(X)$ is or is not defined.

3. Differentiation in polynomial rings. If we think of polynomials as functions, we naturally come upon the following definition. Let

$$f(X) = a_0 X^n + a_1 X^{n-1} + \cdots + a_{n-1} X + a_n$$

be a polynomial of degree n over a field K. By its derivative, we mean the polynomial

$$f'(X) = n a_0 X^{n-1} + (n-1) a_1 X^{n-2} + \cdots + a_{n-1} \quad . \tag{6}$$

If $K = \mathbb{R}$ is the field of real numbers, and $\tilde{f}$ is the polynomial function corresponding to f, then (6) coincides with the usual definition of the derivative as the limit

$$\lim_{\Delta x \to 0} \frac{\tilde{f}(x + \Delta x) - \tilde{f}(x)}{\Delta x} .$$

But in the case of an arbitrary field K there is no sense in speaking of limits of values of a function (for example, what would this mean in the case of the finite field $\mathbb{Z}_p$?), so it is necessary to use the formal definition (6).

In this abstract setting we still have the well-known formulas from calculus:

$$(\alpha f + \beta g)' = \alpha f' + \beta g' , \qquad \alpha, \beta \in K , \tag{7}$$

$$(f g)' = f'g + fg' . \tag{8}$$

Relation (7) follows immediately from (6) and the definition of the sum of two polynomials. If we use (7) and the definition of the product of two polynomials, we can reduce the verification of (8) to the case when $f = X^k$ and $g = X^\ell$, in which case we have:

$$(X^{k+\ell})' = (k + \ell)\, X^{k+\ell-1} = (k X^{k-1})\, X^\ell + X^k(\ell X^{\ell-1}) = (X^k)'\, X^\ell + X^k (X^\ell)' .$$

As a generalization of (8), we have the following formula, which is easy to prove by induction on k:

$$(f_1 f_2 \cdots f_k)' = \sum_{i=1}^{k} f_1 \cdots f_{i-1} f_i' f_{i+1} \cdots f_k .$$

In particular,

$$(f^k)' = k f^{k-1} f' . \tag{9}$$

If we re-write (7) and (8) in terms of the "differentiation operator" $\frac{d}{dX} : f \mapsto f'$, it might occur to us to consider for any ring R (not necessarily a polynomial ring) maps $\mathfrak{D} : R \to R$ having the two properties

$$\mathcal{D}(u+v) = \mathcal{D}u + \mathcal{D}v \quad , \tag{7'}$$

$$\mathcal{D}(uv) = (\mathcal{D}u)v + u(\mathcal{D}u) \quad . \tag{8'}$$

Any such map from a ring R to itself is called a <u>differentiation.</u> The set $\mathrm{Der}(R)$ of such maps is the point of departure for a major branch of mathematics (<u>Lie groups</u> and <u>Lie algebras</u>).

A generalization of (8') is Leibniz' rule

$$\mathcal{D}^m(uv) = \sum_{k=0}^{m} \binom{m}{k} \mathcal{D}^k u \, \mathcal{D}^{m-k} v \quad , \tag{8''}$$

which is proved by induction on m for $m \geq 1$ (since, if we apply $\mathcal{D}$ to (8") and use (8') and the equality $\binom{m}{k-1} + \binom{m}{k} = \binom{m+1}{k}$, we obtain (8") with $m+1$ in place of m).

If $R = K[X]$ and the differentiation $\mathcal{D}$ satisfies $\mathcal{D}(\lambda f) = \lambda \mathcal{D} f$ for $\lambda \in K$, in addition to (7') and (8'), then it is easy to see using these rules that

$$\mathcal{D} f(X) = f'(X) \mathcal{D} X \quad .$$

Thus, any such differentiation of the polynomial ring $K[X]$ is determined once we know the one polynomial $\mathcal{D}X$. If $\mathcal{D}X = 1$, then $\mathcal{D}$ is the usual differential operator $\frac{d}{dX}$.

4. <u>Multiple factors.</u> The result of differentiating $f(X)$ m times is denoted $f^{(m)}(X)$. It is clear that, if $f(X) = a_0 X^n + a_1 X^{n-1} + \cdots + a_n$, then $f^{(n)}(X) = n!\, a_0$ and $f^{(n+1)}(X) = 0$. In addition, if K is a field of characteristic 0, then $\deg f' = \deg f - 1$. But this is not true if K is a field of characteristic p, since, for example,

$$(X^{kp})' = kpX^{kp-1} = 0 \quad .$$

But even in the general case (allowing fields of characteristic p) some information about polynomials can be obtained by looking at their derivatives. If we divide an arbitrary polynomial $f \in K[X]$ by $(X-c)^2$, where c is an element in some field extension

$F \supset K$, and then write the remainder in the form $(X-c)s+r$, where $s, r \in F$, we obtain: $f = (X-c)^2 t + (X-c)s + r$, where $t \in F[X]$, $s, r \in F$. Differentiating gives: $f' = (X-c)[2t + (X-c)t'] + s$. If we substitute $X = c$, we obtain $r = f(c)$, $s = f'(c)$, i.e.,

$$f(X) = (X-c)^2 t(X) + (X-c) f'(c) + f(c) \quad .$$

We have proved the following assertion.

THEOREM 4. Let K be any field, and let F be an extension of K. A polynomial $f \in K[X]$ has $c \in F$ for a multiple root if and only if $f(c) = f'(c) = 0$. □

Example 1. In any field of characteristic p, the polynomial $X^n - 1$ only has simple roots as long as n is not divisible by p, since the roots of the derivative nX^{n-1} (namely, 0) can not at the same time be roots of $X^n - 1$.

Now suppose that K is a field of characteristic zero. (The reader will not sacrifice anything if he takes K to be $\mathbb{Q}$, $\mathbb{R}$, or $\mathbb{C}$.) The monic irreducible polynomial $p_i(X)$ in the factorization

$$f(X) = \lambda p_1(X)^{k_1} \cdots p_r(x)^{k_r} \quad , \qquad \lambda \in K \quad , \tag{10}$$

is called a k_i-fold factor of f (by analogy with k-fold roots). We mentioned earlier that, in practice, it is rather difficult to obtain the factorization (10) of a given polynomial. We shall briefly describe a method, based on the derivative, which allows us to determine whether or not $f(X)$ has multiple factors over a given field K.

THEOREM 5. Let $p(X)$ be a k-fold irreducible factor of a polynomial $f \in K[X]$, where $k \geq 1$ and $\deg p \geq 1$. Then $p(X)$ is a $(k-1)$-fold factor of the derivative $f'(X)$. In particular, if $k = 1$, then f' is not divisible by $p(X)$.

Proof. By assumption, $f(X) = p(X)^k g(X)$, where $\text{g.c.d.}(p(X), g(X)) = 1$, i.e., $g(X)$ is not divisible by $p(X)$. Applying (8) and (9), we obtain:

$$f'(X) = p(X)^{k-1}[kp'(X)g(X) + p(X)g'(X)] \quad .$$

It suffices to show that the polynomial in square brackets is not divisible by $p(X)$. If $p(X)$ did divide this polynomial, then it would also divide $kp'(X)g(X)$, but this is impossible (see the corollaries to Theorems 3 and 4 in §3 Ch. 5), since $g(X)$ is not divisible by $p(X)$, and $\deg kp'(X) < \deg p(X)$. □

Notice that, in proving Theorem 5, it was essential to know that $p(X)$ is irreducible and that $\operatorname{char} K = 0$.

COROLLARY 1. *If $f(X)$ is a polynomial with coefficients in a field K of characteristic zero, the following two conditions are equivalent:*

(i) *in some extension $F \supset K$, f has a root c of multiplicity k;*

(ii) $f^{(j)}(c) = 0$ *for* $0 \le j \le k-1$, *but* $f^{(k)}(c) \ne 0$.

To prove the corollary, apply Theorem 5 k times with K replaced by F and $p(X)$ replaced by $X - c$. □

COROLLARY 2. *If a polynomial $f \in K[X]$ of degree ≥ 1 has factorization (10), then the factorization of the greatest common divisor of f and its derivative f' is*

$$\text{g.c.d.}(f,f') = p_1(X)^{k_1-1}\, p_2(X)^{k_2-1} \cdots p_r(X)^{k_r-1} . \tag{11}$$

(Here we are taking the g.c.d. to be a monic polynomial, as we may always do.)

Proof. By Theorem 5, each of the prime divisors $p_i(X)$ in the factorization of $f(X)$ (see (10)) occurs in the factorization of $f'(X)$ with exponent $k_i - 1$, i.e.,

$$f'(X) = p_1(X)^{k_1-1}\, p_2(X)^{k_2-1} \cdots p_r(X)^{k_r-1} \cdot u(X) ,$$

where $\text{g.c.d.}(u, p_i) = 1$, $1 \le i \le r$ (we take $p_i(X)^0 = 1$). Hence, by the divisibility criterion in Subsection 2 of §3 Ch. 5, we conclude that $\text{g.c.d.}(f, f')$ is given by (11).

□

Using (11) for g. c. d. (f, f'), we can get rid of multiple factors in f by considering the polynomial

$$g(X) = \frac{f(X)}{\text{g. c. d. }(f, f')} = p_1(X)\, p_2(X) \ \dots \ p_r(X) \quad .$$

This polynomial has the same prime divisors as $f(X)$ but with multiplicity one. It is important to note that the polynomial $g(X)$ can be found without knowing the factorization of f, namely, by using the Euclidean algorithm to find g. c. d. (f, f').

Example 2. The polynomial $f(X) = X^5 - 3X^4 + 2X^3 + 2X^2 - 3X + 1$ and its derivative $f'(X) = 5X^4 - 12X^3 + 6X^2 + 4X - 3$ have greatest common divisor $X^3 - 3X^4 + 3X - 1 = (X-1)^3$. (To see this, we can use the division algorithm in $Q[X]$:

$$f(X) = (\frac{1}{5}X - \frac{3}{25})\, f'(X) - \frac{16}{25}(X^3 - 3X^2 + 3X - 1)$$

$$f'(X) = (5X + 3)(X^3 - 3X^2 + 3X - 1) + 0 \quad ,$$

so $X^3 - 3X^2 + 3X - 1$ is the g. c. d. of f and f'.) The "squarefree" polynomial $g(X) = f(X)/(X-1)^3 = X^2 - 1 = (X-1)(X+1)$ has the two roots ± 1. Thus, $f(X) = (X-1)^4 (X+1)$ has 1 as a 4-fold root and -1 as a simple root.

5. Vieta's formulas. When discussing linear systems, we noted the beneficial effect a good system of notation had on the development of the study of linear equations, in particular, leading to the theory of determinants. This was all accomplished by mathematicians of the eighteenth and early nineteenth centuries. But much earlier, at a time when algebra was still considered to be "solving equations", the perfection of algebraic notation by Vieta and Descartes stimulated the development of the theory of polynomials and algebraic equations. After studying special types of equations with numerical coefficients, which impeded the discovery of general principles, it was a bold step to introduce equations with letter coefficients. As often happens, the development of new notation led to new results. Descartes discovered revolutionary new applications of algebra to geometry. We shall discuss a more modest, but nevertheless important, achievement of his predecessor

Vieta.

Suppose that a monic polynomial $f \in K[X]$ of degree n has n roots $c_1, c_2, \ldots, c_n$ in the field K or in some extension of K, where we allow the possibility of multiple roots, i.e., that some of the c_i are the same. Then, by Theorem 2, we have the factorization

$$f(X) = (X - c_1)(X - c_2) \ldots (X - c_n) \quad .$$

We write $f(X)$ in the usual way in powers of X:

$$f(X) = X^n + a_1 X^{n-1} + \cdots + a_k X^{n-k} + \cdots + a_n \quad ,$$

where we find the coefficients by multiplying together all of the $X - c_i$ and combining similar terms. We obtain the following expressions for the coefficients a_i in terms of the roots c_i:

$$\begin{aligned} a_1 &= -(c_1 + c_2 + \cdots + c_n) \quad , \\ &\cdots\cdots\cdots\cdots\cdots \\ a_k &= (-1)^k \sum_{i_1 < i_2 < \cdots < i_k} c_{i_1} c_{i_2} \cdots c_{i_k} \quad , \\ &\cdots\cdots\cdots\cdots\cdots \\ a_n &= (-1)^n c_1 c_2 \cdots c_n \quad . \end{aligned} \tag{12}$$

These formulas are called <u>Vieta's formulas.</u>

If f were not a monic polynomial, i.e., if its leading coefficient were $a_0 \neq 1$, then the same formulas (12) would give expressions for the ratios a_i / a_0.

Vieta's formulas, which give the explicit relationship between the roots and the coefficients of any monic polynomial, have the remarkable property that they do not change under any permutation of the roots $c_1, \ldots, c_n$. This will lead us to introduce the notion of a <u>symmetric function,</u> in much the same way that the example of determinants led us to the notion of skew-symmetric functions. According to the definition given in the corollary to Theorem 3' in Subsection 2 of §2 Ch. 5, an element π of the symmetric group S_n

acts on a function $\widetilde{f}(x_1, \ldots, x_n)$ of n variables by the rule

$$(\pi\widetilde{f})(x_1, \ldots, x_n) = f\left(x_{\pi^{-1}(1)}, \ldots, x_{\pi^{-1}(n)}\right) .$$

We say that $\widetilde{f}$ is <u>symmetric</u> if $\pi\widetilde{f} = \widetilde{f}$ for all $\pi \in S_n$. As an example of symmetric functions we have for each $k = 1, \ldots, n$ the k-th <u>elementary symmetric function</u> s_k:

$$s_k(x_1, \ldots, x_n) = \sum_{1 \le i_1 < i_2 < \ldots < i_k \le n} x_{i_1} x_{i_2} \ldots x_{i_k} . \tag{13}$$

Using these elementary symmetric functions, we can re-write (12) in the form

$$a_k = (-1)^k s_k(c_1, \ldots, c_n), \qquad k = 1, 2, \ldots, n , \tag{12'}$$

so that the coefficient a_k is (up to a sign) equal to the k-th symmetric function evaluated at the roots of the polynomial. Note that, by definition, $a_k \in K$, although the roots $c_1, \ldots, c_n$ may lie in a larger field $F \supset K$. Here we shall not be concerned with the problem of finding a field F over which f decomposes into linear factors. In some situations we will know for various reasons that f splits entirely into linear factors, i.e., has n roots in the field K.

<u>Example.</u> Consider the polynomial $X^{p-1} - 1$ over the finite field $\mathbb{F}_p$. We know that $x^{p-1} = 1$ for all $x \in \mathbb{F}_p^*$, i.e., all non-zero elements are roots of the polynomial $X^{p-1} - 1$. We thus have the factorization

$$X^{p-1} - 1 = (X-1)(X-2) \ldots (X-(p-1)) . \tag{14}$$

(We are supposing enough familiarity with the field $\mathbb{F}_p$ so that the reader will not be confused if we use the same numbers $1, 2, \ldots, p-1$ to denote both ordinary integers and elements of $\mathbb{F}_p \cong \mathbb{Z}/p\mathbb{Z}$, i.e., cosets $\{i\}_p$.) By (12') and (14) we obtain

$$s_k(1, 2, \ldots, p-1) \equiv 0 \pmod p, \qquad k = 1, 2, \ldots, p-2 ,$$

$$s_{p-1}(1, 2, \ldots, p-1) \equiv -1 \pmod p .$$

The last congruence, when written in the form

$$(p - 1)! + 1 \equiv 0 \pmod p \tag{15}$$

is known as Wilson's Theorem. For (15) to hold for an integer p is actually equivalent to p being a prime number. Namely, we just showed that (15) holds if p is prime. Conversely, $p = p_1 p_2 \Rightarrow (p-1)! = p_1 t \Rightarrow (p-1)! + 1 \not\equiv 0 \pmod{p_1} \Rightarrow (p-1)! + 1 \not\equiv$ $\not\equiv 0 \pmod p$. □

EXERCISES

1. Is the ring of polynomial functions over $\mathbb{F}_p$ an integral domain?

2. Let K be an infinite field, and let f be a non-zero polynomial in $K[X_1, \ldots, X_n]$. Using Theorem 3 and induction on n, prove that there exist $a_1, \ldots, a_n \in K$ such that $f(a_1, \ldots, a_n) \neq 0$. Thus, we have an isomorphism of $K[X_1, \ldots, X_n]$ with the ring of polynomial functions in n variables over K.

3. Show that a non-zero polynomial $f \in \mathbb{Z}_p[X_1, \ldots, X_n]$ of degree $< p$ in each of the variables has the property in Exercise 2, i.e., $f(a_1, \ldots, a_n) \neq 0$ for some $a_1, \ldots, a_n \in \mathbb{Z}_p$. Prove that any polynomial $f \in \mathbb{Z}_p[X_1, \ldots, X_n]$ can be written in the form

$$f(X_1, \ldots, X_n) = \sum_{i=1}^{n} g_i(X_1, \ldots, X_n)(X_i^p - X_i) + f^*(X_1, \ldots, X_n) ,$$

where f^* is a reduced polynomial $(\deg_{X_i} f^* < p,\ i = 1, 2, \ldots, n)$ whose total degree is $\leq \deg f$. Conclude from this that the map $f \mapsto \tilde{f} = \tilde{f^*}$ is an epimorphism from the ring $\mathbb{Z}_p[X_1, \ldots, X_n]$ to the ring of polynomial functions in n variables over $\mathbb{Z}_p$, with kernel $L = \sum_{i=1}^{n} (X_i^p - X_i)\,\mathbb{Z}_p[X_1, \ldots, X_n]$.

4. Prove: Theorem (Chevalley). Let $f(X_1, \ldots, X_n)$ be a homogeneous polynomial over $\mathbb{Z}_p$ of degree $r < n$. Then the equation $f(x_1, \ldots, x_n) = 0$ has at

least one non-trivial solution.

By means of a slightly modified argument (computing the sum $\sum_{x_1,\ldots,x_n \in \mathbb{Z}_p} g(x_1, \ldots, x_n)$ in two ways), prove that the total number of solutions to the equation $f(x_1, \ldots, x_n) = 0$ is always divisible by p.

5. Let $f(x_1, \ldots, x_n)$ be a quadratic form with integral coefficients. Chevalley's Theorem (Exercise 4), stated in the language of congruences, tells us that, if $n \geq 3$, then the congruence

$$f(x_1, \ldots, x_n) \equiv 0 \pmod p$$

has a non-zero solution. Show that all solutions of the congruence $x^2 - 2y^2 \equiv 0 \pmod 5$ are trivial, and hence we cannot remove the condition $r < n$ in Chevalley's Theorem.

6. Show that g.c.d. $(f, f') = 1$ if $\operatorname{char} K = 0$, f is an irreducible polynomial over K, and f' is its derivative.

7. Prove that $f' = 0 \Rightarrow f = \text{const}$ if $f(X)$ is a polynomial over a field of characteristic zero, and that $f' = 0 \Rightarrow f(X)$ is of the form $g(X^p)$ for some g, if $f(X)$ is a polynomial over a field of characteristic $p > 0$.

8. From Subsection 3 we know that every differentiation of the polynomial ring $K[X]$ has the form $T_u : f \mapsto uf'$, where $u \in K[X]$. Prove the following assertions:

(i) The set of polynomials which go to zero under the differentiation is the set of constants, which is a subring of $K[X]$.

(ii) The product $T_u T_v$ is not, in general, a differentiation, but, if $\operatorname{char} K = p > 0$, then $(T_u)^p$ is a differentiation.

(iii) The commutator $[T_u, T_v] = T_u T_v - T_v T_u$ is always a differentiation of the form T_w, where $w = uv' - u'v$.

9. If we are dealing with a polynomial ring $K[X_1, \ldots, X_n]$ in n variables, it

is natural to introduce partial differentiation operators with respect to each variable:

$$\frac{\partial}{\partial X_k} : X_1^{i_1} \cdots X_k^{i_k} \cdots X_n^{i_n} \longmapsto i_k X_1^{i_1} \cdots X_k^{i_k - 1} \cdots X_n^{i_n} .$$

(i) Show that the set of "constants" (polynomials which go to zero) for $\frac{\partial}{\partial X_k}$ is the polynomial ring $K[X_1, \ldots, X_{k-1}, X_{k+1}, \ldots, X_n]$ in $n-1$ variables.

(ii) Let $f(X_1, \ldots, X_n)$ be a homogeneous polynomial of degree m. Prove Euler's identity:

$$\sum_{k=1}^{n} X_k \frac{\partial f}{\partial X_k} = m \cdot f(X_1, \ldots, X_n) .$$

Conversely, if $\operatorname{char} K = 0$, show that any polynomial satisfying Euler's identity must be homogeneous of degree m.

10. Show that a polynomial $X^n + a_1 X^{n-1} + \cdots + a_n \in \mathbb{Z}_2[X]$ has no linear factors if and only if $a_n(1 + \Sigma a_i) \neq 0$. Here is a complete list of all irreducible polynomials over $\mathbb{Z}_2$ of degree $n \leq 3$: $X, X+1, X^2+X+1, X^3+X+1, X^3+X^2+1$. Write down all irreducible polynomials over $\mathbb{Z}_2$ of degree 4 and 5 (there are 3 and 6 of them, respectively).

11. Prove that the polynomial $X^5 - X^2 + 1$ is irreducible over $\mathbb{Z}_2$, and hence over $\mathbb{Q}$.

Similarly, use a congruence mod 3 to prove that $X^5 - X - 1$ is irreducible over $\mathbb{Q}$.

§2. Symmetric polynomials

1. The ring of symmetric polynomials. Following the definition of a symmetric function at the end of the last section, we now introduce the analogous notion in the polynomial ring $A[X_1, \ldots, X_n]$, where A is an integral domain. At first glance, it seems that

Theorem 3 of §1 (and its generalization to polynomials and functions of several variables), which allows us to think of polynomials as a subring of the ring of functions, should make it unnecessary to make a new definition. However, in that theorem we had to assume that the integral domain A has infinitely many elements; we would like to give a universal construction, valid for any A.

So let us again return to the corollary of Theorem 3' in Subsection 2 of §2 Ch. 5, and associate to every permutation $\pi \in S_n$ the automorphism $\tilde{\pi} : A[X_1, \ldots, X_n] \to A[X_1, \ldots, X_n]$ which takes a polynomial $f \in A[X_1, \ldots, X_n]$ to the polynomial πf defined by:

$$(\pi f)(X_1, \ldots, X_n) = f\left(X_{\pi^{-1}(1)}, \ldots, X_{\pi^{-1}(n)}\right) .$$

The polynomial f is called <u>symmetric</u> if $\pi f = f$ for all $\pi \in S_n$. As in the case of functions, we introduce the elementary symmetric polynomials s_k:

$$s_k(X_1, \ldots, X_n) = \sum_{1 \le i_1 < i_2 < \cdots < i_k \le n} X_{i_1} X_{i_2} \ldots X_{i_k}, \quad k = 1, 2, \ldots, n. \tag{1}$$

We can see that the s_k are in fact symmetric either directly or by considering the polynomial

$$f(Y) = (Y - X_1)(Y - X_2) \ldots (Y - X_n) = Y^n - s_1 Y^{n-1} + s_2 Y^{n-2} + \cdots + (-1)^n s_n \tag{2}$$

over the ring $A[X_1, \ldots, X_n]$ in the new variable Y and observing that the expression for $f(Y)$ does not change if we permute the linear factors $Y - X_i$; hence the s_k are symmetric.

Note that if we substitute zero in place of X_n in (2), we obtain $(Y - X_1) \ldots (Y - X_{n-1})Y = Y^n - (s_1)_0 Y^{n-1} + \cdots + (-1)^{n-1}(s_{n-1})_0 Y$, where $(s_k)_0$ is the polynomial obtained by substituting $X_n = 0$ in s_k. Canceling Y from both sides (which is justified by Theorem 3 of §4 Ch. 4, applied to $A[X_1, \ldots, X_n, Y]$), we arrive at the identity

$$(Y - X_1)(Y - X_2) \dots (Y - X_{n-1}) = Y^{n-1} - (s_1)_0 Y^{n-2} + \dots + (-1)^{n-1} (s_{n-1})_0 \ . \qquad (3)$$

Comparing (2) and (3), we conclude that $(s_1)_0, \dots, (s_{n-1})_0$ are the elementary symmetric polynomials in the $n - 1$ variables $X_1, \dots, X_{n-1}$.

Next note that, since $\tilde{\pi}$ is an automorphism of the ring $A[X_1, \dots, X_n]$, it follows that any linear combination of symmetric polynomials and any product of symmetric polynomials are symmetric polynomials. This means that the set of all symmetric polynomials is a subring of $A[X_1, \dots, X_n]$. Our next goal is to discover the structure of this subring.

2. The fundamental theorem on symmetric polynomials. It turns out that the most general way of obtaining symmetric polynomials is as follows. Take any polynomial $g \in A[Y_1, \dots, Y_n]$ and substitute $s_1, \dots, s_n$ in place of $Y_1, \dots, Y_n$, respectively. The resulting polynomial

$$f(X_1, \dots, X_n) = g(s_1(X_1, \dots, X_n), \dots, s_n(X_1, \dots, X_n))$$

is, of course, symmetric. We shall prove that any symmetric polynomial f can be obtained in this way.

Notice that a monomial $Y_1^{i_1} \dots Y_n^{i_n}$ in g becomes a homogeneous polynomial of degree $i_1 + 2i_2 + 3i_3 + \dots + ni_n$ in the variables $X_1, \dots, X_n$ after substituting $Y_k = s_k(X_1, \dots, X_n)$, since s_k is homogeneous of degree k. The sum $i_1 + 2i_2 + \dots + ni_n$ is usually called the weight of the monomial $Y_1^{i_1} \dots Y_n^{i_n}$. By the weight of the polynomial $g(Y_1, \dots, Y_n)$, we mean the maximum weight of a monomial occurring in g.

The fundamental fact about symmetric polynomials is

THEOREM 1. *Let $f \in A[X_1, \dots, X_n]$ be a symmetric polynomial of total degree m over an integral domain A. Then there exists a unique polynomial $g \in A[Y_1, \dots, Y_n]$ such that*

$$f(X_1, \ldots, X_n) = g(s_1, \ldots, s_n) \quad .$$

The polynomial g has weight m.

The proof consists of two parts.

I. Proof that such a polynomial g exists. We use induction on both n and m (see §7 of Ch. 1). If $n = 1$, the theorem is obvious, since $s_1 = X_1$, and $f(X_1) = f(s_1)$. Suppose that we know that g exists for polynomials of $\leq n - 1$ variables. We must prove the assertion for polynomials of degree m in n variables. We now use induction on m. There is nothing to prove if $m = 0$, so we suppose that $m > 0$, and that the existence of g has been established for any polynomial of degree $< m$.

Now let $f(X_1, \ldots, X_n)$ be our given polynomial of total degree m. If we set $X_n = 0$, by the induction assumption we can write

$$f(X_1, \ldots, X_{n-1}, 0) = g_1((s_1)_0, \ldots, (s_{n-1})_0) \quad ,$$

where g_1 is some polynomial in $A[Y_1, \ldots, Y_{n-1}]$ of weight $\leq m$ (the weight could be $< m$, since the degree of f might decrease after substituting $X_n = 0$), and $(s_1)_0, \ldots, (s_{n-1})_0$ are the elementary polynomials in the variables $X_1, \ldots, X_{n-1}$ (see (3)). Obviously, $\deg g_1(s_1(X_1, \ldots, X_n), \ldots, s_{n-1}(X_1, \ldots, X_n)) \leq m$. Hence, the polynomial

$$f_1(X_1, \ldots, X_n) = f(X_1, \ldots, X_n) - g_1(s_1, \ldots, s_{n-1}) \tag{4}$$

has total degree in $X_1, \ldots, X_n$ no greater than m, and, since it is the difference of two symmetric polynomials, is symmetric. In addition, $f_1(X_1, \ldots, X_{n-1}, 0) = 0$, so that X_n divides $f_1 : f_1 = X_n \cdot f^\circ$. But, since f_1 is symmetric, we have $f_1 = \pi^{-1} f_1 = X_{\pi(n)} (\pi^{-1} f^\circ)$, $\forall \pi \in S_n$, i.e., f_1 is divisible by each $X_1, \ldots, X_n$, and hence by the product $s_n = X_1 X_2 \ldots X_n$. Thus,

$$f_1(X_1, \ldots, X_n) = s_n \cdot f_2(X_1, \ldots, X_n) \quad , \tag{5}$$

where f_2 is also a symmetric polynomial, having degree $\deg f_2 = \deg f_1 - n \leq m - n$.

By the induction assumption, there exists a polynomial $g_2(Y_1, \ldots, Y_n)$ of weight $\leq m - n$ such that $f_2(X_1, \ldots, X_n) = g_2(s_1, \ldots, s_n)$. Then, by (4) and (5), we obtain the following expression for f:

$$f(X_1, \ldots, X_n) = g_1(s_1, \ldots, s_{n-1}) + s_n g_2(s_1, \ldots, s_n) \quad ,$$

and we have found the desired polynomial $g = g_1(Y_1, \ldots, Y_{n-1}) + Y_n g_2(Y_1, \ldots, Y_n)$ of weight $\leq m$. Since $\deg f = m$, the weight of g cannot be less than m, and so the weight exactly equals m.

II. Proof that g is unique. If we had two distinct polynomials g_1, g_2 such that $f = g_1(s_1, \ldots, s_n) = g_2(s_1, \ldots, s_n)$, then we would have a polynomial $g(Y_1, \ldots, Y_n) = g_1 - g_2 \neq 0$ for which $g(s_1, \ldots, s_n) = 0$. In other words, $s_1, \ldots, s_n$ would be algebraically dependent over A (see the definition in Subsection 2 of §2 Ch. 5). We use induction on n to show that this is impossible. Supposing that polynomials g for which $g(s_1, \ldots, s_n) = 0$ exist, we choose the one of minimal total degree. Considering g as a polynomial in Y_n over the ring $A[Y_1, \ldots, Y_{n-1}]$, we write it in the form

$$g(Y_1, \ldots, Y_n) = g_0(Y_1, \ldots, Y_{n-1}) + \cdots + g_k(Y_1, \ldots, Y_{n-1}) Y_n^k, \quad k = \deg_n g \quad .$$

If $g_0 = 0$, then $g = Y_n h$, where $h \in A[Y_1, \ldots, Y_n]$. By assumption, $s_n h(s_1, \ldots, s_n) = 0$, so that, since $A[X_1, \ldots, X_n]$ is an integral domain (Theorem 1' of §2 Ch. 5), this means that $h(s_1, \ldots, s_n) = 0$. But this is impossible, since $\deg h(Y_1, \ldots, Y_n) = \deg g(Y_1, \ldots, Y_n) - 1$. Thus, $g_0 \neq 0$. Now consider the identity

$$g_0(s_1, \ldots, s_{n-1}) + \cdots + g_k(s_1, \ldots, s_{n-1}) s_n^k = g(s_1, \ldots, s_n) = 0$$

in $A[X_1, \ldots, X_n]$, and substitute 0 in place of X_n. Then all terms except for the first vanish, and we obtain

$$g_0((s_1)_0, \ldots, (s_{n-1})_0) = 0 \quad ,$$

where $(s_1)_0, \ldots, (s_{n-1})_0$ are the elementary symmetric polynomials in the variables

$X_1, \ldots, X_{n-1}$ (see (3)). But by the induction assumption, the $(s_k)_0$, $k = 1, \ldots, n-1$, are algebraically independent over A, which contradicts $g_0 \neq 0$. This contradiction completes the proof of uniqueness, and hence of Theorem 1. □

Note that the proof of the first part (existence of g) was constructive, and so can actually be used to find the polynomial g in a given situation. It is also worth noting that the coefficients of g lie in the subring of A generated by the coefficients of f. For example, if $A = \mathbb{Z}$, the coefficients of both f and g will be integers.

COROLLARY. *Let* $f(X) = X^n + a_1 X^{n-1} + \cdots + a_{n-1} X + a_n$ *be a monic polynomial of degree* n *in one variable* X *over a field* K. *Suppose that* $f(X)$ *has* n *roots* $c_1, \ldots, c_n$ *in some field* F *containing* K. *Let* $h(X_1, \ldots, X_n)$ *be any symmetric polynomial in* $K[X_1, \ldots, X_n]$. *Then the value* $h(c_1, \ldots, c_n)$ *obtained by substituting* c_i *in place of* X_i, $i = 1, \ldots, n$, *lies in* K.

Proof. By the fundamental theorem on symmetric polynomials, there exists a polynomial $g(Y_1, \ldots, Y_n) \in K[Y_1, \ldots, Y_n]$ such that $h(X_1, \ldots, X_n) = g(s_1(X_1, \ldots, X_n), \ldots, s_n(X_1, \ldots, X_n))$. Hence, $h(c_1, \ldots, c_n) = g(s_1(c_1, \ldots, c_n), \ldots, s_n(c_1, \ldots, c_n))$, but, by Vieta's formulas (12) in §1, $s_k(c_1, \ldots, c_n) = (-1)^k a_k \in K$; hence, $h(c_1, \ldots, c_n) = g(-a_1, \ldots, (-1)^n a_n) \in K$. □

3. *The method of undetermined coefficients.* There are many different proofs of the fundamental theorem on symmetric polynomials, and several different methods for finding the expression for a given symmetric polynomial f in terms of the elementary symmetric polynomials are available. In order to describe one of the most commonly used methods, we introduce a new type of symmetric polynomial. To be definite, let us take the integral domain A to be $\mathbb{Z}$ or $\mathbb{R}$. Let $v = X_1^{i_1} \cdots X_n^{i_n}$ be a monomial. We shall call v a *monotonic* monomial if $i_1 \geq i_2 \geq \cdots \geq i_n$. Let $S(v)$ denote the sum of all distinct monomials in the set of $n!$ monomials of the form πv, where π runs through

S_n. In other words,

$$S(v) = \sum_{\pi \in S_n/H} \pi v \quad ,$$

where the notation $\pi \in S_n/H$ means that π runs through a set of left coset representatives of the subgroup $H = \{\tau \in S_n \mid \tau v = v\}$ in S_n (the reader should check that this set H really is a subgroup). An example of such a sum is the so-called <u>power sum</u>

$$p_k(X_1, \ldots, X_n) = S(X_n^k) = X_1^k + X_2^k + \cdots + X_n^k, \qquad k \geq 0 \quad . \tag{6}$$

In this example we clearly have $H = S_{n-1}$. Another example is $S(X_1 X_2 \ldots X_k) = s_k(X_1, \ldots, X_n)$ (what is H in this case?). It is clear that $S(v)$ is a homogeneous symmetric polynomial of the same total degree as v. Since $S(v) = S(\sigma v)$, $\forall \sigma \in S_n$, we need only consider $S(v)$ for monotonic monomials v (since any v can be transformed into a monotonic v by a suitable $\sigma \in S_n$). It is also clear that any symmetric polynomial f over A is a linear combination of polynomials of the type $S(v)$ with coefficients in A:

$$f = \sum a_v S(v) \quad .$$

It is usually possible to see at a glance how to write f in this form; hence, the problem of expressing f in terms of the s_k reduces to the problem of expressing the $S(v)$ in terms of the elementary symmetric polynomials.

Let us agree to arrange the monomials in $S(v)$ in <u>lexicographic</u> ("alphabetical") order, i.e., in such a way that a monomial $v = X_1^{i_1} X_2^{i_2} \ldots X_n^{i_n}$ comes before a monomial $w = X_1^{j_1} X_2^{j_2} \ldots X_n^{j_n}$ ("v is greater than w" : $v > w$) if and only if the sequence $i_1 - j_1, i_2 - j_2, \ldots, i_n - j_n$ is of the form $0, \ldots, 0, t, \ldots,$ where $t > 0$ (there may be negative $i_\ell - j_\ell$ to the right of t). Of course, we can also use lexicographic ordering to arrange the monomial terms in any polynomial $f \in A[X_1, \ldots, X_n]$. Note that the <u>leading</u> (or <u>first</u>) term when $S(v)$ is arranged in lexicographic order is v, since v was assumed to be monotonic. If $v = X_1^{i_1} X_2^{i_2} \dots X_n^{i_n}$ is a monotonic monomial, then we can consider the product

$$g_v = s_1^{i_1-i_2} s_2^{i_2-i_3} \dots s_{n-1}^{i_{n-1}-i_n} s_n^{i_n}, \qquad s_i = s_i(X_1, \dots, X_n) \quad , \tag{7}$$

in which the leading term is simply

$$v = X_1^{i_1-i_2} (X_1 X_2)^{i_2-i_3} \dots (X_1 \dots X_{n-1})^{i_{n-1}-i_n} (X_1 \dots X_n)^{i_n}$$

(since the leading term in a product is the product of the leading terms in each factor). It hence follows that the leading term in the difference $S(v) - g_v$ is less than v. Thus,

$$S(v) - g_v = \sum n'_w S(w) \quad ,$$

where $n'_w \in \mathbb{Z}$, and the summation is taken over the set of monotonic monomials $w < v$. The total degree of all of the w is the same as the total degree of v.

The following method now suggests itself for expressing $S(v)$ in terms of elementary symmetric polynomials. Let $\deg v = m$. Take all "monotonic" partitions

$$m = j_1 + j_2 + \dots + j_n, \quad j_1 \geq j_2 \geq \dots \geq j_n \geq 0 \quad ,$$

of the positive integer m for which $w = X_1^{j_1} X_2^{j_2} \dots X_n^{j_n} < v$. Consider the set M_v of all such monomials w. For each $w \in M_v$ we have the monomial g_w (see (7)). We already know that

$$S(v) = g_v + \sum_{w \in M_v} n_w g_w \quad , \tag{8}$$

where n_w are certain integers. The undetermined coefficients n_w (whence the name: <u>the method of undetermined coefficients</u>) can be found by successively substituting suitable integer values, usually zeros and ones, in place of $X_1, \dots, X_n$ in (8). The values of g_v, g_w and $S(v)$ are known, and so we obtain a compatible system of linear equations for the unknowns n_w.

Example. $v = X_1^3$, $S(v) = p_3(X_1, \ldots, X_n)$, $n \geq 3$, $g_v = s_1^3$,

M_v	$X_1^2 X_2$	$X_1 X_2 X_3$
g_w	$s_1 s_2$	s_3

.

In this case equation (8) has the form

$$p_3 = s_1^3 + a\, s_1 s_2 + b\, s_3 \quad .$$

If $X_1 = X_2 = 1$ and $X_i = 0$ for $i > 2$, then $p_3 = 2$, $s_1 = 2$, $s_2 = 1$, $s_3 = 0$. If we set $X_1 = X_2 = X_3 = 1$ and $X_i = 0$ for $i > 3$, then we have $p_3 = 3$, $s_1 = 3$, $s_2 = 3$, $s_3 = 1$. From the resulting linear system

$$2 = 2^3 + a \cdot 2 \cdot 1 + b \cdot 0 \quad ,$$

$$3 = 3^3 + a \cdot 3 \cdot 3 + b \cdot 1$$

we find: $a = -3$ and $b = 3$, i.e., $p_3 = s_1^3 - 3\, s_1 s_2 + 3\, s_3$.

We have the following convenient formulas, called Newton's formulas, which can be used to express the power sums $p_k(X_1, \ldots, X_n)$ as polynomials in $s_1, s_2, \ldots, s_n$:

$$p_k - p_{k-1} s_1 + p_{k-2} s_2 + \cdots + (-1)^{k-1} p_1 s_{k-1} + (-1)^k k s_k = 0 \tag{9}$$

for $1 \leq k \leq n$; and

$$p_k - p_{k-1} s_1 + p_{k-2} s_2 + \cdots + (-1)^{n-1} p_{k-n+1} s_{n-1} + (-1)^n p_{k-n} s_n = 0 \tag{10}$$

for $k > n$. In order to prove these formulas, we make use of the obvious relations

$$X_i^n - s_1 X_i^{n-1} + \cdots + (-1)^{n-1} s_{n-1} X_i + (-1)^n s_n = 0 \quad ,$$

which are obtained by substituting $Y = X_i$ in (2). We multiply each of these equations by X_i^{k-n} $(k \geq n)$:

$$X_i^k - s_1 X_i^{k-1} + \cdots + (-1)^{n-1} s_{n-1} X_i^{k-n+1} + (-1)^n s_n X_i^{k-n} = 0$$

and then sum over i from 1 to n . This gives us both formula (10) and also (9) when $k = n$ $(p_0 = X_1^0 + \cdots + X_n^0 = n)$. Next, consider the following symmetric homogeneous polynomial $f_{k,n}$ of degree $k \leq n$ (or $-\infty$ if $f_{k,n} = 0$):

$$f_{k,n}(X_1, \ldots, X_n) = p_k - p_{k-1} s_1 + \cdots + (-1)^{k-1} p_1 s_{k-1} + (-1)^k k s_k .$$

Using induction on $r = n - k$, we prove that $f_{k,n}$ is identically zero. We just showed this when $r = 0$. Now set $X_n = 0$ and note that the resulting symmetric polynomials $(s_i)_0$ and $(p_i)_0$ coincide with the polynomials s_i and p_i defined for the $n-1$ variables $X_1, \ldots, X_{n-1}$ (see (3) and (6)). We obtain:

$$f_{k,n}(X_1, \ldots, X_{n-1}, 0) = (p_k)_0 - (p_{k-1})_0 (s_1)_0 + \cdots + (-1)^{k-1} (p_1)_0 (s_{k-1})_0 + (-1)^k k (s_k)_0 =$$

$$= f_{k,n-1}(X_1, \ldots, X_{n-1}) = 0 ,$$

since $n - 1 - k = r - 1 < r$, so that the induction assumption applies. The relation $f_{k,n}(X_1, \ldots, X_{n-1}, 0) = 0$ shows that the polynomial $f_{k,n}$ is divisible by X_n: $f_{k,n} = X_n f_1$. Using the fact that $f_{k,n}$ is symmetric (see the similar argument in the first part of the proof of Theorem 1), we find that

$$f_{k,n}(X_1, \ldots, X_n) = s_n(X_1, \ldots, X_n) \cdot g(X_1, \ldots, X_n) ,$$

which is only possible if $g = 0$, since $\deg s_n = n$ and $\deg f_{k,n} = k < n$. Thus, $f_{k,n} = 0$, and formula (9) is proved.

4. <u>The discriminant of a polynomial.</u> In the ring $K[X_1, \ldots, X_n]$, consider the polynomial

$$\Delta_n = \prod_{1 \leq j < i \leq n} (X_i - X_j) ,$$

which can clearly be written as a Vandermonde determinant

$$\Delta_n = \begin{vmatrix} 1 & 1 & \cdots & 1 \\ X_1 & X_2 & \cdots & X_n \\ \cdot & \cdot & \cdots & \cdot \\ X_1^{n-1} & X_2^{n-2} & \cdots & X_n^{n-1} \end{vmatrix} . \tag{11}$$

Since the determinant is a skew-symmetric function of its columns, it follows that $\pi(\Delta_n) = \epsilon_\pi \Delta_n$, where ϵ_π is the sign of the permutation $\pi \in S_n$. But then Δ_n^2 is a symmetric polynomial, and, by the fundamental theorem, it can be expressed as a polynomial in the elementary symmetric functions:

$$\Delta_n^2 = \prod (X_i - X_j)^2 = \mathrm{Dis}(s_1, \dots, s_n) \quad .$$

The polynomial Dis in the variables $s_1(X_1, \dots, X_n), \dots, s_n(X_1, \dots, X_n)$ is called the <u>discriminant</u> of the n-tuple $X_1, \dots, X_n$. Its coefficients obviously lie in $\mathbb{Z}$. If we substitute $x_i \in F$ in place of X_i, $i = 1, 2, \dots, n$ (where F is an extension of K), we obtain the discriminant of the n-tuple of elements of F. If not all the $x_i \in F$ are distinct, then the discriminant vanishes, since at least one of the factors $x_i - x_j$ is zero. The name "discriminant" comes from the ability of this function to distinguish the case when two or more of the x_i coincide from the case when they are all distinct.

A convenient method for obtaining the discriminant is based on interpreting Δ_n^2 as the product of the determinant (11) and the transpose determinant: $\Delta_n^2 = \Delta_n {}^t\Delta_n$ (recall that $\det {}^tA = \det A$ for any square matrix A). Using the rule for matrix multiplication, we immediately obtain:

$$\mathrm{Dis}(s_1, \dots, s_n) = \begin{vmatrix} n & p_1 & p_2 & \cdots & p_{n-1} \\ p_1 & p_2 & p_3 & \cdots & p_n \\ p_2 & p_3 & p_4 & \cdots & p_{n+1} \\ \cdot & \cdot & \cdot & \cdots & \cdot \\ p_{n-1} & p_n & p_{n+1} & \cdots & p_{2n-2} \end{vmatrix} , \tag{12}$$

where p_k are our familiar power sums (6). If we compute the p_k using the recursive formulas (9) and (10), we obtain an explicit expression for $\mathrm{Dis}(s_1, \dots, s_n)$. For example, $p_1 = s_1$ and $p_2 = s_1^2 - 2s_2$, so that

$$\mathrm{Dis}(s_1, s_2) = \begin{vmatrix} 2 & s_1 \\ s_1 & s_1^2 - 2s_2 \end{vmatrix} = s_1^2 - 4s_2 \quad . \tag{13}$$

Now suppose that we are given a monic polynomial

$$f(X) = X^n + a_1 X^{n-1} + \cdots + a_{n-1} X + a_n \in K[X] ,$$

which has n roots $c_1, \ldots, c_n$ in K or in some extension F of K. We know from Vieta's formulas that $a_k = (-1)^k s_k(c_1, \ldots, c_n)$.

<u>Definition.</u> The discriminant of the n-tuple of roots $c_1, \ldots, c_n$ of a polynomial f, or equivalently, the value of $\mathrm{Dis}(s_1, \ldots, s_n)$ obtained by substituting $(-1)^k a_k$ in place of s_k, is called the <u>discriminant of the polynomial</u> f and is denoted $D(f)$. It is also called the <u>discriminant of the equation</u>

$$f(x) = x^n + a_1 x^{n-1} + \cdots + a_{n-1} x + a_n = 0 . \tag{14}$$

It is clear that $D(f) \in K$ (recall the corollary to Theorem 1). The following fact is also an immediate consequence of the definition of the discriminant:

PROPOSITION. $D(f) = 0$ <u>if and only if the equation (14) has multiple roots (i. e., at least one root of multiplicity</u> $k > 1$). □

Taking into account Corollary 2 of Theorem 5 in §1, we now have two methods of deciding, without leaving the ground field K, whether or not a polynomial $f \in K[X]$ has multiple roots. But this is not the only purpose of the discriminant. For example, when applied to the quadratic polynomial $f(X) = X^2 + aX + b$ with real coefficients a and b, formula (13) gives $D(f) = a^2 - 4b$, which is a familiar expression from elementary algebra. In particular, the sign of $D(f)$ determines whether the equation $x^2 + ax + b = 0$ has two real roots or two complex conjugate roots.

To take another example, let us compute the discriminant of the so-called incomplete cubic equation

$$f(x) = x^3 + ax + b = 0 . \tag{15}$$

Here $s_4 = 0$, and, computing p_k by the recursive formulas, we obtain $p_1 = s_1 = 0$,

$p_2 = s_1^2 - 2s_2 = -2a$, $p_3 = s_1^3 - 3s_1s_2 + 3s_3 = -3b$, $p_4 = s_1^4 - 4s_1^2s_2 + 4s_1s_3 + 2s_2^2 = 2a^2$. Hence, by formula (12), we have

$$D(f) = \begin{vmatrix} 3 & 0 & -2a \\ 0 & -2a & -3b \\ -2a & -3b & 2a^2 \end{vmatrix} = -4a^3 - 27b^2 \quad . \tag{16}$$

$D(f)$ is given by a more complicated expression than (16) if we take the complete cubic equation $x^3 + a_1x^2 + a_2x + a_3 = 0$, but we can avoid the added complication by reducing the complete cubic to the incomplete cubic as follows. Whenever we have a polynomial f in the form (14), we can make a change of variable from x to $y = x + a_1/n$. Substituting $x = y - a_1/n$ in (14) and using the binomial formula, we find that the polynomial

$$g(y) = f\left(y - \frac{a_1}{n}\right) = y^n + ay^{n-2} + \cdots = 0 \quad , \tag{17}$$

has zero coefficient of y^{n-1}. If we know a root y_0 of (17), we immediately know a root $x_0 = y_0 - a_1/n$ of the original equation (14). Hence, without loss of generality we may always assume that $a_1 = 0$.

If we want to find a general formula for the solutions of (15) (this was a major achievement of del Ferro, Cardano, and other mathematicians of the Middle Ages), we inevitably end up making use of the discriminant (16) (see formulas (2) of §2 Ch. 1).

5. The resultant. The basic property of $D(f)$, given in the proposition in the previous subsection, can also be interpreted as a criterion for when a polynomial f and its derivative f' have a common root (or common factor). In the last analysis, this criterion is based on the Euclidean algorithm. This leads one to hope that, more generally, we can find a similar criterion which allows us to determine directly from the coefficients whether two polynomials $f, g \in K[X]$ have a common factor.

Let

$$f(X) = a_0 X^n + a_1 X^{n-1} + \cdots + a_{n-1} X + a_n ,$$

$$g(X) = b_0 X^m + b_1 X^{m-1} + \cdots + b_{m-1} X + b_m$$

be two polynomials with coefficients in K. Here $n > 0$, $m > 0$, and we allow the possibility that $a_0 = 0$ or $b_0 = 0$.

<u>Definition.</u> The <u>resultant</u> $\operatorname{Res}(f, g)$ of f and g is the homogeneous polynomial in the coefficients of f and g (having degree m in $a_0, \ldots, a_n$ and degree n in $b_0, \ldots, b_m$) given by

$$\operatorname{Res}(f,g) = \left|\begin{matrix} a_0 & a_1 & \cdots & a_n & & \\ & a_0 & a_1 & \cdots & a_n & \\ \cdots & \cdots & \cdots & \cdots & \cdots & \cdots \\ & & a_0 & a_1 & \cdots & a_n \\ b_0 & b_1 & \cdots & b_m & & \\ & b_0 & b_1 & \cdots & b_m & \\ \cdots & \cdots & \cdots & \cdots & \cdots & \cdots \\ & & b_0 & b_1 & \cdots & b_m \end{matrix}\right| \begin{matrix} \left.\vphantom{\begin{matrix} a \\ a \\ a \\ a \end{matrix}}\right\} m \text{ rows} \\ \left.\vphantom{\begin{matrix} b \\ b \\ b \\ b \end{matrix}}\right\} n \text{ rows} \end{matrix}$$

Implicit in this definition is an assertion about the degree in the a_i and the b_j of the above determinant. But this follows immediately from the properties of determinants: if we replace a_i by $t a_i$ in the first m rows, then $\operatorname{Res}(tf, g) = t^m \operatorname{Res}(f,g)$, and so the resultant is homogeneous of degree m in the a_i by Exercise 3 of §2 Ch. 5; we similarly show that it is homogeneous of degree n in the b_j.

We now derive the basic properties of the resultant.

Res 1. $\operatorname{Res}(f,g) = 0$ <u>if and only if either</u> $a_0 = 0 = b_0$ <u>or else</u> f <u>and</u> g <u>have a common factor of degree</u> > 0 <u>in</u> $K[X]$.

<u>Proof.</u> We first show that the condition "$a_0 = 0 = b_0$ or else f and g have a common factor in $K[X]$ of degree > 0" holds if and only if there exist polynomials f_1 and g_1 not both zero, such that

$$fg_1 + f_1 g = 0 , \quad \deg f_1 < n , \quad \deg g_1 < m \quad . \tag{18}$$

To see this, let $h = \text{g.c.d.}(f,g)$ have degree $\deg h > 0$. Then $f = hf_1$, $g = -hg_1$, and hence $fg_1 + gf_1 = 0$. In addition, $\deg f_1 < n$ and $\deg g_1 < m$, so that (18) holds. If $a_0 = 0 = b_0$, we can set $f_1 = f$, $g_1 = -g$.

Conversely, suppose that (18) holds. If we had $\text{g.c.d.}(f,g) = 1$, since $K[X]$ is a unique factorization domain (see §3 Ch. 5), we would have the implication $fg_1 = -gf_1 \Rightarrow f|f_1, g|g_1$. Thus, $\deg f < n$ and $\deg g < m$, i.e., $a_0 = 0 = b_0$.

We now prove that (18) is equivalent to $\text{Res}(f,g) = 0$. If we set

$$f_1 = c_0 X^{n-1} + c_1 X^{n-2} + \cdots + c_{n-1} ,$$

$$g_1 = d_0 S^{m-1} + d_1 X^{m-2} + \cdots + d_{m-1}$$

and use the rules for operating with polynomials to compute the coefficients of $fg_1 + gf_1$, which has degree $\leq n + m - 1$, we can write the condition (18) in the form of a square homogeneous system of linear equations with $n + m$ unknowns $d_0, d_1, \ldots, d_{m-1}$, $c_0, c_1, \ldots, c_{n-1}$:

$$\begin{aligned} &a_0 d_0 + \cdots\cdots\cdots + b_0 c_0 \cdots\cdots\cdots = 0 , \\ &a_1 d_0 + a_0 d_1 + \cdots\cdots + b_1 c_0 + b_0 c_1 \cdots\cdots = 0 , \\ &a_2 d_0 + a_1 d_1 + a_0 d_2 \cdots + b_2 c_0 + b_1 c_1 + b_0 c_2 = 0 , \\ &\cdots\cdots\cdots\cdots\cdots\cdots\cdots\cdots\cdots\cdots \\ &\cdots\cdots\cdots\cdots\cdots\cdots\cdots\cdots\cdots\cdots . \end{aligned} \tag{19}$$

The determinant of the system (19) (more precisely, the determinant of the transpose of the matrix of (19)) is exactly $\text{Res}(f,g)$. Thus, (19) has a non-trivial solution if and only if $\text{Res}(f,g) = 0$, and any non-zero solution to (19) gives us a pair of polynomials f_1, g_1 satisfying (18). □

Res 2. Suppose that the polynomials f and g split completely into linear factors in $K[X]$:

$$f(X) = a_0(X - \alpha_1) \ldots (X - \alpha_n) ,$$

$$g(X) = b_0(X - \beta_1) \ldots (X - \beta_m) .$$

<u>Then</u>

$$\mathrm{Res}(f,g) = a_0^m \prod_{i=1}^{n} g(\alpha_i) = (-1)^{mn} b_0^n \prod_{j=1}^{m} f(\beta_j) = a_0^m b_0^n \prod_{i,j} (\alpha_i - \beta_j) .$$

<u>Proof.</u> It is clear that these formulas, if they are true, must be of a "universal" nature, not depending on the particular properties of f and g. This simple "philosophical" principle, which we shall not justify or discuss in detail here, allows us to restrict ourselves to the "general case" when we suppose that all of the $g(\alpha_1), \ldots, g(\alpha_n)$ are distinct, and that all of the $f(\beta_1), \ldots, f(\beta_m)$ are distinct.

Next, since $\mathrm{Res}(g,f) = (-1)^{mn} \mathrm{Res}(f,g)$ (see the definition), it suffices to verify the equality $\mathrm{Res}(f,g) = a_0^m \prod g(\alpha_i)$. To do this we introduce a new variable Y and consider the polynomials $f(X)$ and $g(X) - Y$ over the field of rational functions $K(Y)$. Using the definition of the resultant and replacing b_m by $b_m - Y$, we find that

$$\mathrm{Res}(f, g - Y) = (-1)^n a_0^m Y^n + \cdots + \mathrm{Res}(f,g)$$

is a polynomial of degree n in Y with leading coefficient $(-1)^n a_0^m$ and with constant term $\mathrm{Res}(f,g)$. The polynomials $f(X)$ and $g(X) - g(\alpha_i)$ have the root α_i in common, and so both are divisible by $X - \alpha_i$. Using the property Res 1, we have $\mathrm{Res}(f, g - g(\alpha_i)) = 0$.

By Bezout's Theorem, the polynomial $\mathrm{Res}(f, g - Y)$ must be divisible by $g(\alpha_i) - Y$, $1 \leq i \leq n$. Since we have assumed that all of the $g(\alpha_i)$ are distinct, it follows that $\mathrm{Res}(f, g - Y) = a_0^m \prod_{i=1}^{n} (g(\alpha_i) - Y)$. Setting $Y = 0$, we obtain the required equality. □

We extend the definition of the discriminant in Subsection 4 to the case of non-monic polynomials, by setting

$$D(f) = a_0^{2n-2} \prod_{1 \le j < i \le n} (\alpha_i - \alpha_j)^2 = \left[a_0^{n-1} \prod_{j<i} (\alpha_i - \alpha_j) \right]^2, \quad a_0 \neq 0 \quad .$$

Res 3. The following formula holds:

$$D(f) = (-1)^{\frac{n(n-1)}{2}} a_0^{-1} \operatorname{Res}(f, f') \quad . \tag{20}$$

Proof. According to Res 2,

$$\operatorname{Res}(f, f') = a_0^{n-1} \prod_{i=1}^{n} f'(\alpha_i) \quad .$$

But

$$f'(\alpha_i) = a_0 \prod_{j \neq i} (\alpha_i - \alpha_j) \quad ,$$

as we immediately see by substituting $X = \alpha_i$ in the general expression

$$f'(X) = a_0 \sum_{i=1}^{n} \prod_{j \neq i} (X - \alpha_j) \quad ,$$

which is obtained by differentiating the product $f(X) = a_0 \prod_{j=1}^{n} (X - \alpha_j)$. Thus,

$$\operatorname{Res}(f, f') = a_0^{2n-1} \prod_{i=1}^{n} \prod_{j \neq i} (\alpha_i - \alpha_j) = a_0 (-1)^{\frac{n(n-1)}{2}} a_0^{2n-2} \prod_{j<i} (\alpha_i - \alpha_j)^2 = a_0 (-1)^{\frac{n(n-1)}{2}} D(f).$$

□

Formula (20) gives an explicit formula for the discriminant.

EXERCISES

1. Let p be a prime number. Using Newton's formulas (9) and (10), show that

$$\sum_{i=1}^{p-1} i^m \equiv \begin{cases} -1 \pmod{p}, & \text{if } m \text{ is divisible by } p-1, \\ 0 \pmod{p}, & \text{if } m \text{ is not divisible by } p-1 \end{cases} \quad .$$

2. Let c_1, c_2, c_3 be the complex roots of the polynomial $X^3 - X + 1$. What can be said about the extension $\mathbb{Q}(c_1^{99} + c_2^{99} + c_3^{99})$?

3. A polynomial $f(X_1, \ldots, X_n)$ over a field K of characteristic $\neq 2$ is called skew-symmetric if $(\pi f)(X_1, \ldots, X_n) = \varepsilon_\pi f(X_1, \ldots, X_n)$, $\forall \pi \in S_n$, where ε_π is the sign of the permutation π. Thus, Δ_n is an example of a skew-symmetric polynomial. Show that every skew-symmetric polynomial $f \in K[X_1, \ldots, X_n]$ is of the form $f = \Delta_n \cdot g$, where g is a symmetric polynomial.

4. Using property Res 2 and the existence of a splitting field for a polynomial (see Theorem 2 in the next section), show that

$$\operatorname{Res}(fg,h) = \operatorname{Res}(f,h) \cdot \operatorname{Res}(g,h) \quad .$$

5. Use Exercise 4 and Res 3 to derive the formula:

$$D(fg) = D(f)\, D(g)\, |\operatorname{Res}(f,g)|^2 \quad .$$

6. What is $\operatorname{Res}(f(X), X - a)$ equal to?

7. Show that $D(X^n + a) = (-1)^{\frac{n(n-1)}{2}} n^n a^{n-1}$.

8. Let $f(X) = X^{n-1} + X^{n-2} + \cdots + 1$. Using the relation $X^n - 1 = (X - 1) f(X)$ and the preceding exercises, show that

$$D(f) = (-1)^{\frac{(n-1)(n-2)}{2}} n^{n-2} \quad .$$

§3. $\mathbb{C}$ is algebraically closed

1. Statement of the fundamental theorem. Let K be a field, and let f be a polynomial over K. As noted in Subsection 2 of §1, the behavior of the polynomial function $\tilde{f} : K \to K$ associated to f very much depends on the field K. In particular, we can conclude that $\operatorname{Im} \tilde{f}$ is all of K if $\deg f > 0$ as soon as we know that K

satisfies the following

Definition. A field K is called algebraically closed if every polynomial in $K[X]$ decomposes into a product of linear factors. Equivalently, K is algebraically closed if the only polynomials irreducible over K are those of degree 1 (linear polynomials).

Note that if every polynomial $f \in K[X]$ has at least one root in K, then K is algebraically closed, because then we can write $f(X) = (X - a)h(X)$, where $a \in K$ and $h \in K[X]$. Applying the condition to h, we have $h(X) = (X - b)r(X)$, where $b \in K$ and $r \in K[X]$; continuing this process, we finally obtain $f(X)$ as a product of linear factors. Since this holds for any polynomial f, that means that K is algebraically closed.

It turns out that every field K has an extension $\widetilde{K} \supset K$ which is algebraically closed (Steinitz' Theorem). But at first glance it seems difficult to comprehend not only the construction of $\widetilde{K}$ but even what it means to have such a field. So we are very fortunate to have at our disposal an excellent and important example of an algebraically closed field. It is this fact which is the subject of the so-called "Fundamental Theorem of Algebra":

THEOREM 1. The field of complex number $\mathbb{C}$ is algebraically closed.

We re-state this fundamental fact in terms of roots: Any polynomial $f(X) \in \mathbb{C}[X]$ of degree $n \geq 1$ has exactly n complex roots, counting multiplicity.

The pretentious name "Fundamental Theorem" for Theorem 1 goes back to the days when solving algebraic equations was the foremost activity of algebra. In modern times Theorem 1 is considered to be one of the basic (but not the basic) theorems of algebra.

The first rigorous proof of this theorem was given by Gauss in 1799. Since that time, there have been many different proofs, of varying degrees of "algebraicity". It is always necessary in one form or another to use the continuity properties of the fields $\mathbb{R}$ and $\mathbb{C}$ (in other words, their topology); there is even a completely non-algebraic and rather short proof of the Fundamental Theorem based on a fairly deep fact about analytic functions of a complex variable. Below we shall give a proof which is, in spirit, the most algebraic of those which are accessible to us at this point. Perhaps the most natural proof to give would be one using Galois theory, but we shall have to be satisfied with a different type of algebraic proof.

The non-algebraic part of the proof of Theorem 1 is contained in the following two lemmas.

LEMMA 1. _Let_

$$f(X) = a_0 X^n + a_1 X^{n-1} + \cdots + a_{n-1} X + a_n \tag{1}$$

be a polynomial of degree $n \geq 1$ _with complex coefficients. Then there exists a positive number_ $r \in \mathbb{R}$ _such that for_ $|z| > r$, $z \in \mathbb{C}$, _we have_

$$|a_0 z^n| > |a_1 z^{n-1} + \cdots + a_{n-1} z + a_n| .$$

Proof. Set $A = \max(|a_1|, \ldots, |a_n|)$ and $r = A/|a_0| + 1$. If we take $|z| > r \geq 1$, then we obtain $|a_0| > A/(|z| - 1)$, so that, by the rules on the moduli of complex numbers (see §1 Ch. 5), we have

$$|a_0 z^n| = |a_0||z|^n > \frac{A|z|^n}{|z| - 1} > \frac{A(|z|^n - 1)}{|z| - 1} =$$

$$= A(|z|^{n-1} + \cdots + |z| + 1) \geq |a_1||z|^{n-1} + \cdots$$

$$\cdots + |a_{n-1}||z| + |a_n| = |a_1 z^{n-1}| + \cdots + |a_{n-1} z| + |a_n| \geq |a_1 z^{n-1} + \cdots + a_{n-1} z + a_n| .$$

□

COROLLARY. _Suppose that the polynomial_ (1) _of degree_ $n \geq 1$ _has real coefficients. Then the sign of (the real number)_ $f(x)$ _is the same as the sign of the "leading term"_ $a_0 x^n$ _for all_ $x \in \mathbb{R}$ _which are sufficiently large in absolute value._ □

LEMMA 2. _A polynomial of odd degree with real coefficients has at least one real root._

Proof. Since n is odd, the leading term $a_0 x^n$ of the polynomial function $\tilde{f}: \mathbb{R} \to \mathbb{R}$ takes values with opposite signs for positive and negative $x \in \mathbb{R}$. If we take x sufficiently large in absolute value, by the corollary to Lemma 1, we know that $f(x)$ also has opposite signs for positive and negative x. For example, if $a_0 > 1$, then $f(-r) < 0$ and $f(r) > 0$ if r is the real number in the proof of Lemma 1. We know from calculus (and it is not hard to prove directly) that the polynomial function $\tilde{f}$ is continuous. But a continuous function f has the property that it takes every value between $f(-r)$ and $f(r)$ on the interval $-r \leq x \leq r$. In particular, we have $f(c) = 0$ for some c with $|c| \leq r$. The same argument applies if $a_0 < 0$. □

This geometrically and intuitively clear assertion concludes the non-algebraic part of our proof. We shall give the next step in a context not directly related to $\mathbb{C}$, since it involves a construction which is of independent interest.

2. The splitting field of a polynomial. It often happens that a "glance from the side" at a well-known example makes it possible for us to understand it better and come up with useful generalizations. Recall the realization of $\mathbb{C}$ as the quotient ring $\mathbb{R}[X]/(X^2+1)\mathbb{R}[X]$ (see Theorem 6 in §2 Ch. 5). If we replace $\mathbb{R}$ by an arbitrary field K and replace X^2+1 by an arbitrary polynomial $f \in K[X]$, we obtain the "ring of residue classes modulo (f)", or, equivalently, the quotient ring $K[X]/(f)$, where $(f) = f \cdot K[X]$ is the ideal in $K[X]$. The ideal (f) consists of all polynomials which are divisible by f; according to the corollary to Theorem 5 of §2 Ch. 5, any ideal in $K[X]$ is of this form for some f. Just as we have an analogy between the rings $\mathbb{Z}$ and $K[X]$, we also have an analogy between the residue rings $\mathbb{Z}_n = \mathbb{Z}/(n)$ and $K[X]/(f)$. It is worthwhile for us now to repeat the basic steps in the construction of $\mathbb{Z}_n$ (see §4 Ch.4) for $K[X]$.

The elements of the quotient ring $K[X]/(f)$ are the residue classes $\bar{g} = g + (f)$, each of which can be represented in the form $r + (f)$, where $\deg r < \deg f$. As in the case of $\mathbb{Z}$, this is proved by the division algorithm: if $g = qf + r$, then $g + (f) = = r + qf + (f) = r + (f)$, since $qf \in (f)$. It is easy to see that the elements $\bar{a}$, $a \in K$, form a subring of $K[X]/(f)$ which is isomorphic to the field K. Next, if f is reducible over K, i.e., if we can write $f = f_1 f_2$, where $f_i \in K[X]$ and $0 < \deg f_i < \deg f$, then this means that there are non-trivial zero divisors in $K[X]/(f)$, namely, $\bar{f}_i \neq 0$, $i = 1, 2$, but $\bar{f}_1 \bar{f}_2 = \overline{f_1 f_2} = \bar{f} = \bar{0}$.

Now suppose that f is an irreducible polynomial. If $\deg r < \deg f$ $(r \neq 0)$, then g.c.d. $(r, f) = 1$ and $ur + vf = 1$ for suitable $u, v \in K[X]$ (see Theorem 3 of §3 Ch. 5). In other words,

$$\{r + (f)\}\{u + (f)\} = ru + (f) = 1 - vf + (f) = 1 + (f) \quad ,$$

and hence

$$\bar{r}\bar{u} = \overline{ru} = \bar{1} \quad .$$

Thus, any element $\bar{r} \neq \bar{0}$ has an inverse $\bar{u} = \bar{r}^{-1}$ in $K[X]/(f)$. This shows that, whenever f is an irreducible polynomial, the quotient ring $K[X]/(f)$ is a field containing a subfield isomorphic to K.

One element in $K[X]/(f)$ is $\bar{X}$. For any $a_0, a_1, \ldots, a_m \in K$ we have

$$\sum_{k=0}^{m} \bar{a}_k \bar{X}^k = \sum_k \{a_k + (f)\}\{X + (f)\}^k = \sum_k \{a_k + (f)\}\{X^k + (f)\} = \left\{\sum_k a_k X^k\right\} + (f) = \overline{\sum_k a_k X^k}.$$

Briefly, we can say that, if $g(Y) = \Sigma\, a_k Y^k \in K[Y]$, then $g(\bar{X}) = \overline{g(X)}$. Of course, when we write $g(\bar{X})$, we use the fact that $K[X]/(f)$ contains a field isomorphic to K to think of the coefficients of g as elements $\bar{a}_k$ in $K[X]/(f)$. Applying this to f, we have

$$f(\bar{X}) = \overline{f(X)} = f + (f) = (f) = \bar{0} \quad ,$$

i.e., the element $\bar{X} \in K[X]/(f)$ is a root of the polynomial f.

Thus, we have the following two results.

THEOREM 2. _The ring of residue classes (quotient ring)_ $K[X]/(f)$ _is a field if and only if_ f _is irreducible over_ K. □

COROLLARY. _If_ $f(X)$ _is any irreducible polynomial over_ K, _there exists an extension_ F _of the field_ K _in which_ $f(X)$ _has at least one root. We can take the field_ $K[X]/(f)$ _for_ F. □

It is customary to say that the extension F is obtained by adjoining to K a root c of the polynomial f; we write $F = K(c)$. We then have $f(X) = (X - c)\,g(X)$, where $g \in F[X]$. We now have a real possibility of constructing an extension of K in which f splits completely into a product of linear factors.

Definition. Let K be a field, and let f be a monic polynomial (not necessarily irreducible) of degree n over K. Then a field $F \supset K$ is called a _splitting field_ of f if $f(X) = (X - c_1) \dots (X - c_n)$ in $F[X]$ and $F = K(c_1, \dots, c_n)$, i.e., F is obtained from K by adjoining the roots $c_1, \dots, c_n$ of the polynomial f.

THEOREM 3. _For every monic polynomial_ $f \in K[X]$ _of degree_ $n \geq 1$, _there exists at least one splitting field._

Proof. The condition that f be monic is not really needed, and is included only for convenience. Let

$$f(X) = f_1(X) \dots f_r(X)$$

be the factorization of f into monic irreducible factors in $K[X]$. According to the corollary to Theorem 2, there exists an extension $K_1 \supset K$ containing at least one root of f_1. Of course, this root c_1 will also be a root of f. Suppose we have already found an extension $K_k \supset \dots \supset K_1 \supset K$ over which f has factorization in the form

$$f(X) = (X - c_1) \dots (X - c_k)\,g_1(X) \dots g_s(X)$$

with k (not necessarily distinct) linear factors, where $k < n$. If we again apply the corollary to Theorem 2, this time to the field K_k and the monic irreducible polynomial $g_1 \in K_k[X]$, we obtain a field $K_{k+1} \supset K_k$ in which we can split off another linear factor $X - c_{k+1}$ with $c_{k+1} \in K_{k+1}$. Continuing in this way, we finally obtain a complete decomposition of f into a product of linear factors over some extension $K_n \supset K$. Either K_n or a subfield $F = K(c_1, \ldots, c_n)$ in K_n will be the desired splitting field for f. (We have not ruled out the possibility that F is just K.) □

The proof of Theorem 3 contains too many choices, so we cannot claim to have proved uniqueness of the splitting field of f. Although, in fact, any two splitting fields of the same polynomial must be isomorphic, i.e., the splitting field is unique up to isomorphism, the proof of this is somewhat harder. For now, we do not need this uniqueness property.

Examples. 1) The quadratic field $\mathbb{Q}(\sqrt{d})$ is the splitting field of the polynomial $X^2 - d$.

2) If we adjoin to $\mathbb{Z}_2$ a root θ of the irreducible polynomial $X^2 + X + 1$, we obtain a field $\mathbb{Z}_2(\theta) = \{0, 1, \theta, 1 + \theta\}$ having four elements which is isomorphic to the field $\mathbb{Z}_2[X]/(X^2 + X + 1)$, and also to the field $GF(4)$ in Subsection 6 of §4 Ch. 4. Notice that $X^2 + X + 1 = (X - \theta)(X - \theta^2)$, i.e., $\mathbb{Z}_2(\theta)$ is a splitting field for the polynomial $X^2 + X + 1$.

3) The polynomial $X^2 + 1$ is irreducible not only over $\mathbb{R}$ (over which its splitting field is $\mathbb{C}$), but also over some other fields, for example, over $\mathbb{Z}_3$. Let $\theta^2 = -1$ in $\mathbb{Z}_3$ (more precisely, let θ be the element $X + (X^2 + 1)\mathbb{Z}_3[X]$ in the residue class field $\mathbb{Z}_3[X]/(X^2 + 1)$). Since $X^2 + 1 = (X - \theta)(X + \theta)$, it follows that $\mathbb{Z}_3(\theta) = \{a + b\theta \mid a, b \in \mathbb{Z}_3\}$ is the splitting field of $X^2 + 1$ over $\mathbb{Z}_3$. By the way, $\mathbb{Z}_3(\theta)$ is isomorphic to the field of matrices $\begin{Vmatrix} a & b \\ -b & a \end{Vmatrix}$, $a, b \in \mathbb{Z}_3$, in Exercise 14 of §4 Ch. 4; here is the isomorphism: $a + b\theta \mapsto a \begin{Vmatrix} 1 & 0 \\ 0 & 1 \end{Vmatrix} + b \begin{Vmatrix} 0 & 1 \\ -1 & 0 \end{Vmatrix}$. Notice that

$\mathbb{Z}_3(\theta)^* = \langle \lambda \rangle$, where $\lambda = 1+\theta$, $\lambda^2 = -\theta$, $\lambda^3 = 1-\theta$, $\lambda^4 = -1$, $\lambda^5 = -1-\theta$, $\lambda^6 = \theta$, $\lambda^7 = -1+\theta$, $\lambda^8 = 1$, i.e., the multiplicative group of the field $\mathbb{Z}_3(\theta)$ is not only abelian, but is actually cyclic.

4) According to the Eisenstein criterion, the polynomial $X^3 - 2$ is irreducible over $\mathbb{Q}$. Since not all of its roots are real, it follows that $\mathbb{Q}(\sqrt[3]{2})$ cannot be the splitting field. It turns out that the splitting field is $\mathbb{Q}(\sqrt[3]{2}, \varepsilon)$, where ε is a primitive cube root of 1:

$$X^3 - 2 = (X - \sqrt[3]{2})(X - \varepsilon\sqrt[3]{2})(X - \varepsilon^2\sqrt[3]{2}) \quad .$$

3. Proof of the Fundamental Theorem. All we need from the preceding subsection is Theorem 3.

According to the remark immediately following the definition of an algebraically closed field, we need only prove that any polynomial (1) has at least one complex root. We first suppose that all of the coefficients of f are real. Without loss of generality, we may assume that $a_0 = 1$ and $a_n \neq 0$. Let

$$\deg f = 2^m n_0 \quad ,$$

where n_0 is an odd integer. If $m = 0$, we know by Lemma 2 that f has a root (in fact, a real root). Using induction on m, we suppose that the existence of a root has been proved for all polynomials with real coefficients whose degree has the form $2^{m'} n_0'$ with $m' \leq m - 1$ (there are no restrictions on the odd factor n_0').

We consider the splitting field F over $\mathbb{C}$ of the polynomial $(X^2+1)f(X)$. Let $u_1, u_2, \ldots, u_n$ be the roots of f in F. We consider the following elements of F:

$$v_{ij} = u_i u_j + a(u_i + u_j), \qquad 1 \leq i < j \leq n \quad , \tag{2}$$

where a is a fixed real number. (We should actually write $v_{ij}(a)$, but we shall omit the a in the notation for the sake of brevity.) The number n' of elements of the form

(2) is equal to

$$n' = \binom{n}{2} = \frac{n(n-1)}{2} = \frac{2^m n_0 (2^m n_0 - 1)}{2} = 2^{m-1} n_0' , \tag{3}$$

where n_0' is an odd number. The polynomial

$$f_a(X) = \prod_{1 \le i < j \le n} (X - v_{ij}) = X^{n'} + b_1 X^{n'-1} + \cdots + b_{n'} \in F[X]$$

has degree n', and, by definition, its roots are precisely the elements (2). By Vieta's formulas (12) in §1, the coefficients $b_1, \ldots, b_{n'}$ are plus or minus the elementary symmetric functions $s_1, \ldots, s_{n'}$ of the v_{ij}. If we substitute the expression for v_{ij} in terms of u_i and u_j in $s_k(v_{12}, v_{13}, \ldots, v_{n-1,n})$, we obtain the function

$$h_k(u_1, \ldots, u_n) = s_k(\ldots, u_i u_j + a(u_i + u_j), \ldots), \quad k = 1, \ldots, n' ,$$

which we claim is also a symmetric function. This is true because, for any permutation $\pi \in S_n$ (S_n is the symmetric group on n elements) we have

$$\hat{\pi} v_{ij} = u_{\pi(i)} u_{\pi(j)} + a(u_{\pi(i)} + u_{\pi(j)}) = u_{\pi(i), \pi(j)}$$

(or $v_{\pi(j), \pi(i)}$ if $\pi(i) > \pi(j)$), so that π induces a permutation $\hat{\pi}$ on the set of elements (2). Since $s_k(v_{12}, v_{13}, \ldots, v_{n-1,n})$ is symmetric, it does not change when the arguments are permuted; hence,

$$(\pi h_k)(u_1, \ldots, u_n) = s_k(\hat{\pi} v_{12}, \hat{\pi} v_{13}, \ldots, \hat{\pi} v_{n-1,n}) = s_k(v_{12}, v_{13}, \ldots, v_{n-1,n}) = h_k(u_1, \ldots, u_n) .$$

We note that $h_k(u_1, \ldots, u_n)$ is the value at $X_i = u_i$, $i = 1, \ldots, n$, of a symmetric polynomial $h_k(X_1, \ldots, X_n)$ having real coefficients that depend only on $a \in \mathbb{R}$.

By the basic theorem on symmetric polynomials (Theorem 1 of §2), there exists a polynomial $g_k(Y_1, \ldots, Y_n)$ with real coefficients such that $h_k(X_1, \ldots, X_n) =$ $= g_k(s_1(X_1, \ldots, X_n), \ldots, s_n(X_1, \ldots, X_n))$. Hence,

$$(-1)^k b_k = h_k(u_1, \ldots, u_n) = g_k(s_1(u_1, \ldots, u_n), \ldots, s_n(u_1, \ldots, u_n)) = g_k(-a_1, \ldots, (-1)^n a_n) \in \mathbb{R}$$

(recall that the a_i are the coefficients of our original polynomial $f \in \mathbb{R}[X]$).

Thus, the coefficients b_k of the polynomial $f_a(X)$ are real for any $a \in \mathbb{R}$. Since $\deg f_a = n' = 2^{m-1} n'_0$ (see (3)), it follows by the induction assumption that f_a has at least one complex root, which, of course, must be one of the v_{ij}. Thus, at least one of the v_{ij} is not only in F but in the (perhaps smaller) field $\mathbb{C}$. If we vary the parameter $a \in \mathbb{R}$, we obtain other polynomials $f_a(X)$ with real coefficients, and for each such polynomial there is a pair of indices $i < j$ (depending on a) such that the element $v_{ij} = u_i u_j + a(u_i + u_j) \in F$ is an element of $\mathbb{C}$. Since there are only finitely many pairs of indices $\{i,j\}$ and infinitely many real numbers a, there must be two distinct real numbers a and a' with the same $\{i,j\}$. We may suppose that this pair $\{i,j\}$ is $\{1,2\}$ (re-numbering, if necessary, the $u_1, \ldots, u_n$). Thus,

$$\begin{aligned} u_1 u_2 + a(u_1 + u_2) &= c , \\ u_1 u_2 + a'(u_1 + u_2) &= c' , \qquad a \neq a' , \end{aligned} \tag{4}$$

are both complex numbers. The system of equations (4) implies that

$$u_1 + u_2 = \frac{c - c'}{a - a'} , \qquad u_1 u_2 = c - a \frac{c - c'}{a - a'}$$

also belong to the field $\mathbb{C}$. Hence, u_1 and u_2 are the roots of the quadratic equation

$$(X - u_1)(X - u_2) = X^2 - (u_1 + u_2) X + u_1 u_2$$

with complex coefficients. By the well-known formula for the roots of a quadratic equation, we have

$$u_1, u_2 = \frac{u_1 + u_2}{2} \pm \sqrt{\left(\frac{u_1 + u_2}{2}\right)^2 - u_1 u_2} ,$$

so that u_1 and u_2 are also complex numbers. Thus, we have found a complex root (actually, two complex roots) of the polynomial $f(X)$ under the assumption that f has real coefficients.

Now let

$$f(X) = a_0 X^n + a_1 X^{n-1} + \cdots + a_{n-1} X + a_n$$

be an arbitrary polynomial of degree n with complex coefficients (we may assume that $a_0 = 1$, but this is unimportant). If we replace each a_i by its complex conjugate, we obtain the polynomial

$$\overline{f}(X) = \overline{a}_0 X^n + \overline{a}_1 X^{n-1} + \cdots + \overline{a}_{n-1} X + \overline{a}_n \quad .$$

We now consider the polynomial

$$e(X) = f(X)\overline{f}(X) = e_0 X^{2n} + e_1 X^{2n-1} + \cdots + e_{2n}$$

of degree $2n$ with coefficients

$$e_k = \sum_{i+j=k} a_i \overline{a}_j , \qquad k = 0, 1, \ldots, 2n \quad .$$

Since conjugation $z \mapsto \overline{z}$ is an automorphism of $\mathbb{C}$ of order 2 (see Theorem 1 of §1 Ch. 5), we have $\overline{e}_k = \sum_{i+j=k} \overline{a}_i a_j = e_k$, and this means that $e_k \in \mathbb{R}$. Since we have proved that a polynomial with real coefficients has at least one complex root, it follows that for some $c \in \mathbb{C}$

$$f(c) \cdot \overline{f}(c) = e(c) = 0 \quad .$$

This means that either $f(c) = 0$, in which case the theorem is proved, or else $\overline{f}(c) = 0$, i.e., $\overline{a}_0 c^n + \overline{a}_1 c^{n-1} + \cdots + \overline{a}_{n-1} c + \overline{a}_n = 0$. Applying complex conjugation to both sides of this equation, we obtain $a_0 \overline{c}_1^{\,n} + a_1 \overline{c}^{\,n-1} + \cdots + a_{n-1} \overline{c} + a_n = 0$, i.e., $f(\overline{c}) = 0$. □

The fact that $\mathbb{C}$ is algebraically closed (and also the fact that splitting fields exist) are useful in solving a wide variety of problems.

Example. Let $S_0(f)$ be the set of zeros of a polynomial $f \in \mathbb{C}[X]$, and let

$S_1(f)$ be the set of "ones", i.e., $d \in S_1(f) \Leftrightarrow f(d) = 1$. Now suppose that f and g are two polynomials in $\mathbb{C}[X]$. We claim that

$$S_0(f) = S_0(g), \quad S_1(f) = S_1(g) \Longrightarrow f(X) = g(X) \quad .$$

Since obviously $S_0(f) \cap S_1(f) = \phi$, by the results of §1 it suffices to show that $|S_0(f) \cup S_1(f)| \geq n+1$, where $n = \deg f$, since in that case $f - g$ will be a polynomial with more distinct roots than its degree. By Theorem 1, we have

$$f(X) = a_0 \prod_{i=1}^{\nu} (X - c_i)^{s_i}, \quad f(X) - 1 = a_0 \prod_{j=1}^{\mu} (X - d_j)^{t_j}, \quad c_i, d_j \in \mathbb{C} \quad ,$$

where

$$\sum s_i = n = \sum t_j, \qquad \nu + \mu = |S_0(f) \cup S_1(f)| \quad .$$

According to Theorem 5 of §1, we have

$$f(X)' = (f(X) - 1)' = \prod_{i=1}^{\nu} (X - c_i)^{s_i - 1} \cdot \prod_{j=1}^{\mu} (X - d_j)^{t_j - 1} \cdot h(X) \quad ,$$

so that $(n - \nu) + (n - \mu) = \Sigma\,(s_i - 1) + \Sigma\,(t_j - 1) \leq \deg f(X)' = n - 1$. Hence, $\nu + \mu \geq n + 1$.

§4. Polynomials with real coefficients

1. Factorization in $\mathbb{R}[X]$. It follows from Theorem 1 of §3 that every polynomial f of degree n in $\mathbb{C}[X]$ can be written uniquely (except for the order of the factors) in the form

$$f(X) = a(X - c_1)(X - c_2) \ldots (X - c_n) \quad ,$$

where $a \neq 0$ and $c_1, \ldots, c_n$ are complex numbers. Now let $f(X) = X^n + a_1 X^{n-1} + \cdots + a_{n-1} X + a_n$ be a monic polynomial with real coefficients $a_1, \ldots, a_n$, and let c be a complex root of f, which we write in the form

$c = u + iv$, $u, v \in \mathbb{R}$. If $c \notin \mathbb{R}$, i.e., if $v \neq 0$, and we apply complex conjugation to the equation $f(c) = 0$, as in the proof of Theorem 1 of §3, we find that $f(\bar{c}) = 0$ as well, because $\bar{a}_i = a_i$. Hence, $f(X)$ is divisible by the quadratic polynomial

$$g(X) = (X - c)(X - \bar{c}) = X^2 - (c + \bar{c})X + c\bar{c} = X^2 - 2uX + (u^2 + v^2)$$

with negative discriminant $D(g) = 4u^2 - 4(u^2 + v^2) = -4v^2 < 0$. The condition $D(g) < 0$ is equivalent to the irreducibility of $g \in \mathbb{R}[X]$ over $\mathbb{R}$.

Next, suppose that k is the multiplicity of the root c of $f(X)$, and that $\ell \leq k$ is the multiplicity of the root $\bar{c}$. Then $f(X)$ is divisible by the ℓ-th power of $g(X)$:

$$f(X) = g(X)^{\ell} q(X) \quad .$$

The quotient $q(X)$ of the two polynomials in $\mathbb{R}[X]$ is also a polynomial in $\mathbb{R}[X]$, and if $k > \ell$ the complex number c will be a root of $q(X)$ of multiplicity $k - \ell$, while $\bar{c}$ will not be a root of $q(X)$. But we saw that this is impossible. Hence, $k = \ell$ (the case $\ell \geq k$ is handled similarly). Thus, the complex roots of any polynomial in $\mathbb{R}[X]$ occur in conjugate pairs, where conjugate roots have the same multiplicity. Since $\mathbb{R}[X]$ is a unique factorization domain, we now have the following theorem.

THEOREM 1. Any monic polynomial $f \in \mathbb{R}[X]$ of degree n factors uniquely (except for the order of the factors) into a product of $m \leq n$ linear factors $X - c_i$, corresponding to the real roots $c_1, \ldots, c_m$, and $(n-m)/2$ quadratic factors, which are irreducible over $\mathbb{R}$ and correspond to conjugate pairs of complex roots of f. □

Remarks. 1) Any irreducible polynomial in $\mathbb{R}[X]$ is either a linear polynomial or else a quadratic polynomial with negative discriminant.

2) In the notation of Theorem 1, we have the relation

$$D(f) = (-1)^{\frac{n-m}{2}} |D(f)| \quad ,$$

i.e., the sign of the discriminant is determined by the number of conjugate pairs of roots.

This equality can be obtained either directly from the definition of the discriminant, or else using the formula in Exercise 5 of §2.

3) All primary rational functions in the field $\mathbb{R}(X)$ are of the form given in (9) §4 Ch. 5.

2. <u>The problem of isolating the roots of a polynomial.</u> We shall think of a polynomial $f \in \mathbb{R}[X]$ as a real-valued function $x \mapsto f(x)$ of the real variable x, which we depict by a graph on the xy-plane. The real roots of the polynomial $f(X)$ correspond to the intersection of the graph with the x-axis.

The first important question that often arises in practice is to find bounds for the real roots, i.e., an interval $a < x < b$ which we can determine must contain all of the real roots of a given polynomial f. Actually, from Lemma 1 of §3 we already know that if $|x| > A/|a_0| + 1$ (where a_0 is the leading coefficient and $A = \max\{|a_1|, \ldots, |a_n|\}$), then the function $f(x)$ does not vanish (even if we allow x to be complex). More exact bounds on the roots are given in Exercises 1-4.

A more general problem is to <u>localize the roots</u>, i.e., for each real root to find an interval containing that root and no other root. The first satisfactory (though cumbersome) solution of this problem was given by Sturm in 1829. We shall not give the complete theory, which would also include the problem of isolating complex roots in regions of the complex plane. The simplification of the general results for various special classes of polynomials is a matter of great interest to specialists. We shall not discuss the methods for computing a "localized root" to within a given accuracy. Modern computer science has at its disposal a large arsenal of techniques to do this, but that subject would take us too far afield.

Fortunately, in many situations one is satisfied to know a rough picture of the location of the roots. Important information is furnished by drawing the graph of the function $x \mapsto f(x)$, whose values can be computed, say, at integer values of x. Notice that the roots of the equation $f(x) = 0$ will occur between extremal points (or at those points), i.e., no two roots can occur between adjacent extremal points. These extremal points, in turn, are the roots of the lower degree polynomial $f'(X)$. Looking at the graph can give us an estimate of the number of roots in a given interval -- but only an estimate, since we might have neglected oscillations of the function $x \mapsto f(x)$ in certain small intervals (see the diagram).

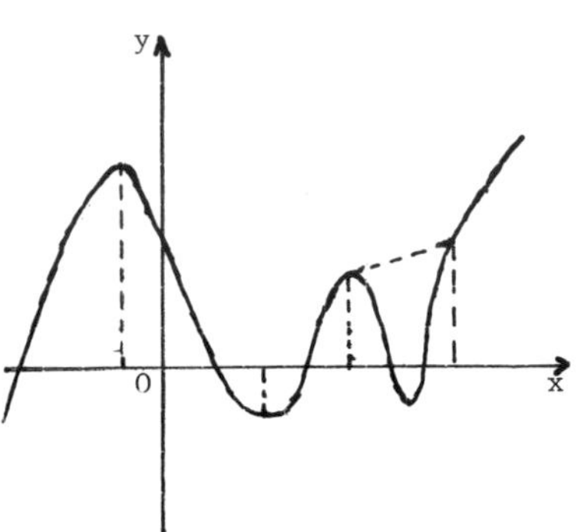

It is a remarkable fact that upper estimates for the number of positive (or negative) roots can be obtained using a very simple observation, which Descartes made in 1637. We introduce the following

<u>Definition.</u> Let

$$a_0, a_{i_1}, a_{i_2}, \dots, a_{i_q} \qquad (0 < i_1 < i_2 < \dots < i_q \leq n) \tag{1}$$

be all of the non-zero coefficients of a polynomial $f(X) = a_0 X^n + a_1 X^{n-1} + \dots + a_n \in \mathbb{R}[X]$, written in the indicated order. If $a_{i_k} a_{i_{k+1}} < 0$, we say that there is a <u>change of sign</u> at the $(k+1)$-st term. We let $L(f)$ denote the total number of changes of sign in the sequence (1).

It is clear that we always have $0 \leq L(f) \leq \deg f$, and also $L(-f) = L(f)$. Further note that $L(f) = L(aX^k + a_{i_1} X^{n-i_1} + \dots)$, where the exponent k need only satisfy the condition $k > n - i_1$, and where $a a_0 > 0$. If $L(f) = 0$, then f obviously does not have any positive roots. But it is possible for f not to have any positive roots even when $L(f) = \deg f$, for example: $f(X) = X^2 - X + 1$. But nevertheless, we shall see that $L(f)$ does have a direct relationship to the number of positive roots of the polynomial f.

LEMMA. *If* $c > 0$, *then* $L((X-c)f) = L(f) + 1 + 2s$, *where* $s \in \mathbb{Z}$, $s \geq 0$.

Proof. We are assuming, of course, that $f \neq 0$, so that $L(f)$ makes sense. If $\deg f = 0$, then $L(f) = 0$, and the lemma holds with $s = 0$. Using induction on $\deg f$, we suppose that the lemma holds for all polynomials of degree $< n$. Let $\deg f = n$, and write

$$f = a_0 X^n + a_k X^{n-k} + \cdots + a_{n-1} X + a_n ,$$

where a_k is the first non-zero coefficient after a_0 if there is any $(k \geq 1)$. Since $L(-f) = L(f)$, without loss of generality we may assume that $a_0 > 0$. Set

$$g(X) = a_k X^{n-k} + \cdots + a_{n-1} X + a_n .$$

We clearly have

$$L(f) = L(g) + \varepsilon , \tag{2}$$

where

$$\varepsilon = \frac{1}{2}\left(1 - \left(\frac{a_0 a_k}{|a_0 a_k|}\right)\right) = 0 \quad \text{or} \quad 1 .$$

If $g = 0$, then the lemma holds trivially for f, so we suppose that $g \neq 0$. For later use, we also set

$$(X - c)\, g(X) = a_k X^{n+1-k} + h(X)$$

(note that if $g \neq 0$, then $h \neq 0$).

Using the induction assumption and (2), we have

$$L((X - c)\, g(X)) = L(g) + 1 + 2t = L(f) + 1 - \varepsilon + 2t . \tag{3}$$

We also have

$$(X - c)f = a_0 X^n (X - c) + (X - c)g = a_0 X^{n+1} - a_0 c X^n + a_k X^{n+1-k} + h(X) .$$

If $k > 1$, then obviously $L((X - c)f) = 2 - \varepsilon + L((X - c)g)$, since $c > 0$ $(2 - \varepsilon$ is the number of changes of sign in the sequence $a_0, -a_0 c, a_k)$. Using (3), we obtain

$$L((X-c)f) = L(f) + 1 + 2s, \quad \text{where} \quad s = t+1-\varepsilon \geq 0 \quad .$$

It remains to consider the case $k = 1$:

$$(X-c)f = a_0 X^{n+1} + (a_1 - a_0 c) X^n + h(X) \quad .$$

If a_1 and $a_1 - a_0 c$ have the same sign, then

$$L((a_1 - a_0 c) X^n + h(X)) = L((X-c)g)$$

and

$$L((X-c)f) = \varepsilon + L((X-c)g) = L(f) + 1 + 2s, \qquad s = t \quad .$$

If a_1 and $a_1 - a_0 c$ have opposite signs, which can only happen if $a_1 > 0$ and $\varepsilon = 0$, then

$$L((a_1 - a_0 c) X^n + h(X)) = L((X-c)g) \pm 1 = L(f) + 1 + 2t \pm 1$$

and

$$L((X-c)f) = 1 + L((a_1 - a_0 c) X^n + h(X)) = L(f) + 1 + 2s \quad ,$$

where $s = t$ or $t+1$. Finally, if $a_1 - a_0 c = 0$, which also can only happen if $a_1 > 0$ and $\varepsilon = 0$, then

$$L((X-c)f) = L(a_0 X^{n+1} + h(X)) = L(a_1 X^n + h(X)) = L((X-c)g) = L(f) + 1 + 2s, \quad s = t \quad .$$

□

Using this lemma, it is easy to prove Descartes' rule of signs.

THEOREM 2. The number of positive roots of a polynomial $f \in \mathbb{R}[X]$ either is equal to $L(f)$ or is less than $L(f)$ by an even number.

Proof. Let $c_1, c_2, \ldots, c_m$ be the positive roots (not necessarily distinct) of the polynomial $f(X) = a_0 X^n + \cdots + a_{n-\nu} X^{\nu}$, where we assume that $a_0 > 0$ and $a_{n-\nu}$ is the last non-zero coefficient. Recalling how f factors in $\mathbb{R}[X]$ (Theorem 1), we may write:

$$f(X) = (X - c_1) \dots (X - c_m)\, g(X) \quad , \tag{4}$$

where $g(X) = a_0 X^{n-m} + \dots + bX^{\nu}$, $a_0 > 0$, $b > 0$ $(\nu \geq 0)$. Since a_0 and b have the same sign, it follows that $L(g) = 2t$ is an even number. Using the lemma and the factorization (4), we obtain the chain of equalities

$$L((X - c_1)g) = 1 + 2(s_1 + t) \quad ,$$

$$L((X - c_2)(X - c_1)g) = 1 + 2(s_1 + t) + 1 + 2s_2 = 2 + 2(s_1 + s_2 + t) \quad ,$$

$$\dots\dots\dots\dots\dots\dots\dots\dots\dots\dots\dots\dots$$

$$L(f) = m + 2(s_1 + s_2 + \dots + s_m + t) \quad .$$

The last equality gives us the theorem. □

Thus, we always have $m \leq L(f)$. We now consider a special case which is important in practice. Suppose we know in advance that all of the roots of f are real. Then we have a more precise fact.

THEOREM 3. _If all of the roots of_ f _are real, and if_ $m(f) = m$ _denotes the number of positive roots, counting multiplicity, then_ $m(f) = L(f)$.

Proof. It is fairly easy to derive Theorem 3 from Theorem 2, but there is a simple and at the same time instructive independent proof which we shall give instead.

By Rolle's Theorem (or the Mean Value Theorem) of calculus, if $a' < b'$ are two roots of our polynomial $f(X)$, then there exists a number $c \in \mathbb{R}$, $a' < c < b'$, such that $f'(c) = 0$. This implies that all of the roots of the derivative $f'(X)$ are also real, and that $m(f') = m(f)$ or $m(f) - 1$. To see this, let $c_1 < c_2 < \dots < c_r$ be the roots of f, and let $n_1, n_2, \dots, n_r$ be their respective multiplicities, so that $n_1 + n_2 + \dots + n_r = \deg f = n$. By Theorem 5 of §1, the derivative f' has the roots $c_1, c_2, \dots, c_r$ with multiplicities $n_1 - 1$, $n_2 - 1, \dots, n_r - 1$; and, by Rolle's

Theorem, in each interval between successive c_i there is at least one root of f', so we also obtain roots $c'_1, c'_2, \ldots, c'_{r-1}$ of f'. In all this gives us $(n_1 - 1) + (n_2 - 1) + + \cdots + (n_r - 1) + r - 1 = n - 1$ real roots of f'. Since $\deg f' = n - 1$, this is all of its roots. Further suppose that $c_{\ell-1} < 0$, while $c_\ell, \ldots, c_r$ are the positive roots of f; thus, $n_\ell + \cdots + n_r = m = m(f)$. The positive roots of $f'(X)$ are the roots $c_\ell, \ldots, c_r$ with multiplicities $n_\ell - 1, \ldots, n_r - 1$, the roots $c'_\ell, \ldots, c'_{r-1}$, and perhaps $c'_{\ell-1}$, i.e., the number of positive roots of f' is $m(f') = m(f) - 1$ or $m(f)$, as claimed. The following formula clearly holds in both cases $m(f') = m(f)$ and $m(f') = m(f) - 1$:

$$m(f) = m(f') + \varepsilon, \qquad \varepsilon = \frac{1}{2}(1 - (-1)^{m(f) + m(f')}) \quad . \tag{5}$$

We further note that, if

$$f(X) = a_0 X^n + \cdots + a_{n-\nu} X^\nu \quad , \tag{6}$$

where $a_{n-\nu}$ is the last non-zero coefficient, and if we write f in the form (4), then $a_{n-\nu} = (-1)^m c_1 c_2 \cdots c_m b$, where $c_k > 0$ and $b > 0$. In other words,

$$(-1)^{m(f)} a_{n-\nu} > 0 \quad . \tag{7}$$

We now use induction on $n = \deg f$ to prove the theorem. Suppose that Theorem 3 holds for all polynomials of degree $< n$. If $\nu > 0$ in (6), i.e., if $a_n = 0$, then $f(X) = X \cdot f_1(X)$, where $m(f) = m(f_1) = L(f_1) = L(f)$, since $m(f_1) = L(f_1)$ by the induction assumption. So we may suppose that $a_n \neq 0$. Let

$$f'(X) = n a_0 X^{n-1} + \cdots + \mu a_{n-\mu} X^{\mu-1}, \qquad a_{n-\mu} \neq 0 \quad .$$

Then

$$L(f) = L(f') + \delta, \qquad \delta = \frac{1}{2}\left(1 - \frac{a_n a_{n-\mu}}{|a_n a_{n-\mu}|}\right) = 0 \text{ or } 1 \quad .$$

But we know (see (7)) that $(-1)^{m(f)} a_n > 0$ and $(-1)^{m(f')} a_{n-\mu} > 0$. Hence

$\delta = \frac{1}{2}(1 - (-1)^{m(f) + m(f')})$; thus, $\delta = \epsilon$. Since $L(f') = m(f')$ by the induction assumption, we conclude that $L(f) = m(f') + \epsilon = m(f)$ by (5). □

COROLLARY. *Suppose that all of the roots of* f *are real. Then the number of roots in the interval* $(a,b]$ *is equal to* $L(f_a) - L(f_b)$, *where*

$$f_a(X) = f(X + a) = \sum_{0 \le k \le n} \frac{f^{(k)}(a)}{k!} X^k ,$$

$$f_b(X) = f(X + b) = \sum_{0 \le k \le n} \frac{f^{(k)}(b)}{k!} X^k ,$$

(*using the Taylor series for* $f(X + a)$, *see Exercise 3 below*).

Proof. By definition, $m(f_a)$ is the number of positive roots of f_a, which is equal to the number of roots of f which are greater than a. Similarly for $m(f_b)$. Thus, the number of roots of f between a and b is equal to the difference $m(f_a) - m(f_b)$, which equals $L(f_a) - L(f_b)$ by Theorem 3. □

3. *Stable polynomials.* A monic polynomial $f(X) = X^n + a_1X^{n-1} + \cdots + a_{n-1}X + a_n$ with real coefficients is called *stable* if all of its roots lie in the left half-plane:

$$f(\lambda) = 0, \quad \lambda = \alpha + i\beta \implies \alpha < 0$$

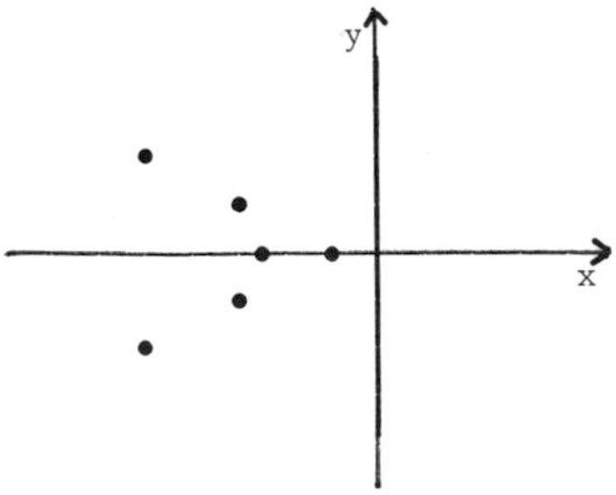

(see Fig. 18). The terminology originates from the theory of differential equations, where one has the following criterion for a physical system (in the broad sense of a mechanical, technological, or economic system) to be asymptotically stable in a neighborhood of an equilibrium position. If f is the polynomial associated to a given n-th order linear differential equation with constant coefficients, then for any root λ we must

Fig. 18

have

$$\lim_{t \to +\infty} e^{\lambda t} = 0 \quad . \tag{8}$$

Since, by Euler's formula (see (15) §1 Ch. 5), we have $e^{\lambda t} = e^{\alpha t} e^{i\beta t} = e^{\alpha t}(\cos \beta t + i \sin \beta t)$, it follows that the dominating term is $e^{\alpha t}$, and the condition (8) is equivalent to: $\alpha < 0$.

This leads to a special type of localization problem, called the <u>Hurwitz-Routh problem</u>, which asks how to determine directly from the coefficients whether or not a polynomial is stable. (Actually, this problem was first stated much earlier, in 1868, by the British physicist Maxwell, and was solved for certain small n by the Russian engineer Vyshnegradskii, who studied the stability problem for regulators in 1876.) The algebraic problem was solved for any n in 1895. The Hurwitz-Routh criterion says: <u>a polynomial f is stable if and only if the following inequalities hold:</u>

$$\Gamma_1 > 0, \quad \Gamma_2 > 0, \ldots, \Gamma_n > 0 , \tag{9}$$

where

$$\Gamma_k = \begin{vmatrix} a_1 & 1 & 0 & 0 & 0 & 0 & \cdots & 0 \\ a_3 & a_2 & a_1 & 1 & 0 & 0 & \cdots & 0 \\ a_5 & a_4 & a_3 & a_2 & a_1 & 1 & \cdots & 0 \\ a_7 & a_6 & a_5 & a_4 & a_3 & a_2 & \cdots & 0 \\ \cdots & \cdots & \cdots & \cdots & \cdots & \cdots & \cdots & \cdots \\ a_{2k-1} & a_{2k-2} & a_{2k-3} & a_{2k-4} & a_{2k-5} & a_{2k-6} & \cdots & a_k \end{vmatrix}$$

(we take $a_s = 0$ for $s > n$).

Without attempting to prove the Hurwitz-Routh theorem (such a proof belongs in other courses), we take note of the fact that the statement of the criterion has such an elegant form thanks to the theory of determinants. We also note that, by Theorem 1, if the conditions in (9) are fulfilled, then the polynomial $f(X)$ is a product of factors of the form $X + u$ and $X^2 + vX + w$ with $u > 0$, $v > 0$, $w > 0$, and this means that all of the coefficients of a stable polynomial are positive:

$$a_1 > 0, \quad a_2 > 0, \ldots, a_n > 0 \quad . \tag{10}$$

Thus, the conditions (10) are necessary in order for $f(X)$ to be stable. Although (10) is not a sufficient condition for stability (i.e., unstable polynomials exist with all positive coefficients), nevertheless the use of (10) allows us to cut in half the number of determinant inequalities in (9). This is of great practical value, since the computation of determinants is a very cumbersome affair.

Example. If $n = 2$, the system of inequalities $\Gamma_1 > 0$, $\Gamma_2 > 0$ is equivalent to the simpler system of inequalities: $a_1 > 0$ and $a_2 > 0$; this criterion can also be seen immediately from the formula for the roots of a quadratic equation.

If $n = 3$, the criterion reduces to the inequalities $a_1 > 0$, $a_2 > 0$, $a_3 > 0$, and $a_1 a_2 > a_3$, since we have: $\Gamma_3 = a_3(a_1 a_2 - a_3)$.

In conclusion, we note that the Hurwitz-Routh criterion does not answer all questions connected with stability, since in practice we are often interested in polynomials and differential equations whose coefficients depend on a parameter. In that case, we want to express the stability conditions in terms of the parameter, and this is a problem of a very different sort.

EXERCISES

1. Let $f(X) = a_0 X^n + a_1 X^{n-1} + \cdots + a_n$ be a polynomial of degree n with real coefficients. Show that knowing an upper bound for the positive roots of the polynomials $f(X)$, $X^n f(1/X)$, $f(-X)$, and $X^n f(-1/X)$ gives both upper and lower bounds for both positive and negative roots of $f(X)$.

2. In the notation of Exercise 1, let $a_0 > 0$, and let m be the lowest index for which $a_m < 0$. Let B be the maximum of the absolute values of the negative

coefficients. Show that

$$c \leq 1 + \sqrt[m]{B/a_0}$$

for every positive real root of $f(X)$.

3. (Taylor's formula). Let K be a field of characteristic zero, and let $a \in K$. Prove that any polynomial $f \in K[X]$ of degree n satisfies the formula

$$f(X) = f(a) + \frac{f'(a)}{1!}(X - a) + \frac{f''(a)}{2!}(X - a)^2 + \cdots + \frac{f^{(n)}(a)}{n!}(X - a)^n \quad .$$

4. Show that, if $f(X) \in \mathbb{R}[X]$ has degree n and positive leading coefficient a_0, and if $f(a) > 0$, $f'(a) > 0, \ldots, f^{(n)}(a) > 0$, then $f(c) = 0$, $c > 0 \Rightarrow c < a$.

5. Using Descartes' rule of signs, find the sign of the discriminant of the polynomials $X^5 - X^2 + 1$ and $X^3 - 6X - 9$ (see the remark at the end of Subsection 1).

6. Can the polynomials $X^5 - X - 1$ and $X^3 + aX + b \in \mathbb{Q}[X]$ have any complex roots in common? Recall that the polynomial $X^5 - X - 1$ is irreducible over $\mathbb{Q}$ (see Exercise 11 of §1).

7. Show that the roots of a polynomial $f(X) = X^5 + uX^4 + vX^3 + w \in \mathbb{R}[X]$ with $w \neq 0$ cannot all be real.

8. It is clear that, if a polynomial $f(X) = a_0 X^n + \cdots + a_n \in \mathbb{Z}[X]$ has a root $c \in \mathbb{Z}$, then c divides the constant term $a_n = f(0)$. Namely, if $f(c) = 0$, then $a_n = c(-a_0 c^{n-1} - \cdots - a_{n-2} c - a_{n-1})$. Show that $c - 1$ divides $f(1) = \Sigma a_i$, and that $c + 1$ divides $f(-1) = (-1)^n \sum (-1)^i a_i$.

9. Show that

$$f(X) = X^n + a_1 X^{n-1} + \cdots + a_n \in \mathbb{Z}[X], \quad f(c) = 0, \quad c \in \mathbb{Q} \Longrightarrow c \in \mathbb{Z} \quad .$$

10. Show that any polynomial $f(X)$ with $f(x) \geq 0$ for all $x \in \mathbb{R}$ can be written in the form

$$f(X) = g(X)^2 + h(X)^2 ,$$

where $g, h \in \mathbb{R}[X]$.

11. Give an independent proof of the stability criterion for $n = 3, 4$. For $n = 4$ write it in the form: $a_1 > 0, \ldots, a_4 > 0$, $a_1 a_2 > a_3$, $a_3(a_1 a_2 - a_3) > a_1^2 a_4$.

"In recent times, the view has become more and more prevalent that many branches of mathematics are nothing but the theory of invariants of special groups".

Sophus Lie, 1893

Part Two
Groups, Rings, Modules

The second part of the book can be viewed as a more sophisticated, but, one hopes, not too abstract continuation of Part I. Relatively few new concepts are introduced. Our old friends from Chapter 4 reappear, and lead us into areas of much greater depth. The reader should pay the closest attention to the examples, which take up at least a quarter of the text (for example, §1 of Ch. 7 and §3 of Ch. 8). Among other things, the examples are chosen in such a way as to provide a bridge between algebra and other branches of mathematics. If they serve to strengthen the reader's feeling for the unity of mathematics, then the author's purpose in Part II can be considered fulfilled.

Further Reading

1. M. F. Atiyah and I. G. Macdonald, Introduction to Commutative Algebra, Addison-Wesley, 1969.

2. T. C. Bartee and G. Birkhoff, Modern Applied Algebra, McGraw-Hill, 1970.

3. Z. I. Borevich and I. R. Shafarevich, Number Theory, Academic Press, 1966.

4. N. Bourbaki, Algebra (Modules, Rings, Forms).

5. J. B. Carrell and J. A. Dieudonné, Invariant Theory Old and New, Academic Press, 1971.

6. P. M. Cohn, Universal Algebra, Harper and Row, 1965.

7. C. C. Faith, Algebra: Rings, Modules and Categories, Springer-Verlag, 1973.

8. M. Hall, The Theory of Groups, Chelsea Pub. Co., 1976.

9. I. N. Herstein, Noncommutative Rings, A. M. S. (J. Wiley), 1968.

10. N. Jacobson, Lie Algebras, Interscience Pub., 1962.

11. A. A. Kirillov, Elements of the Theory of Representations, Springer-Verlag, 1976.

12. A. G. Kurosh, Lectures on General Algebra, Pergamon Press, 1965.

13. G. I. Liubarskiǐ, The Application of Group Theory in Physics, Pergamon Press, 1960.

14. A. I. Mal'tsev, Algebraic Systems, Springer-Verlag, 1973.

15. L. S. Pontryagin, Topological Groups, Gordon and Breach, 1966.

16. M. M. Postnikov, Foundations of Galois Theory, Pergamon Press, 1962.

17. J.-P. Serre, A Course in Arithmetic, Springer-Verlag, 1973.

18. J.-P. Serre, Linear Representations of Finite Groups, Springer-Verlag, 1977.

19. H. Weyl, The Classical Groups; Their Invariants and Representations, Princeton University Press, 1939.

20. D. P. Zhelobenko, Compact Lie Groups and Their Representations, Translated by A. M. S., 1973.

Chapter 7. Groups

This chapter further develops the concept of a group, which was introduced in Chapter 4. In this chapter we emphasize not so much abstract groups as certain natural types of group "actions". It was the concrete realizations of groups which gave the impetus for the development of the general theory and was responsible for its reputation as a valuable instrument for mathematical investigation.

In examining these special (but important) examples we shall see the key role played by (homo-, epi-, iso-) morphisms of groups. These group-theoretic constructions allow us to reduce the study of complicated objects to simpler ones.

§1. Classical groups in low dimensions

1. General definitions. A basic course in linear algebra and geometry supplies us with examples of groups which deserve an especially detailed investigation. The transformations of affine, Euclidean, and Hermitian spaces which leave fixed a given point (say, the origin) lead to the so-called classical groups $GL(n)$, $SL(n)$, $O(n)$, $SO(n)$, $U(n)$, $SU(n)$.

It turns out that these are all examples of so-called Lie groups. One should also include the symplectic group $Sp(n)$, but we do not intend to describe all of the classical groups; there are many books where the reader can find treatments of the symplectic group. If n is small, we speak of the classical groups in low dimensions. We have already encountered the groups $GL(n)$ and $SL(n)$ in Part I. In our definitions of the other groups we shall want to avoid dependence on geometry; once one chooses an orthonormal basis in n-dimensional space, the orthogonal and unitary groups can be defined in terms of matrices:

$$\begin{aligned} O(n) &= \{A \in M_n(\mathbb{R}) \mid {}^tA \cdot A = A \cdot {}^tA = E\} , \\ SO(n) &= \{A \in O(n) \mid \det A = 1\} , \\ U(n) &= \{A \in M_n(\mathbb{C}) \mid A^* \cdot A = A \cdot A^* = E\} , \\ SU(n) &= \{A \in U(n) \mid \det A = 1\} . \end{aligned}$$

Here $A^* = {}^t\bar{A}$ is the matrix obtained from $A = (a_{ij})$ by taking the transpose and then replacing the entries by their complex conjugates. The groups $SL(n)$, $SO(n)$, $SU(n)$ are called the <u>special</u> linear group, the <u>special</u> orthogonal group, and the <u>special</u> unitary group, respectively. In particular,

$$O(1) = \{\pm 1\}, \qquad SO(1) = \{1\} ,$$

$$U(1) = \{e^{i\varphi} \mid 0 \leq \varphi < 2\pi\}, \qquad SU(1) = \{1\} ,$$

$$SO(2) = \left\{ \begin{Vmatrix} \cos\varphi & -\sin\varphi \\ \sin\varphi & \cos\varphi \end{Vmatrix} \;\middle|\; 0 \leq \varphi < 2\pi \right\} \cong U(1) .$$

We have an isomorphism between the groups $SO(2)$ and $U(1)$ given by

$$\begin{Vmatrix} \cos\varphi & -\sin\varphi \\ \sin\varphi & \cos\varphi \end{Vmatrix} \longmapsto e^{i\varphi} .$$

Since the geometric locus of the complex numbers $e^{i\varphi}$, $0 \leq \varphi < 2\pi$, is the unit circle S^1 in the plane, it is also customary to say that the group $SO(2)$ and the circle S^1 are topologically equivalent. The precise meaning of this statement is explained in a geometry or topology course.

There is a remarkable and much less obvious connection between the groups SU(2) and SO(3). We first discuss a geometric realization of SU(2), which will then lead us to a geometric realization of SO(3).

2. Parametrization of SU(2) and SO(3). According to a famous theorem of Euler, every rigid rotation of $\mathbb{R}^3$, i.e., every element of SO(3), is rotation about some fixed axis. For example, the matrices

$$B_\varphi = \begin{Vmatrix} \cos\varphi & -\sin\varphi & 0 \\ \sin\varphi & \cos\varphi & 0 \\ 0 & 0 & 1 \end{Vmatrix}, \qquad C_\theta = \begin{Vmatrix} 1 & 0 & 0 \\ 0 & \cos\theta & -\sin\theta \\ 0 & \sin\theta & \cos\theta \end{Vmatrix} \tag{1}$$

correspond, respectively, to rotation about the z-axis through an angle of φ and rotation about the x-axis through an angle of θ. If we use the parametrization of rotations by the Euler angles φ, θ, ψ, where $0 \le \varphi, \psi < 2\pi$ and $0 \le \theta < \pi$ (for now, we are not concerned with the geometric meaning of these angles), then any matrix $A \in SO(3)$ can be written in the form

$$A = B_\varphi C_\theta B_\psi , \tag{2}$$

where B_φ, C_θ, and B_ψ are the matrices defined in (1).

Now let

$$g = \begin{Vmatrix} \alpha & \beta \\ \gamma & \delta \end{Vmatrix} \in SU(2) .$$

We have

$$g^* = {}^t\bar{g} = \begin{Vmatrix} \bar{\alpha} & \bar{\gamma} \\ \bar{\beta} & \bar{\delta} \end{Vmatrix}, \qquad g^{-1} = \begin{Vmatrix} \delta & -\beta \\ -\gamma & \alpha \end{Vmatrix} .$$

Since $g \in U(2) \Leftrightarrow g^* = g^{-1}$, it follows that $\delta = \bar{\alpha}$ and $\gamma = -\bar{\beta}$. Thus, any matrix g in SU(2) has the form

$$g = \left\| \begin{matrix} \alpha & \beta \\ -\bar{\beta} & \bar{\alpha} \end{matrix} \right\| , \qquad |\alpha|^2 + |\beta|^2 = 1 \quad . \tag{3}$$

Conversely, if g is a matrix of the form (3), then obviously $g \in SU(2)$. Hence, every element of the group $SU(2)$ is uniquely determined by a pair of complex numbers α, β such that $|\alpha|^2 + |\beta|^2 = 1$. If we set $\alpha = \alpha_1 + i\alpha_2$ and $\beta = \beta_1 + i\beta_2$ with $\alpha_k, \beta_k \in \mathbb{R}$ and $i = \sqrt{-1}$, then the condition $|\alpha|^2 + |\beta|^2 = 1$ can be written in the form

$$\alpha_1^2 + \alpha_2^2 + \beta_1^2 + \beta_2^2 = 1 \quad .$$

We are therefore justified in saying that <u>the group</u> $SU(2)$ <u>is topologically equivalent (homeomorphic) to the sphere</u> S^3 <u>in the four-dimensional space</u> $\mathbb{R}^4$.

We now consider the unitary matrices

$$b_\varphi = \left\| \begin{matrix} e^{i\frac{\varphi}{2}} & 0 \\ 0 & e^{-i\frac{\varphi}{2}} \end{matrix} \right\| , \qquad c_\theta = \left\| \begin{matrix} \cos\frac{\theta}{2} & i\sin\frac{\theta}{2} \\ i\sin\frac{\theta}{2} & \cos\frac{\theta}{2} \end{matrix} \right\| \quad . \tag{4}$$

As one proves in a basic course on linear algebra (and as is easy to verify directly in this case), given a unitary matrix g of the form (3), there exists a unitary matrix U such that

$$g = ub_\varphi u^{-1} \tag{5}$$

where $\lambda = e^{i\varphi/2}$ is determined from the quadratic equation

$$\lambda^2 - 2\alpha_1\lambda + 1 = 0 \quad .$$

We further note that any matrix (3) with $\alpha\beta \neq 0$ can be given the form

$$a(\varphi, \theta, \psi) \equiv b_\varphi c_\theta b_\psi = \begin{Vmatrix} \cos\frac{\theta}{2} \cdot e^{i\frac{\varphi+\psi}{2}} & i\sin\frac{\theta}{2} \cdot e^{i\frac{\varphi-\psi}{2}} \\ i\sin\frac{\theta}{2} \cdot e^{i\frac{\psi-\varphi}{2}} & \cos\frac{\theta}{2} \cdot e^{-i\frac{\varphi+\psi}{2}} \end{Vmatrix}, \tag{6}$$

where

$$0 \leq \varphi < 2\pi, \quad 0 \leq \theta < \pi, \quad -2\pi \leq \psi < 2\pi .$$

(We shall later see that φ, θ, ψ are the Euler angles. The unitary matrices g and $-g$ correspond to the same rotation in $\mathbb{R}^3$, so that the range of ψ is restricted to the half-interval $[0, 2\pi)$.) To see this, it suffices to set

$$|\alpha| = \cos\frac{\theta}{2}, \quad \operatorname{Arg}\alpha = \frac{\varphi+\psi}{2}, \quad |\beta| = \sin\frac{\theta}{2}, \quad \operatorname{Arg}\beta = \frac{\varphi-\psi+\pi}{2},$$

and use the fact that any complex number z is given by the two real parameters $|z|$ and $\arg z$ ($\operatorname{Arg} z$ is the principal value of the argument $\arg z$).

We are now ready to resolve the basic problem of this section.

3. The epimorphism $SU(2) \to SO(3)$. To every vector $x = x_1e_1 + x_2e_2 + x_3e_3$ with norm $N(x) = x_1^2 + x_2^2 + x_3^2$ we associate the 2×2 complex matrix

$$H_x = \begin{Vmatrix} x_3 & x_1 + ix_2 \\ x_1 - ix_2 & -x_3 \end{Vmatrix}. \tag{7}$$

The space M_2^+ of matrices of the form (7) consists of all Hermitian matrices with zero trace (i.e., ${}^t\bar{H}_x = H_x$, $\operatorname{tr} H = 0$). The correspondence between vectors $x \in \mathbb{R}^3$ and matrices $H_x \in M_2^+$ is obviously one-to-one. In particular, the basis vectors $e_1, e_2, e_3 \in \mathbb{R}^3$ correspond to basis matrices $h_k = H_{e_k}$:

$$h_1 = \begin{Vmatrix} 0 & 1 \\ 1 & 0 \end{Vmatrix}, \quad h_2 = \begin{Vmatrix} 0 & i \\ -i & 0 \end{Vmatrix}, \quad h_3 = \begin{Vmatrix} 1 & 0 \\ 0 & -1 \end{Vmatrix}; \tag{8}$$

$$H_x = x_1 h_1 + x_2 h_2 + x_3 h_3, \quad M_2^+ = \langle h_1, h_2, h_3 \rangle_{\mathbb{R}} .$$

Note that every linear operator $\Phi^+ : H_x \mapsto H_y$ on M_2^+ with matrix A in the basis (8) completely determines a linear operator $\Phi : x \mapsto y$ on $\mathbb{R}^3$ with the same matrix A in the basis e_1, e_2, e_3, since we have: $H_{\alpha x} = \alpha H_x$ and $H_{x+x'} = H_x + H_{x'}$. Since these bases are the only ones we shall be using, in what follows we shall often identify operators with the corresponding matrices.

Now let g be a fixed element in the group $SU(2)$. We consider the map

$$\Phi_g^+ : H_x \longmapsto gH_x g^{-1} . \tag{9}$$

Since similar matrices have the same trace, it follows that $\operatorname{tr} \Phi_g^+(H_x) = \operatorname{tr} H_x = 0$. In addition, $g^* = {}^t\bar{g} = g^{-1}$, and hence

$$(gH_x g^{-1})^* = (g^{-1})^* H_x^* g^* = gH_x g^{-1}$$

so that $\Phi_g^+(H_x) \in M_2^+$:

$$\Phi_g^+(H_x) = \begin{Vmatrix} y_3 & y_1 + iy_2 \\ y_1 - iy_2 & -y_3 \end{Vmatrix} = H_y ,$$

where $y = (y_1, y_2, y_3) \in \mathbb{R}^3$. It is clear from the defining equations (7) and (9) that

$$\Phi_g^+(H_{\alpha x + \alpha' x'}) = \alpha \Phi_g^+(H_x) + \alpha' \Phi_g^+(H_{x'}) .$$

Thus, Φ_g^+ (respectively, Φ_g) is a linear map on M_2^+ (resp. $\mathbb{R}^3$).

We show that $\Phi_g : \mathbb{R}^3 \to \mathbb{R}^3$ is an orthogonal operator. We have:

$$N(\Phi_g(x)) = N(y) = y_1^2 + y_2^2 + y_3^2 = -\det H_y = -\det \Phi_g^+(H_x) =$$
$$= -\det gH_x g^{-1} = -\det H_x = x_1^2 + x_2^2 + x_3^2 = N(x) ,$$

i.e., Φ_g preserves the norm, and hence also the scalar product. We have not yet established whether or not Φ_g changes the orientation of $\mathbb{R}^3$; this depends on the sign of $\det \Phi_g$. We only know that $\det \Phi_g = \pm 1$.

It follows from the definition that

$$\Phi^+_{g_1}(\Phi^+_{g_2} H_x) = g_1(g_2 H_x g_2^{-1}) g_1^{-1} = (g_1 g_2) H_x (g_1 g_2)^{-1} = \Phi^+_{g_1 g_2}(H_x) \quad ,$$

where Φ^+_E is the orthogonal unit matrix of order 3 corresponding to $E = \begin{Vmatrix} 1 & 0 \\ 0 & 1 \end{Vmatrix} \in SU(2)$. Thus, the map

$$\Phi : g \mapsto \Phi_g \qquad (\text{or } \Phi^+ : g \mapsto \Phi^+_g)$$

is a homomorphism from $SU(2)$ to $O(3)$. The kernel $\operatorname{Ker} \Phi = \operatorname{Ker} \Phi^+$ consists of those unitary matrices g for which $\Phi^+_g = \Phi^+_E$. In other words,

$$\operatorname{Ker} \Phi = \{g \in SU(2) \mid gH = Hg, \ \forall H \in M_2^+\} = \{g \in SU(2) \mid gh_j = h_j g, \ j = 1, 2, 3\} \quad ,$$

where $\{h_1, h_2, h_3\}$ is the basis (8) of the space M_3^+. A direct verification shows that

$$g = \begin{Vmatrix} \alpha & \beta \\ -\bar{\beta} & \bar{\alpha} \end{Vmatrix}, \quad gh_j = h_j g, \quad 1 \leq j \leq 3 \implies g = \pm E \implies \operatorname{Ker} \Phi = \{\pm E\} \quad .$$

We now consider the images of the unitary matrices (4) under the homomorphism Φ. We carry out the calculation for Φ^+ in the basis (8):

$$b_\varphi h_1 b_\varphi^{-1} = (\cos \varphi) h_1 + (\sin \varphi) h_2 \quad ,$$

$$b_\varphi h_2 b_\varphi^{-1} = (-\sin \varphi) h_1 + (\cos \varphi) h_2 \quad ,$$

$$b_\varphi h_3 b_\varphi^{-1} = h_3 \quad .$$

Thus (here we feel free to switch from Φ^+ to Φ and from matrices to operators), $\Phi_{b_\varphi} = B_\varphi$ (see (1)) is rotation of $\mathbb{R}^3$ about the x_3-axis through an angle of φ. If φ and u are chosen so that (5) holds, then, since Φ is a homomorphism, we have

$$\Phi_g = \Phi_u \Phi_{b_\varphi} \Phi_u^{-1} \qquad \text{and} \qquad \det \Phi_g = \det \Phi_u \cdot 1 \cdot (\det \Phi_u)^{-1} = 1 \quad .$$

This shows that Φ is in fact a homomorphism from $SU(2)$ to $SO(3)$.

We can similarly verify that $\Phi_{c_\theta} = C_\theta$ is rotation about the x_1-axis through an angle of θ. Now for any matrix $A \in SO(3)$ we have

$$A = B_\varphi C_\theta B_\psi = \Phi_{b_\varphi} \Phi_{c_\theta} \Phi_{b_\psi} = \Phi_{b_\varphi c_\theta b_\psi} = \Phi_{a(\varphi, \theta, \psi)} \quad .$$

Hence, the image $\operatorname{Im} \Phi$ contains all of $SO(3)$, and we have proved

THEOREM 1. _The group_ $SO(3)$ _is the homomorphic image of_ $SU(2)$ _under the homomorphism_ $\Phi : g \mapsto \Phi_g$ _with kernel_ $\operatorname{Ker} \Phi = \{\pm E\}$. _Each rotation in_ $SO(3)$ _corresponds to precisely two unitary operators_ g _and_ $-g$ _in_ $SU(2)$. □

4. _Geometrical characterization of_ $SO(3)$. Theorem 1 immediately implies the following

COROLLARY. _The group_ $SO(3)$ _is topologically equivalent (homeomorphic) to three-dimensional real projective space_ $\mathbb{R}(P^3)$.

Proof. We saw in Subsection 2 that the elements in $SU(2)$ are in one-to-one correspondence with the points of the sphere S^3 in $\mathbb{R}^4$. The two linear operators g and $-g \in SU(2)$ correspond to diametrically opposite points in S^3, which are glued together (identified) under the homomorphism Φ. We thereby obtain one of the models of the projective space $\mathbb{R}(P^3)$. □

In the usual course on linear algebra and geometry, the projective space $\mathbb{R}(P^n)$ is defined to be the set of straight lines through the origin in $\mathbb{R}^{n+1}$. Each such line intersects the unit sphere S^n (centered at the origin) at precisely two diametrically opposite points. Giving one of these points uniquely determines the line through the origin. But this means that $\mathbb{R}(P^n)$ can be defined as the quotient space of the unit sphere S^n in $\mathbb{R}^{n+1}$

with respect to the equivalence relation that calls two points of S^n equivalent if they are diametrically opposite one another. We are not at this point concerned with giving the topology on $\mathbb{R}(P^n)$.

We have arrived at a remarkable result. The sphere S^3 and the space $\mathbb{R}(P^3)$ have a group structure -- $SU(2)$ in the first case, and $SO(3)$ in the second case. It turns out that any attempt to define a continuous group structure on S^2 or on $\mathbb{R}(P^2)$ is doomed to fail (this fact is tangential to our theme, and will not be proved here).

According to Theorem 1 and its corollary, the group $SO(3)$ is "twice as small" as the group $SU(2)$. Since we have an epimorphism $SU(2) \to SO(3)$, it is natural to ask whether there exists a monomorphism $SO(3) \to SU(2)$. We shall see in Chapter 8 that this question has a negative answer.

EXERCISES

1. Fill in the gaps in the proof of Theorem 1, i.e., go through an actual verification (without alluding to courses in linear algebra and geometry) of all the minor assertions, starting with Equation (2).

2. Using the geometrical characterization of $SU(2)$, show that

$$(0,1,0,0) * (0,0,1,0) = (0,0,0,1) \neq (0,0,1,0) * (0,1,0,0)$$

(taking the product of points on S^3). These same points $(0,1,0,0)$ and $(0,0,1,0)$ commute when considered on $\mathbb{R}(P^3)$.

3. Show that differentiating the entries in the unitary matrices

$$K_1(t) = \begin{Vmatrix} \cos\frac{t}{2} & i\sin\frac{t}{2} \\ i\sin\frac{t}{2} & \cos\frac{t}{2} \end{Vmatrix}, \quad K_2(t) = \begin{Vmatrix} \cos\frac{t}{2} & -\sin\frac{t}{2} \\ \sin\frac{t}{2} & \cos\frac{t}{2} \end{Vmatrix}, \quad K_3(t) = \begin{Vmatrix} e^{i\frac{t}{2}} & 0 \\ 0 & e^{-i\frac{t}{2}} \end{Vmatrix}$$

with respect to t and then setting $t = 0$ leads to the matrices

$$K_1 = \frac{i}{2}\begin{Vmatrix} 0 & 1 \\ 1 & 0 \end{Vmatrix} = \frac{i}{2}h_1, \quad K_2 = \frac{i}{2}\begin{Vmatrix} 0 & i \\ -i & 0 \end{Vmatrix} = \frac{i}{2}h_2, \quad K_3 = \frac{i}{2}\begin{Vmatrix} 1 & 0 \\ 0 & -1 \end{Vmatrix} = \frac{i}{2}h_3 ,$$

which form a basis of the space M_2^- of <u>skew-hermitian matrices</u>

$$K = \begin{Vmatrix} ik_3 & -k_2 + ik_1 \\ k_2 + ik_1 & -ik_3 \end{Vmatrix}, \qquad k_j \in \mathbb{R} ,$$

with zero trace: $K^* = -K$, $\operatorname{tr} K = 0$.

§2. Group actions on sets

1. <u>Homomorphisms</u> $G \to S(\Omega)$. We began group theory in Chapter 4 with examples of transformation groups, i.e., subgroups of the group $S(\Omega)$ of all one-to-one maps of a set Ω to itself. This approach is consistent both with the historical path along which group theory developed and with the importance of transformation groups in other areas of mathematics. The so-called abstract theory of groups, which arose in a later era (the first half of our century), has gone far beyond transformation groups, but many of the concepts in this theory bear the imprint of earlier times. In fact, the most common source of these concepts is the idea of a <u>realization</u> (a <u>representation</u>) of a given group G in $S(\Omega)$, where Ω is some suitably chosen set. By a realization of G in $S(\Omega)$ we mean any homomorphism $\Phi : G \to S(\Omega)$. If Φ_g is the transformation in $S(\Omega)$ corresponding to $g \in G$, then $\Phi_e = e_\Omega$ is the identity map $\Omega \to \Omega$, and we have $\Phi_{gh} = \Phi_g \circ \Phi_h$ for $g, h \in G$. The image $\Phi_g(x)$ of a point (element) $x \in \Omega$ under the transformation Φ_g is often denoted simply gx; we speak of a map $(g, x) \mapsto gx$ from the cartesian product (G, Ω) to Ω. Perhaps we should be more careful and write $g \circ x$ or $g * x$, so as not to confuse this operation with multiplication in G, but there is usually no need to do this; in practice, there is rarely a danger of ambiguity. We can now write the above properties of Φ_g in the form

(i) $ex = x, \qquad x \in \Omega;$

(ii) $(gh)x = g(hx); \qquad g, h \in G \quad .$

Any time we have a map $(g, x) \mapsto gx$ from the cartesian product $G \times \Omega$ to Ω which satisfies (i) and (ii), we say that the group acts (on the left) on the set Ω, and Ω is called a G-set. Conversely, if we have a G-set Ω, then we can use the formula

$$\Phi_g(x) = gx, \qquad x \in \Omega \quad ,$$

for each $g \in G$ to define a map $\Phi_g : \Omega \to \Omega$. It then follows from (i) and (ii) that the map $\Phi : g \mapsto \Phi_g$ is a homomorphism from G to $S(\Omega)$. It is also customary to say (especially when $|\Omega| < \infty$) that we have a representation (Φ, Ω) of G in a permutation group. The kernel $\operatorname{Ker} \Phi$ is called the kernel of the action of G. If Φ is a monomorphism (in other words: if $gx = x$, $\forall x \in \Omega \Rightarrow g = e$), then we say that G acts effectively on the set Ω.

Remark. Any action of G on Ω induces an action of G on $\Omega^k = \Omega \times \dots \times \Omega$ by the obvious rule: $g \cdot (x_1, \dots, x_k) = (gx_1, \dots, gx_k)$. There is also an induced action of G on the set of all subsets $\mathcal{P}(\Omega)$ (see Exercise 4 of §5 Ch. 1). We set $g\phi = \phi$, and, if T is a non-empty subset of Ω, then we set $gT = \{gt \mid t \in T\}$. The properties (i) and (ii) are easily checked. Clearly, T and gT have the same cardinality; hence, G induces an action on the subsets of a given cardinality.

2. The orbit and stationary subgroup of a point. Two points $x, x' \in \Omega$ are said to be G-equivalent, where G is a group acting on Ω, if $x' = gx$ for some $g \in G$. Using (i) and (ii) in Subsection 1, we easily show that we have reflexivity, symmetry, and transitivity, and hence an equivalence relation which divides Ω into disjoint equivalence classes. Each equivalence class is called a G-orbit. The orbit containing $x_0 \in \Omega$ is denoted $G(x_0)$; thus, $G(x_0) = \{gx_0 \mid g \in G\}$. But sometimes other notation is used, depending on the special nature of various actions of groups of sets. The notion of an orbit

arose from geometry. For example, if $G = SO(2)$ is the group of rotations of the plane about the origin, then the orbit of a point P is the circle centered at the origin passing through P, and the set $\Omega = \mathbb{R}^2$ is the union of all of the concentric circles, including the one with zero radius (consisting of one point, the origin). We have encountered orbits before, in the first part of this book. In Chapter 4 we used orbits to write a permutation $\pi \in S_n$ as a product of disjoint cycles. In that case G was the cyclic group $\langle \pi \rangle$.

Let x_0 be a given point in Ω. Consider the set

$$St(x_0) = \{g \in G \mid g x_0 = x_0\} \subset G \quad .$$

Since $e x_0 = x_0$ and $g, h \in St(x_0) \Rightarrow gh^{-1} \in St(x_0)$, it follows that $St(x_0)$ is a subgroup of G. It is called the stationary subgroup (or the stabilizer) in G of the point $x_0 \in \Omega$, and is often denoted G_{x_0}. In the case of the above example of $SO(2)$ acting on $\mathbb{R}^2$, we have $St(\text{origin}) = SO(2)$ and $St(P) = e$ if P is not the origin. We always have

$$g x_0 = g' x_0 \iff g^{-1} g' \in St(x_0) \iff g' \in g\, St(x_0) \quad .$$

Thus, the lefts cosets $g\, St(x_0)$ of the stationary subgroup $St(x_0)$ in G are in one-to-one correspondence with the points in the orbit $G(x_0)$. In particular,

$$\text{Card}\, G(x_0) = \text{Card}\, (G / St(x_0)) = (G : St(x_0)) \quad . \tag{1}$$

Here, as before, $G / St(x_0)$ denotes the quotient set of G by $St(x_0)$, and $(G : St(x_0))$ is the index of the subgroup $St(x_0)$ in G. The cardinality $\text{Card}\, G(x_0)$ is often called the length of the G-orbit of x_0.

From (1) and Lagrange's Theorem it follows that the length of any orbit of a finite group G divides the order of the group.

It should be noted that the point x_0 in the right side of (1) can be replaced by any other point $x_0' \in G(x_0)$. Thus,

$$\mathrm{Card}\, G(x_0) = \mathrm{Card}\, G(x_0') = (G : \mathrm{St}(x_0')) \quad .$$

We now give a stronger statement concerning stationary subgroups. Suppose $x' = g x_0$. Then

$$\mathrm{St}(x_0')\, g x_0 = \mathrm{St}(x_0')\, x_0' = x_0' = g x_0 \quad ,$$

so that

$$g^{-1}\, \mathrm{St}(x_0')\, g x_0 = x_0 , \quad \text{i.e.} \quad g^{-1}\, \mathrm{St}(x_0')\, g \subset \mathrm{St}(x_0) \quad .$$

Similarly,

$$g\, \mathrm{St}(x_0)\, g^{-1} \subset \mathrm{St}(x_0') \quad ,$$

since

$$\mathrm{St}(x_0)\, g^{-1} x_0' = \mathrm{St}(x_0)\, x_0 = x_0 = g^{-1} x_0' \quad .$$

Thus, we have the equality

$$\mathrm{St}(x_0') = g\, \mathrm{St}(x_0)\, g^{-1} = \{ g h g^{-1} | h \in \mathrm{St}(x_0) \} \quad .$$

Two subgroups $H, H' \subset G$ are called <u>conjugate</u> if $H' = g H g^{-1}$ for some $g \in G$ (see Example 1 below). We can then state the following theorem.

THEOREM 1. <u>Suppose that a group</u> G <u>acts on a set</u> Ω. <u>If two points</u> $x_0, x_0' \in \Omega$ <u>lie in the same orbit, then their stationary subgroups are conjugate:</u>

$$x_0' = g x_0 \implies \mathrm{St}(x_0') = g\, \mathrm{St}(x_0)\, g^{-1} \quad .$$

<u>Further, if</u> G <u>is a finite group, and if</u> Ω <u>splits up into finitely many orbits</u>

$$\Omega = \Omega_1 \cup \Omega_2 \cup \ldots \cup \Omega_r$$

<u>with representatives</u> $x_1, x_2, \ldots, x_r$, <u>then</u>

$$|\Omega| = \sum_{i=1}^{r} (G : \mathrm{St}(x_i)) \quad . \qquad \square \tag{2}$$

Many applications of the "orbit method" to finite groups are based on this formula (2).

3. Examples of group actions on sets. We now discuss some examples which relate to ideas from group theory.

Example 1 (the conjugation action). Taking $\Omega = G$, we have the action of G on G defined by

$$x \longmapsto I_g(x) = gxg^{-1}, \qquad \forall x \in G .$$

We could have written $g \circ x = gxg^{-1}$, but we prefer to use our old notation from Subsection 2 of §3 Ch. 4 for the inner automorphism I_g corresponding to $g \in G$.

The action of g given by $I_g \in \mathrm{Inn}(G)$ is called conjugation. The kernel of this action is called the center of the group G:

$$Z(G) = \{z \in G \mid I_g(z) = z, \quad \forall g \in G\} = \{z \in G \mid zg = gz, \quad \forall g \in G\} .$$

The orbit of an element $x \in G = \Omega$, which in the present context we shall denote x^G, is called the conjugacy class of x. If $a, b \in x^G$, we sometimes write $a \underset{G}{\sim} b$. The stationary subgroup $St(x)$, which in this context is called the centralizer of x, is often denoted $C(x)$ (or $C_G(x)$, if it is necessary to indicate the group G in order to avoid confusion).

According to the remark at the end of Subsection 1, the conjugation action carries over to subsets and subgroups of G. Two subsets $H, T \subset G$ are conjugate if $T = gHg^{-1}$ for some $g \in G$. Let H be a subgroup of G. It is customary to call

$$N(H) = St(H) = \{g \in G \mid gHg^{-1} = H\}$$

the normalizer of H in G. In particular, H is a normal subgroup of G (written $H \lhd G$) if $N(H) = G$, in agreement with the definitions in Chapter 4. Because of the relationship (1), the length of the orbit H^G (the number of subgroups conjugate to H) is equal to the index of the normalizer $N(H)$ in G.

Now suppose that G is a finite group, and that $x_1^G, \dots, x_r^G$ are its conjugacy classes, where the first q of them consist of one element:

$$x_i^G = \{x_i\}, \qquad i = 1, \dots, q \quad (x_1 = e) \quad .$$

Then $Z(G) = \{x_1, x_2, \dots, x_q\}$, and we can write relations (1) and (2) in the form

$$|x_i^G| = (G : C(x_i)) \ ; \tag{1'}$$

$$|G| = |Z(G)| + \sum_{i=q+1}^{r} (G : C(x_i)) \quad . \tag{2'}$$

For example, suppose that $G = S_3$. Then $r = 3$ and $q = 1$ (i.e., $Z(S_3) = e$), and we have

$$S_3 = \{e\} \cup \{(12), (13), (23)\} \cup \{(123), (132)\}$$

for the partition of S_3 into conjugacy classes. The sizes of these conjugacy classes (orbit lengths) divide $6 = |S_3|$, as must be the case by (1').

The relation (2') immediately gives us the following interesting fact.

THEOREM 2. <u>Every finite p-group G (i.e., every group of order $p^n > 1$, where p is a prime) has a non-trivial center, i.e.,</u> $Z(G) \neq e$.

<u>Proof.</u> If G is an abelian group, then $G = Z(G)$, and there is nothing to prove. Otherwise, $r > q$, $(G : C(x_i)) = p^{n_i}$, where $n_i > 1$ for $i > q$, and (2) takes the form

$$p^n = |Z(G)| + \sum_{i=q+1}^{r} p^{n_i} \ ,$$

which shows that $|Z(G)|$ is divisible by p. □

It is easy to find examples of non-abelian p-groups. Take the group of upper triangular matrices with entries in the finite field of p elements:

$$P = \left\{ \begin{Vmatrix} 1 & a & c \\ 0 & 1 & b \\ 0 & 0 & 1 \end{Vmatrix} \middle| a, b, c \in Z_p \right\} .$$

This group P is clearly a non-abelian group of order p^3.

Example 2 (translation). The map $L_a : G \to G$ defined by $L_a(g) = ag$, which we used in the proof of Cayley's Theorem (see §3 Ch. 4), is usually called left translation by a. Since $eg = g$ and $(ab)g = a(bg)$, it follows that the left translations give an action of G on itself, which induces an action of the group G on the set of subsets of G. In particular, let H be a subgroup, and let G/H be the set of left cosets gH, $g \in G$. It is clear that the map

$$(x, gH) \mapsto x(gH) = (xg)H$$

gives an action, denoted L^H, of the group G on the set G/H. The kernel of this action is $\operatorname{Ker} L^H = \{x \in G \mid L^H_x(gH) = gH,\ \forall g \in G\} = \{x \in G \mid xgH = gH,\ \forall g \in G\}$. In other words, $x \in \operatorname{Ker} L^H$ if and only if $g^{-1}xg \in H$ for all $g \in G$, i.e., if $x \in gHg^{-1}$, $\forall g \in G$. Thus,

$$\operatorname{Ker} L^H = \bigcap_{g \in G} gHg^{-1}$$

is the largest normal subgroup of G contained in H. The action of G on G/H is effective if and only if there is no non-trivial normal subgroup of G contained in H.

In any case, if H is any subgroup of index n in G, then we obtain a representation $(L^H, G/H)$ of the group G by permutations L^H_x of cosets of H in G. This representation (which may, however, be not a monomorphism) is much more efficient than the one obtained using Cayley's Theorem.

Example 3 (transitive groups). A group $G \subset S_n$ of permutations of the set $\Omega = \{1, 2, \dots, n\}$ is called transitive if the orbit G_i of some element $i \in \Omega$ (and hence of any element of Ω) is all of Ω. In other words, an action $G \times \Omega \to \Omega$ is transitive if for every pair of elements $i, j \in \Omega$ there exists at least one $g \in G$ such that $g(i) = j$.

Let $\Omega^{[k]}$ be the set of all ordered k-element subsets of Ω. The action of G on Ω induces an action on $\Omega^{[k]}$. If this action is transitive on $\Omega^{[k]}$, we say that G acts k-transitively on Ω. For example, the symmetric group S_n is n-transitive on Ω, and the alternating group A_n is (n - 2)-transitive.

Any group G acts transitively on the set G/H of left cosets of H (see Example 2). To see this, let $g_i H$ be two cosets. Then $g_j g_i^{-1}(g_i H) = g_j H$. But, remarkably, very little is known about k-transitive groups for $k > 5$. There is even a century-old conjecture (unproved) of Jordan that there are only two such groups: S_n and A_n.

We shall now obtain some interesting quantitative results concerning transitive groups, which we shall need later. Let G be a group acting transitively on Ω. We let G_i denote the stationary subgroup St(i) of a point $i \in \Omega$. We know (see Theorem 1) that if $i = g_i(1)$, then $G_i = g_i G_1 g_i^{-1}$, $i = 1, 2, \ldots, n$ $(g_1 = e)$. In addition, the elements g_i can be chosen as left coset representatives for G modulo G_1:

$$G = G_1 \cup g_2 G_1 \cup \ldots \cup g_n G_1 \quad . \tag{3}$$

In particular, $|G| = n|G_1|$ (which agrees with our general results on the length of orbits in Subsection 2).

THEOREM 3. Let G be a transitive group on Ω, and for any $g \in G$ let N(g) be the number of points of Ω which g leaves fixed. Then:

(i) $\sum_{g \in G} N(g) = |G|$ (thus, if we divide both sides by $|G|$, we conclude that "on the average" each element of G leaves one point fixed);

(ii) if G is a 2-transitive group, then

$$\sum_{g \in G} N(g)^2 = 2|G| \quad .$$

Proof. (i). We have

$$\sum_{g \in G} N(g) = \sum_{j=1}^{n} \Gamma(j) \quad ,$$

where $\Gamma(j)$ is the number of elements of G leaving the point j fixed, i.e., $\Gamma(j) = |G_j|$. Since G is transitive, we have $|G_j| = |g_j G_1 g_j^{-1}| = |G_1|$, where the g_j are as in (3). Hence,

$$\sum_{g \in G} N(g) = \sum_{j=1}^{n} |G_j| = \sum_{j=1}^{n} |G_1| = n\,|G_1| = |G| \quad .$$

(ii). The 2-transitivity condition means that the stationary subgroup G_1 acts transitively on the set $\Omega_1 = \Omega \setminus \{1\}$, i.e., the G_1-orbits are $\{1\}$ and Ω_1. Let $N'(x)$ be the number of points in Ω_1 which are left fixed by $x \in G_1$. Applying (i) to G_1 and Ω_1, we obtain

$$\sum_{x \in G_1} N'(x) = |G_1| \quad .$$

Since $N(x) = 1 + N'(x)$ for $x \in G_1$ (because of the point 1), we have

$$\sum_{x \in G_1} N(x) = 2\,|G_1| \quad .$$

The same relations hold for all of the other G_j :

$$\sum_{x \in G_j} N(x) = 2\,|G_j| = 2\,|G_1| \quad .$$

Summing over j, we obtain

$$\sum_{j=1}^{n} \sum_{x \in G_j} N(x) = 2n\,|G_1| = 2\,|G| \quad .$$

In the double sum $N(x)$ is counted once for every G_j which contains x. But x leaves fixed precisely $N(x)$ points, and so is contained in $N(x)$ subgroups G_j. This means that each x contributes $N(x)^2$ to the sum. On the other hand, any element $y \in G$

which is not contained in the union $\bigcup_j G_j$ must permute all of the points, so that $N(y) = 0$. We can thus write

$$\sum_{g \in G} N(g)^2 = \sum_{j=1}^{n} \sum_{x \in G_j} N(x) = 2|G| \quad . \qquad \square$$

4. Homogeneous spaces. In geometry it is of particular interest to consider the case when Ω is a topological space (for example, the line $\mathbb{R}$ or the sphere S^2), G is a so-called continuous (or topological) group, and the action $(g, x) \mapsto gx$ satisfies the reasonable requirement:

(iii) $f(g, x) = gx$ is a continuous function of the two variables g and x.

A group G which acts on Ω in such a way that conditions (i) and (ii) of Subsection 1 and condition (iii) above are all satisfied, is called a group of motions of Ω. In many cases these are motions which preserve some metric on Ω. The space Ω is called homogeneous if G acts transitively in the sense of Example 3, i. e., if all of the points of Ω belong to the same G-orbit.

From the discussion in Subsections 1 and 2 it is clear that there is a one-to-one correspondence between the points of the homogeneous space Ω and the cosets in G of one of the stationary subgroups H. To a motion $g \in G$ of Ω we associate the map $g'H \mapsto gg'H$ on the set G/H.

We now consider our example $SO(3)$ in §1 from this new point of view. The group $SO(3)$ acts on the two-dimensional unit sphere S^2. It is obvious that, to any pair of points $P, Q \in S^2$ there corresponds a motion (rotation) taking P to Q, i. e., S^2 is a homogeneous space with group $SO(3)$. The stationary subgroup $St(P)$ of any point $P \in S^2$ leaves fixed the entire axis through P and the origin. Hence, $St(P) \cong SO(2)$, the group of rotations in the plane perpendicular to the axis through P.

Since the elements of $SO(2)$ are identified with the points of the unit circle S^1, it follows that the group $SO(3)$ can be thought of as a "layer cake" of unit circles "indexed" by the points of the sphere $S^2 : SO(3)/S^1 \cong S^2$. In this we call $SO(3) \to S^2$ a <u>fibration</u> with <u>base</u> S^2 and <u>fibre</u> S^1 over each point $P \in S^2$. We shall not go further into this subject, which properly belongs in a course on topology.

EXERCISES

1. Let Φ and Φ' be homomorphisms of the group G into $S(\Omega)$ and $S(\Omega')$, respectively. Then the actions on Ω and Ω' are said to be <u>equivalent</u> if there exists a bijective map $\sigma : \Omega \to \Omega'$ such that the following diagram is commutative for all $g \in G$:

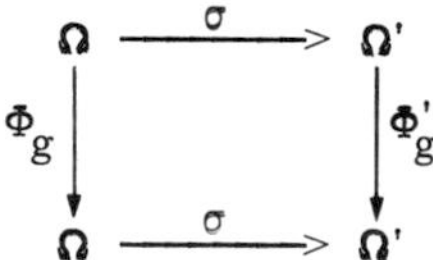

In other words, $\Phi'_g = \sigma \Phi_g \sigma^{-1}$. Prove that any transitive action of a group G is equivalent to the action of G on the left cosets of some subgroup H.

2. Using Theorem 2, prove that any group of order p^2 (where p is a prime) is abelian.

3. Prove that the center of the group P at the end of Example 1 is:

$$Z(P) = \left\{ \begin{Vmatrix} 1 & 0 & c \\ 0 & 1 & 0 \\ 0 & 0 & 1 \end{Vmatrix} \;\middle|\; c \in Z_p \right\} .$$

Find the conjugacy classes in the group P.

4. Let n be a natural number. Write it as a sum $n = n_1 + n_2 + \cdots + n_m$ with $n_1 \geq n_2 \geq \cdots \geq n_m \geq 1$. Let $p(n)$ denote the total number of such partitions for all $m = 1, 2, \ldots$. Thus, $p(3) = 3$, $p(4) = 5$, and so on. Given a permutation $\pi \in S_n$, by writing it as a product of disjoint cycles $\pi = \pi_1 \pi_2 \ldots \pi_m$ (see §2 Ch. 4),

we obtain a corresponding partition of n. Show that the conjugacy classes in the group S_n are in one-to-one correspondence with the partitions of the integer n.

5. Suppose that $\pi \in S_n$ is a product of r cycles of length 1, s cycles of length 2, t cycles of length 3, and so on, so that $n = r + 2s + 3t + \cdots$. Show that the cardinality of the conjugacy class in S_n which contains π is given by the formula

$$\left| {}_{\pi}S_n \right| = \frac{n!}{1^r r!\, 2^s s!\, 3^t t! \cdots} .$$

6. Suppose that a group G acts on a set Ω. We call a subset $\Gamma \subset \Omega$ <u>invariant</u> under G (or G-<u>invariant</u>) if $gx \in \Gamma$ for all $g \in G$ and $x \in \Gamma$. For example, an invariant set for the action $SO(2) \times \mathbb{R}^2 \to \mathbb{R}^2$ is a set of concentric circles about the origin. Prove that any invariant subset of Ω is a union of orbits, and that the G-orbit of an element $x \in \Omega$ is the same thing as the smallest invariant subset containing x.

7. Given a group G and a subgroup H, show that the action $H \times G \to G$ defined by $(h, g) \mapsto hg$ gives the partition of G into right cosets of H.

8. By modifying the proof of Theorem 1, derive the relation

$$r(G:\Omega) = \frac{1}{|G|} \sum_{g \in G} N(g) ,$$

where $r(G:\Omega)$ is the number of orbits for the action of G on Ω.

§3. Some group theoretic constructions

This section, especially the first subsection, is somewhat more difficult, and we shall have to return to it several times, using different concrete examples to solidify our understanding of the abstract concepts.

1. General theorems on group homomorphisms. In §4 of Chapter 4, we saw that, given a normal subgroup K of a group G, it is possible to construct a new group G/K, which is called the quotient group of G by K. For example, in working with the epimorphism $\Phi : SU(2) \to SO(3)$ (see §1), it is natural to introduce the quotient group $SU(2)/\{\pm E\}$ and compare it with the image $\operatorname{Im} \Phi = SO(3)$. It is not hard to see that $SU(2)/\{\pm E\} \cong SO(3)$, but, in order not to have to go through the argument again each time, it is useful to prove some general facts about subgroups, homomorphisms, and quotient groups. First, recall that the notation $K \triangleleft G$ means that K is a normal subgroup of G.

THEOREM 1 (fundamental homomorphism theorem). Let $\varphi : G \to H$ be a group homomorphism with $K = \operatorname{Ker} \varphi$. Then K is a normal subgroup of G, and $G/K \cong \operatorname{Im} \varphi$. Conversely, if $K \triangleleft G$, then there exists a group H (namely, G/K) and an epimorphism $\pi : G \to H$ having kernel K. (π is often called the natural map, natural homomorphism, or natural projection.)

Proof. We already know that $\operatorname{Ker} \varphi = K \triangleleft G$. We define the map $\overline{\varphi} : G/K \to H$ by setting

$$\overline{\varphi}(gK) = \varphi(g) \quad .$$

If $g_1 K = g_2 K$, then $g_1^{-1} g_2 \in K$, $\varphi(g_1^{-1} g_2) = e$, and so $\varphi(g_1) = \varphi(g_2)$; this means that the map $\overline{\varphi}$ is well defined (i.e., does not depend on the choice of coset representative). Since $\overline{\varphi}(g_1 K \cdot g_2 K) = \overline{\varphi}(g_1 g_2 K) = \varphi(g_1 g_2) = \varphi(g_1)\varphi(g_2) = \overline{\varphi}(g_1 K)\overline{\varphi}(g_2 K)$, it follows that $\overline{\varphi}$ is a homomorphism. Actually, $\overline{\varphi}$ is a monomorphism, because if we had $\overline{\varphi}(g_1 K) = \overline{\varphi}(g_2 K)$ it would follow that $\varphi(g_1) = \varphi(g_2)$, so that $\varphi(g_1^{-1} g_2) = e$, $g_1^{-1} g_2 \in K$, and $g_1 K = g_2 K$. It is also clear that $\operatorname{Im} \overline{\varphi} = \operatorname{Im} \varphi$. Hence $\overline{\varphi}$ is the desired isomorphism from G/K to $\operatorname{Im} \varphi$.

Conversely, suppose $K \triangleleft G$. Take π to be the function which associates to any element of G its K-coset, i.e., set $\pi(g) = gK$. It is clear that π has the required properties. $\square$

It should be noted that giving the kernel of a homomorphism does not determine the homomorphism uniquely. For example, the two automorphisms $g \mapsto g$ and $g \mapsto g^{-1}$ of an abelian group of order a prime $p > 2$ are different, but their kernels are the same $(= e)$.

If we have a homomorphism $\rho: G \to G_1$ and a subgroup $H \subset G$, it is natural to consider the restriction $\rho|_H$ and the image of H under this homomorphism. The following theorem greatly simplifies the investigation of all such situations.

THEOREM 2 (first isomorphism theorem). _Let_ G _be a group, and let_ H _and_ K _be subgroups, where_ K _is normal in_ G. _Then_ $HK = KH$ _is a subgroup of_ G _containing_ K. _In addition, the intersection_ $H \cap K$ _is a normal subgroup of_ H, _and the map_

$$\varphi : hK \longmapsto h(H \cap K)$$

is an isomorphism of groups:

$$HK/K \cong H/H \cap K \quad .$$

Proof. The condition $K \lhd G$ can be written in the form $gK = Kg$, $\forall g \in G$; in particular, $hK = Kh$ for all $h \in H$. The set $HK = \{hk | h \in H, k \in K\}$ consists of a certain number of cosets hK, i.e., $HK = \bigcup_{h \in H} hK$. If we replace hK by Kh, we obtain the equality

$$HK = \bigcup_{h \in H} hK = \bigcup_{h \in H} Kh = KH \quad .$$

It is obvious that the identity element e, which is in both H and K, is contained in HK. Next $(hk)^{-1} = k^{-1}h^{-1} = h^{-1}(hkh^{-1})^{-1} \in HK$, so that the inverse of an element in HK lies in HK. Finally, $HK \cdot HK = H \cdot KH \cdot K = H \cdot HK \cdot K = HK$, i.e., the set HK is closed under multiplication. We conclude that $HK \subset G$ is a subgroup of

G.

Since $K \subset HK$ and $K \lhd G \Rightarrow K \lhd HK$, it makes sense to speak of the quotient group HK/K. Let $\pi : G \to G/K$ be the natural epimorphism, and let $\pi_0 = \pi|_H$ be the restriction of π to H. The image $\operatorname{Im} \pi_0$ consists of the cosets hK, $h \in H$, i.e., all cosets of K in G which have a representative in H. In other words, $\operatorname{Im} \pi_0 = HK/K$. Thus, we have an epimorphism

$$\pi_0 : H \longrightarrow HK/K \quad .$$

Its kernel $\operatorname{Ker} \pi_0$ consists of those $h \in H$ for which $\pi_0(h) = hK = K$, the identity in HK/K. But $hK = K \Leftrightarrow h \in H \cap K$, so that $\operatorname{Ker} \pi_0 = H \cap K$. The subgroup $H \cap K$, like any kernel of a homomorphism, is normal in H (this can also be easily verified directly).

By the fundamental homomorphism theorem (Theorem 1), the map $\overline{\pi}_0 : h(H \cap K) \mapsto \pi_0(h) = hK$ gives an isomorphism $H/H \cap K \cong HK/K$. Note that $\varphi = \overline{\pi}_0^{-1} : hK \mapsto h(H \cap K)$ is also an isomorphism, from HK/K to $H/H \cap K$. □

Since we have given a "first isomorphism theorem", the reader may surmise that we shall give a "second isomorphism theorem". This is the case, but we shall give a more special, simplified theorem than what usually goes by the name "second theorem".

THEOREM 3. Suppose that G is a group with subgroups H and K, where $K \lhd G$ and $K \subset H$. Then $\overline{H} = H/K$ is a subgroup of $\overline{G} = G/K$, and $\pi^* : H \mapsto \overline{H}$ is a one-to-one correspondence between the set $\Omega(G, K)$ of subgroups of G which contain K and the set $\Omega(\overline{G})$ of all subgroups of the group $\overline{G}$. If $H \in \Omega(G, K)$, then $H \lhd G \Leftrightarrow \overline{H} \lhd \overline{G}$, and

$$G/H \cong \overline{G}/\overline{H} = (G/K)/(H/K) \quad .$$

Proof. Let $H \in \Omega(G,K)$. From the definition of G/K it immediately follows that H/K is a subgroup of G/K. In order to see that the map $\pi^* : H \mapsto \overline{H}$ is injective, suppose that $H_1/K = H_2/K$, where $H_1, H_2 \in \Omega(G,K)$. Then if $h_1 \in H_1$, we have $h_1K = h_2K$ for some $h_2 \in H_2$, i.e., $h_1 = h_2k$ for some $k \in K$, and, since $K \subset H_2$, we have $h_1 \in H_2$. Thus, $H_1 \subset H_2$; we similarly show that $H_2 \subset H_1$, so that $H_1 = H_2$.

We now show that π^* is surjective. Suppose that $\overline{H} \in \Omega(\overline{G})$, and let H be the set of all elements of G which are contained in the cosets of K in $\overline{H}$ (recall that each element of $\overline{H}$ is a coset of K). Then $K \subset H$; also, $a, b \in H \Rightarrow aK$, $bK \in \overline{H} \Rightarrow abK = aKbK \in \overline{H} \Rightarrow ab \in H$; and finally, $a \in H \Rightarrow aK \in \overline{H} \Rightarrow a^{-1}K = (aK)^{-1} \in \overline{H} \Rightarrow a^{-1} \in H$. Thus, H is a subgroup of G, and $\overline{H} = H/K$ (H is usually called the preimage of $\overline{H}$ in G).

The fact that $H \in \Omega(G,K)$, $H \lhd G \Rightarrow \overline{H} \lhd \overline{G}$, is fairly obvious; namely, $gK \cdot hK \cdot (gK)^{-1} = ghg^{-1}K = h'K \in \overline{H}$ for all $g \in G$ and $h \in H$. By the same argument, $\overline{H} \lhd \overline{G} \Rightarrow ghg^{-1} = K = gK \cdot hK \cdot (gK)^{-1} = h'K \Rightarrow ghg^{-1} \in H \Rightarrow H \lhd G$.

Finally, if H is a normal subgroup of G containing K, then, by what we have proved, we have two natural epimorphisms

$$\pi : G \longrightarrow G/K; \qquad \overline{\pi} : \overline{G} \longrightarrow \overline{G}/\overline{H}$$

($\overline{\pi}(\overline{g}) = \overline{g}\overline{H}$, where $\overline{g} = gK \in \overline{G}$), and we can consider the composition, which is an epimorphism

$$\sigma = \overline{\pi} \circ \pi : G \longrightarrow \overline{G}/\overline{H} \ ,$$

defined by: $\sigma(g) = \overline{\pi}(\overline{g}) = \overline{g}\overline{H}$. We have: $\operatorname{Ker} \sigma = \{g \in G \,|\, \sigma(g) = \overline{H}\} = \{g \in G \,|\, \overline{g} \in \overline{H}\} = \{g \in G \,|\, gK = hK \text{ for some } h \in H\} = H$. Consequently, by the fundamental homomorphism theorem, the map $gH \mapsto \overline{g}\overline{H}$ is an isomorphism between G/H and $\overline{G}/\overline{H}$. □

Example 1. Let $n = dm$ be a natural number with divisor $d > 1$. Obviously, $n\mathbb{Z} \subset d\mathbb{Z}$, and the map $x \mapsto dx + n\mathbb{Z}$ is an epimorphism of additive groups

$$\mathbb{Z} \longrightarrow d\mathbb{Z}/n\mathbb{Z} = \{di + n\mathbb{Z} \,|\, i = 0, 1, \ldots, m-1\}$$

with kernel $m\mathbb{Z}$. By Theorem 1, we have an isomorphism

$$Z_m = \mathbb{Z}/m\mathbb{Z} \cong d\mathbb{Z}/n\mathbb{Z}$$

(this is also easy to see directly). Using Theorem 3, we find

$$\mathbb{Z}/d\mathbb{Z} \cong (\mathbb{Z}/n\mathbb{Z})/(d\mathbb{Z}/n\mathbb{Z}), \quad \text{i.e.} \quad Z_d \cong Z_n/Z_m \quad .$$

If we recall Theorem 5 of §3 Ch. 4, we can conclude that all of the subgroups and quotient groups of a cyclic group are themselves cyclic. Of course, this result could also be obtained without using the homomorphism theorems.

Example 2. Consider the following subgroups of the symmetic group S_4:

$V_4 = \{e, (12)(34), (13)(24), (14)(23)\} \lhd S_4$ (see Exercise 4 of §2),

$S_3 = \{e, (12), (13), (23), (123), (132)\}$

(here S_3 is the stationary subgroup of the point $i = 4$). Since obviously $S_3 \cap S_4 = e$, by Theorem 2 the subgroup $H = S_3 V_4$ has the property that

$$H/V_4 \cong S_3/S_3 \cap V_4 \cong S_3 \quad .$$

In particular, $|H| = |V_4|\,|S_3| = 24$, i.e., $H = S_4$. Thus, in addition to a subgroup isomorphic to S_3, S_4 has a quotient group isomorphic to S_3. Applying Theorem 3, we obtain a description of the set $\Omega(S_4, V_4)$ of subgroups of S_4 which contain V_4:

$$\Omega(S_4, V_4) = \{V_4, \langle(12)\rangle V_4, \langle(13)\rangle V_4, \langle(23)\rangle V_4, A_4 = \langle(123)\rangle V_4, S_4\} \quad .$$

Notice that for each divisor d of 24, S_4 has at least one subgroup of order d. For example, we have exactly four subgroups of order 3 -- $\langle(123)\rangle$, $\langle(124)\rangle$, $\langle(134)\rangle$, and $\langle(234)\rangle$ -- and three subgroups of order 8 -- $\langle(12)\rangle V_4$, $\langle(13)\rangle V_4$, and $\langle(23)\rangle V_4$. (These are the so-called 3-Sylow and 2-Sylow subgroups.) There are in all two proper normal subgroups (i.e., other than e and S_4): V_4 and A_4. To see this, first suppose that $K \lhd S_4$ and $K \cap V_4 \neq e$. Then $K \supset V_4$, since the elements $\neq e$

in V_4 are all conjugate. By considering the set $\Omega(S_4, V_4)$, we see that either $K = V_4$ or $K = A_4$. Now suppose that $K \cap V_4 = e$, $K \neq e$. Then

$$K \lhd S_4, \quad V \lhd S_4 \Longrightarrow KV_4 \lhd S_4 ,$$

and so $KV_4 = A_4$ or S_4. In either case, it is not hard to show that K could not be a normal subgroup if its intersection with V_4 is trivial. We leave the details to the reader.

2. Solvable groups. The expression

$$[x,y] = xyx^{-1}y^{-1}$$

is called the commutator of the elements x and y in the group G. It is the "correction term" needed in order to reverse the order of multiplication of x and y:

$$xy = [x,y]yx \quad .$$

If x and y commute, then $[x,y] = e$. Roughly speaking, one can say that, the more commutators in G are distinct from e, the farther G is from being an abelian group. Let M be the set of all commutators in G. The commutant (or derived subgroup) of G is defined to be the subgroup $G' = G^{(1)} = [G,G]$ generated by the set M (see Subsection 2 of §2, Ch. 4):

$$G' = \langle [x,y] | x, y \in G \rangle \quad .$$

Although $[x,y]^{-1} = yxy^{-1}x^{-1} = [y,x]$ is a commutator, it is not true that the product of two commutators is always a commutator. So we define G' to consist of all products of the form

$$[x_1,y_1][x_2,y_2] \ldots [x_k,y_k] \quad \text{with} \quad x_i, y_i \in G \quad .$$

Of course, when dealing with a specific example it is useful to have a more down-to-earth description of the commutant G'.

Example. Let $G = S_n$. The commutator $[\alpha, \beta] = \alpha\beta\alpha^{-1}\beta^{-1}$ of any two permutations $\alpha, \beta \in S_n$ is obviously an even permutation. Hence, $S_n' \subset A_n$. Furthermore,

$$(ij)(ik)(ij)^{-1}(ik)^{-1} = (ij)(ik)(ij)(ik) = (ijk) \quad ,$$

and, since the 3-cycles (ijk) generate all of the alternating group A_n (see Exercise 8 of §2 Ch. 4), we conclude that $S_n' = A_n$. Note that $S_n' \lhd S_n$, and that the quotient group S_n/S_n' is abelian.

Returning to the general situation, let us consider an arbitrary group homomorphism $\varphi : G \to \overline{G}$. Since

$$\varphi([x,y]) = \varphi(xyx^{-1}y^{-1}) = \varphi(x)\varphi(y)\varphi(x)^{-1}\varphi(y)^{-1} = [\varphi(x), \varphi(y)] \quad ,$$

we have $\varphi(G') \subset (\overline{G})'$, and $\varphi(G') = (\overline{G})'$ if φ is an epimorphism. Now let K be a normal subgroup of G, and let $\varphi = I_a : x \mapsto axa^{-1}$ be an inner automorphism of G, which then induces an automorphism of K. By what we have just shown, $I_a(K') \subset K'$ for any $a \in G$, and this means that

$$K \lhd G \Longrightarrow K' \lhd G \quad . \tag{1}$$

In particular, $G' \lhd G$. We now prove a more general fact, which gives the intrinsic meaning of the concept of the commutant.

THEOREM 4. Any subgroup $K \subset G$ which contains the commutant G' is normal in G. The quotient group G/G' is abelian, and G' is contained in any normal subgroup K for which G/K is abelian. In particular, the maximal order of an abelian quotient group G/K is equal to the index $(G:G')$.

Proof. If $x \in K$, $g \in G$, and $G' \subset K$, then $gxg^{-1} = (gxg^{-1}x^{-1})x = [g,x]x \in G'K = K$, and hence K is normal in G. Next, whenever $G' \subset K$ and $K \lhd G$ (in particular, for $K = G'$), we have

$$[aK, bK] = aK \cdot bK \cdot a^{-1}K \cdot b^{-1}K = aba^{-1}b^{-1}K = [a,b]K = K \quad ,$$

i.e., the commutator of any two elements of the quotient group G/K is the identity element $(= K)$. Hence, G/K is an abelian group. Conversely, if $K \triangleleft G$ and G/K is abelian, then

$$[a,b]K = [aK, bK] = K$$

for all $a, b \in G$. Hence, $[a,b] \in K$, and so $G' \subset K$, since G' is generated by the commutators. □

Remark. We now know two important normal subgroups of any group G: the center $Z(G)$ and the commutant G'. In general, there is only a weak connection between them, but, roughly speaking, the following principle applies: the "nearer" G is to being abelian, the larger $Z(G)$ is and the smaller G' is. Here is an interesting fact:

The quotient group $G/Z(G)$ of a non-abelian group G by its center cannot be cyclic.

Proof. If $G/Z(G)$ is a cyclic group, then $G = \bigcup_i a^i Z(G)$, and any element of G has the form $g = a^i z$, $z \in Z$. In that case, $[g,h] = [a^i z, a^j z'] = a^{i+j-i-j}[z,z'] = e$ for any two elements $g, h \in G$. Thus, $G' = e$, and G is an abelian group, contradicting our assumptions. □

In G' we can consider the commutant $(G')' = G''$, which is called the second derived group (or second commutant) of the group G. Continuing in this manner, we define the k-th derived group $G^{(k)} = (G^{(k-1)})'$. By (1), we have $G^{(k)} \triangleleft G$, and of course $G^{(k)} \triangleleft G^{(k-1)}$. We thus have a sequence of normal subgroups

$$G \triangleright G^{(1)} \triangleright G^{(2)} \triangleright \cdots \triangleright G^{(k)} \triangleright G^{(k+1)} \triangleright \cdots \tag{2}$$

with abelian quotient groups $G^{(k)}/G^{(k+1)}$.

A group G is called solvable if the sequence (2) terminates with the trivial subgroup e, i.e., if $G^{(m)} = e$ for some m. The least such m in that case is called the level of solvability of G. It is obvious that any abelian group (in particular, any cyclic group) is solvable of level 1. Notice that any solvable group G has a normal abelian subgroup $\neq e$, namely $G^{(m-1)}$, if m is the level of solvability. An example of a solvable group is S_4: $S_4' = A_4$, $A_4' = V_4$, $V_4' = e$. Thus, the alternating group A_4 is solvable of level 2, and the symmetric group S_4 is solvable of level 3.

The term "solvable groups" comes from Galois theory, which we alluded to in Subsection 1 of §2 Ch. 1. The solvability of S_4 turns out to imply that algebraic equations of degree $n \leq 4$ can always be solved by radicals. The reader can study these questions in several of the texts listed in the suggestions for further reading at the beginning of Part II.

3. Simple groups. There exist groups other than e which are equal to their commutant, and, in particular, are not solvable. We shall now show, in fact, that there exist non-abelian groups having no non-trivial ($\neq e$ or G) normal subgroups. Such a group is called simple.

LEMMA. *Any normal subgroup K of a group G is a union of some set of conjugacy classes in G.*

Proof. If $x \in K$, then also $gxg^{-1} \in K$ for all $g \in G$, since K is normal. Hence, if x is contained in K, then so is the entire conjugacy class x^G. So we can write $K = \bigcup_{i \in I} x_i^G$. □

THEOREM 5. *The alternating group A_5 is simple.*

Proof. Not counting the identity permutation e, the group A_5 has 15 elements whose square is e, namely the permutations $(ij)(k\ell)$ (there are three such elements in the stationary subgroups of each of the points $1,2,3,4,5$); there are $20 = 2\binom{5}{3}$ elements (ijk) of order 3; and there are $24 = 4!$ elements $(1i_1i_2i_3i_4)$ of order 5. The elements of order 2 are all conjugate in A_5 : they are clearly conjugate in S_5, and, since conjugation by the transposition (ij) leaves $(ij)(k\ell)$ fixed, we can always make two elements of order 2 conjugate using a product of an even number of transpositions, i.e., any two such elements are conjugate in A_5. The same holds for the elements of order 3. But the elements of order 5, though they are conjugate in S_5, split into two conjugacy classes in A_5 with representatives (12345) and (12354). To see this, note that $(45)(12345)(45)^{-1} = (12354)$, and the centralizer (= the stationary subgroup under the conjugation action) of (12345) in A_5 is the cyclic group of order 5 generated by (12345). Thus, we have the following table of the number of elements in each conjugacy class in A_5:

1	15	20	12	12
e	(12)(34)	(123)	(12345)	(12354)

The bottom row gives representatives of the conjugacy classes, and the top row gives the number of elements in each conjugacy class.

Now let K be a normal subgroup of A_5. According to the lemma,

$$|K| = \delta_1 \cdot 1 + \delta_2 \cdot 15 + \delta_3 \cdot 20 + \delta_4 \cdot 12 + \delta_5 \cdot 12 \quad ,$$

where $\delta_1 = 1$ (since $e \in K$) and $\delta_i = 0$ or 1 for $i = 2,3,4,5$. It is not hard to see that, since $|K|$ must be a divisor of $|A_5| = 60$ (by Lagrange's Theorem), we are left with only two possibilities:

a) $\delta_2 = \delta_3 = \delta_4 = \delta_5 = 0$, and $K = e$;

b) $\delta_2 = \delta_3 = \delta_4 = \delta_5 = 1$, and $K = A_5$. □

Using induction on n, it is now possible to establish the following fact, discovered by Galois: <u>all of the groups A_n for $n \geq 5$ are simple.</u> Since any subgroup of a solvable group is solvable ($H \subset G \Rightarrow H^{(k)} \subset G^{(k)}$, $k = 1, 2, \ldots$), Theorem 5 implies, in particular, that the symmetric group S_n is not solvable if $n \geq 5$.

THEOREM 6. <u>The rotation group</u> SO(3) <u>is simple.</u>

<u>Proof.</u> By Theorem 3, it is sufficient to show that any normal subgroup K of $SU(2)$ which contains the kernel $\{\pm E\}$ of the epimorphism $\Phi: SU(2) \to SO(3)$ (see Subsection 3 of §1) and is not $\{\pm E\}$, must be all of $SU(2)$. We can interpret the relation (5) in §1 as saying that every conjugacy class in $SU(2)$ contains a diagonal matrix $d_\varphi = b_{2\varphi} = \operatorname{diag}\{e^{i\varphi}, e^{-i\varphi}\}$. Since, by the lemma, K is the union of some set of conjugacy classes in $SU(2)$, without loss of generality we may assume that $d_\varphi \in K$ for some $\varphi > 0$ for which $\sin\varphi \neq 0$. Then K must also contain any commutator

$$[d_\varphi, g] = d_\varphi(g d_\varphi^{-1} g^{-1}) = \begin{Vmatrix} e^{i\varphi} & 0 \\ 0 & e^{-i\varphi} \end{Vmatrix} \begin{Vmatrix} \alpha & \beta \\ -\bar\beta & \bar\alpha \end{Vmatrix} \begin{Vmatrix} e^{-i\varphi} & 0 \\ 0 & e^{i\varphi} \end{Vmatrix} \begin{Vmatrix} \bar\alpha & -\beta \\ \bar\beta & \alpha \end{Vmatrix} =$$

$$= \begin{Vmatrix} |\alpha|^2 + |\beta|^2 e^{i2\varphi} & * \\ * & |\alpha|^2 + |\beta|^2 e^{-i2\varphi} \end{Vmatrix},$$

where $|\alpha|^2 + |\beta|^2 = 1$ (see (3) in §1). We obtain the following expression for the trace of the matrix $[d_\varphi, g]$:

$$\operatorname{tr}[d_\varphi, g] = 2|\alpha|^2 + |\beta|^2(e^{i2\varphi} + e^{-i2\varphi}) = 2(1 - 2|\beta|^2 \sin^2\varphi) \quad .$$

Here $|\beta|$ can take any value in $[0,1]$, and $\sin\varphi \neq 0$. Again using (5) of §1, we can find a unitary matrix $h \in SU(2)$ such that $h[d_\varphi, g]h^{-1} = d_\psi = \operatorname{diag}\{e^{i\psi}, e^{-i\psi}\}$, where $d_\psi \in K$. Since $e^{i\psi}$ and $e^{-i\psi}$ are the roots of the characteristic equation

$$\lambda^2 + (4|\beta|^2 \sin^2\varphi - 2)\lambda + 1 = 0$$

of the matrix $[d_\varphi, g]$, it follows that, if we let $|\beta|$ run through the values from 0 to 1, then we obtain for ψ any point on the interval $[0, 2\varphi]$. Thus, K contains any such d_ψ, and also the corresponding conjugacy class, as the parameter ψ varies in $0 \leq \psi \leq 2\varphi$. But for every $\sigma > 0$ there is a natural number n such that

$0 < \psi = \frac{\sigma}{n} \leq 2\varphi$. Hence, we may conclude that K contains any given element $d_\sigma = d_\psi^n$. □

It is already clear from Theorems 5 and 6 that the class of simple groups contains many groups, both finite and infinite, which are important in applications. But, surprising as it may seem, no one has been able to find a reasonable description of all finite simple groups, and it is unclear whether such a description will ever be found. (Note: This was true until 1980, but the situation is changing now.)

4. Products of groups. We now consider a construction which allows us to make new groups out of the ones we have. We have already encountered special cases of this construction.

The direct product of two groups A and B is the set $A \times B$ of all ordered pairs (a,b), where $a \in A$ and $b \in B$, with the binary operation

$$(a_1, b_1)(a_2, b_2) = (a_1a_2, b_1b_2) \quad .$$

Strictly speaking, we should write $(a_1, b_1) * (a_2, b_2) = (a_1 \circ a_2, b_1 \Box b_2)$, where $\circ$, $\Box$ and $*$ are the binary operations in A, B and $A \times B$, respectively; but, for simplicity, we shall use a dot (or nothing) to denote all of the operations. If we are using additive notation for abelian groups, then we write the direct sum $A \oplus B$.

The group $A \times B$ contains the subgroups $A \times e$ and $e \times B$, which are isomorphic to A and B, respectively (here we are using the same letter e to denote the identity in both A and B). The map $\varphi : A \times B \to B \times A$ given by $\varphi((a,b)) = (b,a)$ is obviously an isomorphism between $A \times B$ and $B \times A$. If we have three groups A, B, C, then we can speak of the direct products $(A \times B) \times C$ and $A \times (B \times C)$. If we set $\psi(((a,b), c)) = (a, (b,c))$, we easily see that

$$(A \times B) \times C \cong A \times (B \times C) \quad .$$

Because of these properties of "commutativity" and "associativity" of the direct product, we can speak of the direct product of any finite number of groups $G_1, G_2, \ldots, G_n$, and write

$$G_1 \times G_2 \times \ldots \times G_n = \prod_{i=1}^{n} G_i ,$$

without needing to indicate by parentheses in what order the direct product is taken. (Note that this makes the set of all groups into a commutative semigroup, whose elements are groups, and whose binary operation is the direct product.)

THEOREM 7. _Let_ G _be a group with normal subgroups_ A _and_ B. _If_ $A \cap B = e$ _and_ $AB = G$, _then_ $G \cong A \times B$.

Proof. Since $AB = G$, any element $g \in G$ can be written in the form $g = ab$, where $a \in A$ and $b \in B$. If we also had $g = a_1 b_1$, $a_1 \in A$, $b_1 \in B$, then $ab = a_1 b_1$, and so $a_1^{-1} a = b_1 b^{-1} \in A \cap B = e$. Thus, $a_1 = a$, $b_1 = b$, and we conclude that g can be written in the form ab in only one way. Furthermore, $A \lhd G \Rightarrow k = a(ba^{-1}b^{-1}) = aa' \in A$; $B \lhd G \Rightarrow k = (aba^{-1})b^{-1} = b'b^{-1} \in B$, i.e., the commutator k is in $A \cap B = e$, and so $ab = ba$.

We now define a map $\varphi : G \to A \times B$ by setting $\varphi(g) = (a,b)$ for any $g = ab$. We then have $\varphi(gg') = \varphi(aba'b') = \varphi(aa'bb') = (aa', bb') = (a,b)(a',b') =$ $= \varphi(ab)\varphi(a'b') = \varphi(g)\varphi(g')$. Furthermore, $\varphi(ab) = (e,e) \Leftrightarrow a = e$, $b = e$, i.e., $\operatorname{Ker}\varphi = e$. φ is obviously surjective. Thus, φ satisfies all of the requirements for an isomorphism between G and $A \times B$. $\square$

A group G which satisfies the conditions in Theorem 7 is called the direct product of the subgroups A and B. There is a (somewhat pedantic) distinction between this meaning of "direct product" and the earlier meaning, in that now G contains the two groups of which it is the product; strictly speaking, the group $A \times B$, which is the direct product of A and B is the earlier sense, is the direct product of $A \times e$ and $e \times B$

in the new sense. But usually one neglects this distinction, and identifies A with $A \times e$ and B with $e \times B$.

Our next result concerns homomorphisms of direct products.

THEOREM 8. Let $G = A \times B$, and let $A_1 \lhd A$ and $B_1 \lhd B$. Then $A_1 \times B_1 \lhd G$, and $G/(A_1 \times B_1) \cong (A/A_1) \times (B/B_1)$. In particular, $G/A \cong B$.

Proof. Let $\pi : A \to A/A_1$ and $\rho : B \to B/B_1$ be the natural homomorphisms. We define the map $\varphi : G \to (A/A_1) \times (B/B_1)$ by setting $\varphi(ab) = (\pi(a), \rho(b))$. We immediately verify that φ is a homomorphism with kernel $\operatorname{Ker} \varphi = A_1 \times B_1$ and image $(A/A_1) \times (B/B_1)$. □

Just as in the theory of vector spaces, it is easy to prove that, if G is a group with normal subgroups $G_1, \ldots, G_n$, then $G \cong \prod G_i$ if and only if $G = \langle G_1, \ldots, G_n \rangle$ and $G_j \cap \langle G_1, \ldots, G_{j-1}, G_{j+1}, \ldots, G_n \rangle = e$ for all j. This same fact can also be expressed as follows: G is the direct product of the normal subgroups $G_1, \ldots, G_n$ if every element $g \in G$ can be uniquely expressed in the form $g = g_1 \ldots g_n$, $g_i \in G_i$.

The direct product of n copies of a group H is called the n-th cartesian power and is denoted $H^n = H \times \ldots \times H$. One subgroup of H^n which is of special interest is the diagonal $\Delta = \{(h, h, \ldots, h) \mid h \in H\}$, which is a group isomorphic to H.

If we omit the condition $B \lhd G$ in Theorem 7, then we arrive at the notion of a semidirect product: $G = AB$, $A \cap B = e$, $A \lhd G$ (sometimes one write $G = A \rtimes B$). If we define G to be a semidirect product of two groups A and B, that does not uniquely determine G until we describe the action of B on the normal subgroup A, i.e., in any concrete case we must describe how to compute bab^{-1} for $a \in A$, $b \in B$.

Many of the groups we have encountered are direct or semidirect products. For example, S_n is the semidirect product of the normal subgroup A_n and the cyclic subgroup $\langle (12) \rangle$ of order 2. Using the notation in Example 2 of Subsection 1, we can write: $A_4 = V_4 \times \langle(123)\rangle \cong (Z_2 \times Z_2) \rtimes Z_3$; $S_4 = V_4 \rtimes S_3 \cong (Z_2 \times Z_2) \rtimes (Z_3 \rtimes Z_2)$.

One more example: The group $A(1, \mathbb{R})$ of affine transformations $\mathbb{R} \to \mathbb{R}$ (see Exercise 3 of §2 Ch. 4) is the semidirect product of the normal subgroup of translations and the subgroup $GL(1, \mathbb{R})$ of transformations which leave the point $x = 0$ fixed.

5. Generators and defining relations. The subject of systems of generators for a group G was already discussed in §2 Ch. 4. We return to this question in order to look at some of the groups we now know from this point of view. It follows from the results of Chapter 4 that in the case of cyclic groups it is not necessary to make a cumbersome Cayley table. Writing

$$C_n = \langle c \mid c^n = e \rangle \tag{3}$$

gives all possible information about the abstract cyclic group C_n of order n; we have: $C_n = \{e, c, c^2, \dots, c^{n-1}\}$, with $c^s c^t = c^{s+t}$ if $s + t < n$ and $c^s c^t = c^{s+t-n}$ if $s + t \geq n$. We can also say that any cyclic group is a homomorphic image of the single group $(\mathbb{Z}, +)$.

In the same way, the "universal" group for all possible direct products $A = \langle a_1 \rangle \times \dots \times \langle a_r \rangle$ of r cyclic groups is the r-th cartesian power $\mathbb{Z}^r = \mathbb{Z} \oplus \dots \oplus \mathbb{Z}$ (see Subsection 4), which has generators

$$z_i = (0, \dots, 1, \dots, 0), \qquad i = 1, 2, \dots, r,$$

and addition law

$$\sum s_i z_i + \sum t_i z_i = \sum (s_i + t_i) z_i = (s_1 + t_1, \dots, s_r + t_r) .$$

The map $z_i \mapsto a_i$, $1 \leq i \leq r$, extends uniquely to a group homomorphism $\varphi : (s_1, s_2, \dots, s_r) \mapsto a_1^{s_1} a_2^{s_2} \dots a_r^{s_r}$ with kernel $\operatorname{Ker} \varphi = m_1 \mathbb{Z} \oplus \dots \oplus m_r \mathbb{Z}$ (see Theorem 8), where m_i is the order of $\langle a_i \rangle$ if this is finite, and $m_i = 0$ otherwise.

In analogy with (3), we can write

$$A = \langle a_1, \dots, a_r \mid a_1^{m_1} = e, \dots, a_r^{m_r} = e \rangle ,$$

where we always implicitly assume when we write this that the generators $a_1, \dots, a_r$ commute with one another. It is customary to call $a_1^{m_1} = e, \dots, a_r^{m_r} = e$ the defining relations for the abelian group A, and to call $\mathbb{Z}^r$ the free abelian group of rank r (or with r free generators $z_1, \dots, z_r$). It is obvious that

$$A \cong \mathbb{Z}^r \iff (a_1^{s_1} \dots a_r^{s_r} = e \iff s_1 = \dots = s_r = 0) .$$

Now if F_d is any group which is generated by d elements $f_1, \dots, f_d$, then any element $f \in F_d$ can be written (perhaps in many ways) in the form

$$f = f_{i_1}^{s_1} f_{i_2}^{s_2} \dots f_{i_k}^{s_k}; \quad i_j \in \{1, 2, \dots, d\}, \quad s_j \in \mathbb{Z} , \tag{4}$$

where $i_j \neq i_{j+1}$, $j = 1, 2, \dots, k-1$. We can always obtain this latter condition using: $f_i^s f_i^t = f_i^{s+t}$, $f_i^0 = e$, and $f_j e = e f_j = f_j$.

If the condition $f = e \Leftrightarrow s_1 = \dots = s_k = 0$ holds for every f written in the form (4), we say that F_d is a free group, generated by d free generators. The elements of F_d are often called words in the alphabet $\{f_1, f_1^{-1}, \dots, f_d, f_d^{-1}\}$. The irreducible form (4) of the word f and the length $\ell(f) = |s_1| + |s_2| + \dots + |s_k|$ are uniquely determined, since otherwise we would have an empty word $e = ff^{-1}$ (the identity element in F_d) with length > 0. If d is fixed, then two free groups F_d and G_d with d free generators $f_1, \dots, f_d$ and $g_1, \dots, g_d$, respectively, are isomorphic: we need only set $\Phi(f_i) = g_i$, $1 \leq i \leq d$, and, given any word f in the form (4), set

$$\Phi(f) = g_{i_1}^{s_1} g_{i_2}^{s_2} \dots g_{i_k}^{s_k}$$

(the identities in F_d and G_d are denoted by the same symbol). However, if G_d is not a free group, then Φ will only be an epimorphism, whose kernel $\operatorname{Ker}\Phi$ consists of those words f which become the identity element of G_d after the substitution $f_i \mapsto g_i$. This universal property of F_d (the fact that the substitution $f_i \mapsto g_i$ always extends to an epimorphism $\Phi : F_d \to G_d$ whenever G_d is a group with d generators g_i) can actually be taken as the definition of a free group with d generators, but we shall not dwell on this point of view.

So as not to give the impression of free groups as something mystical, we give some concrete realizations of such groups.

$d = 1$. $F_1 \cong (\mathbb{Z}, +)$ is the free abelian group of rank 1, or, equivalently, the infinite cyclic group.

$d = 2$. Let $\mathbb{Z}[t]$ be the ring of polynomials in t with integer coefficients. In the special linear group $SL(2, \mathbb{Z}[t])$ we consider the subgroup F generated by the matrices

$$A = \begin{Vmatrix} 1 & t \\ 0 & 1 \end{Vmatrix} \quad \text{and} \quad B = \begin{Vmatrix} 1 & 0 \\ t & 1 \end{Vmatrix} .$$

We prove that F is a free group. A simple induction on k shows that the element

$$W_k = A^{\alpha_1} B^{\beta_1} \dots A^{\alpha_k} B^{\beta_k}, \qquad \alpha_i, \beta_i \neq 0, \qquad 1 \leq i \leq k ,$$

has the form

$$W_k = \begin{Vmatrix} 1 + \cdots + \sigma_k t^{2k} & t(\dots + \sigma_{k-1} \alpha_k t^{2(k-1)}) \\ t(\dots + \alpha_1^{-1} \sigma_k t^{2(k-1)}) & 1 + \cdots + \alpha_1^{-1} \sigma_{k-1} \alpha_k t^{2(k-1)} \end{Vmatrix} ,$$

where $\sigma_k = \alpha_1 \beta_1 \dots \alpha_k \beta_k$, and the dots denote terms of lower degree in t. It is clear that $W_k \neq E$. Any element of F can be written either in the form $B^{\beta} A^{\alpha}$, which is $\neq E$, or in the form $W = B^{\beta} W_k A^{\alpha}$. If $W = E$, then $W_k = B^{-\beta} A^{-\alpha}$, which, however, is impossible (compare the degrees if $k > 1$, and do a direct verification if $k = 1$). Thus, F is free.

It is also not hard to show by an additional argument that, if we replace t by any integer $m \geq 2$ in the above example (i.e., we take the subgroup of $SL(2, \mathbb{Z})$ generated by $\begin{Vmatrix} 1 & m \\ 0 & 1 \end{Vmatrix}$ and $\begin{Vmatrix} 0 & 0 \\ m & 1 \end{Vmatrix}$), then the resulting group F is still a free group with two generators.

<u>Definition.</u> Let F_d be a free group with d generators $f_1, \ldots, f_d$, let $S = \{w_i, i \in I\}$ be some subset of elements $w_i(f_1, \ldots, f_d) \in F_d$, and let $K = \langle S^{F_d} \rangle$ be the smallest normal subgroup of F_d containing S (i.e., the intersection of all normal subgroups containing S). We say that a group G <u>is given by</u> d <u>generators</u> $a_1, \ldots, a_d \in G$ <u>and the relations</u> $w_i(a_1, \ldots, a_d) = e$, $i \in I$, if there exists an epimorphism $\pi : F_d \to G$ with kernel K such that $\pi(f_k) = a_k$, $1 \leq k \leq d$. In this case we write

$$G = \langle a_1, \ldots, a_d \mid w_i(a_1, \ldots, a_d) = e, \quad i \in I \rangle$$

and we call G a <u>finitely presented</u> group if $\operatorname{Card} I < \infty$. The group F_d itself is "free of any relations" (whence its name). If a group H with d generators $b_1, \ldots, b_d$ has the same relations $w_i(b_1, \ldots, b_d) = e$, $i \in I$, and possibly some other relations as well, then H is a homomorphic image of G. In particular, $|H| \leq |G|$.

<u>Example 1</u> (the dihedral group). The group $G = \langle a, b \mid a^3 = b^2 = abab = e \rangle$ with two generators and three relations has order $|G| \leq 6$, since $ba = a^{-1}b^{-1} = (a^3)^{-1}a^2b(b^2)^{-1} = a^2b$, and it is not hard to see that the elements e, a, a^2, b, ab, a^2b exhaust G. Since the permutations (123) and (12), which generate S_3, satisfy the relation $(123)^3 = (12)^2 = (123)(12)(123)(12) = e$, it follows that the map $\varphi : G \to S_3$ defined by taking $a \mapsto (123)$ and $b \mapsto (12)$ gives an isomorphism $G \cong S_3$. Thus, the symmetric group S_3 is given by two generators and three relations. Recall that S_3 can also be identified with the group of all symmetry transformations of an equilateral triangle.

The full group of symmetry transformations of a regular n-gon P_n is called a <u>dihedral group</u>, and is denoted D_n. The rotation

$$\mathfrak{A} = \begin{Vmatrix} \cos\theta & -\sin\theta \\ \sin\theta & \cos\theta \end{Vmatrix}$$

of the polygon about its center through the angle $\theta = 2\pi/n$ generates a cyclic subgroup $\langle\mathfrak{A}\rangle$ of order n. D_n also contains the reflection $\mathfrak{B} = \begin{Vmatrix} 1 & 0 \\ 0 & -1 \end{Vmatrix}$ of P_n relative to an axis passing through the center and one of the vertices. By definition, we have $\mathfrak{B}^2 = 1$. The $2n$ distinct symmetry transformations

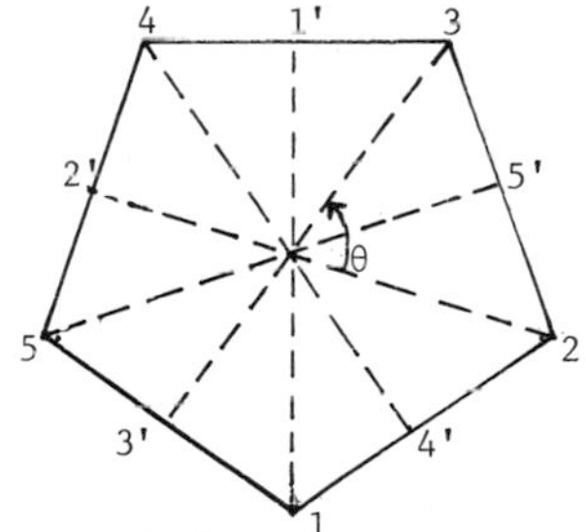

$$e, \mathfrak{A}, \mathfrak{A}^2, \ldots, \mathfrak{A}^{n-1} ; \mathfrak{B}, \mathfrak{A}\mathfrak{B}, \ldots, \mathfrak{A}^{n-1}\mathfrak{B} \qquad (5)$$

exhaust all of D_n. To see this, note that any symmetry transformation is determined by its action on the vertices. If a transformation takes 1 to k, then it must either preserve the same cyclic order of the vertices, in which case it is $\mathfrak{A}^{k-1}$, or else reverse this order, in which case it is $\mathfrak{A}^{k-1}\mathfrak{B}$. Hence, D_n does not have any elements other than those in (5). Note that $\mathfrak{B}\mathfrak{A}$ coincides with $\mathfrak{A}^{n-1}\mathfrak{B}$, since both transformations interchange 1 and n. Thus, we have the relations

$$\mathfrak{A}^n = e, \quad \mathfrak{B}^2 = e, \quad \mathfrak{A}\mathfrak{B}\mathfrak{A}\mathfrak{B} = e \quad .$$

This means that D_n is a homomorphic image of the group

$$G = \langle a, b \mid a^n = e, b^2 = e, (ab)^2 = e \rangle \quad .$$

As in the case $n = 3$, we find that $ba = a^{-1}b^{-1} = a^{n-1}b$, so that any word in the alphabet $\{a, a^{-1}, b, b^{-1}\}$ reduces either to a^i or to $a^i b$, $0 \leq i \leq n-1$. Hence, $|G| \leq 2n$, and, by what we determined before, we must have an isomorphism $G \cong D_n$. We have thereby obtained the dihedral group in terms of generators and relations. We identify G with D_n:

$$D_n = \langle a, b \mid a^n = e, b^2 = e, (ab)^2 = e \} \quad .$$

Since $\langle a \rangle \triangleleft D_n$, and $D/\langle a \rangle$ is a cyclic group (of order 2), it follows by Theorem 4 that the commutant D'_n of D_n is contained in $\langle a \rangle$. But $a^2 = aba^{-1}b^{-1} = [a,b] \in D'_n$, so for odd n we have $D'_n = \langle a \rangle$. For even n we have $D_n/\langle a^2 \rangle = \langle \bar{a}, \bar{b} \rangle \cong V_4$, the direct product of two cyclic groups of order 2, and hence $D'_n = \langle a^2 \rangle$. The center $Z(D_n)$ and the number r of conjugacy classes in D_n also depend on whether n is odd or even. We give the information in the following tables (which are easy to verify), in which the representatives of conjugacy classes are in the top row and the number of elements in a class is given in the bottom row:

$n = 2m$. $D'_n = \langle a^2 \rangle$, $(D_n : D'_n) = 4$, $Z(D_n) = \langle a^m \rangle$, $r = m + 3$

1	1	2	...	2	m	m
e	a^m	a	...	a^{m-1}	b	ab

$n = 2m + 1$. $D'_n = \langle a \rangle$, $(D_n : D'_n) = 2$, $Z(D_n) = e$, $r = m + 2$

1	2	...	2	2	n
e	a	...	a^{m-1}	a^m	b

It should be emphasized that the form of the relations (the w_i in the above notation) very much depends on our choice of generators for the group. For example, the dihedral group D_n for $n = 2m$ is generated by any two reflections about axes which intersect at an angle of π/m. Hence,

$$D_n = \langle g_1, g_2 \mid g_1^2 = g_2^2 = (g_1 g_2)^n = e \rangle \quad .$$

If we use the previous definition, we can set $g_1 = ab$, $g_2 = b$.

Example 2 (the group of quaternions). Unlike in the previous case, we shall define the group of quaternions Q_8 (whose name will be explained in Chapter 9) from the very beginning in terms of generators and relations:

$$Q_8 = \langle a, b \mid a^4 = e, \ b^2 = a^2, \ bab^{-1} = a^{-1} \rangle \quad .$$

Again we have $ba = a^{-1}b = a^3 b$, and, since $b^2 = a^2$, any word in the alphabet $\{a, a^{-1}, b, b^{-1}\}$ can be reduced to the form $a^s b^t$, $0 \leq s \leq 3$, $0 \leq t \leq 1$; thus, $|Q_8| \leq 8$.

Can we assert that $|Q_8| = 8$? Yes we can, but only after exhibiting a group of 8 elements having two generators which are subject to the same relations as a and b in Q_8. We claim that the group generated by the following matrices is such a group:

$$A = \begin{Vmatrix} i & 0 \\ 0 & -i \end{Vmatrix}, \quad B = \begin{Vmatrix} 0 & 1 \\ -1 & 0 \end{Vmatrix} \quad (i = \sqrt{-1}) \quad .$$

In fact,

$$A^4 = E, \quad B^2 = A^2, \quad BAB^{-1} = A^{-1}$$

and

$$\langle A, B \rangle = \left\{ \pm \begin{Vmatrix} 1 & 0 \\ 0 & 1 \end{Vmatrix}, \ \pm \begin{Vmatrix} i & 0 \\ 0 & -i \end{Vmatrix}, \ \pm \begin{Vmatrix} 0 & 1 \\ -1 & 0 \end{Vmatrix}, \ \pm \begin{Vmatrix} 0 & i \\ i & 0 \end{Vmatrix} \right\} \quad .$$

The map $a \mapsto A$, $b \mapsto B$ gives an isomorphism $Q_8 \cong \langle A, B \rangle$. Note that $a^2 \in Z(Q_8)$, and since the quotient group of a non-abelian group by its center cannot be cyclic (see the remark in Subsection 2), it follows that $\langle a^2 \rangle = Z(Q_8)$. All groups of order 4 are abelian, and we have $Q_8 / Z(Q_8) \cong V_4$, the direct product of two cyclic groups of order 2. Thus, the commutant Q_8' coincides with $Z(Q_8)$, and $(Q_8 : Q_8') = 4$. The information concerning conjugacy classes is contained in the following table:

1	1	2	2	2
e	a^2	a	b	ab

Finitely presented groups, some of the simplest of which we have just considered, occur in many areas of mathematics, for example, as the so-called fundamental groups of manifolds. It is not surprising that there are still many questions concerning such groups which remain open.

EXERCISES

1. Recall the definition in Subsection 2 of §3 Ch. 4 of the inner automorphism $I_a : g \mapsto aga^{-1}$ and the group $\mathrm{Inn}(G) \subset \mathrm{Aut}(G)$. Show that $\mathrm{Inn}(G) \lhd \mathrm{Aut}(G)$, and that $\mathrm{Inn}(G) \cong G/Z(G)$, where $Z(G)$ is the center of G. The quotient group $\mathrm{Out}(G) = \mathrm{Aut}(G)/\mathrm{Inn}(G)$ is sometimes called the <u>group of outer automorphisms.</u>

2. Let H and K be subgroups of a group G. Show that $|HK| \cdot |H \cap K| = |H| \cdot |K|$ (this is analogous to a well-known formula in linear algebra). Next show that the set HK is a subgroup if and only if $HK = KH$; if $K \lhd G$, then this condition automatically holds.

3. Show that, if G is a finite solvable group, then there exists a chain of subgroups $E = G_0 \subset G_1 \subset \cdots \subset G_n = G$, where $G_{i-1} \lhd G_i$, $1 \leq i \leq n$, and each index $(G_i : G_{i-1}) = p_i$ is a prime number.

4. Compose for the symmetric group S_4 the table

1	3	6	8	6
e	(12) (34)	(12)	(123)	(1234)

which is analogous to the one used to prove Theorem 5. Using the same reasoning as in the proof of Theorem 5, repeat the description given in Example 2 of the normal subgroups of S_4.

5. Prove that the alternating group A_n, where $n \geq 5$, is simple by filling in

the details in the following outline.

a) If $K \neq e$ is a normal subgroup of A_n, take a permutation $\pi \neq e$ which leaves fixed the maximum possible number k of elements of $\Omega = \{1, 2, \ldots, n\}$. If $k = n - 3$, then $\pi = (ijk)$, and $K = A_n$ (see Exercise 8 of §2 Ch. 4): so suppose $k < n - 3$.

b) If $\pi = (123 \ldots)\ldots$ is the decomposition of π into disjoint cycles, then the fact that π is even and $k < n - 3$ implies that $k \leq n - 5$. It is also possible for $\pi = (12)(34)\ldots$ to be a product of disjoint cycles of length 2.

c) In either case consider the commutator $[\pi, \sigma] = \pi\sigma\pi^{-1}\sigma^{-1} \neq e$ with $\sigma = (123)$, and verify that it leaves more than k points fixed. This contradicts the choice of k, and proves our assertion.

6. Show that $Z(A \times B) = Z(A) \times Z(B)$.

7. If $K_1, K_2 \lhd G$ and $K_1 \cap K_2 = e$, then G is isomorphic to some subgroup of $(G/K_1) \times (G/K_2)$. Is this true?

8. Let $K \lhd G = A \times B$. Prove that either K is abelian, or else one of the intersections $K \cap A$, $K \cap B$ is non-trivial. Give an example of a group $A \times B$ with a non-trivial normal subgroup K for which $K \cap A = e$ and $K \cap B = e$. Thus, $K \lhd A \times B$ does not necessarily imply that $K = (K \cap A) \times (K \cap B)$.

9. Is the quaternion group Q_8 a semidirect product of two of its proper subgroups?

10. Prove that $H \lhd Q_8$ for any proper subgroup $H \subset Q_8$.

11. Show that the groups D_4 and Q_8 are not isomorphic.

12. Show that $\mathrm{Aut}(D_4) \cong D_4$. (Since $|Z(D_4)| = 2$, by Exercise 1 this then means that $|\mathrm{Out}(D_4)| = 2$.)

13. The set of all complex p^i-th roots of 1, $i = 0, 1, 2, \ldots$, forms an infinite group $C(p^\infty)$. It is called quasi-cyclic, since any finite number of elements generate a cyclic group. Verify this, and show that

$$C(p^\infty) = \langle a_1, a_2, a_3, \ldots \mid a_1^p = 1,\ a_{i+1}^p = a_i,\quad i = 1, 2, 3, \ldots \rangle \quad .$$

14. Let $G = \langle a, b \mid aba = ba^2b,\ a^3 = e,\ b^{2n-1} = e \}$, where $n \in \mathbb{N}$. Prove that $n = 1$, i.e., that $b = e$ and so $G = \langle a \mid a^3 = e \rangle$ is actually a cyclic group of order 3.

15. Fill in the details in the following formal definition of the free group F_n with n generators. Take the alphabet $A = \{a_1, a_1^{-1}, \ldots, a_n, a_n^{-1}\}$, which consists of n letters $a_1, \ldots, a_n$, and their "opposites" $a_1^{-1}, \ldots, a_n^{-1}$, and add the symbol e. Let S be the set of all "words" obtained by writing out these $2n+1$ symbols in any order, with any possible repetitions, in a row of finite length. We define the product uv of two words u and v to be the word obtained by juxtaposing the word v after u. By the inverse of $u = a_{i_1}^{\epsilon_1} \ldots a_{i_m}^{\epsilon_m}$, $\epsilon_k = \pm 1$, $k = 1, \ldots, m$, we mean the word $u^{-1} = a_{i_m}^{-\epsilon_m} \ldots a_{i_1}^{-\epsilon_1}$, $e^{-1} = e$. We introduce an equivalence relation $\sim$ on S as follows. Two words are considered to be equivalent if one is obtained from the other by applying finitely many of the following elementary transformations:

$$ee \sim e,$$

$$a_i a_i^{-1} \sim e, \qquad a_i^{-1} a_i \sim e,$$

$$a_i e \sim a_j, \qquad a_i^{-1} e \sim a_i^{-1},$$

$$e a_i \sim a_j, \qquad e a_i^{-1} \sim a_i^{-1} \quad .$$

Each equivalence class has a unique "irreducible" (shortest) word. Multiplication (juxtaposition) of words induces an associative multiplication operation on $\sim$-equivalence classes (and we similarly have inverses defined for equivalence classes). The identity will be the equivalence class of the "empty" word e. We finally define our free group F_n to be the set of equivalence classes with this operation.

Example. A cat with a spool of thread runs in "figure eights" around two columns, each time carrying the thread above the previous loop of thread. (The cat runs in different directions when it comes to the point between the two columns, not necessarily in a continuous figure eight.) The possible paths of the cat starting and ending at the point between the columns are viewed, naturally enough, as elements of a free group F_2 with 2 generators. The irreducible words

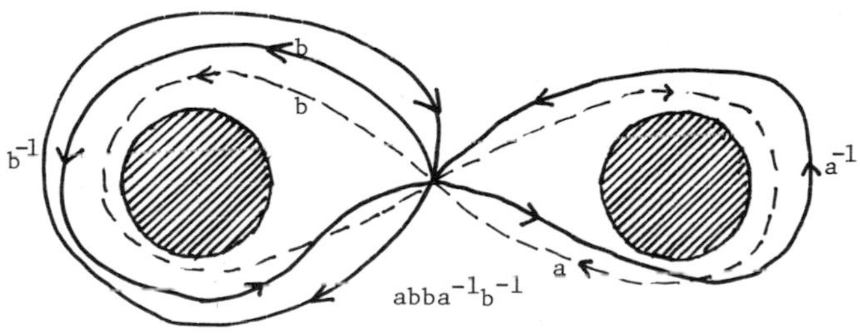

Fig. 19

correspond to paths without "retracing of steps", such as aa^{-1}, $a^{-1}a$, bb^{-1}, $b^{-1}b$. The segments a and a^{-1}, and b and b^{-1} are not drawn right on top of one another in Fig. 19 only for visual clarity. Our example amounts to realizing F_2 as the set of classes of "homotopically equivalent paths" (to use the topological terminology) of the lemniscate. In the same way, the fundamental group of Fig. 21 at the end of §3 Ch. 8 is the free group F_5.

§4. The Sylow theorems

In Subsection 4 of §3 Ch. 5, we observed that a finite group G of order $|G|$ might not have a subgroup of order d even if d divides $|G|$. The simplest example is $G = A_4$, $d = 6$.

Since there can be no subgroup of index 2 in a non-abelian simple group (since such a group would be normal), it follows by Theorem 5 of §3 that the alternating group A_5 has no subgroup of order 30. Actually, neither does A_5 have a subgroup of order 20 or 15. (Why not? Use the arguments in Example 2 of Subsection 3 §2.) Because of these examples, it is especially remarkable that, more than a century ago, the Norwegian mathematician Sylow was able to prove a series of general facts about subgroups which exist in any group. These facts concern p-groups (which we already encountered in §2) which occur as subgroups of a given group G.

Let $|G| = p^n m$, where p is a prime number and m is an integer not divisible by p. A subgroup $P \subset G$ of order $|P| = p^n$ (if such a subgroup exists) will be called a p-*Sylow subgroup* of G. As in Subsection 3 of §2, we let N(P) denote the normalizer of P in G.

THEOREM 1 (the first Sylow theorem). *There always exists a* p-*Sylow subgroup*.

THEOREM 2 (the second Sylow theorem). *Let* P *and* P_1 *be two* p-*Sylow subgroups of* G. *Then there exists an element* $a \in G$ *such that* $P_1 = aPa^{-1}$. *In other words, all* p-*Sylow subgroups are conjugate.*

THEOREM 3 (the third Sylow theorem). *The number* N_p *of* p-*Sylow subgroups in* G *is equal to* (G:N(P)) *and satisfies the congruence* $N_p \equiv 1 \pmod p$.

The proofs of Theorems 1-3 illustrate the general methods and concepts that were presented in §2. We begin with Theorem 2.

Proof of Theorem 2. Suppose that G actually has p-Sylow subgroups, and that

P is one of them. Now let P_1 be any p-subgroup of G, not necessarily a p-Sylow one. We make P_1 act by left translation on the set $G/P = \bigcup_i g_i P$ of left cosets of P in G (this is a restriction of the action of G on G/P given in §2). According to the results of Subsection 2 §2, the length of any orbit of this action divides the order $|P_1| = p^k$, $k \leq n$. Thus,

$$m = \frac{p^n m}{p^n} = \frac{|G|}{|P|} = |G/P| = p^{k_1} + p^{k_2} + \cdots ,$$

where $p^{k_1}, p^{k_2}, \ldots$ are the lengths of the orbits. Since g. c. d. $(m, p) = 1$, it follows that at least one orbit has length $p^{k_1} = 1$, i. e.,

$$P_1 \cdot aP = aP \tag{1}$$

for some element $a = g_i \in G$ (this argument resembles the proof of Theorem 2 of §2). Rewriting (1) in the form

$$P_1 \cdot aPa^{-1} = aPa^{-1} ,$$

we conclude that

$$P_1 \subset aPa^{-1} \tag{2}$$

(since aPa^{-1} is a group). In particular, if P_1 is a p-Sylow subgroup, then $|P| = |P_1|$, and (2) implies that $P_1 = aPa^{-1}$. □

<u>Proof of Theorems 1 and 3</u>. Theorem 1 is a consequence of Theorem 3, since $N_p \equiv 1 \pmod p$ implies $N_p \neq 0$, and hence the set S of all p-Sylow subgroups of G is non-empty. So it suffices to prove Theorem 3.

First, the equality $N_p = (G : N(P))$ follows immediately from the fact that all p-Sylow subgroups are conjugate (Theorem 2) and the general fact about the length of the orbit H^G that was given in §2. We shall arrive at the congruence $N_p \equiv 1 \pmod p$ by considering a somewhat more general situation. Namely, let $|G| = p^s t$, where $s \leq n$ (t may be divisible by p), and let $N_p(s)$ be the total number of subgroups of order p^s in G. We shall see that we always have the congruence $N_p(s) \equiv 1 \pmod p$; in particular, G contains subgroups of any order p^s, $s = 1, 2, \ldots, n$. Note that $N_p(n) = N_p$.

To prove this congruence, we argue as follows. The action of G on itself by left translation induces an action of G on the set

$$\Omega = \{M \subset G \,|\, |M| = p^s\}$$

of all p^s-element subsets $\{g_1, \ldots, g_{p^s}\}$ (see the remark at the end of Subsection 1 §2). Recall that $g \cdot \{g_1, \ldots, g_{p^s}\} = \{gg_1, \ldots, gg_{p^s}\}$. The set Ω splits up into G-orbits $\Omega_i : \Omega = \bigcup_i \Omega_i$, so that

$$|\Omega| = \sum_i |\Omega_i|, \qquad |\Omega_i| = (G : G_i) \quad ,$$

where $G_i = \{g \in G \,|\, gM_i = M_i\}$ is the stationary subgroup of some representative $M_i \in \Omega_i$.

Since $G_i M_i = M_i$, it follows that $M_i = \bigcup_{j=1}^{\nu_i} G_i g_{ij}$ is the union of a certain number of right cosets of G_i. Hence, $p^s = |M_i| = \nu_i |G_i|$, so that $|G_i| = p^{s_i} \leq p^s$. In the case $|G_i| < p^s$, we have $|\Omega_i| = p^{s-s_i} t \equiv 0 \pmod{pt}$; while $|G_i| = p^s$ if and only if $|\Omega_i| = t$. We obtain

$$\binom{|G|}{p^s} = |\Omega| \equiv \sum_{|\Omega_i| = t} |\Omega_i| \pmod{pt} \quad . \tag{3}$$

From what has been said, we have $|\Omega_i| = t \Rightarrow |G_i| = p^s \Rightarrow M_i = G_i a_i$ ($a_i = g_{i1}$ is some element in G), and hence $a_i^{-1} M_i = a_i^{-1} G_i a_i = P_i$ is a subgroup of order p^s.

The orbit Ω_i consists of the various cosets gP_i of P_i.

Conversely, every subgroup $H \subset G$ of order $|H| = p^s$ leads to an orbit $\Omega' = \{gH | g \in G\}$ of length t. Different subgroups H_i with $|H_i| = p^s$ lead to different orbits Ω'_i, since if we had $H_i = gH_j$, then we would have $e = gh_j$ for some $h_j \in H_j$, which gives $g \in H_j$ and $H_i = H_j$. Thus, there is a one-to-one correspondence between subgroups of order p^s and orbits Ω_i of length t. The congruence (3) can now be written in the form

$$\binom{|G|}{p^s} \equiv \sum_{|\Omega_i|=t} |\Omega_i| \equiv tN_p(s) \pmod{pt} \quad . \tag{4}$$

So far we have used nothing about the particular group G except for its order. If we take G to be the cyclic group of order $p^s t$, then the $N_p(s)$ for this G would be 1 (Theorem 5 of §3 Ch. 4), and so we must have

$$\binom{|G|}{p^s} \equiv t \cdot 1 \pmod{pt} \quad . \tag{5}$$

Since the left sides of (4) and (5) coincide (i.e., $|G|$ for our original G is the same as $|G|$ for the cyclic group of order $p^s t$), it follows that

$$t \equiv tN_p(s) \pmod{pt} \quad ,$$

which gives us the desired congruence $N_p(s) \equiv 1 \pmod{p}$. □

Although we have actually proved more than stated in the theorems, we shall not have occasion to use any deeper results, and so refer the interested reader to more specialized books.

Example. Let $G = SL(2, \mathbb{Z}_p)$ be the group of all 2×2 matrices with determinant 1 over the field $\mathbb{Z}_p$ of p elements. If we write the general linear group $GL(2, \mathbb{Z}_p)$ as a union of cosets of $SL(2, \mathbb{Z}_p)$:

$$GL(2, \mathbb{Z}_p) = \bigcup_{k=1}^{p-1} \begin{Vmatrix} k & 0 \\ 0 & 1 \end{Vmatrix} SL(2, \mathbb{Z}_p)$$

we see that

$$|GL(2, \mathbb{Z}_p)| = (p - 1)|SL(2, \mathbb{Z}_p)| \quad . \tag{6}$$

If we think of $GL(2, \mathbb{Z}_p)$ as the group of automorphisms of a two-dimensional vector space V over $\mathbb{Z}_p$, we can easily find $|GL(2, \mathbb{Z}_p)|$. Namely, $GL(2, \mathbb{Z}_p)$ acts on the set of bases $\{v_1, v_2\}$. The image of v_1 can be any non-zero vector $f_1 \in V$ (there are $p^2 - 1$ such f_1), and given a choice of f_1, the image of v_2 can be any vector f_2 in $V \setminus \langle f_1 \rangle$ (there are $p^2 - p$ such vectors). Hence, $|GL(2, \mathbb{Z}_p)| =$ $= (p^2 - 1)(p^2 - p)$, which, combined with (6), gives the formula

$$|SL(2, \mathbb{Z}_p)| = p(p^2 - 1) \quad .$$

We can exhibit at least two p-Sylow subgroups of $SL(2, \mathbb{Z}_p)$ right away:

$$P_1 = \left\{ \begin{Vmatrix} 1 & \alpha \\ 0 & 1 \end{Vmatrix} \,\middle|\, \alpha \in \mathbb{Z}_p \right\}, \quad P_2 = \left\{ \begin{Vmatrix} 1 & 0 \\ \alpha & 1 \end{Vmatrix} \,\middle|\, \alpha \in \mathbb{Z}_p \right\} \quad .$$

By Theorem 3, we have

$$N_p = (G : N(p)) = 1 + kp > 1 \, .$$

Since

$$\begin{Vmatrix} \lambda & 0 \\ 0 & \lambda^{-1} \end{Vmatrix} \begin{Vmatrix} 1 & \alpha \\ 0 & 1 \end{Vmatrix} \begin{Vmatrix} \lambda^{-1} & 0 \\ 0 & \lambda \end{Vmatrix} = \begin{Vmatrix} 1 & \lambda^2 \alpha \\ 0 & 1 \end{Vmatrix}$$

and so the normalizer $N(P)$ contains the subgroup

$$H = \left\{ \begin{Vmatrix} \lambda & \alpha \\ 0 & \lambda^{-1} \end{Vmatrix} \,\middle|\, \alpha, \lambda \in \mathbb{Z}_p , \quad \lambda \neq 0 \right\}$$

of order $p(p - 1)$, it follows that the only possibility is

$$N(P) = H, \qquad N_p = 1 + p \quad .$$

With $p = 2$, we can obtain an isomorphism between the symmetric group S_3 and the group

$$SL(2, \mathbb{Z}_2) = \left\{ \begin{Vmatrix} 1 & 0 \\ 0 & 1 \end{Vmatrix}, \begin{Vmatrix} 1 & 1 \\ 1 & 0 \end{Vmatrix}, \begin{Vmatrix} 0 & 1 \\ 1 & 1 \end{Vmatrix}, \begin{Vmatrix} 0 & 1 \\ 1 & 0 \end{Vmatrix}, \begin{Vmatrix} 1 & 0 \\ 1 & 1 \end{Vmatrix}, \begin{Vmatrix} 1 & 1 \\ 0 & 1 \end{Vmatrix} \right\}$$

by letting

$$(1\ 2\ 3) \longmapsto \begin{Vmatrix} 1 & 1 \\ 1 & 0 \end{Vmatrix}, \qquad (1\ 2) \longmapsto \begin{Vmatrix} 0 & 1 \\ 1 & 0 \end{Vmatrix}$$

(note that both groups are given by the same generators and relations). If $p > 2$, the group $G = SL(2, \mathbb{Z}_p)$ has center $Z(G) = \{\pm E\}$ of order 2. The quotient group $PSL(2, \mathbb{Z}_p) = G/Z(G)$ is called the <u>projective special linear group</u> (it is the group of transformations of the projective line $\mathbb{Z}_p P^1 = P^1(V) = \{0, 1, \ldots, p-1\} \cup \{\infty\}$). It has played an important role in algebra since the time of Galois. It turns out that, when $p > 3$, $PSL(2, \mathbb{Z}_p)$ is a simple group, and provided, along with A_n, one of the earliest known examples of a finite simple group.

We return to the general situation, and give a useful refinement of Sylow's theorems.

THEOREM 4. (i) <u>A</u> p-<u>Sylow subgroup</u> P <u>of</u> G <u>is normal in</u> G <u>if and only if</u> $N_p = 1$.

(ii) <u>A finite group</u> G <u>of order</u> $|G| = p_1^{n_1} \cdots p_k^{n_k}$ <u>is a direct product of the</u> p_i-<u>Sylow subgroups</u> $P_1, \ldots, P_k$ <u>if and only if these subgroups are normal in</u> G.

<u>Proof.</u> (i) By the second Sylow theorem, all of the Sylow subgroups corresponding to a given prime divisor p of $|G|$ are conjugate, and, if P is one such subgroup, then $N_p = 1 \Leftrightarrow xPx^{-1} = P, \forall x \in G \Leftrightarrow P \lhd G$.

(ii) If $G = P_1 \times \ldots \times P_k$ is the direct product of its Sylow subgroups, then $P_i \lhd G$, since it is a direct factor in G. Hence, normality of the Sylow subgroups is a necessary condition.

Now suppose that $P_i \lhd G$, $1 \leq i \leq k$, i.e., $N_{p_i} = 1$. In the first place, we note that

$$x \in P_i \cap P_j, \quad i \neq j \Longrightarrow x^{p_i^s} = e, \quad x^{p_j^t} = e \Longrightarrow x = e \quad .$$

Hence, $P_i \cap P_j = e$, and so for any $x_i \in P_i$, $x_j \in P_j$ we have

$$[x_i, x_j] = \begin{cases} (x_i x_j x_i^{-1}) x_j^{-1} = x_j' x_j^{-1} \in P_j \\ x_i (x_j x_i^{-1} x_j^{-1}) = x_i x_i' \in P_i \end{cases} \Longrightarrow [x_i, x_j] = e \quad ,$$

i.e., the elements x_i and x_j commute with one another.

Suppose that we could write the identity element $e \in G$ in the form $e = y_1 y_2 \ldots y_k$, where $y_i \in P_i$ is an element of order $a_i = p_i^{b_i}$. Setting $a = \prod_{i \neq j} a_i$ and using the fact that the y_i commute with one another, we obtain

$$e = (y_1 y_2 \ldots y_k)^a = y_1^a y_2^a \ldots y_k^a = y_j^a \quad .$$

But, since a and a_j are relatively prime, the fact that $y_j^{a_j} = e$ and $y_j^a = e$ implies that $y_j = e$. This is true for any j, and so $e = y_1 y_2 \ldots y_k$ is only possible if $y_1 = y_2 = \cdots = y_k = e$.

On the other hand, any element $x \in G$ of order $r = r_1 r_2 \ldots r_k$, $r_i = p_i^{s_i}$, can be written in the form

$$x = x_1 x_2 \ldots x_k, \quad |\langle x_i \rangle| = r_i, \quad 1 \leq i \leq k \quad . \tag{7}$$

To see this, it suffices to set $x_i = x^{t_i r_i'}$, where t_i and r_i' are determined from the conditions:

$$r_i' = r / r_i, \quad 1 = \sum_{i=1}^{k} t_i r_i' \quad .$$

Now if $x = x_1' x_2' \ldots x_k'$ is another expression for x in the form (7), then, since the

x_i and x'_i with different subscripts commute, we have

$$e = (x'_1 x'_2 \dots x'_k)(x_1 x_2 \dots x_k)^{-1} = x'_1 x_1^{-1} \cdot x'_2 x_2^{-1} \dots x'_k x_k^{-1} ,$$

which, as shown above, implies that $x'_1 x_1^{-1} = x'_2 x_2^{-1} = \dots = x'_k x_k^{-1} = e$, i.e., $x'_1 = x_1, x'_2 = x_2, \dots, x'_k = x_k$.

Thus, every element in G can be written uniquely in the form (7), i.e., (see §3), we have $G = P_1 \times \dots \times P_k$. □

Remark. A normal p-Sylow group P is invariant under the action of any automorphism $\varphi \in \mathrm{Aut}(G)$, since $|\varphi(P)| = |P|$, and so $\varphi(P)$ is also a p-Sylow subgroup; hence, $\varphi(P) = P$ if $N_p = 1$.

In conclusion, we note that analogs of Sylow subgroups have been studied in algebraic structures which are very different from finite groups.

EXERCISES

1. Find the number of 5-Sylow subgroups in A_5.

2. Verify that the set P of matrices

$$\pm \begin{Vmatrix} 1 & 0 \\ 0 & 1 \end{Vmatrix}, \quad \pm \begin{Vmatrix} 1 & -1 \\ -1 & -1 \end{Vmatrix}, \quad \pm \begin{Vmatrix} -1 & -1 \\ -1 & 1 \end{Vmatrix}, \quad \pm \begin{Vmatrix} 0 & 1 \\ -1 & 0 \end{Vmatrix}$$

over $\mathbb{Z}_3$ is a group isomorphic to the quaternion group Q_8, and that P is a 2-Sylow subgroup of $SL(2, \mathbb{Z}_3)$. Show that $P \triangleleft SL(2, \mathbb{Z}_3)$.

3. Show that the groups S_4 and $SL(2, \mathbb{Z}_3)$ are not isomorphic. Are the groups $PSL(2, \mathbb{Z}_3)$ and A_4 isomorphic?

4. Prove that every group G of order pq (where p and q are prime numbers, $p < q$) is either cyclic, or else non-abelian with a normal q-Sylow subgroup, and that the second possibility can occur if and only if $q - 1$ is divisible by p. For

example, all groups of order 15 are cyclic.

5. Give a new proof of the congruence $(p-1)! + 1 \equiv 0 \pmod p$ for p a prime (see §1 Ch. 6), by making a direct computation of the number N_p of p-Sylow subgroups in the symmetric group S_p.

§5. Finite abelian groups

In an abelian group all subgroups are normal. This obvious fact, together with Theorem 4 §4, immediately imply that any abelian group A of order $|A| = p_1^{n_1} p_2^{n_2} \cdots p_k^{n_k}$ is the direct product of its Sylow subgroups $A(p_i)$:

$$A = A(p_1) \times A(p_2) \times \cdots \times A(p_k) \quad . \tag{1}$$

The factors $A(p_i)$ are often called the primary components of A. The direct product expansion (1) is unique: each component $A(p_i)$ is simply the set of all elements whose order is a power of p_i.

Our purpose is to express the abelian group A as a direct product of the simplest possible groups, i.e., cyclic groups. If we do not make any restrictions on the orders of the cyclic groups, then we can generally do this in many ways; to take the simplest example,

$$A = \langle a \mid a^6 = e \rangle = \langle a^3 \rangle \times \langle a^2 \rangle \quad .$$

But we only have a limited choice, and, despite the non-uniqueness, the final result (Theorem 3) gives us a very useful picture of the nature of finite abelian groups.

1. Primary abelian groups. In what follows we must keep in mind that, if an abelian group A is generated by subgroups B and C, then actually $A = BC$; in addition, $A = B \times C$ if and only if $B \cap C = e$ (see Subsection 4 of §3).

Unlike other cyclic groups, the cyclic group C_{p^n} of prime power order p^n cannot be decomposed, i.e., cannot be written as a direct product of groups of smaller order. To see this, write $C_{p^n} = \langle a \rangle$ and $C_{p^i} = \langle a^{p^{n-i}} \rangle$. Then the chain

$$C_{p^n} \supset C_{p^{n-1}} \supset \cdots \supset C_p \supset e$$

contains all subgroups of C_{p^n}. Any two of them $X \neq e$ and $Y \neq e$ have a non-trivial intersection $X \cap Y \supset C_p$, and so cannot be factors in a direct product expansion.

THEOREM 1. Every finite abelian p-group is a direct product of cyclic groups.

Proof. We use induction, and suppose the theorem true for all abelian p-groups of order $< p^n$. Let A be an abelian group of order $|A| = p^n$, and choose an element $a \neq e$ of maximal order p^m. Consider the quotient group $\overline{A} = A/\langle a \rangle$. Since $|\overline{A}| = p^{n-m} < p^n$, it follows by the induction assumption that

$$\overline{A} = \overline{A}_1 \times \cdots \times \overline{A}_r , \tag{2}$$

where $\overline{A}_i = \langle \overline{b}_i \rangle = \langle b_i \langle a \rangle \rangle = \{ \langle a \rangle , b_i \langle a \rangle , \ldots , b_i^{p^{m_i}-1} \langle a \rangle \}$ is a cyclic group of order p^{m_i}, $1 \leq i \leq r$, $m_1 + \cdots + m_r = n - m$. By definition,

$$\overline{b}_i^{p^{m_i}} = \overline{e} = \langle a \rangle , \quad \text{i.e.,} \quad b_i^{p^{m_i}} = a^{s_i} \in \langle a \rangle , \tag{3}$$

and, although every element $x \in A$ has the form

$$x = b_1^{k_1} \cdots b_r^{k_r} \cdot a^k ,$$

in general this expression is not unique. We must "correct" the elements $b_i \in A$ in such a way that the exponents s_i in (3) vanish. This is not hard to do. If we recall that $m_i \leq m$ and raise both sides of (3) to the power p^{m-m_i}, we obtain

$$e = a^{s_i p^{m-m_i}} ,$$

so that $s_i = t_i p^{m_i}$ (by Theorem 3 of §2 Ch. 4). If we now set $a_i = b_i a^{-t_i}$, then (3) gives

$$a_i^{p^{m_i}} = e , \quad 1 \le i \le r \ (\Longleftrightarrow \langle a_i \rangle \cap \langle a \rangle = e) , \tag{3'}$$

where $\bar{a}_i = a_i \langle a \rangle = b_i \langle a \rangle = \bar{b}_i$, and consequently $\langle \bar{a}_i \rangle = A_i$. Once more, we have

$$x = a_1^{k_1} \cdots a_r^{k_r} a^k$$

for any $x \in A$, and this expression is now unique. To see uniqueness, note that if there were two such expressions for x, we would obtain a relation

$$a_1^{\nu_1} \cdots a_r^{\nu_r} a^{\nu} = e , \quad 0 \le \nu_i < p^{m_i} , \quad 0 \le \nu < p^m ,$$

(not all of the ν_i and ν vanish), which the epimorphism $A \to \bar{A}$ would take to the relation $\bar{a}_1^{\nu_1} \cdots \bar{a}_r^{\nu_r} = \bar{e}$. Because of the direct product (2), this gives $\bar{a}_i^{\nu_i} = \bar{e}$, $1 \le i \le r$, or, equivalently, $a_i^{\nu_i} \in \langle a \rangle$. But, by (3'), this can only happen if all of the $\nu_i = 0$; then also $\nu = 0$.

This contradiction shows that $A = \langle a_1 \rangle \times \dots \times \langle a_r \rangle \times \langle a \rangle$. □

Remark. This proof of Theorem 1 resembles the geometric proof of the theorem on the Jordan normal form of the matrix of a nilpotent linear operator (see the Appendix).

THEOREM 2. If a finite abelian p-group A is expressed in two ways as a direct product of cyclic subgroups:

$$A = A_1 \times \dots \times A_r = B_1 \times \dots \times B_s ,$$

<u>then</u> $r = s$, <u>and the orders of the</u> A_i <u>coincide with the orders of the</u> B_j <u>for a suitable ordering of the</u> B_j.

<u>Proof.</u> The theorem obviously holds if $|A| = p$. We use induction on $|A|$. It is convenient from the very beginning to order the A_i and B_j in such a way that their orders are non-increasing:

$$\begin{gathered} A_i = \langle a_i \rangle, \qquad |\langle a_i \rangle| = p^{m_i}, \\ m_1 \geq m_2 \geq \cdots \geq m_q > m_{q+1} = \cdots = m_r - 1; \end{gathered} \tag{4}$$

$$\begin{gathered} B_j = \langle b_j \rangle, \qquad |\langle b_j \rangle| = p^{n_j}, \\ n_1 \geq n_2 \geq \cdots \geq n_t > n_{t+1} = \cdots = n_s \quad . \end{gathered} \tag{5}$$

The relations

$$(xy)^p = x^p y^p, \qquad (x^p)^{-1} = (x^{-1})^p ,$$

which hold in any abelian group (see (3) in §1 Ch. 4), imply that the set

$$A^p = \{x^p \mid x \in A\}$$

of p-th powers of all elements of A is a subgroup of A, and it does not, of course, depend on the direct product expansion of A. On the other hand, if

$$a_1^{i_1} \ldots a_q^{i_q} \ldots a_r^{i_r} = x = b_1^{j_1} \ldots b_t^{j_t} \ldots b_s^{j_s} ,$$

then, using (4) and (5), we have

$$(a_1^p)^{i_1} \ldots (a_q^p)^{i_q} = x^p = (b_1^p)^{j_1} \ldots (b_t^p)^{j_t} \quad .$$

Hence,

$$\langle \bar{a}_1 \rangle \times \ldots \times \langle \bar{a}_q \rangle = A^p = \langle \bar{b}_1 \rangle \times \ldots \times \langle \bar{b}_t \rangle ,$$

where $\bar{a}_i = a_i^p$, $\bar{b}_j = b_j^p$ are elements of order p^{m_i-1} and p^{n_j-1}, respectively. Since $|A^p| < |A|$, it follows by the induction assumption that $q = t$ and $m_1 - 1 = n_1 - 1, \ldots, m_q - 1 = n_q - 1$; hence, $m_1 = n_1, \ldots, m_q = n_q$. If we further note that

$$|A_{q+1} \times \ldots \times A_r| = p^{r-q}, \qquad |B_{t+1} \times \ldots \times B_s| = p^{s-t}, \qquad q = t ,$$

we find that

$$p^{m_1 + \cdots + m_q} p^{r-q} = |A| = p^{m_1 + \cdots + m_q} p^{s-q} .$$

Hence, $s = r$, and all of the assertions in the theorem have been proved. □

The orders $p^{m_1}, \ldots, p^{m_r}$ of the cyclic factors are called the _invariants_ (or _elementary divisors_) of the finite abelian p-group A. If two abelian p-groups A and B have the same invariants, then

$$A = A_1 \times \ldots \times A_r, \qquad B = B_1 \times \ldots \times B_r, \qquad A_i \cong C_{p^{m_i}} \cong B_i ,$$

and the set of isomorphisms $\varphi_i : A_i \to B_i$ induces an isomorphism $\varphi : A \to B$ given by $\varphi((a_1, \ldots, a_r)) = (\varphi(a_1), \ldots, \varphi_r(a_r))$. Hence, Theorem 2 says that _the group_ A _is determined up to isomorphism by its invariants._ In particular, we have the following

COROLLARY. _The number of non-isomorphic abelian groups of order_ p^n _is equal to the number_ $p(n)$ _of partitions_

$$n = n_1 + n_2 + \cdots + n_r, \qquad n_1 \geq n_2 \geq \cdots \geq 1, \qquad 1 \leq r \leq n .$$ □

We encountered the partition function $p(n)$ when we were describing the conjugacy classes in the symmetric group S_m (see Exercise 4 of §2). An abelian group of order p^r with invariants $p, \ldots, p$ is often called an _elementary abelian group._ An elementary abelian group is characterized by the property that $A^p = e$. If we switch to additive notation, we note that an abelian group A for which $pA = 0$ (where p is a prime) is a vector space over the finite field $\mathbb{F}_p$ of p elements. To see this, think of the elements of $\mathbb{F}_p = \mathbb{Z}_p$ as residue classes $\bar{k}$ modulo p for $k \in \mathbb{Z}$, and set $\bar{k}a = ka$, $a \in A$. This gives an action of $\mathbb{F}_p$ on A which makes A into an $\mathbb{F}_p$-vector space. The action is correctly defined, because, if $\bar{k} = \bar{k'}$, then $(k-k')a$ is of the form $\ell(pa) = 0$. The expansion of A as a direct product of cyclic subgroups corresponds to an expansion of A considered as an $\mathbb{F}_p$-vector space as a direct sum of one-dimensional subspaces (using the basis theorem). Thus,

$$A \cong \mathbb{Z}_p^r = \mathbb{Z}_p \oplus \ldots \oplus \mathbb{Z}_p \quad .$$

The amount of choice in the one-dimensional subspaces even when $r = 2$ is clear from the example in §4: $\mathbb{Z}_p^2$ has $p(p+1)$ different expansions.

2. <u>The structure theorem for finite abelian groups.</u> Using the unique expansion (1), along with Theorems 1 and 2, we immediately arrive at the following fundamental fact about abelian groups.

THEOREM 3. <u>Every finite abelian group</u> A <u>is a direct product of primary cyclic subgroups. Any two such product expansions have the same number of factors of each cyclic order.</u> □

Borrowing the vector space terminology, we say that elements $a_1, \ldots, a_r$ of orders $d_1, \ldots, d_r$, respectively, form a <u>basis</u> of an abelian group A if every element $x \in A$ can be uniquely written in the form

$$x = a_1^{i_1} a_2^{i_2} \ldots a_r^{i_r}, \qquad 0 \leq i_k < d_k, \qquad k = 1, \ldots, r \quad .$$

Of course, in that case we have

$$A = \langle a_1 \rangle \times \ldots \times \langle a_r \rangle, \qquad |A| = d_1 d_2 \ldots d_r \quad . \tag{6}$$

Theorem 3 is equivalent to the statement that every finite abelian group A has a basis whose elements are primary (i.e., their orders d_i are powers of prime divisors of $|A|$), and that the set $\{d_1, d_2, \ldots, d_r\}$ does not depend on the choice of basis. Because the d_i only depend on A, and not on the choice of basis, they are called the <u>invariants</u> or <u>elementary divisors</u> of A, just as in the case of primary groups. We sometimes say that $\{d_1, \ldots, d_r\}$ is the <u>type</u> of the finite abelian group A.

Given a finite abelian group A, we shall write out all of the invariants in rows corresponding to the different prime divisors of $|A|$, as follows:

$$\begin{array}{ll}
p_1^{n_{11}}, p_1^{n_{12}}, p_1^{n_{13}}, \ldots; & n_{11} \geq n_{12} \geq n_{13} \geq \cdots; \\
p_2^{n_{21}}, p_2^{n_{22}}, p_2^{n_{23}}, \ldots; & n_{21} \geq n_{22} \geq n_{23} \geq \cdots; \\
\cdots\cdots\cdots & \cdots\cdots\cdots \\
p_k^{n_{k1}}, p_k^{n_{k2}}, p_k^{n_{k3}}, \ldots; & n_{k1} \geq n_{k2} \geq n_{k3} \geq \cdots \quad .
\end{array}$$

We may take all of the rows to be the same length ℓ if we fill in some of the rows with ones.

The integers

$$m_j = p_1^{n_{1j}} p_2^{n_{2j}} \ldots p_k^{n_{kj}}, \quad j = 1, 2, \ldots, \ell ,$$

are called the <u>invariant factors</u> of A. By construction, we have

$$|A| = m_1 m_2 \ldots m_\ell, \quad m_{j+1} | m_j, \quad j = 1, 2, \ldots, \ell - 1 \quad . \tag{7}$$

The expansion (6), written in the form

$$A = (\langle a_{11} \rangle \times \ldots \times \langle a_{k1} \rangle) \times \ldots \times (\langle a_{1\ell} \rangle \times \ldots \times \langle a_{k\ell} \rangle) ,$$

gives us an expansion

$$A = \langle u_1 \rangle \times \langle u_2 \rangle \times \ldots \times \langle u_\ell \rangle \tag{8}$$

whose cyclic direct factors have order $m_1, m_2, \ldots, m_\ell$. To obtain the expansion (8), it suffices to set

$$u_j = a_{1j} a_{2j} \ldots a_{kj}, \quad 1 \leq j \leq \ell ,$$

and use the proposition at the end of Subsection 3 §2 Ch. 4.

If A is a primary group, then the direct products (6) and (8) are obviously the same, but in general (8) is more economical than (6) $(\ell \leq r \leq k\ell)$. Note that the expansion (8) immediately gives an element u_1 of maximal order m_1. The integer m_1 is called the exponent of the group A. An abelian group A is cyclic if and only if its exponent is equal to $|A|$.

Finally, note that we can always find an abelian group with any given invariant factors $m_1, m_2, \ldots, m_\ell$: we need only take the direct sum of the cyclic groups $\mathbb{Z}_{m_1}, \ldots, \mathbb{Z}_{m_\ell}$.

As an example, let us compute all abelian groups of order 16 and of order 36.

$$|A| = 16 = 2^4, \quad p(4) = 5: \quad \mathbb{Z}_{16}, \quad \mathbb{Z}_8 \oplus \mathbb{Z}_2,$$

$$\mathbb{Z}_4 \oplus \mathbb{Z}_4, \quad \mathbb{Z}_4 \oplus \mathbb{Z}_2 \oplus \mathbb{Z}_2, \quad \mathbb{Z}_2^4 = \mathbb{Z}_2 \oplus \mathbb{Z}_2 \oplus \mathbb{Z}_2 \oplus \mathbb{Z}_2 .$$

$\lvert A\rvert = 36 = 2^2 \cdot 3^2$		elementary divisors	invariant factors
$\mathbb{Z}_4 \oplus \mathbb{Z}_9$	$\cong \mathbb{Z}_{36}$	4, 9	36
$\mathbb{Z}_2 \oplus \mathbb{Z}_2 \oplus \mathbb{Z}_9$	$\cong \mathbb{Z}_{18} \oplus \mathbb{Z}_2$	2, 2, 9	18, 2
$\mathbb{Z}_4 \oplus \mathbb{Z}_3 \oplus \mathbb{Z}_3$	$\cong \mathbb{Z}_{12} \oplus \mathbb{Z}_3$	4, 3, 3	12, 3
$\mathbb{Z}_2 \oplus \mathbb{Z}_2 \oplus \mathbb{Z}_3 \oplus \mathbb{Z}_3$	$\cong \mathbb{Z}_6 \oplus \mathbb{Z}_6$	2, 2, 3, 3	6, 6

Let us consider one more example. We write the group $\mathbb{Z}_{72} \oplus \mathbb{Z}_{84}$ in terms of invariant factors. We first express each cyclic summand in terms of cyclic primary components:

$$\mathbb{Z}_{72} = \mathbb{Z}_8 \oplus \mathbb{Z}_9, \quad \mathbb{Z}_{84} = \mathbb{Z}_4 \oplus \mathbb{Z}_3 \oplus \mathbb{Z}_7 .$$

Next, we gather together all of the primary components:

$$\mathbb{Z}_{72} \oplus \mathbb{Z}_{84} = (\mathbb{Z}_8 \oplus \mathbb{Z}_4) \oplus (\mathbb{Z}_9 \oplus \mathbb{Z}_3) \oplus \mathbb{Z}_7$$

(this is a direct sum of p-Sylow subgroups). Finally, we take the cyclic summand of highest order in each primary component, and repeat this process with the summands that remain:

$$\mathbb{Z}_{72} \oplus \mathbb{Z}_{84} = (\mathbb{Z}_8 \oplus \mathbb{Z}_9 \oplus \mathbb{Z}_7) \oplus (\mathbb{Z}_4 \oplus \mathbb{Z}_3) = \mathbb{Z}_{504} \oplus \mathbb{Z}_{12} \quad .$$

Note that we would get the same result if we had started with $\mathbb{Z}_{36} \oplus \mathbb{Z}_{168}$; thus

$$\mathbb{Z}_{72} \oplus \mathbb{Z}_{84} = \mathbb{Z}_{36} \oplus \mathbb{Z}_{168}$$

(strictly speaking, we should use $\cong$ instead of $=$ here). In particular, note that the exponent of both groups is 504.

EXERCISES

1. Prove Theorem 1 and the first half of Theorem 3 without passing to quotient groups.

2. Obtain an expansion for a finite abelian group A into a direct product of primary components without using the Sylow theorems, and of course without using Theorem 3. In particular, Example 1 in Subsection 1 §3 or the proposition in Subsection 3 §2 Ch. 4 can be used to obtain the expansion

$$\mathbb{Z}_n \cong \mathbb{Z}_{d_1} \oplus \mathbb{Z}_{d_2} \oplus \dots \oplus \mathbb{Z}_{d_k}$$

for $n = d_1 d_2 \dots d_k$, $d_i = p_i^{e_i}$ (where p_i are the prime divisors).

3. Show that, if A is a finite abelian group, then A has at least one subgroup of each order d dividing $|A|$ (this is a converse to Lagrange's Theorem).

4. Show that, for a suitable ordering, the invariants of any subgroup of an abelian

group are divisors of the invariants of the group.

5. Prove that, if $A \oplus A \cong B \oplus B$, where A and B are finite abelian groups, then $A \cong B$.

6. Prove that, if A, B and C are finite abelian groups and $A \oplus C \cong B \oplus C$, then $A \cong B$.

7. Show that an abelian group with invariant factors $m_1, \ldots, m_\ell$ cannot be generated by fewer than ℓ elements.

8. Show that a finite abelian group whose order is not divisible by the square of any integer greater than 1 must be cyclic.

9. List all non-isomorphic abelian groups of order 72.

10. Are the groups $\mathbb{Z}_{12} \oplus \mathbb{Z}_{72}$ and $\mathbb{Z}_{18} \oplus \mathbb{Z}_{48}$ isomorphic?

Chapter 8. Elements of Representation Theory

Before giving the precise definitions of representation theory, we shall discuss two problems which are similar in spirit.

Problem 1. In the $(m+1)$-dimensional vector space V_m consisting of real homogeneous degree m polynomials

$$f(x,y) = a_0 x^m + a_1 x^{m-1} y + \cdots + a_{m-1} y^{m-1} x + a_m y^m$$

(or rather, polynomial functions $(x,y) \mapsto f(x,y)$), we consider the set of solutions of the two-dimensional Laplace equation

$$\frac{\partial^2 f}{\partial x^2} + \frac{\partial^2 f}{\partial y^2} = 0 \qquad (*)$$

(see Exercise 9 in §1 Ch. 6). The Laplace operator $\Delta = \frac{\partial^2}{\partial x^2} + \frac{\partial^2}{\partial y^2}$ is linear:

$$\Delta(\alpha f + \beta g) = \alpha \Delta f + \beta \Delta g, \qquad \forall \alpha, \beta \in \mathbb{R} .$$

Hence, the solutions of equation $(*)$ form a subspace H_m of V_m. We immediately find

$$\Delta f = \sum_{k=0}^{m-2} [(m-k)(m-k-1)a_k + (k+2)(k+1)a_{k+2}]x^{m-2-k}y^k \quad .$$

Consequently,

$$\Delta f = 0 \iff (m-k)(m-k-1)a_k + (k+2)(k+1)a_{k+2} = 0, \quad 0 \le k \le m-2 \quad ,$$

and all of the coefficients a_k can be expressed in terms of two of them, say, a_0 and a_1. Thus, $\dim H_m \le 2$.

But it is possible to give two linearly independent solutions right away. Namely, if we extend the Δ operator by linearity to polynomials with complex coefficients, then we have

$$\Delta(x+iy)^m = m(m-1)(x+iy)^{m-2} + imi(m-1)(x+iy)^{m-2} = 0, \quad i^2 = -1 \quad .$$

Separating the real and imaginary parts, we obtain

$$z_m(x,y) \equiv (x+iy)^m = u_m(x,y) + iv_m(x,y) \quad ,$$

so that

$$\Delta u_m + i\Delta v_m = \Delta z_m = 0 \implies \Delta u_m = 0, \quad \Delta v_m = 0 \quad .$$

Thus,

$$H_m = \langle u_m(x,y), v_m(x,y) \rangle_{\mathbb{R}} \quad .$$

If we now interpret x and y as the coordinates of a vector in the Euclidean space $\mathbb{R}^2$ with a fixed rectangular system of coordinates, we can see what happens under an orthogonal change of coordinates, i.e., when the plane $\mathbb{R}^2$ is rotated about the origin through an angle θ:

$$x' = \Phi_0(x) = x\cos\theta - y\sin\theta \quad ,$$

$$y' = \Phi_0(y) = x\sin\theta + y\cos\theta \quad .$$

The chain rule of calculus (which is easy to verify directly for polynomials) gives

$$\frac{\partial^2 f}{\partial x'^2} = \frac{\partial^2 f}{\partial x^2}\cos^2\theta - 2\frac{\partial^2 f}{\partial x\,\partial y}\cos\theta\cdot\sin\theta + \frac{\partial^2 f}{\partial y^2}\sin^2\theta ,$$

$$\frac{\partial^2 f}{\partial y'^2} = \frac{\partial^2 f}{\partial x^2}\sin^2\theta + 2\frac{\partial^2 f}{\partial x\,\partial y}\cos\theta\cdot\sin\theta + \frac{\partial^2 f}{\partial y^2}\cos^2\theta ,$$

and hence

$$\frac{\partial^2 f}{\partial x'^2} + \frac{\partial^2 f}{\partial y'^2} = \frac{\partial^2 f}{\partial x^2} + \frac{\partial^2 f}{\partial y^2} .$$

This means that the equation (*) is invariant under an orthogonal change of variables, or, to put it another way, under the action of the group $SO(2) = \{\Phi_\theta\}$. In particular, the polynomials $u_m(x', y')$ and $v_m(x', y')$ are solutions of (*), and so can be expressed as a linear combination of $u_m(x,y)$ and $v_m(x,y)$. Thus, the group $SO(2)$ acts on the space of solutions of the Laplace equation. We call this a two-dimensional, real, linear representation

$$\Phi^{(m)} : \Phi_\theta \longmapsto \Phi^{(m)}(\theta)$$

of the group $SO(2)$.

If we return once again to complex polynomials, we notice that

$$x' + iy' = xe^{i\theta} + iye^{i\theta} = e^{i\theta}(x + iy) ,$$

$$(x' + iy')^m = e^{im\theta}(x + iy)^m .$$

Letting the complexified linear operator $\Phi^{(m)}(\theta)$ keep its earlier meaning, we have

$$\Phi^{(m)}(\theta) : z_m \longmapsto z'_m = e^{im\theta} z_m .$$

The so-called one-dimensional unitary representations $\Phi^{(m)} : \Phi_\theta \mapsto e^{im\theta}$, $m \in \mathbf{Z}$, of the group $SO(2)$ play an important role in analysis.

We note that the action Φ induces an action of $SO(2)$ on the whole space V_m; from this point of view, H_m is an invariant subspace of V_m.

Problem 2. Estimating the number of organic compounds, for example, in the chemistry of cyclic hydrocarbons, leads to the following general problem: How many different necklaces of length n can be made from (an unlimited supply of) pearls of q different colors?

Following Pólya, we shall attempt to answer this question by first supposing that the necklaces are oriented, i.e., a necklace and the same necklace turned upside-down are counted separately. Note that the number of possible sections of thread with n pearls is equal to q^n (the number of words of length n in the free group with q generators). The cyclic group $\langle\sigma\rangle$ of order n with generator

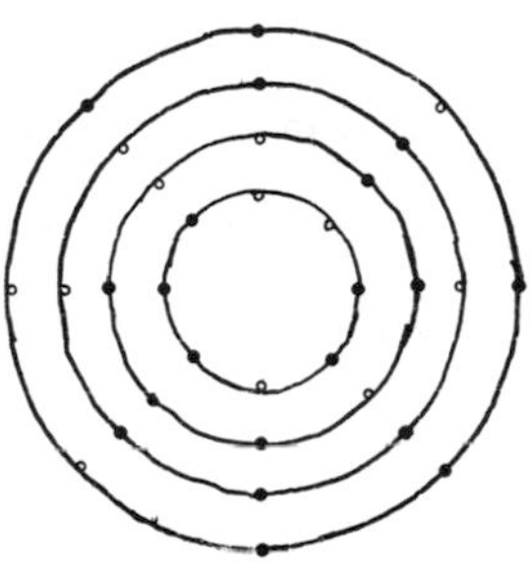

Fig. 20

$\sigma = (12 \ldots n) \in S_n$ acts on the set Ω_n of these pieces of thread by permuting the pearls cyclically on the thread. It is natural to think of a necklace as the $\langle\sigma\rangle$-orbit of a section of thread, or, if we want, as a certain set of concentric circles (see Fig. 20). The second interpretation is easiest to visualize. It is connected with the isomorphism

$$\Phi : \sigma \longmapsto \Phi(\sigma) = \begin{Vmatrix} \cos \frac{2\pi}{n} & -\sin \frac{2\pi}{n} \\ \sin \frac{2\pi}{n} & \cos \frac{2\pi}{n} \end{Vmatrix},$$

which we have encountered before, and which we will soon be calling two-dimensional linear real representation of the group $\langle\sigma\rangle$. The number r of necklaces can now be expressed by the formula in Exercise 8 §2 Ch. 7 :

$$r = \frac{1}{n} \sum_{k=0}^{n-1} N(\sigma^k) \quad .$$

If $d \mid n$, then the element σ^d of order n/d leaves fixed the sections of thread (and necklaces) which can be divided into d periods of length n/d (in this connection, see Exercise 12 of §2 Ch. 4). Hence, $N(\sigma^d) = q^d$, and, more generally, $N(\sigma^k) = q^{\text{g.c.d.}(n,k)}$. Exactly $\varphi(n/d)$ (where φ is the Euler function) of the $N(\sigma^k)$

in $\Sigma N(\sigma^k)$ have g.c.d.$(n,k) = d$. This means that

$$r = \frac{1}{n} \sum_{d|n} \varphi(\frac{n}{d})\, q^d \quad .$$

If we are interested in physically different (un-oriented) necklaces, we must identify elements in Ω_n by means of a two-dimensional linear representation of the dihedral group D_n. We leave this for the reader to do on his own.

Not only in these contrived examples, but in actual physical problems linear representations of groups inevitably arise as the reflection of some symmetry in the problem. The ideas and language of representation theory are very natural. In fact, the examples we shall give in §1 relate to old problems and at first glace do not seem to be anything new. But the very fact that all of these examples have been brought together "under one roof" should suggest that we are dealing with a concept having fundamental importance.

Representation theory has two aims: (1) in pure mathematics, the development of new techniques for investigating various groups, and (2) in applications, as a powerful tool in such areas as crystallography and quantum mechanics. In this chapter our concern will be to say something substantial about representation theory using only the material that is already accessible to us from linear algebra and group theory.

§1. Definitions and examples of linear representations

1. Basic concepts. Strictly speaking, we have already worked with representation theory, when we studied the action of groups on sets (§2 Ch. 7). We now take our set to be a vector space V of dimension n over a field K, and in the group $S(V)$ of all bijective set maps $V \to V$ we consider the subgroup $GL(V)$ of all invertible linear operators on V (i.e., the group of automorphisms of the vector space V). Clearly, given a basis $\{e_1, \dots, e_n\}$ of V, the group $GL(V)$ can be identified with the usual matrix group $GL(n, K)$, the group of automorphisms of the vector space K^n. Then to

every linear operator $\mathcal{a} \in GL(V)$ there corresponds a matrix $A = (a_{ij})$ such that

$$\mathcal{a} e_j = \sum_{i=1}^{n} a_{ij} e_i ; \qquad a_{ij} \in K , \qquad \det A \neq 0 \quad .$$

Definition 1. Let G be a group. Any homomorphism $\Phi : G \to GL(V)$ is called a linear representation of G in the vector space V. A representation is called faithful if the kernel of the representation $\operatorname{Ker} \Phi$ only consists of the identity element of G, and it is called the trivial representation (or the unit representation) if $\Phi(g)$ is the identity operator $\mathcal{E}$ for all $g \in G$. The dimension $\dim_K V$ is called the dimension of the representation. When $K = \mathbb{Q}$, $\mathbb{R}$ or $\mathbb{C}$, we speak of a rational, real, or complex representation, respectively.

Thus, a linear representation is a pair (Φ, V) consisting of a representation space V (also called a G-space) and a homomorphism $\Phi : G \to GL(V)$. By definition, we have

$$\Phi(e) = \text{the identity operator } \mathcal{E} ;$$

$$\Phi(gh) = \Phi(g)\Phi(h) \quad \text{for all} \quad g, h \in G \quad .$$

If we agree to let $g * v$ denote the action of the linear operator $\Phi(g)$ on the vector $v \in V$, then we arrive at the relations

$$\begin{aligned} g * (u + v) &= g * u + g * v , & u, v \in V , \\ g * (\lambda v) &= \lambda (g * v) , & \lambda \in K , \\ e * v &= v , & \\ (gh) * v &= g * (h * v) , & \end{aligned} \tag{1}$$

which imitate the properties of linear operators (the last two are the same as the relations expressed just before using the Φ; compare with (i) and (ii) in §2 Ch. 7). Relations (1) highlight the role of the G-space V in the linear representation (Φ, V); this is often convenient, especially in situations where V is not just an abstract vector space, but has a concrete meaning.

On the other hand, the vector space V need not be explicitly indicated; we can think of a linear representation simply as a homomorphism Φ from G to $GL(n, K)$. As before, we have $\Phi_{gh} = \Phi_g \Phi_h$, but now Φ_g is a non-singular matrix, and $\Phi_e = E$ is the identity matrix. The matrix point of view is usually better for computational purposes, but it is less invariant and lacks the geometrical clarity of the vector space point of view. In practice, it is important to be able to go back and forth freely between the G-space and the matrix interpretations.

In this connection, recall the basic fact from linear algebra that two matrices A and B which correspond to the same operator but in different bases, are similar: $B = CAC^{-1}$ (C is the transfer matrix from one basis to the other). In the situation of representation theory, when we are dealing with groups of linear operators, we take into account the dependence on the choice of basis in the following way.

Definition 2. Two linear representations (Φ, V) and (Ψ, W) of a group G are said to be equivalent (or isomorphic) if there exists an isomorphism of vector spaces $\sigma : V \to W$ such that the diagram

$$\begin{array}{ccc} V & \xrightarrow{\sigma} & W \\ {\scriptstyle \Phi(g)}\downarrow & & \downarrow{\scriptstyle \Psi(g)} \\ V & \xrightarrow{\sigma} & W \end{array}$$

is commutative for all $g \in G$, i.e.,

$$\Psi(g)\sigma = \sigma\Phi(g), \qquad g \in G,$$

or, equivalently,

$$\Psi(g) = \sigma\Phi(g)\sigma^{-1} \tag{2}$$

(compare with the definition of equivalent actions of a group on sets, which was given in Exercise 1 of §2 Ch. 7). We shall sometimes write $\Phi \approx \Psi$ for equivalent representations, $\Phi \not\approx \Psi$ for inequivalent representations.

Here are two variants of Definition 2.

(a) G-space terminology. Let G be a group, and let $V : (g,v) \mapsto g * v$ and $W : (g,w) \mapsto g \square w$ be two G-spaces with $*$ and $\square$ satisfying condition (1). A vector space isomorphism $\sigma : V \to W$ is called an isomorphism of G-spaces if

$$g \square \sigma(v) = \sigma(g * v) \tag{2'}$$

for all $g \in G$ and $v \in V$. In that case we also say that the map σ commutes with the action of G.

(b) Matrix terminology. If $V = \langle v_1, \dots, v_n \rangle$, $W = \langle w_1, \dots, w_n \rangle$, and Φ_g and Ψ_g are the matrices of the linear maps $\Phi(g)$ and $\Psi(g)$ in these bases, then the condition (2) for equivalence can be written in the form

$$\Psi_g = C \Phi_g C^{-1} , \tag{2''}$$

where C is some non-singular matrix which is the same for all $g \in G$. The entries in all of these matrices are in the same field K.

The relation of similarity of matrices, which is expressed by (2''), is an equivalence relation which divides the set $M_n(K)$ into disjoint equivalence classes. In the same way, the representations of a group G divide up into classes of equivalent representations. It will soon be clear that what is important and interesting is precisely the equivalence classes of representations.

Again using linear algebra, we try to give a clearer picture of how a group G acts on a space V. If $\mathfrak{a} : V \to V$ is a linear operator, there may exist an invariant subspace U, i.e., for which $u \in U \Rightarrow \mathfrak{a} u \in U$. If we take an arbitrary basis $\{e_1, \dots, e_k\}$ in U and extend it to a basis for all of $V = \langle e_1, \dots, e_k, e_{k+1}, \dots, e_n \rangle$, we see that the matrix of $\mathfrak{a}$ in the basis $\{e_1, \dots, e_n\}$ has the following triangular block form:

$$A = \begin{Vmatrix} A_1 & A_0 \\ 0 & A_2 \end{Vmatrix} .$$

The block A_1 corresponds to the invariant subspace U, and the block A_2 corresponds to the quotient space V/U. If A_0 happens to be the zero matrix, then $A = A_1 \dotplus A_2$ is the direct sum of the blocks, and $V = U \oplus W$ is a direct sum of invariant subspaces.

We can always find an eigen-vector, i.e., $v \in V$, $v \neq 0$, for which $\mathfrak{A}v = \lambda v$, $\lambda \in K$, if we suppose that K is algebraically closed (see §3 Ch. 6), for example, the field $\mathbb{C}$ of complex numbers. Here λ is a root of the characteristic polynomial

$$f_{\mathfrak{A}}(t) = |tE - A| = t^n - (\operatorname{tr} A)t^{n-1} + \cdots + (-1)^n \det A$$

(A is the matrix of $\mathfrak{A}$ in any basis). Using eigen-vectors, we can easily choose a basis of V with respect to which A has the triangular form

$$A = \begin{Vmatrix} \lambda_1 & & & * \\ & \lambda_2 & & \\ & & \ddots & \\ 0 & & & \lambda_n \end{Vmatrix}$$

with roots $\lambda_1, \lambda_2, \ldots, \lambda_n$ along the diagonal. A more careful analysis allows us to reduce A to the so-called Jordan normal form $J(A)$ (see the Appendix), which is a direct sum of the Jordan cells

$$J_{m,\lambda} = \begin{Vmatrix} \lambda & 1 & 0 & \ldots & 0 \\ 0 & \lambda & 1 & \ldots & 0 \\ \cdot & \cdot & \cdot & \cdots & \cdot \\ 0 & 0 & 0 & \ldots & \lambda \end{Vmatrix}$$

($m \times m$ is the size of the cell, and λ is a root of the characteristic polynomial).

Note that if we have $A^q = E$, then it follows that $J^q_{m,\lambda} = E_m$ is the $m \times m$ identity matrix for each Jordan cell $J_{m,\lambda}$, and this is obviously only possible when $m = 1$ and λ is a q-th root of 1 (let us suppose that $K = \mathbb{C}$). Thus,

$$A^q = E \Longrightarrow CAC^{-1} = \begin{Vmatrix} \lambda_1 & & & 0 \\ & \lambda_2 & & \\ & & \ddots & \\ 0 & & & \lambda_n \end{Vmatrix}, \quad \lambda_i^q = 1, \tag{3}$$

for a suitable invertible matrix C. Alternately, this can be shown using the fact that the characteristic polynomial $f_A(t) = t^q - 1$ has no multiple roots.

These properties of a single linear operator $\mathfrak{a} : V \to V$ should be born in mind when we study a group $\Phi(g)$, $g \in G$, of linear operators.

Definition 3. Let (Φ, V) be a linear representation of a group G. A subspace $U \subset V$ is called G-invariant (or G-stable) if $\Phi(g)\, u \in U$ for all $u \in U$ and all $g \in G$. The zero subspace and the entire space V are called the trivial invariant subspaces. A representation all of whose invariant subspaces are trivial is called irreducible. A representation is called reducible if it has at least one non-trivial invariant subspace.

According to what was said above, if (Φ, V) is a reducible representation and U is an invariant subspace, then V has a basis relative to which

$$\Phi_g = \begin{Vmatrix} \Phi'_g & \Phi^0_g \\ 0 & \Phi''_g \end{Vmatrix} \tag{4}$$

for all $g \in G$. Since $\Phi'_{gh} = \Phi'_g \Phi'_h$, $\Phi'_e = E_k$ and $\Phi'_g(U) \subset U$, it follows that the map $\Phi' : g \mapsto \Phi'_g$ gives a representation on U, which is called a subrepresentation of Φ. In that case we also have a representation on V/U, which is called a quotient representation; it is given by the matrices Φ''_g, $g \in G$.

If it is possible to choose a basis of V in such a way that all of the matrices Φ^0_g in (4) are zero, then we say that Φ is the direct sum of the representations Φ' and Φ'': $\Phi = \Phi' \dotplus \Phi''$. A representation (Φ, V) has a direct sum decomposition if and only if it has an invariant subspace $U \subset V$ for which there is an invariant complement W, i.e., $V = U \oplus W$, where $\Phi(U) \subset U$ and $\Phi(W) \subset W$. In this case Φ' is the restriction $\Phi|_U$ of Φ to U and Φ'' is the restriction of Φ to W. A linear representation (Φ, V) is called indecomposable (and V is called an indecomposable G-space) if it cannot be written as a direct sum of two non-trivial subrepresentations.

If we successively write V, U, W, etc. as direct sums of invariant subspaces (when this is possible), we obtain a direct sum $V = V_1 \oplus \ldots \oplus V_r$ of invariant subspaces (equivalently, we obtain a direct sum $\Phi = \Phi^{(1)} \dotplus \cdots \dotplus \Phi^{(r)}$ of representations). For a suitable choice of basis in V, the matrices of the linear operators are of the form

$$\Phi_g = \begin{Vmatrix} \Phi_g^{(1)} & 0 & \ldots & 0 \\ 0 & \Phi_g^{(2)} & \ldots & 0 \\ \cdot & \cdot & \cdot & \cdot \\ 0 & 0 & \ldots & \Phi_g^{(r)} \end{Vmatrix}$$

Definition 4. A linear representation (Φ, V) of a group G is said to be completely reducible if it is a direct sum of irreducible representations. In that case we also call V a completely reducible G-space.

It is intuitively clear that the irreducible representations play the role of building blocks which are used to construct arbitrary linear representations. The completely reducible representations are obtained from them using the simplest construction -- the direct sum. We shall later see that in many cases this is sufficient for constructing all representations. We should remark that some groups which are important in physics, for example, the Lorenz group, have infinite dimensional irreducible representations. Of course, such representations cannot in any way be reduced to finite-dimensional ones, and they must be studied separately.

2. Examples of linear representations. We have introduced all of the basic concepts of representation theory. In order to acquire a solid understanding of these ideas, it is very useful to start by becoming familiar with (and taking pains to understand in depth) the following examples.

Example 1. By its definition, the general linear group $GL(n, K)$ over a field K has a faithful irreducible n-dimensional linear representation with representation space $V = K^n$. Any linear group $H \subset GL(n, K)$ acts faithfully on this V, but the action may be reducible.

Similar remarks apply to the other classical groups in §1 Ch. 7. For example, the unitary group $U(n)$ acts irreducibly on a Hermitian space, and the orthogonal group $O(n)$ acts on Euclidean space. This all follows immediately from the stronger assertion (proved in a basic linear algebra course) that the groups $U(n)$ and $O(n)$ act transitively (in the sense of Example 3 in Subsection 3 of §2 Ch. 7) on the set of vectors of unit length.

Example 2. If we make $GL(n,K)$ act on the vector space $M_n(K)$ of $n \times n$ matrices by the rule $\Psi_A : X \mapsto AX$ $(A \in GL(n,K),\ X \in M_n(K))$, we easily see that $\Psi_A(\alpha X + \beta Y) = \alpha\Psi_A X + \beta\Psi_A Y$ and $\Psi_{AB} = \Psi_A \Psi_B$. Hence, $(\Psi, M_n(K))$ is an n^2-dimensional linear representation. Let $M_n^{(i)}(K)$ be the subspace of matrices

$$\begin{Vmatrix} 0 & \dots & x_{1i} & \dots & 0 \\ \cdot & \cdot & \cdot & \cdot & \cdot \\ 0 & \dots & x_{ni} & \dots & 0 \end{Vmatrix}$$

with only one non-zero column $X^{(i)}$. It is easy to check that this subspace is invariant under Ψ_A for $A \in GL(n,K)$, is irreducible, and is isomorphic (as a $GL(n,K)$-space) to the natural $GL(n,K)$-space K^n in Example 1. Thus,

$$M_n(K) = M_n^{(1)}(K) \oplus \dots \oplus M_n^{(n)}(K)$$

is a direct sum decomposition of $M_n(K)$ into n isomorphic $GL(n,K)$-subspaces; it corresponds to a direct sum decomposition

$$\Psi = \Psi^{(1)} \dotplus \dots \dotplus \Psi^{(n)}$$

into n equivalent representations. Symbolically, this can be written

$$M_n(K) \cong n M_n^{(1)}(K)\,; \qquad \Psi \approx n\Psi^{(1)} \quad .$$

Example 3. We now define an action Φ of the group $GL(n,K)$ on $M_n(K)$ by setting $\Phi_A : X \mapsto AXA^{-1}$. Again $(\Phi, M_n(K))$ is an n^2-dimensional linear representation. If $X = (x_{ij})$, then, as usual, we let $\operatorname{tr} X = \sum_{i=1}^{n} x_{ii}$ denote the trace of X. It is well known that $\operatorname{tr}(\alpha X + \beta Y) = \alpha \operatorname{tr} X + \beta \operatorname{tr} Y$ (linearity of the trace function) and $\operatorname{tr} \Phi_A(X) = \operatorname{tr} X$. This implies that the set $M_n^0(K)$ of matrices with zero trace is a Φ-invariant subspace. On the other hand, $\Phi_A(\lambda E) = \lambda E$ and $\operatorname{tr} \lambda E = n\lambda$. Therefore, if K is a field of characteristic zero, we have a direct sum decomposition of $GL(n, K)$-subspaces

$$M_n(K) = \langle E \rangle \oplus M_n^0(K) \tag{5}$$

of dimension 1 and $n^2 - 1$, respectively. Note that if $n = p$ and $K = \mathbb{Z}_p$, then there is no decomposition (5), since in that case $\operatorname{tr} E = 0$.

According to the definition, the Jordan normal form $J(X)$ of a matrix X is nothing more nor less than a convenient and simple representative of the $GL(n, \mathbb{C})$-orbit containing X. If we restrict Φ to a subgroup $H \subset GL(n, K)$, we have the natural question of finding similar forms for representatives of the H-orbits.

<u>Example 4.</u> Set $K = \mathbb{R}$ in the previous example, and consider the restriction of Φ to the orthogonal group $O(n)$. Since $A \in O(n) \Leftrightarrow {}^tA = A^{-1}$, we have ${}^tX = \epsilon X$, $\epsilon = \pm 1 \Rightarrow {}^t(AXA^{-1}) = {}^tA^{-1}\,{}^tX\,{}^tA = \epsilon AXA^{-1}$. Hence, the representation space $M_n(\mathbb{R})$ for $O(n)$ can be written as the following sum of $O(n)$-subspaces:

$$M_n(\mathbb{R}) = \langle E \rangle_{\mathbb{R}} \oplus M_n^+(\mathbb{R}) \oplus M_n^-(\mathbb{R})$$

i.e., the sum of the one-dimensional space $\langle E \rangle_{\mathbb{R}}$ of scalar matrices, the $(n+2)(n-1)/2$-dimensional space of symmetric matrices with zero trace, and the $n(n-1)/2$-dimensional space of skew-symmetric matrices. There is a well-known one-to-one correspondence between the symmetric matrices (resp. skew-symmetric matrices) and the symmetric (resp. skew-symmetric) bilinear forms. The action of $O(n)$ on

$\langle E\rangle_{\mathbb{R}} \oplus M_n^+(\mathbb{R})$ and on $M_n^-(\mathbb{R})$ carries over to the spaces of the corresponding forms. The theorem on reducing a quadratic form $q(x)$ to diagonal form is equivalent to saying that in the orbit containing $q(x)$ one can choose a diagonal form $\Sigma\lambda_i x_i^2$ with real λ_i which are uniquely determined up to permuting their order.

If we replace $\mathbb{R}$ by $\mathbb{C}$ and replace $O(n)$ by the unitary group $U(n)$, we obtain the decomposition

$$M_n(\mathbb{C}) = \langle E\rangle_{\mathbb{C}} \oplus M_n^+(\mathbb{C}) \oplus M_n^-(\mathbb{C})$$

into the direct sum of the $U(n)$-subspaces of scalar matrices, hermitian matrices with zero trace, and skew-hermitian matrices. The case $n = 2$ was discussed in detail in §1 Ch. 7.

Example 5. Let G be a group of permutations acting on a set Ω of cardinality $|\Omega| = n > 1$, i.e., $G \subset S_n$. Let

$$V = \langle e_i \mid i \in \Omega\rangle_K$$

be the vector space over a field K of characteristic zero with basis indexed by the elements of the set Ω. We make V into a G-space by setting

$$\Phi(g)\left(\sum_{i\in\Omega} \lambda_i e_i\right) = \sum_{i\in\Omega} \lambda_i \Phi(g) e_i = \sum_{i\in\Omega} \lambda_i e_{g(i)}$$

($i \mapsto g(i)$ is the action of a permutation $g \in G$ on $i \in \Omega$). Since $(gh)(i) = g(h(i))$, we obtain an n-dimensional linear representation of G. It is never irreducible, since

$$V = \langle \sum_{i\in\Omega} e_i\rangle \oplus \left\{ \sum\lambda_i e_i \;\middle|\; \sum\lambda_i = 0,\ \lambda_i \in K\right\} \tag{6}$$

decomposes into the direct sum of a one-dimensional invariant subspace and an $(n-1)$-dimensional subspace. (If $\operatorname{char} K = p > 0$ and $p \mid n$, then we no longer obtain this direct sum.)

We consider two special cases.

(a) $G = S_n$. The monomorphism $S_n \to GL(n, \mathbb{R})$ in Subsection 5 of §3 Ch. 4 coincides with our linear representation Φ if we take the i-th coordinate column $E^{(i)}$ for e_i. The decomposition (6) shows that we have a more economical imbedding $S_n \to GL(n-1, \mathbb{Q})$. We shall later show that this (n - 1)-dimensional linear representation is irreducible (even over the field $\mathbb{C}$).

(b) The regular representation. Let G be any finite group. If we set $\Omega = G$, we obtain the so-called regular G-space $V = \langle e_g \mid g \in G \rangle$ and the corresponding regular representation (ρ, V) of the group $G : \rho(a) e_g = e_{ag}$ for all $a, g \in G$. We already encountered the regular representation in somewhat different notation in the proof of Cayley's theorem (§3 Ch. 4), but at that time we were just interested in the set $\{e_g\}$ and not in the space V. The regular representation of a finite group G is important because it contains all of the irreducible representations of G (up to equivalence), as we shall see in §5.

Example 6. A one-dimensional representation is simply a homomorphism $\Phi : G \to K^*$ from the group G to the multiplicative group of the field K (K is a one-dimensional vector space over itself, and $GL(1, K) = K^*$). Since the multiplicative group of a field is abelian, it follows that $\operatorname{Ker} \Phi \supset G'$, where G' is the commutant of G (Theorem 4 of §3 Ch. 7). Note that equivalence of two one-dimensional representations Φ' and Φ'' (having the same representation space) is the same as equality, since $a\Phi'(g)a^{-1} = \Phi''(g) \Rightarrow \Phi'(g) = \Phi''(g) \Rightarrow \Phi' = \Phi''$. Suppose that $g^n = e$. Then $\Phi(g)^n = \Phi(g^n) = \Phi(e) = 1$, i.e., $\Phi(g)$ is a root of unity. If $K = \mathbb{C}$, we shall see that every cyclic group has a faithful one-dimensional representation. But in general it may happen that even a homomorphism from a cyclic group to K^* always has a non-trivial kernel; for example, let $G = \mathbb{Z}_4$ and $K = \mathbb{Z}_7$, in which case always $\operatorname{Ker} \Phi \supset 2\mathbb{Z}_4$.

(a) $G = (\mathbb{Z}, +)$, $K = \mathbb{C}$. The representation $k \mapsto \lambda^k$ is faithful if $|\lambda| \neq 1$. If $|\lambda| = 1$, then by Euler's formula $\lambda = e^{2\pi i\theta}$, $\theta \in \mathbb{R}$, and the map $k \mapsto e^{2\pi i\theta k}$ has a non-trivial kernel if and only if $\theta \in \mathbb{Q}$.

To find complex representations of the group $\mathbb{Z}$ of arbitrarily high dimension which are indecomposable (but not irreducible), we can use the Jordan normal form of a matrix, and consider the map

$$k \longmapsto J_{m,1}^{k} = \begin{Vmatrix} 1 & 1 & 0 & \dots & 0 & 0 \\ 0 & 1 & 1 & \dots & 0 & 0 \\ & \cdot & \cdot & \cdot & \cdot & \cdot \\ 0 & 0 & 0 & \dots & 1 & 1 \\ 0 & 0 & 0 & \dots & 0 & 1 \end{Vmatrix}^{k} .$$

(b) $G = \langle a \mid a^n = e \rangle$, $K = \mathbb{C}$. Let $\epsilon = e^{2\pi i/n}$ be a primitive n-th root of one. Out of the n one-dimensional representations

$$\Phi^{(m)} : a^k \longmapsto \epsilon^{mk}, \qquad m = 0, 1, \dots, n-1 \,, \tag{7}$$

exactly $\varphi(n)$ of them are faithful. We note the following interesting fact: <u>a cyclic group of order n has exactly n non-equivalent irreducible representations over $\mathbb{C}$. They are all one-dimensional, and they have the form (7).</u> Indeed, it suffices to show that a finite cyclic group has no irreducible complex representations of dimension > 1. But before giving Definition 3 we noted that any linear operator $\Phi(g)$ of finite order is diagonalizable over $\mathbb{C}$. In the present situation, this gives us complete reducibility of Φ. <u>If $\dim \Phi = r$, then Φ decomposes into a direct sum of r one-dimensional representations.</u>

We have thus obtained a description of all complex linear representations of a cyclic group of finite order. Up to equivalence we have

$$\Phi_g = \begin{Vmatrix} \Phi_g^{(i_1)} & & 0 \\ & \ddots & \\ 0 & & \Phi_g^{(i_r)} \end{Vmatrix} ,$$

where $\Phi^{(m)}$ is one of the representations of the form (7).

We would like to establish similar rules for more general cases.

Example 7. We have already noted in the above examples how the properties of a linear representation Φ of a group G can depend strongly on the ground field K. We now clarify this question somewhat.

If we let the cyclic group $G = \langle a \mid a^p = e \rangle$ of prime order p act on the two-dimensional vector space $V = \langle v_1, v_2 \rangle$ over a field K of characteristic p according to the rule $a * v_1 = v_1$, $a * v_2 = v_1 + v_2$, we obtain an indecomposable representation (Φ, V)

$$a^k \longmapsto \Phi_a^k = \begin{Vmatrix} 1 & k \\ 0 & 1 \end{Vmatrix}, \qquad 0 \leq k \leq p-1 .$$

In fact, the matrix Φ_a has characteristic root 1 with multiplicity 2. Hence, if Φ decomposed into a direct sum of two one-dimensional representations, there would exist an invertible matrix C for which $C\Phi_a C^{-1} = \begin{Vmatrix} 1 & 0 \\ 0 & 1 \end{Vmatrix} = E$. But then $\Phi_a = C^{-1} E C = E$, which is false.

Now let $G = \langle a \mid a^3 = e \rangle$ be a cyclic group of order 3, and let $K = \mathbb{R}$. The two-dimensional representation (Φ, V), $V = \langle v_1, v_2 \rangle$ which is defined in this basis by the matrix

$$\Phi_a = \begin{Vmatrix} -1 & -1 \\ 1 & 0 \end{Vmatrix},$$

is irreducible, since the characteristic polynomial $t^2 + t + 1$ of this matrix does not have real roots. But if we consider V over $\mathbb{C}$, then, of course, V decomposes into a sum of one-dimensional G-subspaces

$$V = \langle v_1 + \varepsilon^{-1} v_2 \rangle \oplus \langle v_1 + \varepsilon v_2 \rangle$$

and we have

$$C\Phi_a C^{-1} = \begin{Vmatrix} \varepsilon & 0 \\ 0 & \varepsilon^{-1} \end{Vmatrix} ; \quad \varepsilon = \frac{-1 + \sqrt{-3}}{2} , \quad C = \begin{Vmatrix} 1 & -\varepsilon^{-1} \\ 1 & -\varepsilon \end{Vmatrix} .$$

Thus, we may lose irreducibility of a representation when we extend the field.

In what follows, with rare exceptions, we shall take the ground field K to be the field of complex numbers (which is the most important case from a practical point of view), or else an arbitrary algebraically closed field of characteristic zero.

EXERCISES

1. The group $SO(2)$ is defined by its natural two-dimensional representation

$$\Phi'(\theta) = \begin{Vmatrix} \cos\theta & -\sin\theta \\ \sin\theta & \cos\theta \end{Vmatrix} ,$$

which is irreducible over $\mathbb{R}$. Verify that

$$A\Phi'(\theta)A^{-1} = \begin{Vmatrix} e^{i\theta} & 0 \\ 0 & e^{-i\theta} \end{Vmatrix} \quad \text{for} \quad A = \frac{1}{\sqrt{2}} \begin{Vmatrix} 1 & i \\ i & 1 \end{Vmatrix} \in GL(2, 0) .$$

Hence, Φ' is a direct sum of two non-equivalent (which in this situation simply means distinct) one-dimensional representations.

2. Is the $GL(n, \mathbb{C})$-space $M_n^0(\mathbb{C})$ in the decomposition (5) irreducible when $n = 2$ and 3? (Answer: yes.)

3. Let Φ and Ψ be irreducible complex representations of a cyclic group $\langle a \mid a^n = e \rangle$ of order n. Show that

$$\frac{1}{n} \sum_{k=0}^{n-1} \Phi(a^k)\overline{\Psi(a^k)} = \begin{cases} 1, & \text{if } \Phi \approx \Psi, \\ 0, & \text{if } \Phi \not\approx \Psi . \end{cases}$$

4. Use Exercise 3 to prove the following assertion. Any complex-valued function f on a finite cyclic group $\langle a \mid a^n = e \rangle$ can be expanded "in simple harmonics" as follows:

$$f(a^k) = \sum_{m=0}^{n-1} c_m \varepsilon^{mk} , \qquad \varepsilon = e^{\frac{2\pi i}{n}} .$$

The "Fourier coefficients" c_m are computed according to the formula

$$c_m = \frac{1}{n} \sum_{k=0}^{n-1} f(a^k) \varepsilon^{-mk} .$$

5. Use the formula for the number of necklaces (see the beginning of the chapter) to prove: (a) $q^p - q \equiv 0 \pmod p$ (Fermat's Little Theorem; see §4 Ch. 4); (b) $\sum_{d|n} \varphi(d) = n$.

§2. Unitary and reducible representations

1. Unitary representations. Recall from linear algebra that a non-degenerate form $(u,v) \mapsto (u|v)$ on a vector space V over $\mathbb{C}$ is called hermitian if

$$\begin{aligned} (u|v) &= \overline{(v|u)} , \\ (\alpha u + \beta v|w) &= \alpha(u|w) + \beta(v|w) , \\ (v|v) &> 0 \quad \text{for all} \quad v \neq 0 \end{aligned} \tag{1}$$

(as always, $z \mapsto \bar{z}$ denotes complex conjugation). The space V, considered together with a non-degenerate hermitian form $(u|v)$, is called a hermitian space. The analog over $\mathbb{R}$ is euclidean space with a scalar product given by a non-degenerate symmetric bilinear form. If we take a basis $e_1, \dots, e_n$ for V, we can write the form $(u|v)$ for $u = \Sigma\, u_i e_i$ and $v = \Sigma\, v_i e_i$ as follows:

$$(u|v) = \sum h_{ij} u_i \bar{v}_j .$$

The matrix $H = (h_{ij})$ satisfies the condition $\bar{h}_{ij} = h_{ji}$. Such a matrix is also called hermitian. We have already used this terminology in §1 Ch. 7.

There exists an orthonormal basis (i. e., such that $(e_i | e_j) = \delta_{ij}$) relative to which

$$(u|v) = \sum_{i=1}^{n} u_i \overline{v}_i \quad .$$

A linear operator $\mathcal{A} : V \to V$ which preserves this form, i. e., such that $(\mathcal{A}u | \mathcal{A}v) = (u|v)$, is called a <u>unitary</u> operator. The analogy over $\mathbb{R}$ is the <u>orthogonal</u> operators. In Chapter 7 we already encountered the unitary condition, written in matrix form, i. e., $A \cdot {}^t\overline{A} = E$ with $A = (a_{ij})$, ${}^t\overline{A} = A^* = (\overline{a}_{ij})$. If we let $\mathcal{A}^*$ denote the linear operator with matrix ${}^t\overline{A} = A^*$, then we can express the unitary condition in the form $\mathcal{A} \cdot \mathcal{A}^* = \mathcal{E} = \mathcal{A}^* \cdot \mathcal{A}$.

It is customary to let $U(n)$ denote the group of all unitary matrices (also called the group of unitary operators, or simply the unitary group). By definition, $U(n) \subset GL(n, \mathbb{C})$. <u>If a representation</u> $\Phi : G \to GL(n, \mathbb{C})$ <u>has the property that</u> $\operatorname{Im} \Phi \subset U(n)$, <u>then</u> (Φ, V) <u>is called a unitary representation.</u>

THEOREM 1. <u>If</u> G <u>is a finite group, then every linear representation</u> (Φ, V) <u>of</u> G <u>over</u> $\mathbb{C}$ <u>is equivalent to a unitary representation.</u>

<u>Proof.</u> In the representation space V choose any non-degenerate hermitian form $H : (u,v) \mapsto H(u,v) = \Sigma h_{ij} u_i \overline{v}_j$ (in terms of a basis $f_1, \ldots, f_n$ for V). Consider the form $(u|v)$ obtained from $H(u,v)$ by "averaging over G":

$$(u|v) = |G|^{-1} \sum_{g \in G} H(\Phi(g)u, \Phi(g)v) \quad . \tag{2}$$

The factor $|G|^{-1}$ is not essential, and is only inserted so that, if H is already unitary, we get $(u|v) = H(u,v)$. Since

$$H(\Phi(g)u\,,\ \Phi(g)v) = \overline{H(\Phi(g)v\,,\ \Phi(g)u)}\ ,$$

$$H(\Phi(g)(\alpha u + \beta v)\,,\ \Phi(g)w) =$$

$$= H(\alpha\Phi(g)u + \beta\Phi(g)v\,,\ \Phi(g)w) =$$

$$= \alpha H(\Phi(g)u\,,\ \Phi(g)w) + \beta H(\Phi(g)v\,,\ \Phi(g)w)\ ,$$

$$H(\Phi(g)v\,,\ \Phi(g)v) > 0$$

for $v \neq 0$ and all $g \in G$, it follows that the form (2) satisfies the conditions in (1), and so is a non-degenerate hermitian form. In addition (and this is what is most important),

$$(\Phi(g)u \mid \Phi(g)v) =$$

$$= |G|^{-1} \sum_{h \in G} H(\Phi(h)\Phi(g)u\,,\ \Phi(h)\Phi(g)v) =$$

$$= |G|^{-1} \sum_{h \in G} H(\Phi(hg)u\,,\ \Phi(hg)v) =$$

$$= |G|^{-1} \sum_{t \in G} H(\Phi(t)u\,,\ \Phi(t)v) = (u \mid v)\ ,$$

i.e., for any $g \in G$ the operator $\Phi(g)$ leaves the form $(u|v)$ invariant. Choose a basis $e_1, \ldots, e_n$ in V which is orthonormal relative to the form $(u|v)$. Then the matrices Φ_g of the operators $\Phi(g)$ will be unitary in this basis. □

Remarks. (1) Theorem 1 does not follow automatically from the (much weaker) fact that we knew before which says that each individual matrix Φ_g with $g^m = e$ is similar to a unitary matrix $\operatorname{diag}\{\lambda_1, \ldots, \lambda_n\}$ with $\lambda_i^m = 1$.

(2) In the real case, a completely analogous argument shows that every linear representation of a finite group is equivalent to an orthogonal representation.

(3) For a variety of reasons, unitary representations play an important role in applications of representation theory. Remarkably, Theorem 1 remains true for a much broader class of groups, for example, for $G = U(n)$ or $O(n)$. The proof is the same, except that the summation over the elements of G is replaced by integration (suitably

defined) over the compact group G. Recall that the compact group $SU(2)$ is geometrically indistinguishable from the three-dimensional sphere S^3, and so it makes sense to speak, for example, about its volume. In general, there is a remarkable parallel in representation theory between finite and compact groups, but we cannot dwell on this here. It is clear from Example 6a of §1 that representations of non-compact groups (such as $G = \mathbb{Z}$) need not be unitary.

In conclusion, we note that, while the proof of Theorem 1 is constructive, it would not be very practical to use it to find a unitary realization of a given representation. For example, if G is generated by elements $a_1, \ldots, a_d$, then it would be sufficient to find a representation for which the matrices $\Phi_{a_1}, \ldots, \Phi_{a_d}$ are unitary, since in that case $\operatorname{Im}\Phi = \langle \Phi_{a_1}, \ldots, \Phi_{a_d} \rangle \subset U(n)$.

Example 1. The symmetric group $S_3 = \langle (12), (123) \rangle$ has a two-dimensional representation Φ which is a direct summand in the natural three-dimensional representation (see Example 5 of §1). Namely, if $\Phi(\pi)e_i = e_{\pi(i)}$, $i = 1,2,3$, and $f_1 = e_1 - e_3$, $f_2 = e_2 - e_3$, then

$$\Phi((12))f_1 = e_2 - e_3 = f_2, \qquad \Phi((12)f_2 = e_1 - e_3 = f_1,$$

$$\Phi((123))f_1 = e_2 - e_1 = -f_1 + f_2, \qquad \Phi((123))f_2 = e_3 - e_1 = -f_1.$$

Since $\pi = (123)^i(12)^j$, where $i = 0, 1$, or 2 and $j = 0$ or 1, we easily obtain all of the matrices

$$e \longmapsto \begin{Vmatrix} 1 & 0 \\ 0 & 1 \end{Vmatrix}, \quad (12) \longmapsto \begin{Vmatrix} 0 & 1 \\ 1 & 0 \end{Vmatrix}, \quad (13) \longmapsto \begin{Vmatrix} -1 & -1 \\ 0 & 1 \end{Vmatrix},$$

$$(23) \longmapsto \begin{Vmatrix} 1 & 0 \\ -1 & -1 \end{Vmatrix}, \quad (123) \longmapsto \begin{Vmatrix} -1 & -1 \\ 1 & 0 \end{Vmatrix}, \quad (132) \longmapsto \begin{Vmatrix} 0 & 1 \\ -1 & -1 \end{Vmatrix}.$$

Since $\det \begin{Vmatrix} -1 & -1 \\ 1 & 0 \end{Vmatrix} = 1$ and $(123)^3 = e$, it follows that

$$C\begin{Vmatrix} -1 & -1 \\ 1 & 0 \end{Vmatrix}C^{-1} = \begin{Vmatrix} \varepsilon & 0 \\ 0 & \varepsilon^{-1} \end{Vmatrix}, \qquad \varepsilon = \frac{-1+\sqrt{-3}}{2},$$

for some non-singular matrix C. If we conjugate $\begin{Vmatrix} 0 & 1 \\ 1 & 0 \end{Vmatrix}$ by C, we do not lose the unitary property of this matrix. Solving the linear equations

$$C\begin{Vmatrix} 0 & 1 \\ 1 & 0 \end{Vmatrix} = \begin{Vmatrix} 0 & 1 \\ 1 & 0 \end{Vmatrix}C, \quad C\begin{Vmatrix} -1 & -1 \\ 1 & 0 \end{Vmatrix} = \begin{Vmatrix} \varepsilon & 0 \\ 0 & \varepsilon^{-1} \end{Vmatrix}C, \quad C = \begin{Vmatrix} \alpha & \beta \\ \gamma & \delta \end{Vmatrix}$$

for the entries of C, we obtain:

$$C = \begin{Vmatrix} 1 & -\varepsilon^2 \\ -\varepsilon^2 & 1 \end{Vmatrix}.$$

We can now write out a table of all of the unitary representations of S_3 which we know: the trivial representation $\Phi^{(1)}$, the representation $\Phi^{(2)}: \pi \mapsto \operatorname{sgn}\pi \in \{\pm 1\}$, and the two-dimensional representation $\Phi^{(3)}$ which we just found. The following table is convenient for future reference:

Φ \ g	e	(12)	(13)	(23)	(123)	(132)
$\Phi^{(1)}$	1	1	1	1	1	1
$\Phi^{(2)}$	1	-1	-1	-1	1	1
$\Phi^{(3)}$	$\begin{Vmatrix} 1 & 0 \\ 0 & 1 \end{Vmatrix}$	$\begin{Vmatrix} 0 & 1 \\ 1 & 0 \end{Vmatrix}$	$\begin{Vmatrix} 0 & \varepsilon \\ \varepsilon^{-1} & 0 \end{Vmatrix}$	$\begin{Vmatrix} 0 & \varepsilon^{-1} \\ \varepsilon & 0 \end{Vmatrix}$	$\begin{Vmatrix} \varepsilon & 0 \\ 0 & \varepsilon^{-1} \end{Vmatrix}$	$\begin{Vmatrix} \varepsilon^{-1} & 0 \\ 0 & \varepsilon \end{Vmatrix}$

Example 2. The epimorphism $\Phi: SU(2) \to SO(3)$ that was constructed in §1 Ch. 7 can be thought of as a natural orthogonal representation of the infinite group $SU(2)$.

2. Complete reducibility. The following fact is fundamental, as should be clear from the remarks and definitions in §1.

THEOREM 2. Every linear representation of a finite group G over a field K of characteristic zero or of characteristic not dividing $|G|$, is completely reducible.

Recall that this means that the representation (Φ, V) can be decomposed into a direct sum of irreducible representations. Actually, the classical version of Theorem 2 is as follows:

Every G-invariant subspace $U \subset V$ has a G-invariant complement W:

$$V = U \oplus W \quad . \tag{3}$$

It is this assertion which we shall prove. Theorem 2 will then follow immediately, since either (Φ, V) is irreducible, in which case there is nothing to prove, or else there exists a proper G-invariant subspace U, in which case (3) holds for some G-subspace W. Then $\dim U < \dim V$ and $\dim W < \dim V$. Applying the same argument to U and W, and using induction on the dimension of the representation, we obtain the required decomposition into irreducible components.

Thus, it suffices to prove that every G-invariant U has a G-invariant complement. As usual, we are most interested in the case $K = \mathbb{C}$, so it is useful to give two independent proofs.

First proof $(K = \mathbb{C})$. By Theorem 1, there exists a non-degenerate hermitian form $(u|v)$ on the representation space V which is invariant with respect to the linear operators $\Phi(g)$. For every subspace $U \subset V$ there exists an orthogonal complement

$$U^{\perp} = \{v \in V \,|\, (u|v) = 0, \quad \forall u \in U\} \quad ,$$

and, as is well known from linear algebra, we have

$$V = U \oplus U^{\perp} \quad ,$$

and also $(U^{\perp})^{\perp} = U$. Now suppose that U is a G-subspace of V, i.e., that $\Phi(g)U \subset U$ for all $g \in G$. Since $\Phi(g)\big|_U$ is an automorphism, it follows that any element $u \in U$ can be written in the form $u = \Phi(g)u'$, $u' \in U$. We now use the invariance of the form $(u|v)$:

$$v \in U^{\perp} \Longrightarrow (u|\Phi(g)v) = (\Phi(g)u'|\Phi(g)v) = (u'|v) = 0 \quad .$$

Thus, $v \in U^{\perp} \Rightarrow \Phi(g)v \in U^{\perp}$. Setting $W = U^{\perp}$, we obtain (3). □

Second proof. As before, let U be a subspace of V which is invariant under the action of G. Consider the direct sum

$$V = U \oplus U' \quad ,$$

where U' is any complement of U. In general, U' is not G-invariant. Consider the projection $P : V \to U'$, which is defined by $Pv = u'$ for every vector $v = u + u'$. We have

$$v - Pv \in U \,, \qquad P(U) = 0 \,, \qquad P^2 = P \quad . \tag{4}$$

We now introduce the linear "averaging" operator

$$P_G = |G|^{-1} \sum_{h \in G} \Phi(h)\, P\, \Phi(h^{-1})$$

(by our assumption regarding char K, we are allowed to divide by $|G|$). We have

$$\Phi(g)\, P_G = P_G\, \Phi(g) \,, \quad \forall g \in G \quad . \tag{5}$$

To establish (5), we verify that

$$\Phi(g)\, P_G\, \Phi(g^{-1}) = |G|^{-1} \sum_{h \in G} \Phi(g)\, \Phi(h)\, P\, \Phi(h^{-1})\, \Phi(g^{-1}) =$$

$$= |G|^{-1} \sum_{h \in G} \Phi(gh)\, P\, \Phi((gh)^{-1}) = |G|^{-1} \sum_{t \in G} \Phi(t)\, P\, \Phi(t^{-1}) = P_G \quad ,$$

as required. We set

$$W = P_G(V) = \{P_G v \,|\, v \in V\} \quad .$$

According to (5), we have $\Phi(g)w = \Phi(g)P_G v = P_G \Phi(g) v = P_G v' = w' \in W$ for every $w \in W$, so that the vector subspace $W \subset V$ is actually a G-subspace.

It remains to show that $V = U \oplus W$. Since $\Phi(h^{-1})v - P\Phi(h^{-1})v \in U$ (see (4)), it follows that $v - \Phi(h)P\Phi(h^{-1})v = \Phi(h)\{\Phi(h^{-1})v - P\Phi(h^{-1})v\} \in \Phi(h)U = U$ (by the invariance of U). Consequently,

$$v - P_G v = |G|^{-1} \sum_{h \in G} (v - \Phi(h)P\Phi(h^{-1})v) = u \in U \quad ,$$

and we obtain $v = u + w$ with $w = P_G v \in W$, i.e., $V = U + W$.

Next, we have $\Phi(h^{-1})U \subset U \Rightarrow P\Phi(h^{-1})U = 0$ (by (4)) $\Rightarrow \Phi(h)P\Phi(h^{-1})U = 0 \Rightarrow P_G(U) = 0$. Hence, $v - P_G v = u \in U \Rightarrow P_G(v - P_G v) = 0$, so that $P_G v = P_G^2 v$ for all $v \in V$. This means that P_G is projection along U onto W:

$$P_G(U) = 0 , \qquad P_G^2 = P_G \quad . \tag{6}$$

Now $v \in U \cap W \Rightarrow P_G v = 0$, since $v \in U$, and $v = P_G v'$, since $v \in W = P_G(V)$. Using (6), we obtain $0 = P_G v = P_G(P_G v') = P_G^2 v' = P_G v' = v \Rightarrow U \cap W = 0$. □

We would not be justified in making the stronger assertion that the decomposition into irreducible components (irreducible G-spaces) is unique. For example, if $\Phi(g) = \mathcal{E}$ is the identity for all $g \in G$, then any direct sum decomposition of V into one-dimensional subspaces is a decomposition into irreducible G-spaces, and there are infinitely many such decompositions. But suppose that we group together all of the isomorphic irreducible components, and write

$$V = U_1 \oplus \ldots \oplus U_s \quad .$$

Since we do not distinguish between isomorphic G-spaces, we may suppose that

$$\begin{aligned} U_1 &= V_1 \oplus V_1 \oplus \ldots \oplus V_1 = n_1 V_1 \quad , \\ &\ldots\ldots\ldots\ldots\ldots\ldots \\ U_s &= V_s \oplus V_s \oplus \ldots \oplus V_s = n_s V_s \quad , \end{aligned}$$

where n_i is the <u>multiplicity</u> with which the irreducible component V_i occurs in V. We shall see that these multiplicities are uniquely determined.

EXERCISES

1. Every one-dimensional continuous representation of the group $(\mathbb{R}, +)$ (i.e., such that nearby numbers correspond to nearby operators) has the form $\Phi^{(\alpha)} : t \mapsto e^{i\alpha t}$, where α is a complex number. Show that $\Phi^{(\alpha)}$ is unitary if and only if $\alpha \in \mathbb{R}$.

2. The kernel of the homomorphism $f : t \mapsto \begin{Vmatrix} \cos t & -\sin t \\ \sin t & \cos t \end{Vmatrix}$ from the group $(\mathbb{R}, +)$ to $SO(2)$ consists of the numbers $t = 2\pi m$, $m \in \mathbb{Z}$. Thus, $SO(2) \cong \mathbb{R}/2\pi\mathbb{Z}$, and to every irreducible unitary representation Φ of the group $SO(2)$ (by the results in §4, such a representation must be one-dimensional) there corresponds an irreducible unitary representation $\widetilde{\Phi} : t + 2\pi m \mapsto \Phi(t)$, $0 \leq t < 2\pi$, of the group $(\mathbb{R}, +)$, for which $\widetilde{\Phi}(2\pi) = \widetilde{\Phi}(0) = 1$. Use Exercise 1 to show that $\widetilde{\Phi} = \Phi^{(n)}$ for some $n \in \mathbb{Z}$. Together with Remark 3) in Subsection 1, this means that every irreducible representation of $SO(2)$ has the form $\Phi^{(n)}(t) = e^{int}$, $n \in \mathbb{Z}$. Verify that

$$\frac{1}{2\pi} \int_0^{2\pi} e^{ikt} \cdot \overline{e^{i\ell t}}\, dt = \delta_{k\ell}$$

(compare with the relation in Exercise 3 of §1 : the order n has been replaced by the "volume" 2π of the group $SO(2)$). In analysis, the set of functions $\{e^{int}\}$ is the classic example of a complete orthonormal system of periodic functions (i.e., functions on the circle $S^1 \sim SO(2)$). This is the point of departure for the theory of Fourier series.

3. Use Theorem 2 to prove that any faithful two-dimensional complex representation of a finite non-abelian group is irreducible.

§3. Finite rotation groups

In this section we shall be concerned with finite subgroups of the group $SO(3)$. In the process of determining these groups, we shall obtain the irreducible orthogonal representations of such groups as A_4, S_4, A_5 in an easily remembered geometrical setting. Subsection 1 and the proof of Theorem 2 can be omitted in a first reading, but the reader who really wants to be sure of having a firm grasp of the idea of "group actions" (§2 Ch. 7) would be well-advised to become familiar with the entire section.

1. The orders of finite subgroups of SO(3). According to Euler's theorem of linear algebra, every element $\mathcal{A} \in SO(3)$, $\mathcal{A} \neq \mathcal{E}$, is a rotation in $\mathbb{R}^3$ about some axis. In other words, there are precisely two points on the two-dimensional unit sphere S^2 which are left fixed by $\mathcal{A}$, namely, the points of intersection of the sphere and the axis. These two points are called the poles of the rotation $\mathcal{A}$.

Now let G be a finite subgroup of $SO(3)$, and let S be the set of poles of all of the rotations (besides the identity) in G. It is clear that G acts like a permutation group on the set S. If x is a pole for some rotation $\mathcal{A} \neq \mathcal{E}$, $\mathcal{A} \in G$, then for any $\mathcal{B} \in G$ we have

$$(\mathcal{B}\mathcal{A}\mathcal{B}^{-1})\mathcal{B}x = \mathcal{B} \cdot \mathcal{A}x = \mathcal{B}x ,$$

i.e., $\mathcal{B}x$ is a pole for $\mathcal{B}\mathcal{A}\mathcal{B}^{-1}$, and so $\mathcal{B}x \in S$. We let Ω denote the set of all ordered pairs $(\mathcal{A}, x)$, where $\mathcal{A} \in G$, $\mathcal{A} \neq \mathcal{E}$, and x is a pole for $\mathcal{A}$. Further, let G_x be the stationary subgroup (stabilizer) of the point x, i.e., the subgroup of all elements of G which leave x fixed. If

$$G = G_x \cup g_2 G_x \cup \dots \cup g_{m_x} G_x$$

is the partition of G into left cosets of G_x, then the G-orbit of x is the set

$$G(x) = \{x, g_2 x, \dots, g_{m_x} x\}$$

containing $|G(x)| = m_x$ elements. By Lagrange's theorem, $N = m_x n_x$, where $N = |G|$ and $n_x = |G_x|$ (we are using somewhat different notation from that in §1 Ch. 7). Note that n_x is the order of a cyclic subgroup of G, each of whose elements is a rotation about the axis through x. We say that n_x is the multiplicity of the pole x, or that x is an n_x-pole.

Every element $\mathfrak{a} \neq \mathcal{E}$ in G has two poles; hence $|\Omega| = 2(N-1)$.

On the other hand, for each pole x there are $n_x - 1$ elements (besides the identity) in G which leave x fixed. Consequently, the number of pairs $(\mathfrak{a}, x)$ is equal to the sum

$$|\Omega| = \sum_{x \in S} (n_x - 1) \quad .$$

If we let $\{x_1, \ldots, x_k\}$ be a set of poles taken one from each orbit, set $n_i = n_{x_i}$ and $m_i = m_{x_i}$, and note that $n_x = n_{x_i} = n_i$ for all $x \in G(x_i)$, we obtain

$$|\Omega| = \sum_{x \in S} (n_x - 1) = \sum_{i=1}^{k} m_i(n_i - 1) = \sum_{i=1}^{k} (N - m_i) \quad .$$

Thus,

$$2N - 2 = \sum_{i=1}^{k} (N - m_i) \quad .$$

Dividing through by N, we obtain

$$2 - \frac{2}{N} = \sum_{i=1}^{k} \left(1 - \frac{1}{n_i}\right) \quad . \tag{1}$$

We suppose that $N > 1$, so that $1 \leq 2 - \frac{2}{N} < 2$. Since $n_i \geq 2$, we have $1/2 \leq 1 - \frac{1}{n_i} < 1$, and so k must equal 2 or 3.

Case 1. $k = 2$. Then

$$2 - \frac{2}{N} = \left(1 - \frac{1}{n_1}\right) + \left(1 - \frac{1}{n_2}\right) \quad ,$$

or, equivalently,

$$2 = \frac{N}{n_1} + \frac{N}{n_2} = m_1 + m_2 \quad ,$$

so that $m_1 = m_2 = 1$ and $n_1 = n_2 = N$. Hence, G has precisely one axis of rotation, and $G = C_N$ is a cyclic group of order N.

Case 2. $k = 3$. To be definite, suppose that $n_1 \leq n_2 \leq n_3$. If we had $n_1 \geq 3$, then we would have

$$\sum_{i=1}^{3}\left(1 - \frac{1}{n_i}\right) \geq \sum_{i=1}^{3}\left(1 - \frac{1}{3}\right) = 2 \quad ,$$

which is impossible. Thus, $n_1 = 2$, and equation (1) can be written in the form

$$\frac{1}{2} + \frac{2}{N} = \frac{1}{n_2} + \frac{1}{n_3} \quad .$$

Obviously, $n_2 \geq 4 \Rightarrow \frac{1}{n_2} + \frac{1}{n_3} \leq \frac{1}{2}$, a contradiction. Hence, $n_2 = 2$ or 3.

If $n_2 = 2$, then $n_3 = N/2 = m$ (N must be even), and $m_1 = m_2 = m$, $m_3 = 2$. These data correspond to the dihedral group D_m (see Example 1 in Subsection 5 §3 Ch. 7).

If $n_2 = 3$, then

$$\frac{1}{6} + \frac{2}{N} = \frac{1}{n_3} \quad ,$$

and we only have three possibilities:

2') $n_3 = 3$, $N = 12$, $m_1 = 6$, $m_2 = 4$, $m_3 = 4$;

2'') $n_3 = 4$, $N = 24$, $m_1 = 12$, $m_2 = 8$, $m_3 = 6$;

2''') $n_3 = 5$, $N = 60$, $m_1 = 30$, $m_2 = 20$, $m_3 = 12$.

We collect all of this information in the following table:

N	Number of orbits	$\lvert S \rvert$	orders of the stabilizers		
n	2	2	n	n	—
2 m	3	2 m + 2	2	2	m
12	3	14	2	3	3
24	3	26	2	3	4
60	3	62	2	3	5

(2)

We have proved the following fact.

THEOREM 1. Let G be a finite subgroup of $SO(3)$ which is not cyclic or dihedral. Then there are only three possibilities for $N = |G|$: $N = 12, 24, 60$. Other conditions satisfied by G are given in the table (2). □

2. Symmetry groups for regular polyhedra. It is not hard to prove the existence of groups of order 12, 24, 60 (which are not cyclic or dihedral) which are contained in $SO(3)$. There are only five regular convex polyhedra in $\mathbb{R}^3$ (up to similarity). They have been known since antiquity; they are: the tetrahedron Δ_4, the cube $\square_6$, the octahedron Δ_8, the dodecahedron $⬠_{12}$, and the icosahedron Δ_{20}:

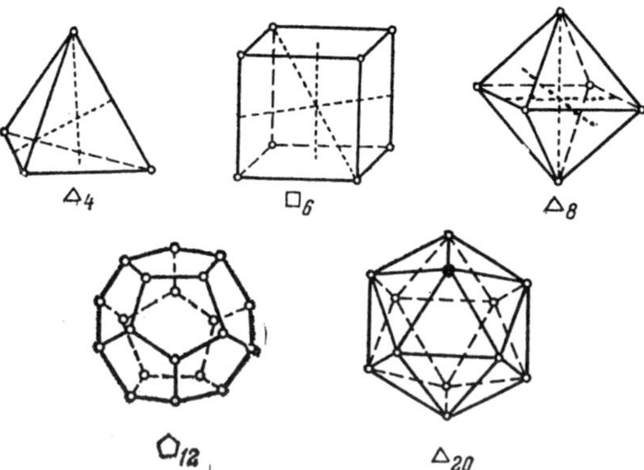

If the center of a regular polyhedron M is placed at the origin in $\mathbb{R}^3$, then the rotations in SO(3) which take M to itself form a finite subgroup. But instead of five, we only obtain three different (i.e., non-isomorphic) groups, since the cube and octahedron, and also the dodecahedron and icosahedron, lead to the same group. This is very easy to see geometrically. If we join the centers of adjacent faces of the cube with line segments, then these line segments are the edges of an octahedron inscribed in the cube. Every rotation in $\mathbb{R}^3$ which takes the cube to itself also takes the inscribed octahedron to itself, and conversely. A similar observation applies to the dodecahedron and icosahedron. In the table below, N_0 is the number of vertices of the polyhedron, N_1 is the number of edges, N_2 is the number of faces, μ is the number of sides (edges) in each face, and ν is the number of faces which meet at a vertex. As before, N is the order of the corresponding group.

	N_0	N_1	N_2	μ	ν	N
Tetrahedron ..... Δ_4	4	6	4	3	3	12
Cube .... $\square_6$	8	12	6	4	3	24
Octahedron ..... Δ_8	6	12	8	3	4	24
Dodecahedron ... ⬠$_{12}$	20	30	12	5	3	60
Icosahedron ..... Δ_{20}	12	30	20	3	5	60

According to Euler's theorem on polyhedra, we have $N_0 - N_1 + N_2 = 2$. The total number of poles is equal to $N_0 + N_1 + N_2 = 2N_1 + 2$. Under any rotation which takes the polyhedron to itself, a given edge a_1b_1 can go to any other edge a_ib_i or b_ia_i; thus, $N = 2N_1$. We also note that $\{\mu, \nu\} = \{n_2, n_3\}$, where n_2 and n_3 are the multiplicities of the poles, which we introduced in Subsection 1.

Further, let T be the group of the tetrahedron, O be the group of the cube (or octahedron), and I be the group of the icosahedron (dodecahedron).

The elements of T are the rotations through multiples of $\pi/2$ around the four axes connecting the vertices with the centers of the opposite faces, the rotations through π

around each of the three axes connecting the midpoints of opposite edges, and the identity rotation.

Besides the identity, the group O consists of the rotations through $\pi/2$, π and $3\pi/2$ around the three axes connecting the centers of opposite faces of the cube, the rotations through $2\pi/3$ and $4\pi/3$ around the four axes connecting diametrically opposite vertices, and the rotations through π around each of the six axes connecting the midpoints of diametrically opposite edges.

The regular tetrahedron can be inscribed in the cube, and then it remains invariant under some of the rotations of order 3 and 2 in O. There are 12 such rotations (including the identity), and they make up all of the group T. Consequently, $T \subset O$, and, since $|O:T| = 2$, it follows that $T \lhd O$.

To each element of O there corresponds exactly one permutation of the set consisting of the four principal diagonals of the cube. Since $|O| = |S_4| = 24$, it follows that $O \cong S_4$.

Similarly, $T \cong A_4$.

In Exercise 2 below, we see that $I \cong A_5$.

Returning to the proof of Theorem 1, we note that when $n_1 = 2$ and $n_2 = n_3 = 3$ there are two four-element orbits of poles $G(p_1) = \{p_1, p_2, p_3, p_4\}$ and $G(q_1) = \{q_1, q_2, q_3, q_4\}$, where p_i and q_i are opposite points on S^2. If Δ_4^0 is the tetrahedron with vertices p_i, then its symmetry group T^0 contains G. Since $|G| = 12$, it follows that Δ_4^0 is a regular tetrahedron, i.e., $\Delta_4^0 = \Delta_4$, and $T^0 = G = T$.

When $n_2 = 3$ and $n_3 = 4$, we take the six-element orbit of poles $G(p_1) = \{p_1, \ldots, p_6\}$. These poles divide up into pairs, since $i \neq 3 \Rightarrow n_i \neq 4$. We take these three pairs of points on S^2 as the three pairs of opposite vertices of an octahedron Δ_8^0. As in the previous case, since $|G| = 24$, we have $\Delta_8^0 = \Delta_8$ (i.e., Δ_8^0 is a regular octahedron), and $O^0 = G = O$.

Finally, when $n_1 = 2$, $n_2 = 3$ and $n_3 = 5$, we construct an icosahedron Δ_{20}^0 whose vertices p_i are taken from the orbit $G(p_1) = \{p_1, \ldots, p_{20}\}$. Again, since $|G| = 60$, it follows that Δ_{20}^0 is regular, and $I^0 = G = I$.

It remains to note that any two regular polyhedra of the same type which are inscribed in the sphere S^2 can be obtained from one another by a rotation (or by a change of coordinates). This shows that the isomorphic finite subgroups of SO(3) are conjugate to one another. We gather together our results in the form of a theorem.

THEOREM 2. _All of the finite subgroups of_ SO(3) _are up to isomorphism one of the groups_ $C_n, D_n, n \in \mathbb{N}$; $T \cong A_4$, $O \cong S_4$, _or_ $I \cong A_5$. _Any two isomorphic finite subgroups are conjugate in_ SO(3). □

COROLLARY. _The isomorphisms in Theorem 2 give irreducible three-dimensional orthogonal representations of the groups_ A_4, S_4, _and_ A_5. □

Using Theorem 2 and the epimorphism $\Phi : SU(2) \to SO(3)$ (Theorem 1 of §1 Ch. 7), we easily obtain a description of all of the finite subgroups of SU(2) (one can also go the other way, first finding the finite subgroups of SU(2), and then of SO(3)). Any such group G^* which is not cyclic is the preimage of a finite subgroup $G \subset SO(3)$. This gives the so-called _binary groups:_

$$D_n^* = \Phi^{-1}(D_n), \qquad T^* = \Phi^{-1}(T), \qquad O^* = \Phi^{-1}(O), \qquad I^* = \Phi^{-1}(I)$$

-- the binary dihedral group, the binary tetrahedral group, the binary octahedral group, and the binary icosahedral group. Like the orthogonal representation $\Phi : SU(2) \to SO(3)$ itself, the binary groups arise in a natural way when one describes the states of a physical system of particles with spin.

EXERCISES

1. Besides the trivial subgroup, the icosahedral group I contains 15 conjugate cyclic subgroups of order 2, 10 conjugate cyclic subgroups of order 3, and 6 conjugate cyclic subgroups of order 5. Prove that I is a simple group.

2. Construct an isomorphism between the groups I and A_5.

3. Show that, if H is a finite subgroup of odd order in SU(2) or SO(3), then H is cyclic.

4. Show that, if a finite subgroup $H \subset SU(2)$ is not the preimage of any subgroup $G \subset SO(3)$, then $|H| \equiv 1 \pmod 2$.

5. Show that, up to conjugation

$$D_3^* = \left\langle \begin{Vmatrix} 0 & 1 \\ -1 & 0 \end{Vmatrix}, \begin{Vmatrix} \varepsilon & 0 \\ 0 & \varepsilon^{-1} \end{Vmatrix} \;\middle|\; \varepsilon^2 + \varepsilon + 1 = 0 \right\rangle .$$

6. What do the following two groups have in common: the binary icosahedral group I^* and the group

$$SL(2, Z_5) = \left\{ \begin{Vmatrix} a & b \\ c & d \end{Vmatrix} \;\middle|\; ad - bc = 1; \quad a, b, c, d \in Z_5 \right\} ?$$

7. Suppose that atoms of q different sorts $(q < 200)$ can be placed in any possible way (we are neglecting chemical bonds) at the vertices of a regular polyhedron M. We do not distinguish between the "molecules" which can be obtained from one another by a rotation around some axis. Let $f(M, q)$ be the number of different "molecules". Derive the formulas:

$$f(\Delta_4, q) = \frac{q^2}{12}(q^2 + 11) ,$$

$$f(\Box_6, q) = \frac{q^2}{24}(q^6 + 17q^2 + 6) ,$$

$$f(\Delta_8, q) = \frac{q^2}{24}(q^4 + 3q^2 + 12q + 8) .$$

8. Show that, if we compute the number of ways of coloring the faces of M with q sorts of colors, in the case of the tetrahedron Δ_4 we obtain the same formula as in Exercise 7, and in the case of the cube and octahedron the formulas are interchanged.

§4. Characters of linear representations

1. Schur's lemma and corollary. At the base of every fundamental mathematical theory one usually finds several relatively simple (but subtle) ideas. Onc of the corner-stones of representation theory is the following fact.

THEOREM 1 (Schur's lemma). *Suppose that* (Φ, V) *and* (Ψ, W) *are two irreducible complex representations of a group* G, *and suppose that* $\sigma : V \to W$ *is a linear map such that*

$$\Psi(g)\sigma = \sigma\Phi(g) , \quad \forall g \in G . \tag{1}$$

Then:

(i) *if the representations* Φ *and* Ψ *are not equivalent, it follows that* $\sigma = 0$;

(ii) *if* $V = W$ *and* $\Phi = \Psi$, *then* $\sigma = \lambda\mathcal{E}$ *for some scalar* λ.

Proof. If $\sigma = 0$, there is nothing to prove. So suppose that $\sigma \neq 0$, and set $V_0 = \operatorname{Ker} \sigma \subset V$.

Since $\sigma\Phi(g)v_0 = \Psi(g)\sigma v_0 = 0$ for any $v_0 \in V_0$, it follows that $\Phi(g)V_0 = V_0$, i.e., the subspace V_0 is G-invariant. Since (Φ, V) is irreducible, we have $V_0 = 0$ or V. But we cannot have $V_0 = V$, because $\sigma \neq 0$. Hence, $\operatorname{Ker} \sigma = 0$.

Similarly, if we set $W_1 = \operatorname{Im} \sigma \subset W$, we have $w_1 \in W_1 \Rightarrow \Psi(g)w_1 = \Psi(g)\sigma(v_1) = \sigma(\Phi(g)v_1) = w_1' \in W_1$, so that W_1 is an invariant subspace of W. Again $\sigma \neq 0 \Rightarrow W_1 \neq 0$, and, since (Ψ, W) is an irreducible representation, the only possibility is that $W_1 = W$.

(i) Since $\operatorname{Ker} \sigma = 0$ and $\operatorname{Im} \sigma = W$, it follows that $\sigma : V \to W$ is an isomorphism, and condition (1) is neither more nor less than the definition of equivalence of two representations Φ and Ψ (see Definition 2 in §1). This proves assertion (i).

(ii) By assumption, $\sigma : V \to V$ is a linear operator on V. Since $\mathbb{C}$ is algebraically closed, it has an eigen-value; let λ be an eigen-value of σ. The linear operator $\sigma_0 = \sigma - \lambda \mathcal{E}$ has a non-trivial kernel (since it contains an eigen-vector for λ), and it satisfies the equality $\Psi(g)\sigma_0 = \sigma_0 \Phi(g)$. By what was proved before, this means that $\sigma_0 = 0$, i.e., $\sigma = \lambda \mathcal{E}$. □

COROLLARY. *Let* (Φ, V) *and* (Ψ, W) *be two irreducible complex representations of a finite group* G *of order* $|G|$, *and let* $\sigma : V \to W$ *be any linear map. Then the "averaging" map*

$$\tilde{\sigma} = \frac{1}{|G|} \sum_{g \in G} \Psi(g)\, \sigma\, \Phi(g)^{-1}$$

has the following properties:

(i) $$\Phi \not\sim \Psi \Longrightarrow \tilde{\sigma} = 0 ;$$

(ii) $$V = W, \quad \Phi = \Psi \Longrightarrow \tilde{\sigma} = \lambda \mathcal{E}, \quad \lambda = \frac{\operatorname{tr} \sigma}{\dim V} .$$

Proof. We have

$$\Psi(g)\widetilde{\sigma}\Phi(g)^{-1} = |G|^{-1} \sum_{h \in G} \Psi(g)\Psi(h)\sigma\Phi(h)^{-1}\Phi(g)^{-1} =$$

$$= |G|^{-1} \sum_{n} \Psi(gh)\sigma\Phi(gh)^{-1} = |G|^{-1} \sum_{t \in G} \Psi(t)\sigma\Phi(t)^{-1} = \widetilde{\sigma} ,$$

so that $\Psi(g)\widetilde{\sigma} = \widetilde{\sigma}\Phi(g)$, $\forall g \in G$. Schur's lemma immediately gives us both assertions, and the precise formula for λ follows from the relations

$$(\dim V)\lambda = \operatorname{tr} \lambda \mathcal{E} = \operatorname{tr} \widetilde{\sigma} = |G|^{-1} \sum_{g \in G} \operatorname{tr} \Phi(g)\sigma\Phi(g)^{-1} = |G|^{-1} \sum_{g \in G} \operatorname{tr} \sigma = \operatorname{tr} \sigma .$$

Here we have used the well-known property of the trace function: $\operatorname{tr} CAC^{-1} = \operatorname{tr} A$. □

We shall need the <u>matrix version</u> of this corollary. To formulate this, we choose any bases in V and W: $V = \langle e_i \mid i \in I \rangle$, $W = \langle f_j \mid j \in J \rangle$. We write our maps in these bases, and identify the maps with the corresponding matrices:

$$\Phi_g = (\varphi_{ii'}(g)), \qquad \Psi_g = (\psi_{jj'}(g)) ,$$

$$\sigma = (\sigma_{ji}), \qquad \widetilde{\sigma} = (\widetilde{\sigma}_{ji}); \qquad i, i' \in I, \quad j, j' \in J .$$

By the definition of $\widetilde{\sigma}$, we have

$$\widetilde{\sigma}_{ji} = |G|^{-1} \sum_{g \in G,\, i' \in I,\, j' \in J} \psi_{jj'}(g)\,\sigma_{j'i'}\,\varphi_{i'i}(g^{-1}) . \tag{2}$$

Our map $\sigma : V \to W$ is completely arbitrary. We can take, for example,

$$\sigma_{ji} = 0, \quad \forall (j,i) \neq (j_0, i_0); \quad \sigma_{j_0 i_0} = 1 . \tag{3}$$

Part (i) of the corollary then corresponds to the relation

$$|G|^{-1} \sum_{g \in G} \psi_{jj_0}(g) \cdot \varphi_{i_0 i}(g^{-1}) = 0, \quad \forall i, i_0, j, j_0 \tag{4}$$

(Φ and Ψ are inequivalent representations).

Now if $V = W$ and $\Phi = \Psi$, then

$$\operatorname{tr}\sigma = \sigma_{ii} = \sum_{i',j'} \delta_{j'i'}\sigma_{j'i'} ,$$

$$\tilde{\sigma} = \frac{\operatorname{tr}\sigma}{\dim V}\,\mathcal{E} \Longrightarrow \tilde{\sigma}_{ji} = \delta_{ji}\frac{\operatorname{tr}\sigma}{\dim V} = \frac{\delta_{ji}}{\dim V}\sum_{i',j'} \delta_{j'i'}\sigma_{j'i'} .$$

Comparing this expression with (2), we obtain

$$|G|^{-1} \sum_{g\in G,\, i',\, j'} \varphi_{jj'}(g)\,\sigma_{j'i'}\,\varphi_{i'i}(g^{-1}) = \frac{1}{\dim V}\sum_{i',j'} \delta_{ji}\,\delta_{j'i'}\,\sigma_{j'i'} ,$$

from which, because of the arbitrariness in the choice of σ (see (3)), we conclude that part (ii) of the corollary corresponds to the relation

$$|G|^{-1} \sum_{g\in G} \varphi_{jj_0}(g)\,\varphi_{i_0 i}(g^{-1}) = \begin{cases} \dfrac{\delta_{ji}}{\dim V}, & \text{if } j_0 = i_0 , \\ 0 & \text{otherwise} . \end{cases} \tag{5}$$

Relations (4) and (5) contain the information we shall need.

2. Characters of representations. To each complex finite-dimensional linear representation (Φ, V) of a group G we associate the function

$$\chi_\Phi : G \longrightarrow \mathbb{C} ,$$

defined by setting

$$\chi_\Phi(g) = \operatorname{tr}\Phi(g) , \qquad g \in G ;$$

this function is called the character of the representation. It can also be denoted χ_V or simply χ if it is clear what representation is being discussed.

Let $\Phi_g = (\varphi_{ij}(g))$ be the matrix corresponding to the operator $\Phi(g)$ in some basis of the space V, and let $\lambda_1, \dots, \lambda_n$ $(n = \dim V)$ be the characteristic roots of this matrix, counted with multiplicity. (The λ_i, of course, depend on g.) By definition, we have

$$\chi_{\Phi}(g) = \chi_V(g) = \sum_{i=1}^{n} \varphi_{ii}(g) = \sum_{i=1}^{n} \lambda_i \quad .$$

If C is any invertible matrix, then

$$\operatorname{tr} C \Phi_g C^{-1} = \operatorname{tr} \Phi_g \quad .$$

But we know that every representation Ψ which is equivalent to Φ has the form $g \mapsto C \Phi_g C^{-1}$. Hence <u>the characters of isomorphic (equivalent) representations coincide.</u> In other words, the notion of the character of a representation is well-defined, i.e., depends only on the equivalence class of the representation.

We note some more elementary properties of the characters of representations.

PROPOSITION. <u>Let</u> χ_{Φ} <u>be the character of a complex linear representation</u> (Φ, V) <u>of a group</u> G. <u>Then:</u>

(i) $\chi_{\Phi}(e) = \dim V$;

(ii) $\chi_{\Phi}(hgh^{-1}) = \chi_{\Phi}(g)$, $\forall g, h \in G$, <u>i.e.,</u> χ_{Φ} <u>is a function which is constant on conjugacy classes of elements of</u> G;

(iii) $\chi_{\Phi}(g^{-1}) = \overline{\chi_{\Phi}(g)}$ <u>for any element</u> $g \in G$ <u>of finite order (the bar denotes complex conjugation)</u>;

(iv) <u>the direct sum</u> $\Phi = \Phi' \dotplus \Phi''$ <u>of two representations has character</u> $\chi_{\Phi} = \chi_{\Phi'} + \chi_{\Phi''}$.

<u>Proof.</u> First, $\chi_{\Phi}(e) = \operatorname{tr} \Phi(e) = \operatorname{tr} \mathcal{E} = \dim V$. Next, $\chi_{\Phi}(hgh^{-1}) = \operatorname{tr} \Phi(hgh^{-1}) = \operatorname{tr} \Phi(h)\Phi(g)\Phi(h)^{-1} = \operatorname{tr} \Phi(g) = \chi_{\Phi}(g)$. To prove (iii), we note that

$$g^m = e \Longrightarrow \Phi(g)^m = \mathcal{E} \quad ,$$

and, if $\lambda_1, \ldots, \lambda_n$ are the characteristic roots of the operator $\Phi(g)$, then $\lambda_1^k, \ldots, \lambda_n^k$ are the characteristic roots of the operator $\Phi(g)^k$. In particular, $\lambda_i^m = 1$,

$1 \le i \le n$, and hence $|\lambda_i| = 1$, $\bar{\lambda}_i = \lambda_i^{-1}$. Thus,

$$\chi_\Phi(g^{-1}) = \operatorname{tr}\Phi(g^{-1}) = \operatorname{tr}\Phi(g)^{-1} = \sum_i \lambda_i^{-1} = \sum_i \bar{\lambda}_i = \overline{\left(\sum_i \lambda_i\right)} = \overline{\chi_\Phi(g)} .$$

Finally, if $\Phi = \Phi' \dot{+} \Phi''$, we know that for a suitable choice of basis in the representation space V all of the matrices Φ_g, $g \in G$, have the form

$$\Phi_g = \begin{Vmatrix} \Phi'_g & 0 \\ 0 & \Phi''_g \end{Vmatrix} ,$$

and hence $\operatorname{tr}\Phi_g = \operatorname{tr}\Phi'_g + \operatorname{tr}\Phi''_g$. But this means that $\chi_\Phi(g) = \chi_{\Phi'}(g) + \chi_{\Phi''}(g)$. □

Note that if $n = \dim V = 1$, then $\chi_\Phi(g) = \Phi(g)$, but that for $n > 1$ the character χ_Φ is not a homomorphism from G to $\mathbb{C}$.

Example 1. We consider the natural two-dimensional representation of the group $SU(2)$. Let χ be the corresponding character. According to (5) in §1 Ch. 7, any matrix $g \in SU(2)$ is conjugate to a matrix of the form

$$b_\varphi = \begin{Vmatrix} e^{i\frac{\varphi}{2}} & 0 \\ 0 & e^{-i\frac{\varphi}{2}} \end{Vmatrix}, \qquad 0 \le \varphi < 2\pi ,$$

so that the conjugacy classes of elements of $SU(2)$ are parametrized by the real numbers φ in the interval $[0, 2\pi)$. According to property (ii) of characters, we have:

$$\chi(g) = \chi(U b_\varphi U^{-1}) = \chi(b_\varphi) = e^{i\frac{\varphi}{2}} + e^{-i\frac{\varphi}{2}} = 2\cos\frac{\varphi}{2} .$$

Under the canonical representation $\Phi : SU(2) \to SO(3)$, the matrix b_φ goes to the matrix

$$B_\varphi = \begin{Vmatrix} \cos\varphi & -\sin\varphi & 0 \\ \sin\varphi & \cos\varphi & 0 \\ 0 & 0 & 1 \end{Vmatrix} ,$$

which is also a convenient choice of conjugacy class representative in the group SO(3). It is obvious that

$$\chi_\Phi(B_\varphi) = 1 + 2\cos\varphi \quad . \tag{6}$$

We shall later make use of the formula (6).

The set $\mathbb{C}^G = \{G \to \mathbb{C}\}$ of all functions from G to $\mathbb{C}$ has a natural vector space structure over $\mathbb{C}$: if $\alpha_1, \alpha_2 \in \mathbb{C}$ and $\chi_1, \chi_2 \in \mathbb{C}^G$, then by $\alpha_1\chi_1 + \alpha_2\chi_2$ we mean the function with values

$$(\alpha_1\chi_1 + \alpha_2\chi_2)(g) = \alpha_1\chi_1(g) + \alpha_2\chi_2(g) \quad .$$

A function in $\mathbb{C}^G$ is called central if it is constant on each conjugacy class of the group G. The central functions obviously form a vector subspace of $\mathbb{C}^G$, which we denote $X_\mathbb{C}(G)$. Generally speaking, $X_\mathbb{C}(G)$ is an infinite dimensional space, but if G has only finitely many conjugacy classes $C_1, C_2, \ldots, C_r$ (as is always the case if G is finite), then the space $X_\mathbb{C}(G)$ is finite dimensional. For example,

$$X_\mathbb{C}(G) = \langle \Gamma_1, \Gamma_2, \ldots, \Gamma_r \rangle_\mathbb{C} , \tag{7}$$

where

$$\Gamma_i(g) = \begin{cases} 1 , & \text{if} \quad g \in C_i , \\ 0 , & \text{if} \quad g \notin C_i \quad . \end{cases}$$

By what we have proved (part (ii) of the proposition), the characters of the group G belong to the space $X_\mathbb{C}(G)$. We shall see that the space spanned by the characters actually is all of $X_\mathbb{C}(G)$, at least when G is a finite group.

We now suppose that G is finite. We make $\mathbb{C}^G$ into a hermitian space by introducing the scalar product

$$(\sigma, \tau)_G = \frac{1}{|G|} \sum_{g \in G} \sigma(g)\overline{\tau(g)}, \qquad \sigma, \tau \in \mathbb{C}^G \quad . \tag{8}$$

It is easily verified that the form $(\sigma, \tau) \mapsto (\sigma, \tau)_G$ satisfies all of the properties of a non-degenerate hermitian form. Its restriction to the subspace $X_{\mathbb{C}}(G) \subset \mathbb{C}^G$ is a very useful tool, especially for studying the characters of linear representations.

THEOREM 2. *Let* Φ *and* Ψ *be irreducible complex representations of a finite group* G. *Then*

$$(\chi_\Phi, \chi_\Psi)_G = \begin{cases} 1, & \text{if} \quad \Phi \approx \Psi, \\ 0, & \text{if} \quad \Phi \not\approx \Psi \quad . \end{cases} \tag{9}$$

Proof. In matrix notation we have

$$\chi_\Phi(g) = \sum_{i=1}^{n} \varphi_{ii}(g), \qquad \chi_\Psi(g) = \sum_{i=1}^{n} \psi_{ii}(g) \quad .$$

Setting $i_0 = i$ and $j_0 = j$ in (4) and then summing over i and j (in the appropriate range), we obtain

$$0 = |G|^{-1} \sum_{g,i,j} \psi_{jj}(g)\varphi_{ii}(g^{-1}) = |G|^{-1} \sum_g \left(\sum_j \psi_{jj}(g)\right)\left(\sum_i \varphi_{ii}(g^{-1})\right) =$$

$$= |G|^{-1} \sum_{g \in G} \chi_\Psi(g)\chi_\Phi(g^{-1}) = |G|^{-1} \sum_{g \in G} \chi_\Psi(g)\overline{\chi_\Phi(g)} = (\chi_\Psi, \chi_\Phi)_G$$

for any non-equivalent irreducible representations Φ and Ψ of the group G.

We now use (5) (for $i_0 = i$, $j_0 = j$):

$$1 = \left(\sum_{j,i} \delta_{ji}\right)/\dim V = |G|^{-1} \sum_{g \in G} \left(\sum_j \varphi_{jj}(g)\right)\left(\sum_i \varphi_{ii}(g^{-1})\right) =$$

$$= |G|^{-1} \sum_{g \in G} \chi_\Phi(g)\chi_\Phi(g^{-1}) = (\chi_\Phi, \chi_\Phi)_G \quad .$$

Since the characters of isomorphic representations coincide, it follows that we have $(\chi_\Phi, \chi_\Psi)_G = 1$ when $\Phi \approx \Psi$. □

The relation (9) is called the (first) orthogonality relation for characters.

COROLLARY. *Let*

$$V = V_1 \oplus \dots \oplus V_k \tag{10}$$

be a decomposition of the complex G-space V into a direct sum of irreducible G-spaces V_i. *If* W *is an irreducible G-space with character* χ_W, *then the number of terms* V_i *in (10) which are isomorphic to* W *is equal to* $(\chi_V, \chi_W)_G$ *and does not depend on the choice of decomposition (10). This number is called the multiplicity with which* W *occurs in* V. *Two representations (two G-spaces) with the same character are isomorphic.*

Proof. As we have already noted (part (iv) of the proposition), we have $\chi_V = \chi_{V_1} + \dots + \chi_{V_k}$, and hence

$$(\chi_V, \chi_W)_G = (\chi_{V_1}, \chi_W)_G + \dots + (\chi_{V_k}, \chi_W)_G \quad .$$

By Theorem 2, the sum on the right consists of k zeros and ones, and the number of ones is equal to the number of G-subspaces V_i which are isomorphic to W. But the scalar product $(\chi_V, \chi_W)_G$ does not depend on any direct sum decomposition (see the definition (8)), so that we have also shown that the multiplicity of W in V is a well-defined number, depending only on W and V.

Suppose that two G-spaces V and V' have the same character $\chi = \chi_V = \chi_{V'}$. Given any irreducible G-space W, V and V' contain W the same number of times, namely $(\chi, \chi_W)_G$. Hence, if we decompose V and V' into direct sums of irreducible G-spaces:

$$V = \bigoplus_{i=1}^{k} V_i , \qquad V' = \bigoplus_{j=1}^{\ell} V'_j$$

we have $\ell = k$ and $V'_i \cong V_i$, $1 \le i \le k$ (for a suitable ordering of the V_i). Hence, V and V' are isomorphic G-spaces. □

The remarks following the proof of Theorem 2 of §2 and the above corollary allow us to express the character χ_Φ of any complex linear representation (Φ, V) of a finite group G as a linear combination with integer coefficients

$$\chi_\Phi = \sum_{i=1}^{s} m_i \chi_i \quad .$$

Here m_i is the multiplicity with which the irreducible representation (Φ_i, V_i) occurs in (Φ, V), so that we assume that $\Phi_i \not\approx \Phi_j$ for $i \neq j$. Using the orthogonality relation (9), we may write:

$$(\chi_\Phi, \chi_\Phi)_G = \sum_{i=1}^{s} m_i^2 \quad . \tag{11}$$

We conclude: <u>the scalar square</u> $(\chi_\Phi, \chi_\Phi)_G$ <u>of the character</u> χ_Φ <u>of an arbitrary complex representation</u> Φ <u>is always an integer; it equals</u> 1 <u>if and only if</u> Φ <u>is irreducible.</u>

We have arrived at a remarkable result. The characters, i.e., the "traces of representations", which contain the scantiest information about each separate linear operator $\Phi(g)$, somehow in their totality express all of the essential properties of the totality $\{\Phi(g) \mid g \in G\}$, i.e., the properties of the representation Φ.

<u>Example 2.</u> We show that the representations of the groups A_4, S_4 and A_5 by rotations of three-dimensional space are irreducible over $\mathbb{C}$. To do this we must return to the corollary of Theorem 2 §3 and make use of the formulas (6) and (11). According to the description of the representation Φ in §3, if σ is a permutation of order q, then $\Phi(\sigma)$ is rotation about some axis through an angle of $k2\pi/q$, where

g.c.d. $(k,q) = 1$. Hence, the values of the character $\chi = \chi_\Phi$ can be computed directly from formula (6):

$$\chi(\sigma) = 1 + 2\cos k\frac{2\pi}{q} = 3\,,\ -1\,,\ 0\,,\ 1\,,\ \frac{1+\sqrt{5}}{2}\,,\ \frac{1-\sqrt{5}}{2}\,,$$

if $q = 1, 2, 3, 4, 5$ $(k = \pm 1)$, 5 $(k = \pm 2)$, respectively. We note that

$$\frac{1+\sqrt{5}}{2} = \mathrm{tr}\begin{Vmatrix} \varepsilon & 0 & 0 \\ 0 & \varepsilon^{-1} & 0 \\ 0 & 0 & 1 \end{Vmatrix} = \varepsilon + \varepsilon^{-1} + 1\,, \quad \frac{1-\sqrt{5}}{2} = \varepsilon^2 + \varepsilon^{-2} + 1\,, \quad \varepsilon = e^{\frac{2\pi i}{5}}\,.$$

In Corollary 1 of Theorem 4 §2 Ch. 4, we described now to compute the order of π from its decomposition into disjoint cycles. The elements are divided into conjugacy classes in the tables there (see Exercise 8 §2 Ch. 7 for A_4, Exercise 4 §3 Ch. 7 for S_4, and the proof of Theorem 5 §3 Ch. 7 for A_5). Here are the same tables, with the values of χ filled in:

A_4	1	3	4	4
	e	(12)(34)	(123)	(132)
χ	3	-1	0	0

S_4	1	3	6	8	6
	e	(12)(34)	(12)	(123)	(1234)
χ	3	-1	-1	0	1

A_5	1	15	20	12	12
	e	(12)(34)	(123)	(12345)	(12354)
χ	3	-1	0	$(1+\sqrt{5})/2$	$(1-\sqrt{5})/2$

The relations

$$(\chi,\chi)_{A_4} = \frac{1}{12}\left\{1\cdot 3^2 + 3(-1)^2 + 4\cdot 0^2 + 4\cdot 0^2\right\} = 1 ,$$

$$(\chi,\chi)_{S_4} = \frac{1}{24}\left\{1\cdot 3^2 + 3(-1)^2 + 6(-1)^2 + 8\cdot 0^2 + 6\cdot 1^2\right\} = 1 ,$$

$$(\chi,\chi)_{A_5} = \frac{1}{60}\left\{1\cdot 3^2 + 15(-1)^2 + 20\cdot 0^2 + 12\left(\frac{1+\sqrt{5}}{2}\right)^2 + 12\left(\frac{1-\sqrt{5}}{2}\right)^2\right\} = 1$$

show that the representation Φ with character χ is irreducible over $\mathbb{C}$ (see (11)).

EXERCISES

1. Let Φ and Ψ be irreducible complex representations of a finite group G. Derive the following generalization of Theorem 2:

$$|G|^{-1} \sum_g \chi_\Psi(hg)\overline{\chi_\Phi(g)} = \delta_{\Phi,\Psi} \frac{\chi_\Phi(h)}{\chi_\Phi(e)} .$$

Here h is any element of G; $\delta_{\Phi,\Psi} = 1$ or 0 depending on whether or not $\Phi \approx \Psi$.

2. Apply the irreducibility criterion in terms of characters to the representation $\Phi^{(3)}$ of S_3 in Example 1 of Subsection 1 §2.

3. Using Schur's lemma, prove that all irreducible complex representations of an abelian group G are one-dimensional.

4. If τ is an automorphism of the group G, then for every linear representation (Φ, V) of G we have another representation (Φ^τ, V), which is defined by the rule: $\Phi^\tau(g) = \Phi(\tau(g))$. Verify that this is a representation, and that Φ^τ is irreducible whenever Φ is. It usually happens that $\Phi^\tau \approx \Phi$, but there are cases when we obtain a new representation. What happens if τ is an inner automorphism?

Let $G = A_5$, and let Φ be the representation in Example 2. The map $\tau : \pi \mapsto (12)\pi(12)^{-1}$ is an (outer) automorphism of A_5 which interchanges the conjugacy classes of (12345) and (12354). The sets of values of the characters χ and χ^τ are obtained from one another by switching $(1+\sqrt{5})/2$ and $(1-\sqrt{5})/2$. Show that the characters χ and χ^{τ} are non-equivalent.

5. Let $\Phi : G \to U(n)$ and $\Psi : G \to U(n)$ be equivalent irreducible unitary representations of a finite group G. Prove that there exists a unitary matrix U such that $U\Phi_g U^{-1} = \Psi_g$, $\forall g \in G$.

§5. Irreducible representations of finite groups

1. The number of irreducible representations. In the case of finite groups, the above ideas allow us to answer the basic questions of representation theory. One of these fundamental facts is the following

THEOREM 1. The number of irreducible pair-wise non-equivalent representations of a finite group G over $\mathbb{C}$ is equal to the number of conjugacy classes in G.

The proof of this theorem is contained in Lemmas 1 and 2, if we note that the number r of conjugacy classes in G can be interpreted as the dimension of the space $X_{\mathbb{C}}(G)$ of complex-valued central functions of G (see (7) §4). Since the characters of linear representations are central functions, they span a subspace of $X_{\mathbb{C}}(G)$ of some dimension $s \leq r$. By Theorem 2 §4, the characters of irreducible representations form an orthonormal basis (in the metric $(*, *)_G$) for this subspace. Hence, the number of irreducible representations is s. We know that $s \leq r$, and so it remains to prove that $s = r$.

LEMMA 1. Let Γ be a central function on a group G, and let (Φ, V) be a complex irreducible representation with character χ_{Φ}. Then for the linear operator

$$\Phi_{\Gamma} = \sum_{h \in G} \overline{\Gamma}(h)\, \Phi(h) : V \longrightarrow V$$

we have $\Phi_{\Gamma} = \lambda \mathcal{E}$, where

$$\lambda = \frac{|G|}{\chi_\Phi(e)} (\chi_\Phi, \Gamma)_G$$

($\overline{\Gamma}$ is the central function defined by setting $\overline{\Gamma}(g) = \overline{\Gamma(g)}$.)

<u>Proof.</u> Since Γ is a central function, we have

$$\Phi(g)\Phi_\Gamma\Phi(g)^{-1} = \sum_{h\in G} \overline{\Gamma}(h)\Phi(g)\Phi(h)\Phi(g^{-1}) =$$

$$= \sum_{h\in G} \overline{\Gamma}(ghg^{-1})\Phi(ghg^{-1}) = \sum_{t\in G} \overline{\Gamma}(t)\Phi(t) = \Phi_\Gamma \quad .$$

Thus, $\Phi_\Gamma\Phi(g) = \Phi(g)\Phi_\Gamma$, $\forall g \in G$. Shur's lemma (Theorem 1 §4), applied to $\sigma = \Phi_\Gamma$, shows that $\Phi_\Gamma = \lambda\mathcal{E}$. Computing the trace of the operators on both sides of this equality, we find that $\lambda\chi_\Phi(e) = \lambda \dim V = \operatorname{tr}\lambda\mathcal{E} = \operatorname{tr}\Phi_\Gamma = \sum_{h\in G} \overline{\Gamma}(h) \operatorname{tr}\Phi(h) =$

$$= |G| \left\{ |G|^{-1} \sum_{h\in G} \chi_\Phi(h)\overline{\Gamma(h)} \right\} = |G|(\chi_\Phi, \Gamma)_G . \quad \square$$

LEMMA 2. <u>The characters</u> $\chi_1, \ldots, \chi_s$ <u>of all of the pair-wise non-equivalent irreducible representations of</u> G <u>over</u> $\mathbb{C}$ <u>form an orthonormal basis for the space</u> $X_\mathbb{C}(G)$.

<u>Proof.</u> By Theorem 2 of §4, the set $\chi_1, \ldots, \chi_s$ is orthonormal and can be included in an orthonormal basis for $X_\mathbb{C}(G)$. Let Γ be any central function which is orthogonal to all of the $\chi_i : (\chi_i, \Gamma)_G = 0$. Then, by Lemma 1, the linear operator $\Phi_\Gamma^{(i)}$ corresponding to the representation $\Phi^{(i)}$ with character χ_i is equal to zero.

By Theorem 2 §2, every complex representation Φ can be decomposed into a direct sum

$$\Phi = m_1\Phi^{(1)} \dotplus \cdots \dotplus m_s\Phi^{(s)}$$

with certain multiplicities $m_1, \ldots, m_s$. For the operator Φ_Γ defined by the relation

$$\Phi_\Gamma = \sum_{h \in G} \bar{\Gamma}(h)\, \Phi(h) \quad ,$$

we have the corresponding decomposition

$$\Phi_\Gamma = m_1 \Phi_\Gamma^{(1)} + \cdots + m_s \Phi_\Gamma^{(s)} = 0 \quad .$$

In particular, this holds for the linear operator ρ_Γ, where ρ is the regular representation (see Example 5 of §1). But in that case we have (here we temporarily let 1 denote the unit element of G, so as to avoid writing e_e):

$$0 = \rho_\Gamma(e_1) = \sum_{h \in G} \bar{\Gamma}(h)\, \rho(h) e_1 = \sum_{h \in G} \bar{\Gamma}(h) e_h \Longrightarrow \bar{\Gamma}(h) = 0 , \qquad \forall h \in G \quad ,$$

from which $\bar{\Gamma} = 0$, and hence $\Gamma = 0$. □

<u>Example.</u> In the case of the symmetric group S_3, Theorem 1 says that this group has exactly three irreducible complex representations. But we already know which ones these are: the table at the end of Subsection 1 §2 contains all of the necessary information. We note, in passing, that the squares of the dimensions of the representations $\Phi^{(1)}$, $\Phi^{(2)}$, $\Phi^{(3)}$ satisfy the relation $1^2 + 1^2 + 2^2 = 6 = |S_3|$. We shall now see that a similar equality holds in general.

2. <u>The degrees of the irreducible representations.</u> We consider the regular representation $(\rho, \langle e_g \mid g \in G \rangle_{\mathbb{C}})$ in somewhat greater detail. Let R_h denote the matrix of the linear operator $\rho(h)$ in the given basis $\{e_g \mid g \in G\}$. Since $\rho(h) e_g = e_{hg}$, it follows that all of the diagonal elements of R_h for $h \neq e$ are zero, and $\operatorname{tr} R_h = 0$. Hence,

$$\chi_\rho(e) = |G| , \qquad \chi_\rho(h) = 0 , \qquad \forall h \neq e \quad .$$

Now let (Φ, V) be an arbitrary irreducible complex representation of G. By the corollary to Theorem 2 §4, the multiplicity with which Φ occurs in ρ is equal to the scalar product $(\chi_\rho, \chi_\Phi)_G$. According to (1),

$$(\chi_\rho, \chi_\Phi)_G = |G|^{-1} \sum_{h \in G} \chi_\rho(h)\overline{\chi_\Phi(h)} = |G|^{-1} \chi_\rho(e)\overline{\chi_\Phi(e)} =$$

$$= |G|^{-1} |G| \chi_\Phi(e) = \dim V \quad . \qquad (2)$$

We see that every irreducible representation (considered up to equivalence) occurs in the regular representation with multiplicity equal to its degree. By Theorem 1, there are r pair-wise non-equivalent irreducible representations

$$\Phi^{(1)}, \Phi^{(2)}, \ldots, \Phi^{(r)}$$

(where r is the number of conjugacy classes in G) having characters

$$\chi_1, \chi_2, \ldots, \chi_r; \qquad \chi_i = \chi_{\Phi^{(i)}} ,$$

and degrees

$$n_1, n_2, \ldots, n_r; \qquad n_i = \chi_i(e) \quad .$$

We usually take $\Phi^{(1)}$ to be the trivial representation, so that $\chi_1(g) = 1$, $\forall g \in G$. By (2), we have

$$\rho = n_1 \Phi^{(1)} + \cdots + n_r \Phi^{(r)} ,$$

and hence

$$\chi_\rho = n_1 \chi_1 + \cdots + n_r \chi_r \quad .$$

In particular,

$$|G| = \chi_\rho(e) = n_1 \chi_1(e) + \cdots + n_r \chi_r(e) = n_1^2 + \cdots + n_r^2 \quad .$$

We have obtained the following theorem.

THEOREM 2. <u>Each irreducible representation</u> $\Phi^{(i)}$ <u>occurs in the regular</u>

<u>representation</u> ρ <u>with multiplicity equal to its degree</u> n_i. <u>The order of the group</u> G <u>and the degrees</u> $n_1, \dots, n_r$ <u>of all of its non-equivalent irreducible representations are connected by the equality:</u>

$$\sum_{i=1}^{r} n_i^2 = |G| \quad . \qquad \square \qquad (3)$$

For groups of small order, the elegant equality (3) is sufficient for determining all of the degrees $n_1, \dots, n_r$, although, in general, we of course need additional information.

It is convenient to write the information about the characters of the irreducible representations (also called: the irreducible characters) in the form of a table

	e	g_2	g_3	$\dots$	g_r
χ_1	n_1	$\chi_1(g_2)$	$\chi_1(g_3)$	$\dots$	$\chi_1(g_r)$
χ_2	n_2	$\chi_2(g_2)$	$\chi_2(g_3)$	$\dots$	$\chi_2(g_r)$
. .	. . .	. . .	. . .	. . .	. . .
χ_r	n_r	$\chi_r(g_2)$	$\chi_r(g_3)$	$\dots$	$\chi_r(g_r)$

which is called the <u>character table.</u> The first row of the character table contains representatives of all of the r conjugacy classes g_i^G. For example, the character table for the group S_3 is:

	e	(12)	(123)
χ_1	1	1	1
χ_2	1	-1	1
χ_3	2	0	-1

(compare with the table at the end of Subsection 1 §2).

As usual, we let $C(g) = C_G(g)$ denote the centralizer of the element g in G.

We know that $|C(g)|\,|g^G| = |G|$ (see Subsection 2 §2 Ch. 7). Hence, if we rewrite the first orthogonality relation (9) §4 in the form

$$\sum_{j=1}^{r} \frac{\chi_i(g_j)}{\sqrt{|C(g_j)|}} \overline{\frac{\chi_k(g_j)}{\sqrt{|C(g_j)|}}} = \frac{1}{|G|} \sum_{j=1}^{r} \frac{|G|}{|C(g_j)|} \chi_i(g_j)\overline{\chi_k(g_j)} =$$

$$= \frac{1}{|G|} \sum_{j=1}^{r} |g_j^G|\, \chi_i(g_j)\overline{\chi_k(g_j)} = \frac{1}{|G|} \sum_{g \in G} \chi_i(g)\overline{\chi_k(g)} = (\chi_i, \chi_k)_G = \delta_{ik} ,$$

we see that the $r \times r$ matrix

$$M = \left(\frac{\chi_i(g_j)}{\sqrt{|C(g_j)|}} \right)$$

is row-unitary. But a matrix is row-unitary if and only if it is column-unitary (since $M \cdot {}^t\overline{M} = E = {}^t\overline{M} \cdot M$), so that

$$\sum_i \frac{\chi_i(g_j)}{\sqrt{|C(g_j)|}} \overline{\frac{\chi_i(g_k)}{\sqrt{|C(g_k)|}}} = \delta_{jk} ,$$

or, written out in more detail:

$$\sum_{i=1}^{r} \chi_i(g)\overline{\chi_i(h)} = \begin{cases} |C_G(g)| , & \text{if } g \text{ and } h \text{ are conjugate ,} \\ 0 & \text{otherwise .} \end{cases} \tag{4}$$

The equality (4) is called the <u>second orthogonality relation</u> for characters.

3. <u>Representations of abelian groups.</u> We can now generalize the description in Example 6 of §1 of the irreducible representations of cyclic groups to all finite abelian groups.

THEOREM 3. <u>Every complex irreducible representation of a finite abelian group</u> A <u>is one-dimensional. The number of pair-wise non-equivalent representations is equal to</u> $|A|$. <u>Conversely, if every irreducible representation of a group</u> A <u>is one-dimensional,</u>

then A is abelian.

Proof. Since the number r of conjugacy classes in an abelian group A is equal to $|A|$, the first two assertions of the theorem follow from Theorem 2 (see also Exercise 3 §4). Next, suppose that all of the n_i in (3) are equal to 1. This means that $r = |A|$, which implies A is abelian. □

Definition. Let A be an abelian group. The set

$$\hat{A} = \mathrm{Hom}(A, \mathbb{C}^*)$$

of homomorphisms from the group A to the multiplicative group of the field of complex numbers, considered along with point-wise multiplication

$$(\chi_1 \chi_2)(a) = \chi_1(a)\chi_2(a)$$

$(\chi_i \in \hat{A},\ a \in A)$, is called the character group of A over $\mathbb{C}$ $(\chi^{-1} = \overline{\chi})$.

THEOREM 4. The groups A and $\hat{A}$ are isomorphic.

Proof. From Theorem 3 we know that, in any case, $|A| = |\hat{A}|$. According to the results of §5 Ch. 7, the group A has a decomposition

$$A = A_1 \times A_2 \times \dots \times A_k$$

into a direct product of cyclic groups $A_i = \langle a_i \rangle$ (we do not care whether they are primary or not; we shall write the group operation in A multiplicatively). If $|A_i| = s_i$ and ε_i is a primitive s_i-th root of 1, then each element $a = a_1^{t_1} a_2^{t_2} \dots a_k^{t_k}$ in A corresponds to the character $\chi_a \in \hat{A}$ defined by:

$$\chi_a\left(a_1^{r_1} a_2^{r_2} \dots a_k^{r_k}\right) = \varepsilon_1^{r_1 t_1} \varepsilon_2^{r_2 t_2} \dots \varepsilon_k^{r_k t_k} .$$

It is clear from the definition that $\chi_a \chi_{a'} = \chi_{aa'}$. If

$$a = a_1^{t_1} a_2^{t_2} \cdots a_k^{t_k} \neq a_1^{t'_1} a_2^{t'_2} \cdots a_k^{t'_k} = a' ,$$

then there exists an index i with $t_i \neq t'_i$. Then

$$\chi_a(a_i) = \epsilon_i^{t_i} \neq \epsilon_i^{t'_i} = \chi_{a'}(a_i) .$$

Consequently, the characters χ_a are pair-wise distinct, and the map $a \mapsto \chi_a$ gives the required isomorphism between A and $\hat{A}$. □

The method of proof of Theorem 4 gives an explicit construction of all of the irreducible representations of an abelian group.

Example. Let V_{2^n} be an elementary abelian group of order 2^n, and let χ be an irreducible complex character which is non-trivial, i.e., $\chi(a) \neq 1$ for some $a \in V_{2^n}$. Since every element in V_{2^n} is assumed to have order 2, we have $\operatorname{Ker} \chi = B \cong V_{2^{n-1}}$, and so we can write $V_{2^n} = B \cup aB$ as the union of cosets of B; thus,

$$\chi(a^i b) = (-1)^i , \qquad i = 0, 1 .$$

For example, the Klein four-group V_4, whose representations were discussed in Problem 2 of §2 Ch. 1, has the following character table:

	e	a	b	ab
χ_1	1	1	1	1
χ_2	1	-1	1	-1
χ_3	1	1	-1	-1
χ_4	1	-1	-1	1

The results concerning representations of abelian groups allow us to obtain some information on the representations of arbitrary finite groups.

THEOREM 5. <u>The one-dimensional representations of a finite group</u> G <u>over</u> $\mathbb{C}$ <u>are in one-to-one correspondence with the irreducible representations of the quotient group</u> G/G' (<u>where</u> G' <u>is the commutant of</u> G). <u>The number of such representations is equal to the index</u> $(G:G')$.

<u>Proof.</u> We first make a general remark. Suppose that G is a group and K is a normal subgroup. If Φ is a representation of G with kernel $\operatorname{Ker}\Phi \supset K$, then we can define a representation $\bar{\Phi}$ of the quotient group G/K by setting

$$\bar{\Phi}(gK) = \Phi(g), \qquad g \in G \quad .$$

This is obviously a well-defined representation (see the proof of Theorem 1 §3 Ch. 7). Furthermore, $\operatorname{Ker}\bar{\Phi} = \operatorname{Ker}\Phi / K$. In particular, if $K = \operatorname{Ker}\Phi$, then we obtain a faithful representation $\bar{\Phi}$.

Conversely, any linear representation Ψ of a group H induces a representation Φ of any group G which maps epimorphically onto H: $\pi : G \to H$. It suffices to set

$$\Phi(g) = \Psi(\pi(g)) \quad .$$

Since π is an epimorphism, it follows that $\Phi(G) = \Psi(H)$, and Φ and Ψ are either both reducible or both irreducible. By Theorem 3 §3 Ch. 7, $\operatorname{Ker}\Phi = \pi^{-1}(\operatorname{Ker}\Psi)$. Given any one-dimensional representation Φ of a group G, we have the associated abelian (actually cyclic) group $\operatorname{Im}\Phi$, so that $\operatorname{Ker}\Phi \supset G'$. We now obtain the theorem as a simple consequence of Theorem 3, the above remark, and Theorem 4 §3 Ch. 7. □

4. <u>Representations of certain special groups.</u> Although, in principle, to obtain all of the irreducible representations of a finite group G it suffices to decompose the regular representation (Theorem 2), in practice this is not usually easy to do, and so one tries to find other methods. It is usually simplest first to construct the character table, and then the representations themselves (in this connection see §1 of Chapter 9). In any case, in the relatively simple examples below there is no need to resort to any subtle tricks.

Example 1. Let G be an arbitrary 2-transitive group of permutations of the set $\Omega = \{1, 2, \ldots, n\}$, $n > 2$ (see Example 3 §2 Ch. 7). Further let Φ be the natural representation of G in the space $V = \langle e_1, e_2, \ldots, e_n \rangle$ with action defined by $\Phi(g)e_i = e_{g(i)}$ (see Example 5 §1). It is not hard to see that the value $\chi_\Phi(g)$ coincides with the number $N(g)$ of points $i \in \Omega$ (i.e., the number of basis vectors e_i) which are left fixed by g. By Theorem 3 §2 Ch. 7, we have

$$\sum_{g \in G} \chi_\Phi(g)\overline{\chi_\Phi(g)} = \sum_{g \in G} \chi_\Phi(g)^2 = \sum_{g \in G} N(g)^2 = 2|G| ,$$

which can obviously be rewritten in the form

$$(\chi_\Phi, \chi_\Phi)_G = 2 . \tag{5}$$

Comparing (5) with the relation (11) of §4, we conclude that Φ is a direct sum of two irreducible representations (since $2 = 1 + 1$ is the only way of writing 2 as a sum of squares of natural numbers). But we also know that $\Phi = \Phi^{(1)} \dot{+} \Psi$, where $(\Phi^{(1)}, U)$ is the unit (trivial) representation and Ψ is the $(n-1)$-dimensional representation on the space $W = \langle e_1 - e_n, e_2 - e_n, \ldots, e_{n-1} - e_n \rangle$. If we could decompose W in the direct sum $V = U \oplus W$, then there would be more than two irreducible terms. We have thereby obtained the following non-trivial fact.

The natural complex linear representation (Φ, V) _of a_ 2_-transitive permutation group_ G _is the sum of the unit representation and one other irreducible representation._ □

In particular, each of the groups S_n, $n > 2$; A_n, $n > 3$, _has an_ $(n-1)$_-dimensional irreducible complex representation_ Ψ, _whose character_ χ_Ψ _is given by the formula_

$$\chi_\Psi(g) = N(g) - 1 . \quad \square \tag{6}$$

As shown before in the case of S_3 (see Example 1 of Subsection 1 §2), the matrices Ψ_g can be easily determined. To compute $\chi_\Psi(g)$ using formula (6) it is enough to know the cyclic structure of the permutation g. Here are a few examples:

A_4	e	(12) (34)	(123)	(132)
χ_Ψ	3	-1	0	0

S_4	e	(12) (34)	(12)	(123)	(1234)
χ_Ψ	3	-1	1	0	-1

A_5	e	(12) (34)	(123)	(12345)	(12354)
χ_Ψ	4	0	1	-1	-1

Example 2. The irreducible representations of the alternating group A_4. Let us gather together all of the facts we know. The group A_4 has four conjugacy classes. Representatives of these classes and the number of elements in them are given in the first two rows of the following table:

	1	3	4	4
	e	(12) (34)	(123)	(132)
χ_1	1	1	1	1
χ_2	1	1	ε	ε^{-2}
χ_3	1	1	ε^{-1}	ε
χ_4	3	-1	0	0

The commutant $A_4' = \{e, (12)(34), (13)(24), (14)(23)\} \cong V_4$ has index 3 in A_4, and so A_4 has three one-dimensional representations $\Phi^{(1)} = \chi_1$, $\Phi^{(2)} = \chi_2$, $\Phi^{(3)} = \chi_3$ (with kernel A_4' and with $\varepsilon^3 = 1$, $\varepsilon \neq 1$) and one three-dimensional representation $\Phi^{(4)}$ $(12 = 1^2 + 1^2 + 1^2 + 3^2)$. Comparing the tables for A_4 in

Example 1 and in Example 2 of §4, we see that the representation $\Phi^{(4)}$ with character χ_4 is equivalent to the representation Φ of the group A_4 by rotations (the tetrahedral group) and to the representation Ψ which is connected to the 2-transitive group A_4.

Example 3. The irreducible representations of the symmetric group S_4. The first two rows of the table

	1	3	6	8	6
	e	(12) (34)	(12)	(123)	(1234)
χ_1	1	1	1	1	1
χ_2	1	1	-1	1	-1
χ_3	2	2	0	-1	0
χ_4	3	-1	-1	0	1
χ_5	3	-1	1	0	-1

are taken from Exercise 4 §3 Ch. 7. The representation $\Phi^{(1)} = \chi_1$ is the unit representation. The representation $\Phi^{(2)} = \chi_2$ is given by taking the sign of a permutation in S_4. Since $(S_4 : S_4') = 2$ (the example in Subsection 2 §3 Ch. 7), it follows that there are no other one-dimensional representations. The two-dimensional representation $\Phi^{(3)}$ with character χ_3 and with kernel $V_4 \lhd S_4$ is obtained from the considerations in the proof of Theorem 5 and in Example 2 of Subsection 1 §3 Ch. 7. The representation $\Phi^{(4)}$ with character χ_4 corresponds to the rotations of the cube (see the table for S_4 in Example 2 §4). The representation $\Phi^{(5)} = \Psi$ with character χ_5 (see the table in Example 1) is the representation which is always connected with a 2-transitive group. It is also equivalent to the representation coming from all of the symmetry transformations of the tetrahedron Δ_4 (rotations and reflections; it is these transformations which are of importance in describing the oscillations of the phosphorus molecule -- see problem 2 of §2 Ch. 1).

Example 4. The irreducible representations of the quaternion group Q_8. Everything we need to know about Q_8 was discussed in Example 2 of Subsection 5 §3 Ch. 7. In particular, we described the two-dimensional irreducible representation $\Phi^{(5)}$ with character χ_5 (of course, at that time we did not use our present terminology).

	1	1	2	2	2
	e	a^2	a	b	ab
χ_1	1	1	1	1	1
χ_2	1	1	-1	-1	1
χ_3	1	1	-1	1	-1
χ_4	1	1	1	-1	-1
χ_5	2	-2	0	0	0

The four one-dimensional representations have the commutant $\langle a^2 \rangle$ as their kernel, and are determined from the table in the example in Subsection 3.

EXERCISES

1. Derive the relation (4) by explicitly writing out the expression $t_{ij} = (\Gamma_i, \chi_j)_G$ for the coefficients in the expansion $\Gamma_i = \Sigma_j t_{ij} \chi_j$ of the central function Γ_i (see (7) §4) in terms of the irreducible characters.

2. Recall the isomorphism between a vector space V and the dual space V^*, and also the natural identification of V with the double dual $(V^*)^*$. Verify that the map $\tau : A \to \hat{\hat{A}}$ defined by

$$a^\tau(\chi) = \chi(a) ,$$

gives an isomorphism between the abelian groups A and $\hat{\hat{A}}$. This exercise, along with Theorem 4, gives part of the so-called <u>duality law for finite abelian groups.</u> A similar, but much deeper duality law for topological abelian groups, which led to important consequences, was established in the 1930's by L. S. Pontryagin.

3. Prove that, if a finite abelian group A has a faithful complex irreducible representation, then A is a cyclic group.

4. Let A be a finite abelian group, and let B be a subgroup. Prove that any character of B extends to a character of A, and that the number of possible extensions is equal to the index $(A:B)$.

5. Justify the sentence before the parentheses at the end of Example 3 in Subsection 4.

6. What is the average value $\frac{1}{|G|}\sum_g \chi(g)$ of a complex character χ on the elements of a finite group G?

7. Gather together the various tables concerning A_5 (see Example 2 in Subsection 2 §4, Exercise 4 §4, and Example 1), and make up a character table:

	1	15	20	12	12
	e	$(12)(34)$	(123)	(12345)	(12354)
χ_1	1	1	1	1	1
χ_2	3	-1	0	$\frac{1}{2}(1+\sqrt{5})$	$\frac{1}{2}(1-\sqrt{5})$
χ_3	3	-1	0	$\frac{1}{2}(1-\sqrt{5})$	$\frac{1}{2}(1+\sqrt{5})$
χ_4	4	0	1	-1	-1
χ_5	*	*	*	*	*

Describe the irreducible representations with characters χ_1, χ_2, χ_3, and χ_4. Fill in the last row of the table by using the second orthogonality relation (4) for characters.

8. Let $P = \{A^iB^jC^k\,;\ 0 \le i,j,k \le p-1\}$ be the group of order p^3 in Exercise 3 §2 Ch. 7. Let $V = \langle e_0, e_1, \ldots, e_{p-1}\rangle_{\mathbb{C}}$ be a complex vector space of

dimension p; let ε be a primitive p-th root of 1; and let $\mathcal{A}, \mathcal{B}_k, \mathcal{C}_k$ be the linear operators on V defined by the relations

$$\mathcal{A} e_i = e_{i+1}, \quad \mathcal{B}_k e_i = \varepsilon^{-k} e_i, \quad \mathcal{C}_k e_i = \varepsilon^k e_i, \quad 0 \le i \le p-1$$

(the subscripts of the basis elements are taken modulo p). Show that the map

$$\Phi^{(k)} : A \longmapsto \mathcal{A}, \quad B \longmapsto \mathcal{B}_k, \quad C \longmapsto \mathcal{C}_k$$

defines an irreducible linear representation of the group P. The representations $\Phi^{(1)}, \dots, \Phi^{(p-1)}$ are pair-wise non-equivalent and, together with the p^2 one-dimensional representations (p^2 is the index of the commutant $P' = \langle C \rangle$ in P), exhaust all of the complex irreducible representations of the group P.

9. Carry out the computations necessary to complete the following argument. Let $D_n = \langle a, b \mid a^n = e,\ b^2 = e,\ bab^{-1} = a^{-1} \rangle$ be the dihedral group of order $2n$, whose properties (including a description of the conjugacy classes) were given in Example 1 of Subsection 5 §3 Ch. 7. Since $\langle a \rangle \lhd D_n$, it follows that the maps $a \mapsto 1$, $b \mapsto 1$ and $a \mapsto 1$, $b \mapsto -1$ give two one-dimensional representations. Let $\varepsilon = e^{2\pi i/n}$ be a primitive n-th root of 1. Then the map

$$\Phi^{(j)} : a \longmapsto \begin{Vmatrix} \varepsilon^j & 0 \\ 0 & \varepsilon^{-j} \end{Vmatrix}, \quad b \longmapsto \begin{Vmatrix} 0 & 1 \\ 1 & 0 \end{Vmatrix}$$

defines a representation of degree 2. The representation $\Phi^{(j)}$ is irreducible for $j = 1, 2, \dots, [(n-1)/2]$ (where $[\alpha]$ denotes the greatest integer less than or equal to a real number α). If $n = 2m$, the representation $\Phi^{(m)}$ splits into a direct sum of two one-dimensional representations: $a \mapsto -1$, $b \mapsto 1$ and $a \mapsto -1$, $b \mapsto -1$. This agrees with the fact that the commutant D'_{2m} has index 4 in D_{2m} and $D_{2m}/D'_{2m} \cong \mathbb{Z}_2 \times \mathbb{Z}_2$. All of these representations are irreducible, and they make up a complete set of complex irreducible representations of the dihedral group. Find a realization of the representation $\Phi^{(j)}$ over the reals. Give an explicit isomorphism (showing equivalence) $\Phi^{(k)} \approx \Phi^{(j)}$ for $k > m$, for suitably chosen $j \leq m$.

10. The crystallographic groups (see Problem 2 in §2 Ch. 1). Let E be n-dimensional Euclidean space, and let V be the associated vector space with the usual scalar product. Every rigid motion d of E corresponds to an orthogonal linear transformation $\bar{d} \in O(n)$, and we have $\overline{d_1 d_2} = \bar{d}_1 \bar{d}_2$. A group D of rigid motions is called a crystallographic group if the D-orbit of any point is discrete (i.e., does not have any limit points), and if there exists a compact set $M \subset E$ for which $D(M) = \bigcup_{d \in D} d(M) = E$. The Schoenflies-Bieberbach theorem states that for any crystallographic group D there exist n independent affine transformations which generate a normal subgroup L in D, where $\bar{D} \cong D/L$ is a finite group (a crystallographic point group). When $n = 3$, there are in all 32 geometrically different crystallographic point groups. Of course, among these groups are the groups containing reflections (improper rigid motions). It follows from the conditions defining a crystallographic group that every proper rotation in $\bar{D}$ is given by a matrix which is similar to

$$A = \begin{Vmatrix} \cos\theta & -\sin\theta & 0 \\ \sin\theta & \cos\theta & 0 \\ 0 & 0 & 1 \end{Vmatrix}$$

and has $\operatorname{tr} A = 1 + 2\cos\theta \in \mathbb{Z}$. Using Theorem 2 §3 and the preceding observation, show that for $n = 3$ the only crystallographic point groups without reflections are the cyclic groups C_1, C_2, C_3, C_4, C_6, the dihedral groups D_2, D_3, D_4, D_6, the tetrahedral group T and the octahedral group O.

§6. Representations of SU(2) and SO(3)

The concrete images of representations of SO(3) play a key role in the study of the physical world, since the action of SO(3) reflects the symmetry of many problems of physics. From a mathematical point of view, the action of SO(3) is of special interest in part because it induces an action on the space of solutions of the differential equation $\Delta f = 0$, where $\Delta = \frac{\partial^2}{\partial x^2} + \frac{\partial^2}{\partial y^2} + \frac{\partial^2}{\partial z^2}$ is the Laplace differential operator. The two-dimensional analog of this was considered at the very beginning of the chapter (Problem 1).

Any element in $SO(3)$ is a product of various operators B_φ and C_θ of the form (1) in §1 Ch. 7. But B_φ does not act on z, and C_θ does not act on x. Hence, the invariance of the equation $\Delta f = 0$ relative to B_φ and C_θ follows from the computations which were made in the two-dimensional case. We conclude that the equation $\Delta f = 0$ is invariant relative to all of $SO(3)$, i.e.,

$$\Delta f = 0 \Longrightarrow \Delta(\Phi_g f) = 0, \quad \forall g \in SO(3),$$

where $\Phi_g f$ is the function defined by:

$$(\Phi_g f)(x,y,z) = f(g^{-1}(x), g^{-1}(y), g^{-1}(z)) \quad . \tag{1}$$

By assumption, if g^{-1} is an orthogonal transformation with matrix $(a_{ij})_1^3$, then the column of the new variables has the form

$$\begin{Vmatrix} g^{-1}(x) \\ g^{-1}(y) \\ g^{-1}(z) \end{Vmatrix} = \begin{Vmatrix} a_{11} & a_{12} & a_{13} \\ a_{21} & a_{22} & a_{23} \\ a_{31} & a_{32} & a_{33} \end{Vmatrix} \begin{Vmatrix} x \\ y \\ z \end{Vmatrix} .$$

According to (1),

$$\begin{aligned} (\Phi_g(\Phi_h f))(x,y,z) &= (\Phi_h f)(g^{-1}(x), g^{-1}(y), g^{-1}(z)) = \\ &= f(h^{-1}(g^{-1}(x)), h^{-1}(g^{-1}(y)), h^{-1}(g^{-1}(z))) = \\ &= f((gh)^{-1}(x), (gh)^{-1}(y), (gh)^{-1}(z)) = (\Phi_{gh} f)(x,y,z) \quad . \end{aligned}$$

Hence,

$$\Phi_g \Phi_h = \Phi_{gh} ,$$

i.e., the linear operators Φ_g, $g \in SO(3)$, act on functions in such a way that the map

$\Phi : g \mapsto \Phi_g$ is a representation of the group SO(3). This is a very natural method for constructing representations (which we actually used before when considering symmetric functions under the action of the group S_n); in principle, this method is suitable for a large class of groups, and it is typical of the techniques used in functional analysis. Starting from certain concrete conditions, one need only choose a suitable space of functions and then decompose it into irreducible invariant subspaces (this is a problem in harmonic analysis).

In the case of the group SO(3), when all of the irreducible representations are finite dimensional (this is true generally for compact groups, but we shall not treat the theory of compact groups in this exposition), we take for our space of functions the homogeneous polynomials

$$f(x,y,z) = \sum_{s,t} a_{s,t}\, x^s y^t z^{m-s-t}$$

of fixed degree m $(m = 1,2,3,\ldots)$. These polynomials form a space P_m of dimension $\binom{m+2}{2}$ (see Exercise 4 §2 Ch. 5). Since $\Delta f \in P_{m-2}$, it follows that the condition $\Delta f = 0$ is equivalent to $\binom{m}{2}$ linear conditions on the coefficients $a_{s,t}$. The solutions $f \in P_m$ of the equation $\Delta f = 0$ are called homogeneous harmonic polynomials of degree m. Since the operator Δ is linear, the harmonic polynomials form a subspace H_m of dimension equal to $\binom{m+2}{2} - \binom{m}{2} = 2m+1$ (at this point we can only say that its dimension is $\leq 2m+1$, but in fact we have equality). By what we have said, H_m is invariant under the action $\Phi = \Phi^{(m)}$ of the group SO(3). It turns out that we have the following fact: the space H_m of the representation $\Phi^{(m)}$ is irreducible over $\mathbb{C}$, and any complex irreducible representation of SO(3) is equivalent to one of the representations $(\Phi^{(m)}, H_m)$ having odd dimension $2m+1$. Rather than prove this theorem, we shall proceed to the group SU(2), where it is somewhat easier to obtain a family of irreducible representations. Because we have a natural epimorphism SU(2) → SO(3) whose kernel consists of the matrices $\pm E$ (see §1 Ch. 7), every representation Ψ of SO(3) can also be considered as a representation of SU(2) (see the proof of Theorem 5 §5), which satisfies the so-called parity condition: $\Psi_{-E} = \Psi_E$. Of course, this means that we also have $\Psi_{-g} = \Psi_g$ for all $g \in SU(2)$. Conversely, any representation Ψ of $SU(2)$ which satisfies the parity condition can be considered as a representation of $SO(3)$. The "double-valued" representations of $SO(3)$, i.e., the representations of $SU(2)$ which do not satisfy the parity condition, also have physical meaning. For example, the usual two-dimensional (spinor) representation is of this type.

We further note that any irreducible representation of $SO(3)$ other than the trivial one is faithful, as immediately follows from the fact that $SO(3)$ is simple (Theorem 6 §3 Ch. 7).

THEOREM 1. *Let* $V_n = \langle x^k y^{n-k} \mid k = 0, 1, \dots, n\rangle_{\mathbb{C}}$ *be the space of homogeneous polynomials of degree* n *in two complex variables with an action* $\Psi^{(n)}$ *of* $SU(2)$ *defined by the rule*

$$(\Psi_g^{(n)} f)(x,y) = f(\bar{\alpha}x - \beta y, \bar{\beta}x + \alpha y)$$

for every element

$$g = \begin{Vmatrix} \alpha & \beta \\ -\bar{\beta} & \bar{\alpha} \end{Vmatrix}, \qquad |\alpha|^2 + |\beta|^2 = 1 .$$

Then $(\Psi^{(n)}, V_n)$ *is an irreducible* $(n+1)$*-dimensional representation of* $SU(2)$. *If* n *is even, then* $(\Psi^{(n)}, V_n)$ *is also an irreducible representation of the group* $SO(3)$.

Proof. Suppose that the polynomial

$$f(x,y) = \sum_{k=0}^{n} a_k x^k y^{n-k} \neq 0$$

is contained in some invariant subspace $U \subset V_n$. Then also

$$\sum_{k=0}^{n} (e^{-i\varphi})^k a_k x^k y^{n-k} = e^{-in\frac{\varphi}{2}} \left(\Psi_{b_\varphi}^{(n)} f\right)(x,y) \in U ,$$

where b_φ is an element of $SU(2)$ of the form (4) §1 Ch. 7. Since φ is an arbitrary real number in the interval $(0, 2\pi)$, we can make up a linear system with Vandermonde determinant, from which it follows that

$$f(x,y) \in U \Longrightarrow x^k y^{n-k} \in U \qquad (2)$$

for any monomial with coefficient $a_k \neq 0$. But if $x^k y^{n-k} \in U$ for some k, then also

$$\bar{\alpha}^k \bar{\beta}^{n-k} x^m + \cdots = (\bar{\alpha} x - \beta y)^k (\bar{\beta} x + \alpha y)^{n-k} = \Psi_g^{(n)}(x^k y^{n-k}) \in U \quad .$$

Taking a g with $\alpha\beta \neq 0$, from (2) we conclude that $x^n \in U$, which, in turn, gives us

$$\sum_{s=0}^{n} \binom{n}{s} \bar{\alpha}^s (-\beta)^{n-s} x^s y^{n-s} \in U \quad .$$

Since $\binom{n}{s} \bar{\alpha}^s (-\beta)^{n-s} \neq 0$, we have $x^s y^{n-s} \in U$, $s = 0, 1, \ldots, n$. Thus, $U = V_n$, and we have proved that $(\Psi^{(n)}, V_n)$ is irreducible.

Next, we have

$$\Psi_{-E}^{(n)}(x^k y^{n-k}) = (-x)^k (-y)^{n-k} = (-1)^n x^k y^{n-k} \quad ,$$

so that the parity condition holds for $n = 2m$ (see the remark above), and $(\Psi^{(2m)}, V_{2m})$ can be considered as an irreducible $(2m+1)$-dimensional representation of $SO(3)$. □

Actually, $\Psi^{(2m)}$ is equivalent to the representation $\Phi^{(m)}$ of $SO(3)$ on the space of homogeneous harmonic polynomials of degree m, but we shall not prove that here; nor shall we show how to choose a basis of V_n in which the representation $\Psi^{(n)}$ becomes unitary (this can be done). We shall only note that, borrowing the terminology of tensor analysis, the representation $\Psi^{(n)}$ of $SU(2)$ can also be realized in the class of coinvariant symmetric tensors of rank n. A complete and transparent theory of the representations of compact groups, including $SU(2)$ and $SO(3)$, is usually given using the infinitesimal method, based on the correspondence between Lie groups and Lie algebras.

EXERCISES

1. Construct $2m+1$ linearly independent homogeneous harmonic polynomials of degree m.

2. Show that any homogeneous polynomial $f \in P_m$ can be written as a linear combination of harmonic polynomials of degree $m, m-2, m-4, \ldots$ with coefficients which are a function of $x^2+y^2+z^2$.

3. Use Exercise 2 to show that every polynomial function $\tilde{g} : (X,Y,Z) \mapsto g(x,y,z)$ on the sphere $S^2 : x^2+y^2+z^2 = 1$ can be decomposed into <u>spherical functions</u> -- the restrictions of harmonic polynomials to S^2.

4. Without using the complete description of the irreducible representations of $SO(3)$, show that the only homomorphism $\tau : SO(3) \to SU(2)$ is the trivial one.

§7. Tensor products of representations

1. <u>The dual representation.</u> Let (Φ, V) be a complex representation of a group G. We consider the dual space V^* (the space of linear functions on V) and set

$$(\Phi^*(g) \cdot f)(v) = f(\Phi(g^{-1})v) ; \qquad f \in V^*, \quad v \in V \quad . \tag{1}$$

We immediately verify that $\Phi^*(g)$ is a linear operator. Next, we choose dual bases of V and V^*:

$$V = \langle e_1, \ldots, e_n \rangle , \quad V^* = \langle f_1, \ldots, f_n \rangle , \quad f_i(e_j) = \delta_{ij} \quad .$$

The matrix of the linear operator $\Phi^*(g)$ in the basis $f_1, \ldots, f_n$ is the transpose of the matrix of the linear operator $\Phi(g^{-1})$ in the basis $e_1, \ldots, e_n$:

$$\Phi^*_g = {}^t\Phi_{g^{-1}} \quad . \tag{2}$$

Since

$$\Phi^*_{gh} = {}^t\Phi_{(gh)^{-1}} = {}^t\Phi_{h^{-1}g^{-1}} = {}^t\left(\Phi_{h^{-1}}\Phi_{g^{-1}}\right) = {}^t\Phi_{g^{-1}}\,{}^t\Phi_{h^{-1}} = \Phi^*_g\Phi^*_h ,$$

it follows that the relation (2) (or (1)) defines another linear representation (Φ^*, V^*) of G; it is called the <u>dual</u> (or <u>contragradient</u>) representation corresponding to (Φ, V). It is not hard to see (for example, from (2)) that $(\Phi^*)^* \approx \Phi$. It is possible for representations which are dual to one another to be equivalent. For example, if (Φ, V) is a real orthogonal representation, then $\Phi^*_g = {}^t\Phi_g^{-1} = \Phi_g$. But the representations Φ and Φ^* are usually not equivalent, as we can see from the simplest example:

$$C_3 = \langle a \mid a^3 = e \rangle ; \quad \Phi(a) = \varepsilon, \quad \Phi^*(a) = \varepsilon^{-1} \quad (\varepsilon^2 + \varepsilon + 1 = 0) .$$

If G is a finite group, then we can obtain a precise criterion for a representation and its dual to be equivalent in terms of characters. Since the characteristic polynomials of the matrices A and ${}^tA$ coincide:

$$\det(\lambda E - {}^tA) = \det{}^t(\lambda E - A) = \det(\lambda E - A) ,$$

it follows from the elementary properties of characters (the proposition in §4) that

$$\chi_{\Phi^*}(g) = \overline{\chi_\Phi(g)} .$$

In particular, a representation Φ whose character takes only real values is equivalent to Φ^*. Note that, since we always have

$$(\chi_{\Phi^*}, \chi_{\Phi^*})_G = (\chi_\Phi, \chi_\Phi)_G ,$$

it follows that Φ and Φ^* are either both reducible or both irreducible.

2. <u>Tensor products of representations.</u> The following fact is normally proved in a course in linear algebra or algebraic geometry (see also Exercise 1 below).

THEOREM 1. <u>Let</u> V <u>and</u> W <u>be vector spaces over a field</u> K. <u>Then there</u>

exist a vector space T over K and a bilinear map $\tau : V \times W \to T$ satisfying the following conditions:

(T1) if $v_1, \dots, v_k \in V$ are linearly independent and $w_1, \dots, w_k \in W$, then

$$\sum_{i=1}^{k} \tau(v_i, w_i) = 0 \Rightarrow w_1 = 0, \dots, w_k = 0 ;$$

(T2) if $w_1, \dots, w_k \in W$ are linearly independent, then $\sum_i \tau(v_i, w_i) = 0 \Rightarrow v_1 = 0, \dots, v_k = 0$;

(T3) τ is surjective, i.e.,

$$T = \langle \tau(v,w) \mid v \in V, \ w \in W \rangle_K \quad .$$

In addition, the pair (τ, T) is universal in the sense that, if (τ', T') is a pair consisting of a vector space T' and a bilinear map $\tau' : V \times W \to T'$, then there exists a unique linear map $\sigma : T \to T'$ such that $\tau'(v,w) = \sigma(\tau(v,w))$, $v \in V$, $w \in W$. □

If we had two such universal pairs (τ, T) and (τ', T'), then we would find that the linear maps $\sigma : T \to T'$ and $\sigma' : T' \to T$ would actually be mutually inverse isomorphisms: $\sigma' \circ \sigma = e_T$, $\sigma \circ \sigma' = e_{T'}$. Thus, in that case $T \cong T'$, and the isomorphism $\sigma : T \to T'$ has the property in the theorem.

The pair (τ, T), which is uniquely determined up to isomorphism once V and W are given, is called the tensor product of V and W. We write $T = V \otimes_K W$, or simply $T = V \otimes W$, but we must also keep in mind that the space T is accompanied with a bilinear map $(v,w) \mapsto v \otimes w$ from $V \times W$ to T which satisfies conditions (T1)-(T3). Thus, the elements of the tensor product $V \otimes W$ are the formal linear combinations of ordered pairs $v \otimes w$ $(v \in V, w \in W)$ with coefficients in K. Here the following relations are fulfilled:

$$
\begin{aligned}
(v_1 + v_2) \otimes w - v_1 \otimes w - v_2 \otimes w &= 0 , \\
v \otimes (w_1 + w_2) - v \otimes w_1 - v \otimes w_2 &= 0 , \qquad (3)\\
\lambda v \otimes w - v \otimes \lambda w &= 0 , \qquad \lambda \in K
\end{aligned}
$$

$$(\lambda (v \otimes w) = \lambda v \otimes w = v \otimes \lambda w) \quad .$$

It is immediately clear from Theorem 1 that the bijective maps $v \otimes w \mapsto w \otimes v$, $(u \otimes v) \otimes w \mapsto u \otimes (v \otimes w)$, and $v \otimes \lambda \mapsto \lambda \otimes v \mapsto \lambda v$ give isomorphisms (called canonical isomorphisms) between the vector spaces:

$$
\begin{aligned}
V \otimes W &\cong W \otimes V , \\
(U \otimes V) \otimes W &\cong U \otimes (V \otimes W) , \\
V \otimes K &\cong K \otimes V \cong V \quad .
\end{aligned}
$$

We also have the following distributive laws:

$$
\begin{aligned}
(U \oplus V) \otimes W &\cong (U \otimes W) \oplus (V \otimes W) , \\
U \otimes (V \oplus W) &\cong (U \otimes V) \oplus (U \otimes W) \quad .
\end{aligned}
$$

In tensor analysis, where the above ideas originated, one studies tensor products of the following special form:

$$\underbrace{V^* \otimes \ldots \otimes V^*}_{p} \otimes \underbrace{V \otimes \ldots \otimes V}_{q} \quad .$$

The elements in such a tensor product are called tensors of type (p,q), p times covariant and q times contravariant. If we choose dual bases $e_1, \ldots, e_n$ in V and $e^1, \ldots, e^n$ in V^*, then the elements $e^{i_1} \otimes \ldots \otimes e^{i_p} \otimes e_{j_1} \otimes \ldots \otimes e_{j_q}$ form a basis for the space of tensors of type (p,q). We usually think of a tensor as simply the set of coordinates $\left\{ t_{i_1 \ldots i_p}^{j_1 \ldots j_q} \right\}$ in this basis, where we have rules for change of coordinates when passing from one basis to another. In this way one obtains the interpretation of such notions as a bilinear form and a linear operator in the language of tensors (in effect, in the language of matrices). We shall not dwell further on such matters, since our purpose is to discuss representations in the general situation of tensor products.

Let $\mathfrak{A} : V \to V$ and $\mathfrak{B} : W \to W$ be linear operators. By their tensor product we mean the linear operator

$$\mathfrak{A} \otimes \mathfrak{B} : V \otimes W \longrightarrow V \otimes W \quad ,$$

which acts according to the rule

$$(\mathfrak{A} \otimes \mathfrak{B})(v \otimes w) = \mathfrak{A}v \otimes \mathfrak{B}w \tag{4}$$

(and extends to all of $V \otimes W$ by linearity: $(\mathfrak{A} \otimes \mathfrak{B})(\Sigma\, v_i \otimes w_i) = \Sigma\, \mathfrak{A}v_i \otimes \mathfrak{B}w_i)$. This definition is clearly compatible with the relations (3). For example,

$$\mathfrak{A}(v_1 + v_2) \otimes \mathfrak{B}w - \mathfrak{A}v_1 \otimes \mathfrak{B}w - \mathfrak{A}v_2 \otimes \mathfrak{B}w = (\mathfrak{A}v_1 + \mathfrak{A}v_2) \otimes \mathfrak{B}w - \mathfrak{A}v_1 \otimes \mathfrak{B}w - \mathfrak{A}v_2 \otimes \mathfrak{B}w = 0.$$

Hence the action of $\mathfrak{A} \otimes \mathfrak{B}$ on $V \otimes W$ is correctly defined. We also note the following relations, which follow directly from the definition (4):

$$\begin{aligned}
(\mathfrak{A} \otimes \mathfrak{B})(\mathfrak{C} \otimes \mathfrak{D}) &= \mathfrak{A}\mathfrak{C} \otimes \mathfrak{B}\mathfrak{D} , \\
(\mathfrak{A} + \mathfrak{C}) \otimes \mathfrak{B} &= \mathfrak{A} \otimes \mathfrak{B} + \mathfrak{C} \otimes \mathfrak{B} , \\
\mathfrak{A} \otimes (\mathfrak{B} + \mathfrak{C}) &= \mathfrak{A} \otimes \mathfrak{B} + \mathfrak{A} \otimes \mathfrak{C} , \\
\mathfrak{A} \otimes \lambda\mathfrak{B} = \lambda\mathfrak{A} \otimes \mathfrak{B} &= \lambda(\mathfrak{A} \otimes \mathfrak{B}) \quad .
\end{aligned}$$

We leave their verification to the reader.

As before, let $V = \langle e_1, \ldots, e_n \rangle$ and $W = \langle f_1, \ldots, f_m \rangle$. We obtain an $nm \times nm$ matrix for the operator $\mathfrak{A} \otimes \mathfrak{B}$, which we denote $A \otimes B$, in the basis

$$\{e_1 \otimes f_1, \ldots, e_1 \otimes f_m, e_2 \otimes f_1, \ldots, e_2 \otimes f_m, \ldots, e_n \otimes f_1, \ldots, e_n \otimes f_m\}$$

if we note that

$$\mathfrak{A} e_i = \sum_{i'} \alpha_{i'i} e_{i'}, \qquad \mathfrak{B} f_j = \sum_{j'} \beta_{j'j} f_{j'}$$

$$(\mathfrak{A} \otimes \mathfrak{B})(e_i \otimes f_j) = \sum_{i',j'} \alpha_{i'i} \beta_{j'j} e_{i'} \otimes f_{j'} \quad .$$

Hence for $A = (\alpha_{i'i})$ and $B = (\beta_{j'j})$ we have

$$A \otimes B = (\alpha_{i'i} \beta_{j'j}) = \begin{Vmatrix} \alpha_{11} B & \alpha_{12} B & \dots & \alpha_{1n} B \\ \alpha_{21} B & \alpha_{22} B & \dots & \alpha_{2n} B \\ \dots & \dots & \dots & \dots \\ \alpha_{n1} B & \alpha_{n2} B & \dots & \alpha_{nn} B \end{Vmatrix} \quad .$$

In particular, we have the trace formula

$$\operatorname{tr} A \otimes B = \alpha_{11} \operatorname{tr} B + \alpha_{22} \operatorname{tr} B + \cdots + \alpha_{nn} \operatorname{tr} B = \operatorname{tr} A \cdot \operatorname{tr} B \quad . \tag{5}$$

We note in passing that

$$\det A \otimes B = \det(A \otimes E_m)(E_n \otimes B) = \det(A \otimes E_m) \cdot \det(E_n \otimes B) = (\det A)^m (\det B)^n \quad ,$$

Now let (Φ, V) and (Ψ, W) be two linear representations of a group G with characters χ_Φ and χ_Ψ, respectively. We define the representation $(\Phi \otimes \Psi, V \otimes W)$ in the natural way, by setting

$$(\Phi \otimes \Psi)(g) = \Phi(g) \otimes \Psi(g), \qquad \forall g \in G \quad .$$

The general properties of the tensor product of linear operators, along with (5), imply that the map $\Phi \otimes \Psi$ actually gives a representation of G with representation space $V \otimes W$ and character

$$\chi_{\Phi \otimes \Psi} = \chi_\Phi \chi_\Psi \quad . \tag{6}$$

We shall call $(\Phi \otimes \Psi, V \otimes W)$ the <u>tensor product of the representations</u> (Φ, V) and (Ψ, W). If $\Psi = \Phi$ and $W = V$, we call it the <u>tensor square</u> of the representation (Φ, V). On the right in (6) we have the usual point-wise product of the central functions

χ_Φ and χ_Ψ.

If U is a G-invariant subspace of V, then it is obvious that $U \otimes W$ is a G-invariant subspace of $V \otimes W$. The analogous remark applies to G-invariant subspaces of W. But irreducibility of V and W by no means implies irreducibility of $V \otimes W$, as we can see by the example of the tensor square $\Phi^{(3)} \otimes \Phi^{(3)}$ of the two-dimensional representation of S_3 (see the table in Subsection 2 §5). In fact, $\dim_{\mathbb{C}} \Phi^{(3)} \otimes \Phi^{(3)} = 4$, and the maximum possible dimension of an irreducible representation of S_3 is 2.

The problem of effectively describing the irreducible representations contained in $\Phi \otimes \Psi$, or more generally in $\Phi^{(1)} \otimes \Phi^{(2)} \otimes \ldots \otimes \Phi^{(p)}$, is fundamental, since many important and very natural group representations arise as tensor products. It is from this point of view that one should consider the representations of the groups $SU(2)$ and $SO(3)$ (see §6), and also Examples 3 and 4 of Subsection 2 §1. The invariant subspaces of symmetric and skew-symmetric covariant (or contravariant) tensors occur constantly in various geometrical applications. This problem is especially attractive when we have a complete reducibility theorem for the representations under consideration.

3. <u>The ring of characters.</u> For simplicity we limit ourselves to the case of complex representations of a finite group G. Let $\Phi^{(1)}, \Phi^{(2)}, \ldots, \Phi^{(r)}$ be a complete set of pair-wise non-equivalent complex irreducible representations of G, and let $\chi_1, \chi_2, \ldots, \chi_r$ be the corresponding characters (r is the number of conjugacy classes in G). We know that the representation $\Phi \otimes \Psi$, like any representation, has a decomposition

$$\Phi \otimes \Psi \approx m_1 \Phi^{(1)} \dot{+} \cdots \dot{+} m_r \Phi^{(r)} ,$$

where the multiplicities m_i only depend on the representations Φ and Ψ whose tensor product we are studying. By (6), we have

$$\chi_\Phi \chi_\Psi = m_1 \chi_1 + \cdots + m_r \chi_r \quad .$$

Let $X_{\mathbb{Z}}(G)$ be the set of all possible integral linear combinations of the characters $\chi_1, \ldots, \chi_r$. Earlier we proved that $\chi_1, \ldots, \chi_r$ form an orthonormal basis for the space $X_{\mathbb{C}}(G)$; hence, in any case, $X_{\mathbb{Z}}(G) \subset X_{\mathbb{C}}(G)$ is a free abelian group isomorphic to $\mathbb{Z}^r$ with generators $\chi_1, \ldots, \chi_r$. We call the elements of this free abelian group the <u>generalized characters</u> of the group G. The only true characters are the linear combinations $\Sigma\, m_i \chi_i$ with all of the m_i nonnegative.

From all this it is clear that tensor product of representations induces a binary operation on $X_{\mathbb{Z}}(G)$ which is commutative and associative and satisfies the distributive laws. To summarize, we have the following

THEOREM 2. <u>The generalized characters form a commutative associative ring $X_{\mathbb{Z}}(G)$ whose unit is the trivial character</u> χ_1.

We say that $X_{\mathbb{C}}(G)$ is a commutative associative algebra of dimension r over $\mathbb{C}$. The structure of the ring $X_{\mathbb{Z}}(G)$ or the algebra $X_{\mathbb{C}}(G)$ is completely determined by the so-called <u>structure constants</u> -- the integers m_{ij}^k in

$$\chi_i \chi_j = \sum m_{ij}^k \chi_k \quad . \tag{7}$$

In particular, the equalities $m_{ij}^k = m_{ji}^k$ and $m_{1j}^k = \delta_{kj}$ reflect the properties that $X_{\mathbb{Z}}(G)$ is commutative and χ_1 is its identity element. According to (7), we have

$$\chi_i(g)\, \chi_j(g) = \sum m_{ij}^k \chi_k(g)\,, \qquad \forall g \in G \quad .$$

If we multiply both sides of this relation by $(|G|)^{-1}\, \overline{\chi_s(g)}$, sum over $g \in G$ and use the first orthogonality relation for characters, we obtain

$$m_{ij}^s = \frac{1}{|G|} \sum_{g \in G} \chi_i(g)\, \chi_j(g)\, \overline{\chi_s(g)} \quad . \tag{8}$$

Thus, the structure constants can be expressed in terms of the characters themselves.

From (8) we can derive the following simple fact:

$$m_{ij}^{1} = \frac{1}{|G|}\sum_{g} \chi_i(g)\chi_j(g)\overline{\chi_1(g)} = \frac{1}{|G|}\sum_{g} \chi_i(g)\chi_j(g) =$$

$$= \frac{1}{|G|}\sum_{g} \chi_i(g)\overline{\overline{\chi_j(g)}} = (\chi_i, \chi_j^*)_G \quad ,$$

where $\chi_j^* = \chi_{\Phi^{(j)*}}$ is the character of the dual representation of $\Phi^{(j)}$ (see Subsection 1). Thus, <u>the unit (trivial) representation occurs in</u> $\Phi^{(i)} \otimes \Phi^{(j)}$ <u>if and only if</u> $\Phi^{(i)}$ <u>is equivalent to the representation</u> $\Phi^{(j')} = \Phi^{(j)*}$ (since otherwise $m_{ij}^{1} = (\chi_i, \chi_j^*)_G = 0$).

We further note that <u>the tensor product of a one-dimensional representation</u> $\Phi^{(i)}$ <u>and an arbitrary irreducible representation</u> $\Phi^{(j)}$ <u>is always an irreducible representation having the same dimension as</u> $\Phi^{(j)}$. This can be seen in many ways, for example, from the criterion for irreducibility in terms of characters. We write $\chi = \chi_{\Phi^{(i)} \otimes \Phi^{(j)}} = \chi_i \chi_j$, where $\chi_i(g)$, since it is a root of 1, satisfies $\chi_i(g)\overline{\chi_i(g)} = 1$; hence,

$$(\chi, \chi)_G = \frac{1}{|G|}\sum_{g} \chi_i(g)\chi_j(g)\overline{\chi_i(g)}\,\overline{\chi_j(g)} = \frac{1}{|G|}\sum_{g} \chi_j(g)\overline{\chi_j(g)} = (\chi_j, \chi_j)_G = 1 \quad .$$

<u>Example 1.</u> $G = S_3$ (see the tables in Subsection 1 §2 and Subsection 2 §5):

$$\Phi^{(1)} \otimes \Phi^{(3)} \approx \Phi^{(2)} \otimes \Phi^{(3)} \approx \Phi^{(3)} \quad .$$

<u>Example 2.</u> $G = S_4$ (see Example 3 in Subsection 4 §5):

$$\Phi^{(2)} \otimes \Phi^{(4)} \approx \Phi^{(5)}, \qquad \Phi^{(2)} \otimes \Phi^{(5)} \approx \Phi^{(4)} \quad .$$

Finally, we prove the following curious theorem, which serves as a generalization of Theorem 2 §5 on the decomposition of the regular representation.

THEOREM 3. <u>Let</u> $\chi = \chi_\Phi$ <u>be the character of a faithful representation</u> (Φ, V) <u>of a finite group</u> G <u>over the field of complex numbers</u> $\mathbb{C}$. <u>Suppose that</u> χ <u>takes precisely</u> m <u>distinct values on</u> G. <u>Then every irreducible character</u> χ_k <u>occurs with</u>

<u>non-zero coefficient in the decomposition of at least one of the characters</u> $\chi^0 = \chi_1, \chi, \chi^2, \ldots, \chi^{m-1}$. <u>In other words, every irreducible representation occurs in at least one of the tensor powers</u> $\Phi^{\otimes i} = \Phi \otimes \ldots \otimes \Phi$, $0 \leq i \leq m-1$, <u>where</u> Φ <u>is a faithful representation.</u>

<u>Proof.</u> Let $\omega_j = \chi(g_j)$, $j = 0, 1, \ldots, m-1$, be the distinct values taken by χ on G, where $\omega_0 = \chi(e) = \deg \Phi$. Further let

$$G_j = \{g \in G \mid \chi(g) = \chi(g_j) = \omega_j\} \quad .$$

Since Φ is faithful, we have

$$G_0 = \operatorname{Ker} \Phi = \{e\} \quad .$$

Let χ_k be an irreducible character of G which does not occur in the decomposition of any of the characters χ^i. Then

$$0 = |G| (\chi^i, \chi_k)_G = \sum_{j=0}^{m-1} (\chi(g_j))^i \sum_{g \in G_j} \overline{\chi_k(g)} = \sum \omega_j^i T_j, \qquad 0 \leq i \leq m-1 \quad ,$$

is a homogeneous system of linear equations in the $T_j = \sum_{g \in G_j} \overline{\chi_k(g)}$ with determinant

$$\det(\omega_j^i) = \begin{vmatrix} 1 & 1 & \ldots & 1 \\ \omega_0 & \omega_1 & \ldots & \omega_{m-1} \\ \ldots & \ldots & \ldots & \ldots \\ \omega_0^{m-1} & \omega_1^{m-1} & \ldots & \omega_{m-1}^{m-1} \end{vmatrix} \quad ,$$

which is non-zero (since it is a Vandermonde determinant). Thus, $T_j = 0$, $j = 0, 1, \ldots, m-1$, i.e.,

$$\sum_{g \in G_j} \chi_k(g^{-1}) = 0, \qquad j = 0, 1, \ldots, m-1 \quad .$$

In particular,

$$0 = \sum_{g \in G_0} \chi_k(g^{-1}) = \chi_k(e) \quad .$$

This contradiction proves the theorem. □

In the case of the regular representation we obviously have $m = 2$.

4. Invariants of linear groups. As usual, by a linear group of degree (or dimension) n we mean a subgroup of $GL(n, K)$, where K is a field. In what follows we shall take $K = \mathbb{R}$ or $\mathbb{C}$. If G is an abstract group and $\Phi : G \to G(n, \mathbb{C})$ is a linear representation, then we shall also call the pair (G, Φ) a linear group. The linear transformations Φ_g act on columns:

$$\left\| \begin{matrix} \Phi_g(x_1) \\ \vdots \\ \Phi_g(x_n) \end{matrix} \right\| = \Phi_g \left\| \begin{matrix} x_1 \\ \vdots \\ x_n \end{matrix} \right\| \quad .$$

These operators take any form (i.e., homogeneous polynomial) f of degree m into another form of degree m:

$$(\widetilde{\Phi}_g f)(x_1, \ldots, x_n) = f\left(\Phi_{g^{-1}}(x_1), \ldots, \Phi_{g^{-1}}(x_n)\right) \quad .$$

We have already encountered some special cases of this action (see §6). The map $\widetilde{\Phi}$ gives a representation of the group G in the space P_m of forms of degree m over $\mathbb{C}$ (i.e., in the space of covariant symmetric tensors of rank m).

Definition. A form $f \in P_m$ which remains fixed under the action of Φ (i.e., $\widetilde{\Phi}_g f = f$, $\forall g \in G$) is called an (integral) invariant of degree m of the linear group (G, Φ).

Actually, in the general theory of invariants one takes a degree m polynomial with coefficients in "generic form" which remains fixed under the action of $\widetilde{\Phi}(G)$; but for simplicity we shall use the above definition. If we take f to be a rational function, then we arrive at the notion of a <u>rational invariant</u>. We also have the important concept of a <u>relative invariant</u> f, where we have

$$\widetilde{\Phi}_g f = \omega_g f ,$$

with $\omega_g \in \mathbb{C}$ a factor which depends on the element $g \in G$.

It is clear that any set of invariants $\{f_1, f_2, \ldots\}$ of a linear group (G, Φ) generates a subring of invariants $\mathbb{C}[f_1, f_2, \ldots]$ in $\mathbb{C}[x_1, \ldots, x_n]$.

We now consider a few examples.

<u>Example 1.</u> The quadratic form $x_1^2 + x_2^2 + \cdots + x_n^2$, along with all polynomials in this form, are integral invariants of the orthogonal group $O(n)$.

<u>Example 2.</u> The elementary symmetric polynomials $s_1(x_1, \ldots, x_n), \ldots, s_n(x_1, \ldots, x_n)$ are integral invariants of the symmetric group S_n, which we consider along with its canonical monomorphism $\Phi : S_n \to GL(n)$. The fundamental theorem on symmetric polynomials states that the invariants $s_1, \ldots, s_n$ of degrees $1, \ldots, n$, respectively, are algebraically independent, and the polynomial functions (rational functions) in these invariants exhaust all of the integral (respectively, rational) invariants of the group (S_n, Φ).

The skew-symmetric polynomials are relative invariants of the linear group (S_n, Φ): $\Phi_\pi f = (\det \Phi_\pi) f = \varepsilon_\pi f$. We have seen (Exercise 3 §2 Ch. 6) that any skew-symmetric polynomial f has the form $f = \Delta_n \cdot g$, where $\Delta_n = \prod_{j<i} (x_i - x_j)$, and g is an arbitrary symmetric polynomial, i.e., an absolute invariant.

<u>Example 3.</u> For the representation $\Phi_A : X \mapsto AXA^{-1}$ of degree n^2 of the general linear group $GL(n, K)$ with representation space $M_n(K)$ (see Example 3 §1) we have the following set of n algebraically independent invariants: the coefficients of the characteristic polynomial of the matrix $X = (x_{ij})$. In particular, this set includes the well-known invariants $\operatorname{tr} X = \Sigma\, x_{ii}$ and $\det X$.

Example 4. The orthogonal group $O(n)$ acts as follows on a quadratic form $f(x_1, \dots, x_n) = \Sigma\, a_{ij} x_i x_j$, which we write in the form $f(x_1, \dots, x_n) = {}^tXAX$, $A = (a_{ij}) = {}^tA$, $X = [x_1, \dots, x_n]$:

$$C \in O(n) \Longrightarrow (C^{-1}f)(x_1, \dots, x_n) = {}^t(CX)A(CX) = {}^tX\,{}^tCACX = {}^tX(C^{-1}AC)X \quad .$$

In this case it is customary to speak of the invariants of the quadratic form f relative to $O(n)$: $\operatorname{tr} A, \dots, \det A$. In the case of the binary quadratic form $ax^2 + 2bxy + cy^2$, the invariants $a + c$ and $ac - b^2$, which distinguish second degree curves which are metrically different, are well known from basic analytic geometry.

Example 5. We consider the symmetric group S_3 as a linear group of degree 2 by using the following representation Γ, which is equivalent to $\Phi^{(3)}$ in the table at the end of Subsection 1 §2 :

$$\Gamma_{(123)} = \begin{Vmatrix} \varepsilon & 0 \\ 0 & \varepsilon^{-1} \end{Vmatrix}, \quad \Gamma_{(23)} = \begin{Vmatrix} 0 & 1 \\ 1 & 0 \end{Vmatrix}, \quad \varepsilon^2 + \varepsilon + 1 = 0$$

(we obtain the equivalence by means of the conjugation

$$\left. \begin{Vmatrix} \varepsilon & 0 \\ 0 & 1 \end{Vmatrix} \Phi_\sigma^{(3)} \begin{Vmatrix} \varepsilon^{-1} & 0 \\ 0 & 1 \end{Vmatrix} = \Gamma_\sigma \right) \quad .$$

Let u and v be independent variables which Γ_σ transforms linearly as follows:

$$\Gamma_{(123)}(u) = \varepsilon u, \quad \Gamma_{(123)}(v) = \varepsilon^{-1}v; \quad \Gamma_{(23)}(u) = v, \quad \Gamma_{(23)}(v) = u \quad .$$

Since

$$\widetilde{\Gamma}_{(123)}(uv) = \Gamma^{-1}_{(123)}(u)\,\Gamma^{-1}_{(123)}(v) = \varepsilon^{-1}u \cdot \varepsilon v = uv,$$

$$\widetilde{\Gamma}_{(23)}(uv) = vu = uv,$$

$$\widetilde{\Gamma}_{(123)}(u^3 + v^3) = (\varepsilon^{-1}u)^3 + (\varepsilon v)^3 = u^3 + v^3,$$

$$\widetilde{\Gamma}_{(23)}(u^3 + v^3) = v^3 + u^3 = u^3 + v^3 \quad ,$$

it follows that the group (S_3, Γ) has as invariants the forms

$$I_1 = uv, \qquad I_2 = u^3 + v^3 \tag{9}$$

of degrees 2 and 3.

Next, S_3 acts naturally on polynomials $f(x_1, x_2, x_3)$ in three independent variables:

$$(\sigma f)(x_1, x_2, x_3) = f\left(x_{\sigma^{-1}(1)}, x_{\sigma^{-1}(2)}, x_{\sigma^{-1}(3)}\right) .$$

If we set

$$u = x_1 + \epsilon x_2 + \epsilon^2 x_3, \qquad v = x_1 + \epsilon^2 x_2 + \epsilon x_3 , \tag{10}$$

we see that

$$\Gamma_\sigma(u) = x_{\sigma^{-1}(1)} + \epsilon x_{\sigma^{-1}(2)} + \epsilon^2 x_{\sigma^{-1}(3)} .$$

In particular,

$$\Gamma_{(123)}(u) = x_3 + \epsilon x_1 + \epsilon^2 x_2 = \epsilon u, \qquad \Gamma_{(23)}(u) = x_1 + \epsilon x_3 + \epsilon^2 x_2 = v,$$

$$\Gamma_{(123)}(v) = x_3 + \epsilon^2 x_1 + \epsilon x_1 = \epsilon^{-1} v, \qquad \Gamma_{(23)}(v) = x_1 + \epsilon^2 x_3 + \epsilon x_2 = u ,$$

i.e., the action of Γ_σ on u and v and the action of σ on x_1, x_2, x_3 are compatible. If we perform the substitution (10) in the invariants (9), those invariants become symmetric functions in the variables x_1, x_2, x_3, which, by Theorem 1 of §2 Ch. 6, can be expressed in terms of the elementary symmetric functions $s_i = s_i(x_1, x_2, x_3)$. It is a simple exercise to show that

$$I_1 = x_1^2 + x_2^2 + x_3^2 + (\epsilon + \epsilon^2)(x_1 x_2 + x_1 x_3 + x_2 x_3) = s_1^2 - 3s_2 ,$$

$$I_2 = 2(x_1^3 + x_2^3 + x_3^3) - 3(x_1^2 x_2 + x_1^2 x_3 + x_1 x_2^2 + x_1 x_3^2 + x_2^2 x_3 + x_2 x_3^2) + 12 x_1 x_2 x_3 = 2s_1^3 - 9s_1 s_2 + 27 s_3 .$$

We specialize I_1 and I_2 by taking x_1, x_2, x_3 to be the three roots of the cubic equation

$$x^3 + px + q = 0 \quad .$$

Then $s_1 = 0$, $s_2 = p$, and $s_3 = -q$, so that

$$I_1 = -3p , \qquad I_2 = -27q \quad . \tag{11}$$

But it follows from (9) that

$$v = \frac{I_1}{u}, \qquad I_2 = u^3 + \frac{I_1^3}{u^3}, \qquad u = \sqrt[3]{\frac{I_2}{2} \pm \sqrt{\frac{I_2^2}{4} - I_1}} \quad .$$

The radicals are chosen in such a way that, after substituting the values (11), we obtain the formulas

$$u = \sqrt[3]{-\frac{27}{2}q + \frac{3}{2}\sqrt{-3D}}, \qquad v = \sqrt[3]{-\frac{27}{2}q - \frac{3}{2}\sqrt{-3D}}, \qquad uv = -3p ,$$

in which $D = -4p^3 - 27q^2$ is the discriminant of our cubic equation (see (16) in §2 Ch. 6). Since we now know u and v, we can find the roots themselves from the linear system

$$\begin{aligned} x_1 + \epsilon x_2 + \epsilon^2 x_3 &= u , \\ x_1 + \epsilon^2 x_2 + \epsilon x_3 &= v , \\ x_1 + x_2 + x_3 &= 0 \quad . \end{aligned}$$

We have obtained in a natural way the formulas of Cardano which were mentioned in Problem 1 of §2 Ch. 1.

The connection in this last example between the invariants of S_3, which is the Galois group of the general cubic equation, and Cardano's formulas is no accident. To a certain extent Galois theory is concerned with the study of invariants of fields (and their corresponding groups) which are generated by roots of algebraic equations.

We mention some facts about generators of the ring of invariants. Let w be an arbitrary form in n independent variables $x_1, \dots, x_n$. A finite group G with an n-dimensional linear representation Φ acts as a permutation group on the set

$$\Omega = \{\widetilde{\Phi}_g(w) \mid g \in G\} \quad .$$

It is clear that any homogeneous symmetric function of $|G|$ (or a divisor of $|G|$) variables which take values in Ω is an invariant of the linear group (G, Φ). If we now let w be the variable x_i, then x_i is a root of the algebraic equation

$$\prod_{g \in G} (X - \Phi_g(x_i)) = 0 \quad ,$$

whose coefficients are invariants of (G, Φ). Thus, each variable x_i is an (algebraic) function of the invariants. If there were fewer than n algebraically independent invariants, then we would be able to express $x_1, \dots, x_n$ in terms of fewer than n algebraically independent quantities, and this is impossible. We have thereby proved (if the reader accepts our rather bold use of the properties of algebraic independence) the following important theorem of invariant theory.

THEOREM 4. A finite linear group of degree n always has a set of n algebraically independent invariants. □

The forms (9) are such a set of invariants for the group (S_3, Γ).

We could have included in Theorem 4 the fact that the full ring of integral invariants of a finite group of degree n is generated by n algebraically independent invariants $f_1, \dots, f_n$ and, as a rule, one more invariant f_{n+1} (which is an algebraic function of the first n invariants). In other words, all integral invariants are polynomials in $f_1, \dots, f_n, f_{n+1}$. This fact holds for many other linear groups, both discrete and continuous.

The general theory of invariants, which developed in the middle of the XIX century in the work of Cayley, Sylvester, Jacobi, Hermite, and others, and then experienced a second birth in several fundamental works of David Hilbert, has in modern times become a part of algebraic geometry and the theory of algebraic groups. The continual interest in the theory of invariants is partly explained by its wide applicability in many areas of physics and mechanics.

EXERCISES

1. Prove Theorem 1, following the outline below (the notation is as in the statement of the theorem).

(a) If $V = \langle e_1, \ldots, e_n \rangle_K$ and $W = \langle f_1, \ldots, f_m \rangle_K$, then (T1)-(T3) combined are equivalent to the following single condition: the vectors $\tau(e_i, f_j)$, $1 \leq i \leq n$, $1 \leq j \leq m$, form a basis for the space T.

(b) For any nm-dimensional vector space T over K a map τ can be defined by setting $\tau(v,w) = \Sigma\, \alpha_i \beta_j g_{ij}$, where g_{ij}, $1 \leq i \leq n$, $1 \leq j \leq m$, form a basis of T. According to (a), the pair (τ, T) satisfies (T1)-(T3), and all such pairs are obtained in this way.

(c) Given any pair (τ', T') with a bilinear map $\tau' : V \times W \to T'$, we define a linear map $\sigma : T \to T'$ by setting $\sigma(\Sigma\, \gamma_{ij} g_{ij}) = \Sigma\, \gamma_{ij} \tau'(e_i, f_j)$.

According to (b) and (c), we have $\tau'(v,w) = \Sigma\, \alpha_i \beta_j \tau'(e_i, f_j) = \sigma(\Sigma\, \alpha_i \beta_j g_{ij}) = \sigma(\tau(v,w))$. Conversely, if $\sigma(\tau(v,w)) = \tau'(v,w)$, then $\sigma(g_{ij}) = \sigma(\tau(e_i, f_j)) = \tau'(e_i, f_j)$.

2. Show that one of the conditions (T1) or (T2) can be omitted from the set (T1)-(T3), and that, if we assume in advance that $\dim T = nm$, then only one of the three conditions is needed in the definition of the tensor product.

3. Prove the relation $\det(A \otimes B) = (\det A)^m (\det B)^n$ for an $n \times n$ matrix A and an $m \times m$ matrix B having complex coefficients, by using reduction to triangular form.

4. Using formula (8) and the tables in Subsection 1 §2, Subsection 2 §5, and Subsection 4 §5, verify the decomposition

$$\Phi^{(3)} \otimes \Phi^{(3)} \approx \Phi^{(1)} \dot{+} \Phi^{(2)} \dot{+} \Phi^{(3)}$$

for the tensor square of the two-dimensional representation $\Phi^{(3)}$ of S_3 and the decomposition

$$\Phi^{(5)} \otimes \Phi^{(5)} \approx \Phi^{(1)} \dot{+} \Phi^{(2)} \dot{+} \Phi^{(3)} \dot{+} \Phi^{(4)}$$

for the tensor square of the two-dimensional representation $\Phi^{(5)}$ of the quaternion group Q_8.

5. <u>Representations of the direct product of groups.</u> Suppose that we have two groups G and H with linear representations (Φ, V) and (Ψ, W), respectively. Then, setting

$$(\Phi \otimes \Psi)(g \cdot h) = \Phi(g) \otimes \Psi(h) \quad ,$$

where $g \cdot h$ is an element of the direct product $G \times H$, we make $G \times H$ act on the tensor product $V \otimes_{\mathbb{C}} W$; as usual,

$$(\Phi(g) \otimes \Psi(h))(v \otimes w) = \Phi(g)v \otimes \Psi(h)w \quad .$$

Verify that the map defined in this way

$$\Phi \otimes \Psi : G \times H \longrightarrow GL(V \otimes W)$$

is a representation of the group $G \times H$ with character $\chi_{\Phi \otimes \Psi} = \chi_\Phi \chi_\Psi$. Prove the following fact. Let $\Phi^{(1)}, \ldots, \Phi^{(r)}$ (respectively, $\Psi^{(1)}, \ldots, \Psi^{(s)}$) be all of the irreducible representations of G (resp. H). Then the representations $\Phi^{(i)} \otimes \Psi^{(j)}$ of the group $G \times H$ are irreducible, and the representations of that form for $1 \leq i \leq r$ and $1 \leq j \leq s$ exhaust all of the irreducible representations of $G \times H$.

6. The forms xy and $x^n + y^n$ are invariants of the two-dimensional linear dihedral group

$$(D_n, \Phi) = \left\langle \left\| \begin{matrix} \varepsilon & 0 \\ 0 & \varepsilon^{-1} \end{matrix} \right\|, \left\| \begin{matrix} 0 & 1 \\ 1 & 0 \end{matrix} \right\| \right\rangle, \qquad \varepsilon^n = 1$$

(see Exercise 9 §5). Prove that any other integral invariant of (D_n, Φ) has the form of a polynomial in xy and $x^n + y^n$.

7. Show that the quaternion group, considered in its irreducible two-dimensional representation, does not have quadratic or cubic invariants. What can be said about the forms x^2y^2 and $x^4 + y^4$?

Chapter 9. Toward a Theory of Fields, Rings and Modules

A second look at some of the algebraic structures already studied is motivated by the following considerations. In the first place, it seems worthwhile to fill out our supply of facts on fields and rings, using, whenever necessary, a solid group-theoretic foundation. In the second place, the results of Chapter 8 on group representations fit in a natural way into the general theory of modules over a ring, and it would be a shame not to go into this at least briefly. The fundamental concept of a module is important in its own right, and merits much deeper study, but for this we refer the reader to other sources.

§1. Finite field extensions

1. Primitive elements and the degree of an extension. If F is a field which contains a subfield K, then F is called an extension of the field K (see §4 Ch. 4). We shall consider the simplest case, when the extension $F = K(\theta)$ is obtained by adjoining a single element θ. We say that θ is a primitive element for the extension F of K. By its definition, $K(\theta)$ is the field of fractions of the integral domain $K[\theta]$. The element

θ is said to be transcendental over K (see §2 Ch. 5) if and only if $K(\theta)$ is isomorphic to the field of rational functions in one variable over K. On the other hand, if θ is an algebraic element, then $K(\theta) \cong K[X]/(f(X))$ (see (9) in §2 Ch. 5 and the corollary of Theorem 5 §2 Ch. 5). Here $f(X)$ is an irreducible polynomial of degree $n > 0$ having θ as a root. Conversely, if $f \in K[X]$ is an irreducible polynomial, then in a canonical way (see §3 Ch. 6) we construct a field F in which f has at least one root θ.

We claim that in the algebraic case the field $F = K(\theta) = K[X]/(f(X))$ can be identified with the set of elements of the form

$$a_0 + a_1\theta + \cdots + a_{n-1}\theta^{n-1}, \qquad a_i \in K, \qquad n = \deg f \quad .$$

This is clear for elements of $K[\theta]$ (if $g(X) \in K[X]$, simply divide $g(X)$ by $f(X)$ to obtain a remainder $r(X)$ of degree $< n$; then $g(\theta) = r(\theta)$). We can divide in $K[\theta]$ as follows: if $g(X) = a_0 + a_1 X + \cdots + a_{n-1} X^{n-1}$, then, since f is irreducible, we have g.c.d.$(f,g) = 1$, and there exist polynomials $u(X)$ and $v(X)$ of degrees $< n$ such that $fu + gv = 1$; hence $g(\theta)v(\theta) = 1$, and $1/g(\theta) = v(\theta)$. The number n is the dimension of the vector space

$$F = \langle 1, \theta, \ldots, \theta^{n-1} \rangle_K$$

over the field K with basis elements $1, \theta, \ldots, \theta^{n-1}$.

If $F \supset K$ is an arbitrary field extension (not necessarily algebraic), we can still consider F as a vector space over K. We let $[F:K]$ denote the dimension $\dim_K F$ (which may be infinite), and we call it the degree of the extension F over K. If $F = K(\theta)$, then $[F:K]$ is also called the degree of the primitive element θ. Clearly, if $\theta \in F$ is transcendental, then the set $1, \theta, \theta^2, \ldots$ is linearly independent over K, and so $[K(\theta):K] = \infty$. On the other hand, from what was said above we have the following fact.

THEOREM 1. *Let F be an extension of the field K. An element $\theta \in F$ is algebraic over K if and only if $[K(\theta):K] < \infty$. In addition, if θ is algebraic, then*

$K(\theta) = K[\theta]$. □

If L, F, and K are fields with $L \supset F \supset K$, we call them a (two-step) <u>tower of extensions.</u> We can then speak of the following three vector spaces: L/K (L as a vector space over K), L/F (L as a vector space over F), and F/K (F as a vector space over K). The dimensions of these vector spaces are connected by a relation which is reminiscent of the analogous formula for the index of one group in another.

THEOREM 2. <u>In a tower of extensions</u> $L \supset F \supset K$ <u>the degree</u> $[L:K]$ <u>is finite if and only if both</u> $[L:F]$ <u>and</u> $[F:K]$ <u>are finite. In that case the following relation holds:</u>

$$[L:K] = [L:F][F:K] \quad .$$

<u>Proof.</u> First suppose that $[L:F]$ and $[F:K]$ are finite. Choose a K-basis $f_1, \dots, f_m$ for F over K and an F-basis $e_1, \dots, e_n$ for L over F. Then any element $x \in L$ can be written in the form $x = \sum \alpha_j e_j$ with $\alpha_j \in F$. Then, in turn, $\alpha_j = \sum_i \beta_{ij} f_i$ with $\beta_{ij} \in K$. Consequently, $x = \sum_{i,j} \beta_{ij} f_i e_j$, and we see that the mn elements $f_i e_j$ span L over K. Now suppose that there is a linear dependence relation $\sum_{i,j} \beta_{ij} f_i e_j = 0$ for some $\beta_{ij} \in K$. Then

$$0 = \sum_{i,j} \beta_{ij} f_i e_j = \sum_j \left(\sum_i \beta_{ij} f_i \right) e_j \Longrightarrow \sum_i \beta_{ij} f_i = 0, \quad \forall j \Longrightarrow \beta_{ij} = 0, \quad \forall i,j \quad ,$$

where we have used the linear independence of the e_j over F and the linear independence of the f_i over K. Thus, the mn elements $f_i e_j$ form a basis of the vector space L over K, and we have $[L:K] = mn = [L:F][F:K]$.

Conversely, suppose that $[L:K]$ is finite. Then $[F:K]$ is finite, since F is a subspace of L. Furthermore, if $\{a_1, \dots, a_r\}$ is a K-basis for L, then any $x \in L$ is a linear combination of the a_i with coefficients in K, and all the more with coefficients in F. The number of linearly independent elements of $\{a_1, \dots, a_r\}$ over F may even be less than r. We thereby see that $[L:F] < \infty$. □

COROLLARY. Let F be an extension of the field K, and let A be the set of all elements of F which are algebraic over K. Then A is a subfield of F which contains K.

Proof. Every element $t \in K$ is a root of a linear polynomial $X - t \in K[X]$; hence $K \subset A$. Next, suppose that $u, v \in A$. Then, by Theorem 1, we have $[K(u):K] < \infty$. Since v is algebraic over K, it is also algebraic over $K(u)$, i.e., $[K(u,v):K(u)] = [K(u)(v):K(u)] < \infty$. According to Theorem 2, we have $[K(u,v):K] = [K(u,v):K(u)][K(u):K] < \infty$. Since $u-v$ and uv are in $K(u,v)$, it follows, again by Theorem 1, that $u-v$ and uv are in A, i.e., A is a subring of F. A is a field because, if $u \in A$ is non-zero, then $[K(u^{-1}):K] = [K(u):K] < \infty$. □

An extension $F \supset K$ is said to be algebraic over K if all of the elements in F are algebraic over K. Every element α of an algebraic extension is a root of some non-zero monic (i.e., leading coefficient 1) polynomial $f \in K[X]$, which depends on α. If $f(\alpha) = 0$ and $g(\alpha) \neq 0$ for all non-zero $g \in K[X]$ with $\deg g < \deg f$, then we say that $f = f_\alpha$ is the minimal polynomial of α. The minimal polynomial is uniquely determined, it is irreducible, and its degree is the same as the degree of the element α. (Often a polynomial obtained from the minimal polynomial by multiplication by a constant is also called a minimal polynomial.) The various roots of the polynomial f_α are called the conjugates of α. The justification for this terminology is given by Theorem 3 below. If $\operatorname{char} K = 0$, then the number of distinct roots coincides with $\deg f_\alpha$ (see §1 Ch. 6), but otherwise this is not always the case (see Exercises 4 and 5 below).

In agreement with the above results, we call an extension $F \supset K$ a finite algebraic extension if the degree $[F:K]$ is finite, i.e., if F is obtained from K by adjoining finitely many algebraic elements $\alpha_1, \dots, \alpha_m$. Note that any extension F which is obtained from K by adjoining finitely many algebraic elements must be of finite degree, since an element α_k which is algebraic over K is all the more algebraic over $K(\alpha_1, \dots, \alpha_{k-1})$; hence $[K(\alpha_1, \dots, \alpha_k) : K(\alpha_1, \dots, \alpha_{k-1})] < \infty$, and, by Theorem 2,

$$[F:K] = [K(\alpha_1, \dots, a_m):K] = \prod_{k=1}^{m} [K(\alpha_1, \dots, \alpha_k) : K(\alpha_1, \dots, \alpha_{k-1})] < \infty \quad .$$

In many cases (always when $\operatorname{char} K = 0$; see Exercise 13 below), a finite field extension can always be obtained by adjoining a single primitive element. In the cases we consider, we will always be able to show directly the existence of a primitive element.

Example. The field $F = \mathbb{Q}(\sqrt{2}, \sqrt{3})$, as a vector space over $\mathbb{Q}$, is four-dimensional: $F = \langle 1, \sqrt{2}, \sqrt{3}, \sqrt{6} \rangle_{\mathbb{Q}}$, in other words, every element $\alpha \in F$ can be written as a linear combination $\alpha = a + b\sqrt{2} + c\sqrt{3} + d\sqrt{6}$ with a, b, c, d rational. On the other hand, we also have $F = \langle 1, \theta, \theta^2, \theta^3 \rangle_{\mathbb{Q}}$, where $\theta = \sqrt{2} + \sqrt{3}$. In fact, we have $\sqrt{2} = -\frac{9}{2}\theta + \frac{1}{2}\theta^3$, $\sqrt{3} = \frac{11}{2}\theta - \frac{1}{2}\theta^3$, $\sqrt{6} = -\frac{5}{2} + \frac{1}{2}\theta^2$. The primitive element θ has minimal polynomial $f_\theta(X) = X^4 - 10X^2 + 1$ with roots

$$\theta^{(1)} = \theta = \sqrt{2} + \sqrt{3}, \ \theta^{(2)} = \sqrt{2} - \sqrt{3}, \ \theta^{(3)} = -\sqrt{2} + \sqrt{3}, \ \theta^{(4)} = -\sqrt{2} - \sqrt{3} \quad .$$

Notice that in this case F is already the splitting field for the polynomial $f_\theta(X)$; we have

$$F = \mathbb{Q}(\theta^{(1)}, \theta^{(2)}, \theta^{(3)}, \theta^{(4)}) = \mathbb{Q}(\theta^{(i)}), \qquad i = 1, 2, 3, 4 \quad .$$

In Galois theory such a field is called normal. The diagram of subfields of F:

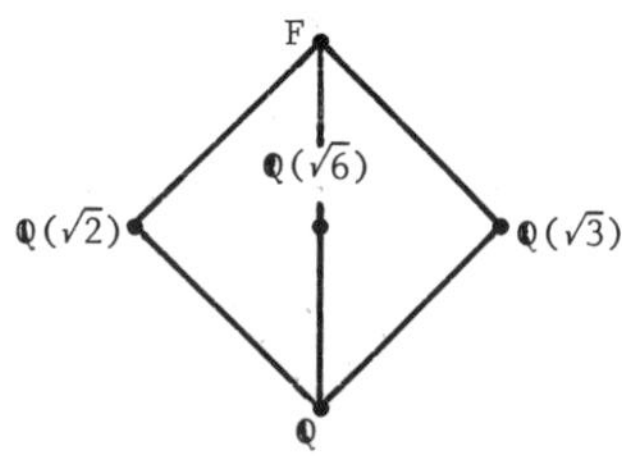

resembles the diagram of subgroups of the Klein four-group V_4, and this is no accident. If we consider any automorphism $\Phi : F \to F$ (see Subsection 5 §4 Ch. 4), then the relations $\Phi(x+y) = \Phi(x) + \Phi(y)$, $\Phi(xy) = \Phi(x)\Phi(y)$, $\forall x, y \in F$, imply that Φ is completely determined by its action on the primitive element θ. Since $\Phi(a) = a$, $\forall a \in \mathbb{Q}$, we have

$$\Phi(\theta)^4 - 10\Phi(\theta)^2 + 1 = \Phi(\theta^4 - 10\theta^2 + 1) = \Phi(0) = 0 \quad .$$

Hence, $\Phi(\theta)$ is one of the roots $\theta^{(i)}$, $i = 1, 2, 3, 4$, and we conclude that the group $\mathrm{Aut}(F/\mathbb{Q})$ of all automorphisms, which is also called the <u>Galois group</u> $G(F/\mathbb{Q})$, has order $4 = [F:\mathbb{Q}]$. There are only two groups of order 4, up to isomorphism: the cyclic group $\mathbb{Z}_4$, and $\mathbb{Z}_2 \times \mathbb{Z}_2 \cong V_4$. Direct computations show that $\mathrm{Aut}(F/\mathbb{Q}) \cong V_4$.

The easiest way to see this is to consider the representation of $\mathrm{Aut}(F/\mathbb{Q})$ by permutations of the set $\Omega = \{1, 2, 3, 4\}$ which indexes the roots $\theta^{(i)}$. If, for example, $\Phi(\theta^{(1)}) = \theta^{(2)}$, then $\theta^{(1)}\theta^{(2)} = -1 \Rightarrow \theta^{(2)}\Phi(\theta^{(2)}) = -1 \Rightarrow \Phi(\theta^{(2)}) = \theta^{(1)}$ and $\Phi(\theta^{(3)}) = -\Phi(\theta^{(2)}) = -\theta^{(1)} = \theta^{(4)}$, i.e., $\Phi \approx (12)(34) = \sigma$. We similarly obtain the automorphisms $(13)(24) = \tau$ and $(14)(23) = \sigma\tau$.

It remains to observe that the cyclic subgroup $\langle\sigma\rangle$ leaves fixed every element of the intermediate field $\mathbb{Q}(\sqrt{2})$, and $\langle\sigma\rangle$ is the group $G(F/\mathbb{Q}(\sqrt{2}))$ of all automorphisms of the field F relative to the subfield $\mathbb{Q}(\sqrt{2})$. Similarly, the "fixed fields" for the subgroups $\langle\tau\rangle$ and $\langle\sigma\tau\rangle$ are $\mathbb{Q}(\sqrt{3})$ and $\mathbb{Q}(\sqrt{6})$, respectively, and the Galois groups $G(F/\mathbb{Q}(\sqrt{3}))$ and $G(F/\mathbb{Q}(\sqrt{6}))$ are $\langle\tau\rangle$ and $\langle\sigma\tau\rangle$. In this special case we have verified that there is a one-to-one correspondence between subfields of a normal field F and subgroups of its automorphism group.

2. <u>Isomorphism of splitting fields.</u> In §3 Ch. 6, where we defined and constructed the splitting field F over K for a monic polynomial $f(X) = X^n + a_1X^{n-1} + \cdots + a_n \in K[X]$, we noted that there are various choices which are made in the construction. But it turns out that all splitting fields over K of a given

polynomial f are isomorphic. In order to make a more precise statement, we consider a somewhat more general situation.

According to Theorem 3 §2 Ch. 5, any isomorphism φ between a field K and another field $\widetilde{K}$ extends uniquely to an isomorphism from $K[X]$ onto $\widetilde{K}[X]$, so that

$$f(X) = X^n + a_1 X^{n-1} + \cdots + a_n \mapsto \widetilde{f}(X) = \varphi_X f = X^n + \varphi(a_1)X^{n-1} + \cdots + \varphi(a_n) .$$

THEOREM 3. Suppose that $\varphi : K \to \widetilde{K}$ is an isomorphism of fields; $f \in K[X]$ is a monic polynomial of degree $n > 0$, $\widetilde{f} = \varphi_X f$ is its image under the isomorphism φ_X; F and $\widetilde{F}$ are splitting fields of the polynomials f and $\widetilde{f}$ over K and $\widetilde{K}$, respectively. Then φ can be extended to an isomorphism $\Phi : F \to \widetilde{F}$ in $k \leq [F:K]$ ways, where $k = [F:K]$ if all of the roots of the polynomial $\widetilde{f}(X)$ are distinct.

Proof. Step I. We first consider the case of arbitrary extensions $L \supset K$, $\widetilde{L} \supset \widetilde{K}$. Let $\theta \in L$ be an algebraic element with minimal polynomial $g = g_\theta \in K[X]$. We claim that the isomorphism $\varphi : K \to \widetilde{K}$ extends to a monomorphism $\rho : K(\theta) \to \widetilde{L}$ precisely when $\widetilde{g}$ has a root in $\widetilde{L}$, and the number of extensions is equal to the number of distinct roots of $\widetilde{g}$ in $\widetilde{L}$.

In fact, it follows from the existence of ρ that the element $\rho(\theta)$ must be a root of $\widetilde{g}$: $g(\theta) = 0 \Rightarrow \widetilde{g}(\rho(\theta)) = \rho(g(\theta)) = 0$. Conversely, if $\widetilde{g}(\omega) = 0$, then $\operatorname{Ker} \psi \supset g(X)\,K[X]$, where $\psi : K[X] \to \widetilde{L}$ is the homomorphism defined by taking $u(X)$ to $\widetilde{u}(\omega)$. As in the case of groups, ψ induces a homomorphism $\overline{\psi} : K[X]/g(X)\,K[X] \to \widetilde{L}$ (given by $(u(X) + g(X)\,K[X]) \mapsto \widetilde{u}(\omega)$; if this is not completely clear, the reader should refer to the results below). Note that, since $g(X)$ is irreducible, the quotient ring $K[X]/g(X)\,K[X]$ is a field, and hence $\overline{\psi}$ is a monomorphism. In exactly the same way we define the isomorphism of fields $\overline{\sigma} : K[X]/g(X)\,K[X] \to K(\theta)$ (given by $u(X) + g(X)\,K[X] \mapsto u(\theta)$). The composite map $\rho = \overline{\psi} \circ \overline{\sigma}^{-1}$ is a monomorphism from $K(\theta)$ to $\widetilde{L}$ (since $\rho(u(\theta)) = \widetilde{u}(\omega)$). Since $K(\theta)$ is generated by θ over K, it follows that ρ is the only extension of φ which takes θ to ω. But this means that the number of distinct monomorphisms ρ with restriction $\rho|_K = \varphi$ is equal to the number of distinct roots of $\tilde{g}(X)$ in $\tilde{L}$.

Step II. The splitting field was constructed by successively adjoining roots of irreducible polynomials. We now use induction on the degree $[F:K]$.

If $[F:K] = 1$, then the polynomial f splits into linear factors already in $K[X]$: $f(X) = (X - c_1) \ldots (X - c_n)$. In this case, $\tilde{f}(X) = (\varphi_X f)(X) = (X - \tilde{c}_1) \ldots (X - \tilde{c}_n)$. The roots $\tilde{c}_1, \ldots, \tilde{c}_n$ of the polynomial $\tilde{f}$ are contained in $\tilde{K}$, and, since $\tilde{F}$ is generated over $\tilde{K}$ by these roots, we have $\tilde{F} = \tilde{K}$, so that $\Phi = \varphi_X$ is the only extension.

If $[F:K] > 1$, we factor $f(X)$ over K into monic irreducible polynomials, where there must be at least one factor with degree $m > 1$. Let $g(X)$ be such a factor. Since

$$f(X) = g(X)\,h(X) \Longrightarrow \tilde{f}(X) = (\varphi_X f)(X) = \tilde{g}(X)\,\tilde{h}(X) \quad ,$$

we have the following factorization of polynomials over the splitting fields F and $\tilde{F}$:

$$g(X) = (X - \theta_1) \ldots (X - \theta_m),$$

$$\tilde{g}(X) = (X - \omega_1) \ldots (X - \omega_m), \quad m \leq n \quad .$$

Since it is irreducible, $g(X)$ is the minimal polynomial of the element θ_1 over K, and we have $[K(\theta_1):K] = m$.

If there are ℓ distinct elements among the $\omega_1, \ldots, \omega_m$, then, by Step I, we can find ℓ monomorphisms $\rho_1, \ldots, \rho_\ell$ from the extension $K_1 = K(\theta_1)$ to $\tilde{F}$ with $\rho_i|_K = \varphi$. Because of the way splitting fields are constructed, we may consider F as a splitting field over K_1 of the polynomial $f \in K_1[X]$, and we may consider $\tilde{F}$ as a splitting field over $\rho_i(K_1)$ of the polynomial $\tilde{f}(X)$ for any $i = 1, 2, \ldots, \ell$. By Theorem 2, we have the inequality $[F:K_1] = [F:K]/m < [F:K]$, so that, by the induction assumption, each of the ρ_i can be extended to an isomorphism $\Phi_{i,j}: F \to \tilde{F}$, and the number of such extensions (the number of indices j) does not exceed $[F:K_1]$, and it is equal to $[F:K_1]$ if all of the roots of $\widetilde{f}$ in $\widetilde{F}$ are distinct. Since $\Phi_{i,j}|_{K_1} = \rho_i$, $1 \leq j \leq [F:K_1]$, and $\rho_i|_K = \varphi$, it follows that $\Phi_{i,j}$ is an extension of φ, and $\rho_i \neq \rho_s \Rightarrow \Phi_{i,j} \neq \Phi_{s,t}$ for $i \neq s$. Hence, altogether we obtain $k \leq m[F:K_1] = [F:K]$ extensions of the isomorphism φ. This inequality becomes an equality if all of the roots of $\widetilde{f}$ are distinct.

Step III. Finally, suppose that $\Phi : F \to \widetilde{F}$ is an arbitrary extension of φ. As in Step II, the restriction $\Phi|_{K_1}$, which is a monomorphism from K_1 to $\widetilde{F}$, coincides with one of the ρ_i, and in this case Φ coincides with one of the $\Phi_{i,j}$. □

COROLLARY 1. *Any two splitting fields* F *and* $\widetilde{F}$ *over* K *of a polynomial* $f \in K[X]$ *are isomorphic.*

In fact, it suffices to set $\widetilde{K} = K$ in Theorem 3 and take φ to be the identity map from K to itself. □

COROLLARY 2. *The group* $\mathrm{Aut}(F/K)$ *of automorphisms of any splitting field* F *over* K *of a polynomial* $f \in K[X]$ *is finite and has order* $\leq [F:K]$. *If all of the roots of* $f(X)$ *are distinct, then* $|\mathrm{Aut}(F/K)| = [F:K]$.

This corollary is an immediate consequence of Theorem 3. □

Remark. Although the splitting field F over $\mathbb{Q}$ (or over any other number field) of a polynomial $f \in \mathbb{Q}[X]$ can be considered imbedded in the complex numbers $\mathbb{C}$, and so is uniquely determined, Corollary 2 shows that, even in this case, it is worthwhile to have Theorem 3 (despite its unpleasant proof).

3. Finite fields. In addition to $\mathbb{Z}_p = \mathbb{Z}/p\mathbb{Z}$, we have encountered other examples of finite fields (see §4 Ch. 4). It is now time to incorporate them into a general

theory.

The first obvious remark concerns an arbitrary finite extension $K \supset F$ of a finite field F : _if_ $|F| = q$ _and_ $[K:F] = n$, _then_ $|K| = q^n$. To see this, choose a basis for the vector space K over F . Then K can be identified with the space F^n of rows $(\alpha_1, \dots, \alpha_n)$ of length n . Since all of the coordinates α_i take on any of the q possible values in F independently of one another, we have $|K| = |F^n| = q^n$, as claimed.

Our second remark is that _any finite field_ F _has finite characteristic_ p (p _is a prime), and_ $|F|$ _is a power of_ p . In fact, since F is finite, the prime field $P \subset F$ must be isomorphic to one of the fields $\mathbb{Z}_p = \mathbb{Z}/p\mathbb{Z}$. According to the first remark, the finite extension $F \supset P$ has cardinality $|F| = p^m$ (where m is the degree of F over P).

THEOREM 4. _For every finite field_ F _and every positive integer_ n _there exists one and (up to isomorphism) only one extension_ $K \supset F$ _of degree_ $[K:F] = n$.

Proof. (a) _Uniqueness._ Let $K \supset F$ be an extension of degree n . We know that $|F| = q \Rightarrow q = p^m$, where p is a prime, and $|K| = q^n$. Consequently, the multiplicative group $K^* = K \backslash \{0\}$ has order $q^n - 1$, and, by Lagrange's theorem, the order of any element in this multiplicative group divides $q^n - 1$: $t^{q^n - 1} = 1$, $\forall t \neq 0$. This means that all of the elements of K (including $t = 0$) are distinct roots of the polynomial $X^{q^n} - X$, and we have the factorization

$$X^{q^n} - X = \prod_{t \in K} (X - t) \quad .$$

It is impossible to have such a factorization into linear terms over any proper subfield of K with fewer than q^n elements; hence, K is the splitting field of the polynomial $X^{q^n} - X$. By Corollary 1 of Theorem 3 , we have the required uniqueness of K .

(b) Existence. The argument in part (a) suggests a possible method for constructing K. We take K to be the splitting field over $P \cong \mathbb{Z}_p$ of the polynomial $X^{q^n} - X$. Since $q = p^m$, we have $q \cdot 1 = 0$ in K. Hence $f'(X) = q^n \cdot 1 \cdot X^{q^n - 1} - 1 = = -1$, and by the well-known criterion (Theorem 4 §1 Ch. 6), $f(X)$ does not have multiple roots. This means that the subset $K_f \subset K$ of roots of $f(X)$ has cardinality $|K_f| = q^n$. We claim that $K_f = K$.

Since $K_f \subset K$ and $\operatorname{char} K = p$, it follows by Exercise 8 §4 Ch. 4 that $(x+y)^{p^s} = x^{p^s} + y^{p^s}$ for any $x, y \in K_f$ and $s = 0, 1, 2, \ldots$. In particular,

$$x, y \in K_f \Longrightarrow (x \pm y)^{q^n} = x^{q^n} \pm y^{q^n} = x \pm y \Longrightarrow x \pm y \in K_f \quad .$$

In addition,

$$1 \in K_f ; \ (xy)^{q^n} = x^{q^n} y^{q^n} = xy \Longrightarrow xy \in K_f ;$$

$$0 \neq x \in K_f \Longrightarrow (x^{-1})^{q^n} = x^{-1} \Longrightarrow x^{-1} \in K_f \quad .$$

Thus, K_f is a subfield of K which contains F (since all elements of F are clearly roots of $f(X)$) and all of the roots of $f(X)$. Because of the definition of the splitting field, we must have $K_f = K$. We have $[K:F] = n$, because $q^{[K:F]} = |K| = |K_f| = q^n$. □

COROLLARY. For every prime p and every positive integer n, there exists one and (up to isomorphism) only one field with p^n elements.

This is merely the special case of Theorem 4 when $|F| = p$. □

As we noted in §4 Ch. 4, it is customary to let $\mathbb{F}_p$ (or sometimes, in honor of Galois, $GF(p^n)$) denote the finite field with $q = p^n$ elements. We now prove some facts about finite fields.

THEOREM 5. (i) The multiplicative group $\mathbb{F}_q^*$ of the finite field $\mathbb{F}_q$ is a cyclic group of order $q-1$.

(ii) The group $\mathrm{Aut}(\mathbb{F}_q)$ of automorphisms of the finite field $\mathbb{F}_q$ with $q = p^n$ elements is cyclic of order n, where

$$\mathrm{Aut}(\mathbb{F}_q) = \langle \Phi \mid \Phi(t) = t^p, \quad \forall t \in \mathbb{F}_q \rangle \quad .$$

(iii) If $\mathbb{F}_{p^d}$ is a subfield of $\mathbb{F}_{p^n}$, then $d \mid n$. Conversely, to every divisor d of n there corresponds precisely one subfield $\{t \in \mathbb{F}_{p^n} \mid \Phi^d(t) = t\} = \mathbb{F}_{p^d}$. The automorphisms of $\mathbb{F}_{p^n}$ which leave fixed all of the elements of this subfield $\mathbb{F}_{p^d}$ form a group $\mathrm{Aut}(\mathbb{F}_{p^n}/\mathbb{F}_{p^d}) = \langle \Phi^d \rangle$. Thus, there is a one-to-one correspondence between the subfields of $\mathbb{F}_q$ and the subgroups of its automorphism group (the correspondence of Galois theory).

(iv) If $q = p^n$ and $\mathbb{F}_q^* = \langle \theta \rangle$, then θ is a primitive element of the field $\mathbb{F}_q$ whose minimal polynomial $h(X)$ over $\mathbb{F}_p$ has degree n. $\mathbb{F}_q$ is the splitting field of $h(X)$ over $\mathbb{F}_p$.

(v) For any natural number m there exists at least one irreducible polynomial of degree m over $\mathbb{F}_q$.

Proof. (i) We shall prove a more general fact. Let F be an arbitrary field, and let A be a finite subgroup of the multiplicative group F^*. We can apply the results of §5 Ch. 7 to the abelian group A. In particular, we know that A is cyclic if and only if $|A|$ coincides with the exponent m of A, i.e., the least natural number such that $a^m = 1$, $\forall a \in A$. If $m < |A|$, then the polynomial $X^m - 1$ would have more than m roots in F, and this is impossible. Hence, the group A is cyclic.

(ii) We shall regard $\mathbb{F}_q$ as a finite extension of degree n of its prime field

$\mathbb{F}_p \cong \mathbb{Z}_p$. Since $\mathbb{F}_q$ is the splitting field of the polynomial $X^q - X$, all of whose roots are distinct, it follows by Corollary 2 of Theorem 3 that $|\mathrm{Aut}(\mathbb{F}_q)| = n$. Because of the relations $(x+y)^p = x^p + y^p$, $(xy)^p = x^p y^p$, and $1^p = 1$, which we noted during the proof of Theorem 4, we see that the map $\Phi : t \mapsto t^p$ is an automorphism of the field $\mathbb{F}_q$ (the finiteness of $\mathbb{F}_q$ is essential here, for surjectivity). If $\Phi^s : t \mapsto t^{p^s}$ is the identity automorphism, then $t^{p^s} - t = 0$ for all $t \in \mathbb{F}_q$, which means that $s \geq n$. But we do obtain the identity automorphism when $s = n$; hence $|\langle\Phi\rangle| = n$, and $\langle\Phi\rangle = \mathrm{Aut}(\mathbb{F}_q)$.

(iii) According to our first remark concerning finite fields at the beginning of the subsection, we have $p^n = (p^d)^r$, where r is the degree of the extension $\mathbb{F}_{p^n} \supset \mathbb{F}_{p^d}$. Hence $n = dr$. Conversely, for any $d \mid n$ we consider the subset $F = \{t \in \mathbb{F}_{p^n} \mid t^{p^d} = 1\}$. Since $n = dr \Rightarrow p^n - 1 = (p^d)^r - 1 = (p^d - 1)s$ for some integer s, it follows that

$$X^{p^n - 1} - 1 = X^{(p^d - 1)s} - 1 = (X^{p^d - 1} - 1)\, g(X) ,$$

$$X^{p^n} - X = (X^{p^d} - X)\, g(X) .$$

Because $\mathbb{F}_{p^n}$ is the splitting field of the polynomial $X^{p^n} - X$, precisely p^d elements in $\mathbb{F}_{p^n}$ are roots of the polynomial $X^{p^d} - X$. This is our set F, which we can now identify with $\mathbb{F}_{p^d}$. This argument also shows uniqueness of the subfield with p^d elements.

We note that, by construction,

$$\mathbb{F}_{p^d} = \{t \in \mathbb{F}_{p^n} \mid \Phi^d(t) = t\}$$

is the set of all elements which remain fixed under the action of $\langle\Phi^d\rangle$. Since the group

$\mathrm{Aut}(\mathbb{F}_{p^n}/\mathbb{F}_p) = \langle\Phi\rangle$ is cyclic, it is immediately clear that any automorphism Φ^ℓ not in $\langle\Phi^d\rangle$ does not act on $\mathbb{F}_{p^d}$ as the identity (simply apply Φ^ℓ to a generator of the group $\mathbb{F}^*_{p^d}$). But this means that the group $\mathrm{Aut}(\mathbb{F}_{p^n}/\mathbb{F}_{p^d})$ of relative automorphisms coincides with $\langle\Phi^d\rangle$. The reference to Galois theory at the end of (iii) has the same sense as in the example in Subsection 1.

(iv) It is obvious that $\mathbb{F}_q = \mathbb{F}_p(\theta)$, $q = p^n$. Let $h(X) = X^n + a_1X^{n-1} + \cdots + a_n$ be the minimal polynomial of the primitive element θ. Since the elements of the prime field $\mathbb{F}_p$ are fixed under all automorphisms, and $a_i \in \mathbb{F}_p$, it follows that the roots of $h(X)$ are $\theta, \theta^p, \theta^{p^2}, \ldots, \theta^{p^{n-1}}$. They are all contained in our field, and $\mathbb{F}_p(\theta, \ldots, \theta^{p^{n-1}}) = \mathbb{F}_p(\theta) = \mathbb{F}_{p^n}$ is the splitting field of $h(x)$ over $\mathbb{F}_p$.

(v) Using Theorem 4, we construct the extension $K \supset \mathbb{F}_q$ of degree m. According to (i), K^* is a cyclic group. If $K^* = \langle\theta\rangle$ and $h(X)$ is the minimal polynomial of the primitive element θ over the field $\mathbb{F}_q$, then $K = \mathbb{F}_q(\theta)$ and $\deg h(X) = [\mathbb{F}_q(\theta):\mathbb{F}_q] = [K:\mathbb{F}_q] = m$. By definition, the minimal polynomial is irreducible (over $\mathbb{F}_q$), so it satisfies the requirement of (v). □

After some simple number-theoretic preliminaries, we shall obtain an exact formula for the number of irreducible polynomials of degree m over $\mathbb{F}_q$.

4. <u>The Möbius inversion formula and its applications.</u> The function μ on the positive integers which is defined by the formulas

$$\mu(n) = \begin{cases} 1, & \text{if } n = 1, \\ (-1)^k, & \text{if } n = p_1 \cdots p_k \text{ is a product of } k \text{ distinct primes}, \\ 0, & \text{if } n \text{ is divisible by a square greater than } 1, \end{cases}$$

is called the <u>Möbius function.</u> The function μ is clearly <u>multiplicative</u> in the sense that, if n and m are relatively prime, then $\mu(nm) = \mu(n)\mu(m)$. It is also clear that, if $n = p_1^{m_1} \cdots p_r^{m_r}$, then $\sum_{d|n} \mu(d) = \sum_{d|n_0} \mu(d)$, where $n_0 = p_1 \cdots p_r$ is the greatest square-free divisor of n. Note that the number of divisors $d = p_{i_1} \cdots p_{i_s}$ of n_0 with fixed s is equal to $\binom{r}{s}$. Thus, for $n > 1$ we have:

$$\sum_{d|n} \mu(d) = \sum_{d|n_0} \mu(d) = \sum_{s=0}^{r} \binom{r}{s}(-1)^s = (1-1)^r = 0$$

(the summation on the left is over all divisors $d \geq 1$ of the integer n). We finally obtain:

$$\sum_{d|n} \mu(d) = \begin{cases} 1, & \text{if } n = 1, \\ 0, & \text{if } n > 1 . \end{cases} \tag{1}$$

The following modification is also useful:

$$\sum_{n,\, d|n|m} \mu\left(\frac{m}{n}\right) = \begin{cases} 1, & \text{if } d = m, \\ 0, & \text{if } d|m \text{ and } d < m \end{cases} \tag{2}$$

(the summation is over n dividing m and divisible by d). If we set $m = dt$ and $n = d\ell$ and make ℓ run through the divisors of t, we easily derive (2) from (1).

It would be possible to use formula (1) (or (2)) to define the Möbius function by induction. The value of this formula for us is contained in the following fact. <u>Let</u> f <u>and</u> g <u>be any two functions from</u> $\mathbb{N}$ <u>to</u> M (where $M = \mathbb{Z}$, $\mathbb{R}$, $\mathbb{F}[X]$, etc.) <u>which are connected by the relation</u>

$$f(n) = \sum_{d|n} g(d) \quad . \tag{3}$$

<u>Then</u>

$$g(n) = \sum_{d|n} \mu(\tfrac{n}{d})\, f(d) \quad . \tag{4}$$

To prove this, we multiply both sides of (3) by $\mu(m/n)$ and sum over n dividing m, using (2). We obtain

$$\sum_{n|m} \mu(\tfrac{m}{n})\, f(n) = \sum_{n|m} \mu(\tfrac{m}{n}) \cdot \sum_{d|n} g(d) = \sum_{d|m} g(d) \cdot \sum_{n,\, d|n|m} \mu(\tfrac{m}{n}) = g(m) \quad .$$

A simple change of notation gives us (4), which is called the Möbius inversion formula. It is possible in a similar way to derive (3) from (4). □

There is also a multiplicative analog of the Möbius inversion formula. If

$$f(n) = \prod_{d|n} g(d) \quad ,$$

then

$$g(n) = \prod_{d|n} f(d)^{\mu(\frac{n}{d})} \quad . \tag{5}$$

This is proved by making the same type of formal computations:

$$\prod_{n|m} f(n)^{\mu(\frac{m}{n})} = \prod_{n|m} \prod_{d|n} g(d)^{\mu(\frac{m}{n})} = \prod_{d|m} \prod_{n,\, d|n|m} g(d)^{\mu(\frac{m}{n})} = \prod_{d|m} g(d)^{\sum_{d|n|m} \mu(\frac{m}{n})} = g(m),$$

and then making a change of notation.

We shall give three examples of how the Möbius inversion formula is applied.

Example 1. Euler's φ function. By definition, $\varphi(n)$ is the number of integers $0, 1, \dots, n-1$ which are prime to n, or, equivalently, $\varphi(n) = |U(\mathbb{Z}_n)|$ is the order of the group of invertible elements in the ring $\mathbb{Z}_n = \mathbb{Z}/n\mathbb{Z}$. From Exercise 5 of §1 Ch. 8 we know the relation

$$n = \sum_{d|n} \varphi(d) \quad . \tag{6}$$

Using (4), we immediately obtain

$$\varphi(n) = \sum_{d|n} \mu(n/d)d = \sum_{d|n} \mu(d)\, n/d = n \sum_{d|n} \frac{\mu(d)}{d} \quad .$$

If $n = p_1^{m_1} \cdots p_r^{m_r}$, then

$$\sum_{d|n} \frac{\mu(d)}{d} = 1 - \sum_i \frac{1}{p_i} + \sum_{i<j} \frac{1}{p_i p_j} - \cdots + (-1)^r \frac{1}{p_1 p_2 \cdots p_r} = (1 - \frac{1}{p_1})(1 - \frac{1}{p_2}) \cdots (1 - \frac{1}{p_r}).$$

Thus,

$$\varphi(n) = n(1 - \frac{1}{p_1})(1 - \frac{1}{p_2}) \cdots (1 - \frac{1}{p_r})$$

a formula which we already gave in Exercise 3 of §8 Ch. 1 and which immediately implies the multiplicativity of the Euler φ function.

Example 2. Cyclotomic polynomials. The splitting field Γ_n of $X^n - 1$ over $\mathbb{Q}$ is called the cyclotomic field of n-th roots of unity. Since the n-th roots of 1 form a cyclic group of order n, it follows that this cyclotomic field has the form $\Gamma_n = \mathbb{Q}(\zeta)$, where ζ is any primitive n-th root ($\zeta \in \mathbb{C}$). We would like to determine $[\Gamma_n : \mathbb{Q}]$ and find the minimal polynomial of ζ over $\mathbb{Q}$.

Let P_n denote the set of primitive n-th roots of 1, which has cardinality $|P_n| = \varphi(n)$. The subgroups of the cyclic group of order n are in one-to-one correspondence with the divisors d of n (Theorem 6 of §3 Ch. 4), and each root ζ^i falls in exactly one of the sets P_d. Hence, we have the following partition into disjoint sets:

$$\{1, \zeta, \zeta^2, \ldots, \zeta^{n-1}\} = \bigcup_{d|n} P_d \tag{7}$$

(note that if we take the cardinality of both sides, we again obtain (6)). The cyclotomic polynomial corresponding to Γ_n is the polynomial

$$\Phi_n(X) = \prod_{\varepsilon \in P_n} (X - \varepsilon)$$

of degree $\varphi(n)$. Corresponding to (7) we have the factorization:

$$X^n - 1 = \prod_{i=0}^{n-1} (X - \zeta^i) = \prod_{d|n} \{ \prod_{\varepsilon \in P_d} (X - \varepsilon)\} = \prod_{d|n} \Phi_d(X) \quad . \tag{8}$$

If we apply the multiplicative version of the Möbius inversion formula to (8), we obtain an explicit formula for Φ_n:

$$\Phi_n(X) = \prod_{d|n} (X^d - 1)^{\mu(\frac{n}{d})} \quad . \tag{9}$$

For the first few values of n we have

$$\Phi_1(X) = X - 1, \quad \Phi_2(X) = X + 1, \quad \Phi_3(X) = X^2 + X + 1,$$

$$\Phi_4(X) = X^2 + 1, \quad \Phi_6(X) = X^2 - X + 1, \quad \Phi_8(X) = X^4 + 1,$$

$$\Phi_9(X) = X^6 + X^3 + 1, \quad \Phi_{10}(X) = X^4 - X^3 + X^2 - X + 1,$$

$$\Phi_{12}(X) = X^4 - X^2 + 1 \quad .$$

Note that

$$\Phi_n(X) \in \mathbb{Z}[X] \quad \text{and} \quad \Phi_n(0) = 1 \quad \text{for} \quad n > 1 \quad . \tag{10}$$

To prove (10), we can either use (9) or else apply induction. The proof by induction is as follows. We have verified (10) for small n. By the induction assumption,

$$g(X) = \prod_{d|n,\, d \neq n} \Phi_d(X)$$

is a monic polynomial with integral coefficients and constant term -1. Using the division algorithm (Theorem 5 of §2 Ch. 5), we obtain uniquely determined polynomials $q, r \in \mathbb{Z}[X]$ such that $X^n - 1 = q(X)\,g(X) + r(X)$, $\deg r(X) < \deg g(X)$. But $X^n - 1 = \Phi_n(X)\,g(X)$ in $\mathbb{Q}[X]$, and so $\Phi_n(X) = q(X) \in \mathbb{Z}[X]$, and $\Phi_n(X)$ is monic with constant term 1.

We can say more: <u>$\Phi_n(X)$ is irreducible over $\mathbb{Q}$, and so $\Gamma_n = \mathbb{Q}(\zeta)$ is an extension of degree $\varphi(n)$ with minimal polynomial $\Phi_n(X)$ for ζ</u>. We shall not prove this, but recall that at the end of §3 Ch. 5 we established the irreducibility of

$\Phi_p(X) = (X^p - 1)/(X - 1) = X^{p-1} + X^{p-2} + \cdots + 1$, where p is any prime number.

It should be noted that the cyclotomic fields, which played a key role in the development of algebraic number theory, are still the subject of active research by many mathematicians.

Example 3. Irreducible polynomials over $\mathbb{F}_q$. Let $\Psi_d(q)$ be the total number of monic irreducible polynomials of degree d over $\mathbb{F}_q$, $q = p^n$, and let $f(X)$ be one such polynomial. Its splitting field over $\mathbb{F}_q$ is isomorphic both to the quotient ring $\mathbb{F}_q[X]/f(X)\mathbb{F}_q[X]$ and to the splitting field of the polynomial $X^{q^d} - X$ (see the corollary to Theorem 4). Since the polynomials $X^{q^d} - X$ and $f(X)$ have a common root θ, while $f(X)$ is irreducible, it follows that $X^{q^d} - X$ is divisible by $f(X)$. Since $X^{q^d} - X$ divides the polynomial $X^{q^m} - X$ for any $m = rd$, and since $X^{q^m} - X$ does not have multiple roots, we may conclude that each of the monic irreducible polynomials

$$f_{d,1}(X), \quad f_{d,2}(X), \ldots, f_{d,\Psi_d(q)}(X)$$

of degree $d|m$ occurs exactly once in the factorization of $X^{q^m} - X$ over $\mathbb{F}_q$:

$$X^{q^m} - X = \prod_{d|m} \left\{ \prod_{k=1}^{\Psi_d(q)} f_{d,k}(X) \right\} . \tag{11}$$

Now taking the degrees of the polynomials in (11), we obtain the relation

$$q^m = \sum_{d|m} d\Psi_d(q) ,$$

from which we find an expression for $\Psi_m(q)$ by applying the Möbius inversion formula (4):

$$\Psi_m(q) = \frac{1}{m} \sum_{d|m} \mu(\tfrac{m}{d})\, q^d \quad . \tag{12}$$

For example, let $q = 2$. Then

$$\Psi_2(2) = \frac{1}{2}(2^2 - 2) = 1, \qquad \Psi_3(2) = \frac{1}{3}(2^3 - 2) = 2,$$

$$\Psi_4(2) = \frac{1}{4}(2^4 - 2^2) = 3, \qquad \Psi_5(2) = \frac{1}{5}(2^5 - 2) = 6,$$

$$\Psi_6(2) = \frac{1}{6}(2^6 - 2^3 - 2^2 + 2) = 9$$

(compare with Exercise 10 §1 Ch. 6). The formula (12) shows that a randomly chosen monic polynomial of degree m over $\mathbb{F}_q$ has a probability of about $1/m$ of being irreducible. But there are no satisfactory criteria for determining in a concrete case whether or not a given polynomial is irreducible. For example, what can be said about irreducibility of the trinomial $X^m + X^k + 1$ over $\mathbb{F}_2$? Questions of this sort constantly arise in algebraic coding theory (see Problem 3 of §2 Ch. 1) and in the construction of pseudo-random sequences.

Example 4. Constructions by ruler and compass. Let $K \subset CS$ be a constructive number field (see p. 219) which is a finite extension of $\mathbb{Q}$. We first suppose that K is real, i.e., its elements are real numbers. In particular, K has a primitive element Θ (see Exercise 13) which is real and can be constructed (as the length of a line segment) in finitely many steps by ruler and compass. This means that Θ is an element of a field $\mathbb{Q}(\Theta_1, \Theta_2, \ldots, \Theta_r)$, where the degree $[\mathbb{Q}(\Theta_1, \ldots, \Theta_k) : \mathbb{Q}(\Theta_1, \ldots, \Theta_{k-1})]$ is at most 2 for each k. This is because Θ_k is a solution to two equations with coefficients in $\mathbb{Q}(\Theta_1, \ldots, \Theta_{k-1})$ either for two lines, for a line and a circle, or for two circles. Now the results of subsection 1 concerning the degrees of algebraic extensions in towers show that $[\mathbb{Q}(\Theta_1, \ldots, \Theta_r) : \mathbb{Q}] = 2^m$, where $m \leq r$. Since $\mathbb{Q}(\Theta) \subset \mathbb{Q}(\Theta_1, \ldots, \Theta_r)$, it follows from Theorem 2 that the degree $[\mathbb{Q}(\Theta):\mathbb{Q}]$ is a power of two.

Turning to the case of K not necessarily real, we again write it in the form $K = \mathbb{Q}(\Theta)$. Now the primitive element $\Theta = a+ib$ is a complex number whose real components a and b are constructive. Namely, if $f(X)$ is the minimal polynomial (with rational coefficients) for Θ, then $f(\Theta) = 0$ and $f(\bar{\Theta}) = 0$, where $\bar{\Theta} = a - ib$. Then clearly $\mathbb{Q}(\Theta,\bar{\Theta})$ is a finite algebraic extension of $\mathbb{Q}$. Its elements $a = (\Theta + \bar{\Theta})/2$ and $ib = (\Theta - \bar{\Theta})/2$ are algebraic over $\mathbb{Q}$, and so is $b = ib / i$ (see the Corollary to Theorem 2), since of course $i^2 + 1 = 0$.

Thus, $\mathbb{Q}(a,b)$ is a finite real algebraic extension of $\mathbb{Q}$ with a and b constructive. According to the above, we have $[\mathbb{Q}(a,b):\mathbb{Q}] = 2^m$. Because $X^2 + 1$ is irreducible over $\mathbb{Q}(a,b) \subset \mathbb{R}$, it follows that $[\mathbb{Q}(a,b)(i):\mathbb{Q}(a,b)] = 2$, and so $[\mathbb{Q}(a,b)(i):\mathbb{Q}] = 2^{m+1}$. Since $K = \mathbb{Q}(\Theta) \subset \mathbb{Q}(a,b,i)$, the degree $[\mathbb{Q}(\Theta):\mathbb{Q}]$ must divide 2^{m+1}. We have proved the following important fact.

<u>If a constructive number field</u> K <u>is a finite algebraic extension of</u> $\mathbb{Q}$, <u>then</u> $[K:\mathbb{Q}] = 2^n$ <u>for some nonnegative integer</u> n.

This enables one to answer various questions that were raised by mathematicians in ancient times.

a) Is it possible to construct (using ruler and compass) the edge of a cube having volume 2 (the Indian problem of doubling the cube)? Here it is assumed that we are given a cube of unit volume. The polynomial $X^3 - 2$, a root of which is the length of the desired side, is irreducible over $\mathbb{Q}$; hence $[\mathbb{Q}(\sqrt[3]{2}):\mathbb{Q}] = 3 \neq 2^n$. Therefore, this question has a negative answer.

b) Is it possible to divide any angle into three equal parts using ruler and compass (the problem of trisecting an angle)? The answer is negative even for the specific angle 60°. Namely, to construct $\phi = 20^\circ$ would mean we could construct

$\cos\phi$ and $2\cos\phi$. But by de Moivre's theorem, $\frac{1}{2} = \cos 60^{\circ} = \cos 3\phi = 4\cos^3\phi - 3\cos\phi$, so that $\Theta = 2\cos\phi$ is a root of the polynomial $f(X) = X^3 - 3X - 1$. Since 1 and -1 are not roots of $f(X)$, it follows that the polynomial $f(X)$ is irreducible over $\mathbb{Q}$ (see Exercise 8 of §4 Ch. 6), and so $[\mathbb{Q}(\Theta):\mathbb{Q}] = 3 \neq 2^n$.

c) Similar arguments show that for various values of n one cannot construct a regular n-gon by ruler and compass. For example, if $n = 7$ it is not hard to see that the number $\Theta = 2\cos\frac{360^{\circ}}{7}$ is a root of the polynomial $X^3 + X^2 - 2X - 1$, which is irreducible over $\mathbb{Q}$.

The great Gauss, at the very beginning of his mathematical career, found necessary and sufficient conditions on n in order that a regular n-gon can be constructed by ruler and compass. In particular, for n a prime he found that it must be a Fermat prime $n = 2^{2^k} + 1$. The complete solution of this problem is connected with the study of the Galois group of cyclotomic fields (see Example 2).

EXERCISES

1. Show that an extension $F \supset K$ of prime degree does not have any subfields besides F and K.

2. Find a primitive element for the extension $\mathbb{Q}(\sqrt{p}, \sqrt{q})$, where p and q are prime numbers.

3. Find the degree over $\mathbb{Q}$ of the splitting field of the polynomial $X^p - 2$.

4. Show that over a field K of characteristic $p > 0$ there are only two possibilities for the polynomial $X^p - a$: either it is irreducible or it is the p-th power of a linear polynomial.

5. Let $\mathbb{Z}_p(Y)$ be the field of rational functions over $\mathbb{Z}_p$, which has characteristic p. Show that $X^p - Y$ is an irreducible polynomial over $\mathbb{Z}_p(Y)$ all of whose roots coincide.

6. Prove that for any $d \mid n$, $d < n$, we have the relation $X^n - 1 = (X^d - 1)\Phi_n(X)h_d(X)$, where $h_d \in \mathbb{Z}[X]$.

7. Let q be a positive integer > 1. According to (10), $\Phi_n(q) \in \mathbb{Z}$. Show that $\Phi_n(q) \mid (q-1) \Rightarrow n = 1$.

8. Verify that the cyclotomic polynomial $\Phi_{15}(X) = X^8 - X^7 + X^5 - X^4 + X^3 - X + 1$, considered over $\mathbb{F}_2$, is the product of the two irreducible polynomials $X^4 + X^3 + 1$ and $X^4 + X + 1$. Using this fact, prove that $\Phi_{15}(X)$ is irreducible over $\mathbb{Q}$ (compare with Exercise 11 of §1 Ch. 6).

9. Starting with the chain of natural inclusions

$$GF(p) \subset GF(p^{2!}) \subset GF(p^{3!}) \subset \ldots$$

introduce the so-called limiting field $\Omega_p = GF(p^{\infty!})$, by setting $\alpha \in \Omega_p \Leftrightarrow \alpha \in GF(p^{n!})$ for n sufficiently large. Using the basic properties of finite fields, prove that Ω_p is an algebraically closed field. Along with $\mathbb{C}$, which has characteristic 0, these fields Ω_p provide examples of algebraically closed fields of any characteristic.

10. Let $q = p^n$. Show that if $p = 2$ all of the elements of $\mathbb{F}_q$ are squares, while if $p > 2$ the squares $\mathbb{F}_q^{*2}$ form a subgroup of index 2 in $\mathbb{F}_q^*$, and $\mathbb{F}_q^{*2} = \operatorname{Ker}(t \mapsto t^{(q-1)/2})$.

11. (M. Aschbacher). Prove the following fact for $p > 5$. Let $\mathbb{F}_q$ be a finite field with an odd number $q = p^n$ of elements. If $q \neq 3$ or 5, then the "circle" $x^2 + y^2 = 1$ contains a point with coordinates $x, y \in \mathbb{F}_q^*$.

12. Is every primitive element of the field $\mathbb{F}_q$ a generator of the multiplicative group $\mathbb{F}_q^*$?

13. (Theorem on the primitive element). Let $F = K(\Theta_1,\Theta_2,\dots,\Theta_r)$ be a finite algebraic extension of a field K of characteristic zero. Show that $F = K(\Theta)$ for some element Θ algebraic over K. (<u>Hint</u>. Use induction on r to reduce to the case $F = K(\alpha,\beta)$, where α and β are algebraic over K with distinct minimal polynomials $f(X)$ and $g(X)$. Let L be the splitting field of the polynomial $f(X)g(X)$, so that $f(X) = (X-\alpha_1)(X-\alpha_2)\cdots(X-\alpha_n)$, $g(X) = (X-\beta_1)(X-\beta_2)\cdots(X-\beta_m)$, where $\alpha_i, \beta_j \in L$, $\alpha_1 = \alpha$, $\beta_1 = \beta$. Irreducibility of f and g, together with the condition $\operatorname{char} K = 0$, guarantees that the elements α_i, β_j are pair-wise distinct (see subsection 4 §1 Ch. 6), and we can consider the elements $(\beta_j - \beta)/(\alpha - \alpha_i) \in L$, where $i, j \neq 1$. Take any rational number $c \neq 0$ different from all of these ratios (again using the condition $\operatorname{char} K = 0$!), and set $\Theta = \beta + c\alpha$.

Clearly $K(\Theta) \subset K(\alpha,\beta) = F$. The polynomials $f(X)$ and $h(X) = g(\Theta - cX) \in K(\Theta)[X]$ have common root α. If α_i is also a common root for some $i > 1$, then $0 = h(\alpha_i) = g(\Theta - c\alpha_i)$, so that $\Theta - c\alpha_i = \beta_j$ for some $j \geq 1$. But this means that either $c(\alpha - \alpha_i) = 0$ or else $c = (\beta_j - \beta)/(\alpha - \alpha_i)$, both of which possibilities are ruled out by our choice of c. Thus, $X - \alpha$ is the greatest common divisor of the polynomials $f, h \in L[X]$. But actually $f, h \in K(\Theta)[X]$, and so (see subsection 3 §3 Ch. 5) we have $\text{g.c.d.}(f,h) \in K(\Theta)[X]$. Therefore, $X - \alpha \in K(\Theta)[X]$, i.e., $\alpha \in K(\Theta)$ and then $\beta = \Theta - c\alpha \in K(\Theta)$. This means $K(\alpha,\beta) \subset K(\Theta)$, i.e., $K(\alpha,\beta) = K(\Theta)$.

14. The picture

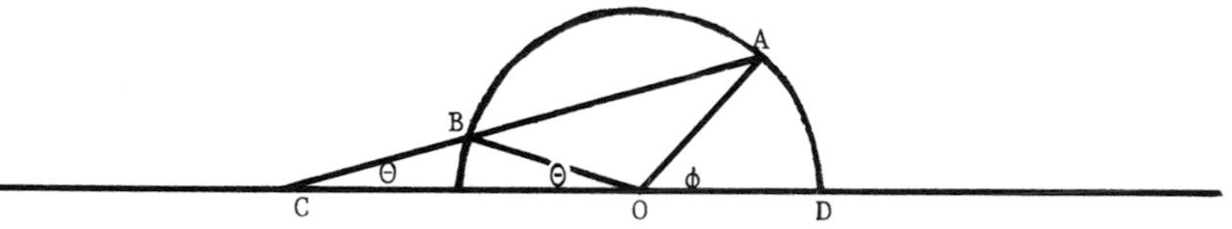

illustrates one way to trisect an angle: $\Theta = \phi/3$. The segments OB and CB have length 1. But how can one construct the point C given the point A?

15. By constructing specific constructive number fields K, show that the degree $[CS : \mathbb{Q}]$ is infinite.

§2. Various results about rings

This section is intended as a small but useful addendum to Chapters 4 and 5.

1. More examples of unique factorization domains. In §3 Ch. 5 we proved that a Euclidean ring has unique factorization. Such rings include the ring $\mathbb{Z}$ and the polynomial ring $K[X]$. Below we give another example of a Euclidean ring, and also an example of a unique factorization domain which is not Euclidean.

Example 1. The ring of Gaussian integers. This is the ring

$$\mathbb{Z}[i] = \{m + in \mid m, n \in \mathbb{Z}\} ,$$

which is contained in the quadratic number field $\mathbb{Q}(i) \subset \mathbb{Q}$, $i^2 + 1 = 0$. Geometrically, this ring can be thought of as the grid (lattice points) in the complex plane with integer coordinates. $\mathbb{Z}[i]$ is clearly an integral domain. On the set $\mathbb{Z}[i]^*$ of non-zero elements of $\mathbb{Z}[i]$ we define a map $\delta : \mathbb{Z}[i]^* \to \mathbb{N} \cup \{0\}$ by setting $\delta(m + in) = |m + in|^2 = m^2 + n^2$ (in other words, $\delta(a) = N(a)$ is the norm of a in $\mathbb{Q}(i)$ in the sense of Subsection 5 of §1 Ch. 5). We know that $\delta(ab) = \delta(a)\delta(b) \geq \delta(a)$ for all $a, b \in \mathbb{Z}[i]^*$, so that property (E1) in the definition of a Euclidean ring (see Subsection 3 §3 Ch. 5) is automatically fulfilled. In order to see (E2), we write the fraction ab^{-1} with $b \neq 0$ in the form $ab^{-1} = \alpha + i\beta$ with $\alpha, \beta \in \mathbb{Q}$ and we take the closest integers k and ℓ to α and β, so that $\alpha = k + \nu$ and $\beta = \ell + \mu$ where $|\nu| \leq \frac{1}{2}$, $|\mu| \leq \frac{1}{2}$. Then

$$a = b[(k + \nu) + i(\ell + \mu)] = bq + r ,$$

where $q = k + i\ell \in \mathbb{Z}[i]$ and $r = b(\nu + i\mu)$. Since $r = a - bq$, we have $r \in \mathbb{Z}[i]$, and

$$\delta(r) = |r|^2 = |b|^2(\nu^2 + \mu^2) \leq \delta(b)(\tfrac{1}{4} + \tfrac{1}{4}) = \tfrac{1}{2}\delta(b) < \delta(b) .$$

Hence, $\mathbb{Z}[i]$ is a Euclidean ring. □

The ring of Gaussian integers is convenient for illustrating the methods of algebraic number theory in a simple setting. For this reason we shall go into a little more detail on the properties of $\mathbb{Z}[i]$. We first make some general remarks.

1) An integral domain R all of whose ideals are principal, i.e., have the form xR, is called a principal ideal domain. Every Euclidean ring is a principal ideal domain. We have already established this in the case of $\mathbb{Z}$ and $K[X]$ (see the corollary to Theorem 5 §2 Ch. 5), and the proof is completely analogous in the general case: if J is an ideal in a Euclidean ring R, then $J = aR$ if we choose $a \in J$ so that $\delta(a) \leq \delta(x)$ for all non-zero $x \in J$.

2) Let R be an arbitrary Euclidean ring with function δ (see Subsection 3 §3 Ch. 5), and let $U(R)$ be its group of invertible elements. Then

$$u \in U(R) \Leftrightarrow \delta(u) = \delta(1) \Leftrightarrow \delta(ux) = \delta(x) \text{ for all } x \in R^* . \tag{1}$$

To see this, using (E1) we have $\delta(x) = \delta(1 \cdot x) \geq \delta(1)$ for all $x \in R^*$, and if $u \in U(R)$, then $\delta(1) = \delta(u \cdot u^{-1}) \geq \delta(u)$, so that $\delta(u) = \delta(1)$. Conversely, by Remark 1), we have $\delta(ux) = \delta(x)$, $\forall x \in R^* \Rightarrow uxR = xR \Rightarrow x = uxv \Rightarrow uv = 1 \Rightarrow$ $\Rightarrow u \in U(R)$.

Applied to $\mathbb{Z}[i]$, the criterion (1) above means that $m + in \in U(\mathbb{Z}[i]) \Leftrightarrow$ $\Leftrightarrow m^2 + n^2 = 1$. Thus, $U(\mathbb{Z}[i])$ is the cyclic multiplicative group of order 4 generated by i.

3) An ideal J in a ring R is called maximal if $J \neq R$ and if every ideal T containing J coincides with either J or R. In a Euclidean ring R, an element $p \in R$ is prime if and only if the ideal pR is maximal. To see this, first let p be a prime element, and suppose that $pR \subset T \subset R$, where T is an ideal of R. By Remark 1), we have $T = aR$, and since $p \in T$, that means that $p = ab$, where one of the elements a or b must be invertible (since p is prime). If $a \in U(R)$, then $T = aR = R$. If $b \in U(R)$, then $T = aR = abR = pR$. Conversely, suppose that

the ideal pR is maximal, and $p = ab$, where $a \notin U(R)$. Then $aR \neq R$, and $pR \subset aR$, so that $pR = aR$. But then $a = pu = abu$, and hence $bu = 1$ and $b \in U(R)$. This means that p is a prime element, and the proof of Remark 3) is complete.

We now look at what happens to a rational prime $p \in \mathbb{Z}$ when considered in the ring $\mathbb{Z}[i]$. It may happen that p remains a prime element in $\mathbb{Z}[i]$, but if this is not the case let $p = \prod_{k=1}^{r} p_k$ be its (unique by Theorem 4 §3 Ch. 5) factorization into prime elements p_k of $\mathbb{Z}[i]$, where $r > 1$. According to Remark 2), we have $\delta(p_k) > 1$ for each k, so that the equation $p^2 = \delta(p) = \prod \delta(p_k)$ implies that $r = 2$, $p = p_1 p_2$, $\delta(p_1) = \delta(p_2) = p$. If $p_1 = m + in$, then $p = \delta(p_1) = m^2 + n^2 = (m + in)(m - in) \Rightarrow$ $\Rightarrow p_2 = m - in$. Thus, <u>if a prime</u> $p \in \mathbb{Z}$ <u>has a non-trivial factorization in</u> $\mathbb{Z}[i]$, <u>then</u>

$$p = (m + in)(m - in) = m^2 + n^2 , \tag{2}$$

<u>where</u> $m + in$ <u>and</u> $m - in$ <u>are prime elements in</u> $\mathbb{Z}[i]$.

For example, $2 = (1 + i)(1 - i)$ is not a prime element in $\mathbb{Z}[i]$. Also note that $t^2 \equiv 0$ or $1 \pmod 4$ for any $t \in \mathbb{Z}$. Hence, if p is an odd prime which is not a prime element in $\mathbb{Z}[i]$, the criterion (2) tells us that

$$p = m^2 + n^2 \equiv 0, 1 \text{ or } 2 \pmod 4 \Rightarrow p \text{ is of the form } 4k + 1 .$$

We now let p be of the form $4k + 1$, and claim that p does not remain prime in $\mathbb{Z}[i]$. Set $t = (2k)!$. Since clearly $t = (-1)^{2k}(2k)! = (-1)(-2) \ldots (-2k) \equiv$ $\equiv (p - 1) \times (p - 2) \ldots (p - 2k) \equiv ((p + 1)/2) \ldots (p - 2)(p - 1) \pmod p$, it follows that

$$t^2 \equiv (2k)!\,((p + 1)/2) \ldots (p - 2)(p - 1) \equiv (p - 1)! \pmod p ,$$

or, by Wilson's theorem (see the end of §1 Ch. 6), we have $t^2 + 1 \equiv 0 \pmod p$. Now if p were a prime element in $\mathbb{Z}[i]$, then it would follow by Theorem 1 §3 Ch. 5 together with the relation $(t + i)(t - i) = t^2 + 1 = \ell p$ for some $\ell \in \mathbb{Z}$ that p must divide either $t + i$ or $t - i$. But comparing imaginary parts in a relation of the form

$t \pm i = p(m + in)$ gives: $\pm 1 = pn$, $n \in \mathbb{Z}$, which is absurd. Thus, we have proved the following fact:

A prime number $p \in \mathbb{Z}$ remains prime in $\mathbb{Z}[i]$ if and only if $p = 4k - 1$. Every prime of the form $p = 4k + 1$ can be written in the form $m^2 + n^2$, where $m, n \in \mathbb{Z}$. □

It is now relatively easy to prove a general number-theoretic theorem about when a given integer can be written as the sum of two squares.

THEOREM 1. A number $t \in \mathbb{Z}$ can be represented as the sum of the squares of two integers m and n if and only if every prime number $p = 4k - 1$ in the prime factorization of t occurs with an even exponent.

Proof. In addition to the facts we already know, it suffices to show that: g.c.d. $(m, n) = 1$, $m^2 + n^2 \equiv 0 \pmod{p} \Rightarrow mn \not\equiv 0 \pmod{p} \Rightarrow m^{p-1} \equiv 1 \pmod{p}$, $n^2 \equiv -m^2 \pmod{p} \Rightarrow (m^{p-2}n)^2 = m^{2p-4}n^2 \equiv -m^{2p-2} \equiv -1 \pmod{p}$. Thus, there exists an integer $s \in \mathbb{Z}$ such that $s^2 \equiv -1 \pmod{p}$, $s^4 \equiv 1 \pmod{p}$. Hence, the order $p - 1$ of the multiplicative group $\mathbb{Z}_p^*$ is divisible by 4, and p is of the form $4k + 1$. □

By Remark 3), the fact that $p = 4k - 1$ is prime in $\mathbb{Z}[i]$ is equivalent to maximality of the ideal $p\mathbb{Z}[i]$, which, in turn, is equivalent to the quotient ring $\mathbb{Z}[i]/p\mathbb{Z}[i]$ being a field of p^2 elements (in this connection, see the isomorphism theorems in Subsection 2 and also Exercise 14 of §4 Ch. 4). This is not surprising, since when $p = 4k - 1$ the polynomial $X^2 + 1$ is irreducible over $\mathbb{Z}_p$.

Example 2. Polynomial rings over unique factorization domains. We shall now show that the polynomial ring $\mathbb{Z}[X_1, \dots, X_n]$ and $K[X_1, \dots, X_n]$ (where K is a field) are unique factorization domains for any n. This important fact follows immediately from the following theorem.

THEOREM 2. _If_ R _is a unique factorization domain, then so is the polynomial ring_ $R[X]$.

Proof. The proof is based on the properties of polynomial rings related to Gauss's lemma (see §3 Ch. 5). Namely, we shall need the following two properties:

(a) _Two primitive polynomials_ $f, g \in R[X]$ _which are associated in_ $Q(R)[X]$ ($Q(R)$ is the field of fractions of R) _are associated in_ $R[X]$ (this is an easy exercise using Gauss's lemma).

(b) _A polynomial_ $f \in R[X]$ _of positive degree which is irreducible over_ R _is also irreducible over_ $Q(R)$ (the proof given in §3 Ch. 5 for $R = \mathbb{Z}$ also works in the general case).

We now proceed to the proof of the theorem. If $f \in R[X]$ is a polynomial of positive degree, we write $f = d(f) f_0$, where $d(f)$ is the content of f and f_0 is a primitive polynomial. Using induction on the degree of a primitive polynomial, we obtain a decomposition of f_0 into a product $f_0 = f_1 \dots f_s$ of primitive polynomials $f_1, \dots, f_s$ which are irreducible over R. Suppose that $f_0 = g_1 \dots g_t$ is another such factorization. Then, by property (b) above, the f_i and g_j are irreducible over $Q(R)$, and, since the ring $Q(R)[X]$ is a unique factorization domain (see the corollary to Theorem 4 §3 Ch. 5), we have $s = t$ and, with a suitable ordering of the f's and g's, f_i is associated with g_i in $Q(R)[X]$. Consequently, by property (a), they are also associated in $R[X]$. If the content $d(f)$ is not invertible in R, we factor it in R: $d(f) = p_1 \dots p_r$, and finally arrive at a factorization of f. This factorization is unique (in the usual sense), because we have just seen that the factorization of f_0 is unique, and the same holds for the factorization of $d(f)$ because R is a unique factorization domain. □

We have strict inclusions

$$\left\{\begin{matrix}\text{Euclidean}\\ \text{rings}\end{matrix}\right\} \subset \left\{\begin{matrix}\text{principal ideal}\\ \text{domains}\end{matrix}\right\} \subset \left\{\begin{matrix}\text{unique factorization}\\ \text{domains}\end{matrix}\right\} \quad . \tag{3}$$

We have already established the first inclusion (see Remark 1)). There are examples (we shall not give them) which show that it is a strict inclusion. To prove the second inclusion, we let R be a principal ideal domain and consider an increasing sequence of ideals $(d_1) \subset (d_2) \subset \dots$ in R. We immediately see that $D = \bigcup_i (d_i)$ is an ideal in R. Hence, $D = (d)$, $d \in D$. By the definition of D, we must have $d \in (d_m)$ for some m, and so $(d_m) = (d_{m+1}) = \dots$. Since we have just shown that an increasing sequence of ideals must stabilize at a finite distance out, it follows that a sequence of non-invertible divisors $d_1, d_2, d_3, \dots$ with $d_i | d_{i-1}$ must also stabilize; hence, any element in R factors into a product of indecomposable elements. We see unique factorization using the same type of argument: $(a, b) = aR + bR = dR = (d) \Rightarrow d = \text{g.c.d.}(a, b) = ax + by$. The rest of the argument is the same as in the proof of Theorem 3 (ii) in §3 Ch. 5 .

The ideal $(2, X)$ in $\mathbb{Z}[X]$ and the ideal (X, Y) in $\mathbb{R}[X, Y]$ are not principal ideals (see the example in Subsection 3 §2 Ch. 5). But, by Theorem 2, the rings $\mathbb{Z}[X]$ and $\mathbb{R}[X, Y]$ are unique factorization domains. Thus, the second inclusion in (3) is strict. □

Principal ideal domains are interesting from a purely algebraic viewpoint, since they are characterized by the properties of very natural objects -- the kernels of homomorphisms. On the other hand, Euclidean rings are more useful to work with because they have the division algorithm.

2. Ring theoretic constructions. We already have at our disposal a significant arsenal of types of rings and methods for constructing new rings from ones we start with. For example, we have matrix rings $M_n(R)$, fields of fractions $Q(R)$, and polynomial rings $R[X_1, \dots, X_n]$, where R is a commutative ring (an integral domain in the case of $Q(R)$). It is worthwhile to discuss, at least briefly, the ring-theoretic analogs of the

general facts about homomorphisms which were established for groups in Chapter 7. As a rule, the proofs will be no different from the case of groups, and so will be left to the reader as exercises.

To the fundamental theorem on ring homomorphisms (Theorem 2 §4 Ch. 4), we add two isomorphism theorems.

THEOREM 3. Let R be a ring, S be a subring, and J be an ideal of R. Then $S + J = \{x + y \mid x \in S, y \in J\}$ is a subring of R containing J as an ideal, and $S \cap J$ is an ideal in S. The map

$$\varphi : x + J \longmapsto x + S \cap J, \quad x \in S,$$

gives a ring isomorphism

$$(S + J)/J \cong S/(S \cap J) \quad .$$

Proof. The first two assertions are obvious. As for the last one, consider the restriction $\pi_0 = \pi|_S$ of the natural epimorphism $\pi : R \to R/J$. The image $\operatorname{Im} \pi_0$ consists of the cosets $x + J$ with $x \in S$, i.e., $\operatorname{Im} \pi_0 = (S + J)/J$. The kernel $\operatorname{Ker} \pi_0$ of the epimorphism $\pi_0 : S \to (S + J)/J$ consists of the elements $x \in S$ for which $x + J = J$. That is, $\operatorname{Ker} \pi_0 = S \cap J$. By the fundamental homomorphism theorem, the correspondence $\overline{\pi}_0 : x + S \cap J \mapsto \pi_0(x) = x + J$ gives an isomorphism $S/(S \cap J) \cong (S + J)/J$. It remains to note that $\varphi = \overline{\pi}_0^{-1}$. □

We have gone through the details of a proof which is essentially copied from the proof of Theorem 2 §3 Ch. 7 in order to emphasize the complete parallelism with group theory.

THEOREM 4. Let R be a ring, S be a subring, and $J \subset S$ be an ideal of R. Then $\overline{S} = S/J$ is a subring of R/J, and $\pi^* : S \mapsto \overline{S}$ is a bijective map from the set $\Omega(R, J)$ of all subrings of R containing J to the set $\Omega(\overline{R})$ of all subrings of $\overline{R}$. If $S \in \Omega(R, J)$, then S is an ideal in R if and only if $\overline{S}$ is an ideal in $\overline{R}$, and

$$R/S \cong \overline{R}/\overline{S} = (R/J)/(S/J) \quad .$$

Proof. This is an easy exercise, following the proof of Theorem 3 §3 Ch. 7. □

COROLLARY. Let R be a commutative ring with unit 1. An ideal J is maximal in R if and only if the quotient ring R/J is a field. □

The following operations are defined on the set of ideals of a ring R:

sum: $$J_1 + J_2 = \{x_1 + x_2 \mid x_k \in J_k\} \quad ,$$

intersection: $$J_1 \cap J_2 = \{x \mid x \in J_1 , x \in J_2\} \quad ,$$

product: $$J_1 J_2 = \left\{\sum_i x_{1i} x_{2i} \mid x_{ki} \in J_k\right\} \subset J_1 \cap J_2 \quad .$$

One can also speak of the sum, product, or intersection of a finite number of ideals, and we have the following fact.

PROPOSITION. If R is a ring with unit, and the equalities

$$J + J_k = R , \qquad k = 1, \dots, n \quad ,$$

hold for ideals $J, J_1, \dots, J_n$, then the following equalities also hold:

$$J + J_1 \cap J_2 \cap \dots \cap J_n = R = J + J_1 J_2 \dots J_n \quad .$$

Proof. Since $J_1 J_2 \dots J_n \subset J_1 \cap J_2 \cap \dots \cap J_n$, it suffices to prove that $J + J_1 J_2 \dots J_n = R$. If $n = 1$, this is true by assumption. If $n = 2$, we have

$$1 = 1^2 = (x_1 + y_1)(x_2 + y_2) = x + y_1 y_2 \quad ,$$

where $x_1, x_2, x \in J$, $y_i \in J_i$. Hence, $1 \in J + J_1 J_2$, and $R = J + J_1 J_2$. We now use an obvious argument by induction on n. □

Let $R_1, \ldots, R_n$ be a finite set of rings, and let $R = R_1 \times \ldots \times R_n$ be the cartesian product of them as sets. We introduce a ring structure on R by defining addition and multiplication component by component:

$$(x_1, \ldots, x_n) + (y_1, \ldots, y_n) = (x_1 + y_1, \ldots, x_n + y_n) \ ;$$

$$(x_1, \ldots, x_n) \cdot (y_1, \ldots, y_n) = (x_1 y_1, \ldots, x_n y_n) \quad .$$

This gives us the so-called (external) <u>direct sum</u> $R = R_1 \oplus \ldots \oplus R_n$ of the rings R_i. Each of the components R_i is the image of the natural epimorphism $\pi_i : R \to R_i$ which takes $(x_1, \ldots, x_n)$ to $x_i \in R_i$. Furthermore, if $J_i = \{(0, \ldots, 0, x_i, 0, \ldots, 0) \mid x_i \in R_i\}$, then $J_i \cong R_i$, J_i is an ideal in R, and $R = J_1 + \cdots + J_n$.

Now suppose that R is a ring with ideals $J_1, \ldots, J_n$, where $R = J_1 + \cdots + J_n$ and $J_k \cap \left(\sum_{j \neq k} J_j \right) = 0$, $1 \leq k \leq n$. Then $R = J_1 \oplus \ldots \oplus J_n$ is the (internal) <u>direct sum of its ideals</u> J_k. As in group theory, the difference between external and internal direct sums is rather pedantic, and there is no need to distinguish between the two in our notation.

3. <u>Number theoretic applications.</u> The universal property of direct sums is the following: <u>if</u> $S = R_1 \oplus \ldots \oplus R_n$, <u>and</u> R <u>is any ring having homomorphisms</u> $\varphi_i : R \to R_i$, <u>then there exists a unique homomorphism</u> $\varphi = (\varphi_1, \ldots, \varphi_n) : R \to S$ <u>with kernel</u> $\operatorname{Ker} \varphi = \cap \varphi_i$ <u>which makes the triangular diagram</u>

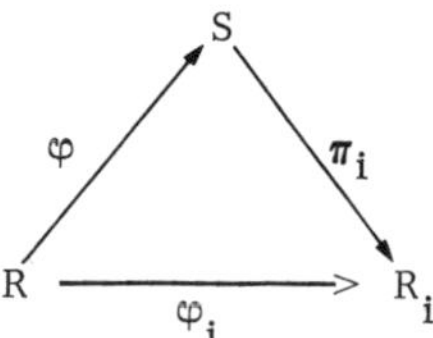

<u>commutative for</u> $i = 1, \ldots, n$. We apply this obvious fact to a ring R having unit 1 and ideals $J_1, \ldots, J_n$ and to the direct sum

$$S = R/J_1 \oplus \dots \oplus R/J_n \quad .$$

Setting $\varphi_i : R \to R/J_i = R_i$, we obtain a homomorphism

$$\varphi : x \longmapsto (x + J_1, \dots, x + J_n) \tag{4}$$

from R to S with kernel $\operatorname{Ker}\varphi = J_1 \cap \dots \cap J_n$.

THEOREM 5 (The Chinese Remainder Theorem). Under the above conditions, if R is a ring with unit and $J_i + J_j = R$ for $1 \leq i \neq j \leq n$, then the map φ in (4) above is an epimorphism.

Proof. We must check that, given any elements $x_1, \dots, x_n \in R$, there exists an $x \in R$ such that $x_i + J_i = x + J_i$, i.e., $x - x_i \in J_i$, for $i = 1, 2, \dots, n$. If $n = 1$, this is obvious; if $n = 2$, we take elements $a_1 \in J_1$ and $a_2 \in J_2$ such that $a_1 + a_2 = 1$, and we set $x = x_1 a_2 + x_2 a_1$. Then

$$x - x_1 - (x_1 a_2 + x_2 a_1) - x_1(a_1 + a_2) = (x_2 - x_1) a_1 \in J_1 \quad ,$$

$$x - x_2 = (x_1 a_2 + x_2 a_1) - x_2(a_1 + a_2) = (x_1 - x_2) a_2 \in J_2 \quad .$$

We now use induction on n. Suppose that we have already found an element y such that $y - x_i \in J_i$, $i = 1, 2, \dots, n-1$. Since by assumption $J_i + J_n = R$, $1 \leq i \leq n-1$, it follows by the proposition in Subsection 2 that $J_1 \cap \dots \cap J_{n-1} + J_n = R$. Now apply the case $n = 2$, which was already proved, using the two ideals $J_1 \cap \dots \cap J_{n-1}$ and J_n and the two elements y and x_n. We find $x \in R$ with $x - y \in J_1 \cap \dots \cap J_{n-1}$ and $x - x_n \in J_n$. But then $x - y \in J_i$, $1 \leq i \leq n-1$, and, by our choice of y,

$$x - x_i = (x - y) + (y - x_i) \in J_i, \qquad 1 \leq i \leq n-1 \quad .$$

Thus, the element x satisfies all of the requirements. □

In Theorem 5 and the arguments preceding it, the ring R was not assumed to be commutative. Now suppose that R is an integral domain and $a_1, \dots, a_n$ are pair-wise relatively prime elements, i.e., $a_i R + a_j R = R$ if $i \neq j$. (In a unique factorization domain, this agrees with the definition that elements are relatively prime if their factorizations contain disjoint sets of primes.) We write the relation $x - x_i \in a_i R$ as a congruence modulo the principal ideal $a_i R$: $x \equiv x_i \pmod{a_i}$.

COROLLARY 1. Let R be an integral domain, and let $a_1, \ldots, a_n$ be pairwise relatively prime elements. Then for any $x_1, \ldots, x_n \in R$ there exists $x \in R$ such that $x \equiv x_i \pmod{a_i}$ for $i = 1, \ldots, n$. □

COROLLARY 2. Let n be a prime number with prime factorization $n = p_1^{m_1} \cdots p_r^{m_r}$; let $\mathbb{Z}_n = \mathbb{Z}/n\mathbb{Z}$ be the ring of residue classes modulo n, and let $U(\mathbb{Z}_n)$ be the multiplicative group of invertible elements in this ring. Then:

(i) $\mathbb{Z}_n \cong \mathbb{Z}_{p^{m_1}} \oplus \cdots \oplus \mathbb{Z}_{p^{m_r}}$ (as a direct sum of rings);

(ii) $U(\mathbb{Z}_n) \cong U\left(\mathbb{Z}_{p^{m_1}}\right) \times \cdots \times U\left(\mathbb{Z}_{p^{m_r}}\right)$ (as a direct product of groups).

Proof. (i) Replacing n by r in (4), and setting $R = \mathbb{Z}$, $J_i = p_i^{m_i}\mathbb{Z}$ and $S = \mathbb{Z}_{p_1^{m_1}} \oplus \cdots \oplus \mathbb{Z}_{p_r^{m_r}}$, we obtain a homomorphism $\varphi : \mathbb{Z} \to S$ with kernel $\operatorname{Ker}\varphi = \cap J_i = n\mathbb{Z}$. Theorem 5 implies that φ is epimorphic, since g.c.d. $(p_i, p_j) = 1$ for $i \neq j$.

(ii) Since the components kill one another in any direct sum of rings $R = R_1 \oplus \cdots \oplus R_r$, in other words $R_i R_j = 0$ for $i \neq j$, it follows immediately from the definition of invertible elements and part (i) that $U(R) = U(R_1) \times \cdots \times U(R_r)$. □

Remark. Part (ii) of Corollary 2 immediately implies that $\varphi(n) = \prod_{i=1}^{r} \varphi(p^{m_i})$, and, since $\varphi(p^m) = p^{m-1}(p-1)$, we again obtain the formula for the Euler function (see Example 1 in Subsection 4 of §1). Since the order of an element in a finite group divides the order of the group, we find that

$$a^{\varphi(n)} \equiv 1 \pmod{n}$$

<u>for any integer</u> a <u>prime to</u> n. (This generalization of Fermat's Little Theorem is sometimes known as Euler's Theorem.)

By Corollary 2, in order fully to understand the structure of the group $U(\mathbb{Z}_n)$, it suffices to consider the case $n = p^m$.

THEOREM 6. *Let* m *be a positive integer.*

(i) *If* p *is an odd prime, then* $U(\mathbb{Z}_{p^m})$ *is a cyclic group.*

(ii) *The groups* $U(\mathbb{Z}_2)$ *and* $U(\mathbb{Z}_4)$ *are cyclic of order* 1 *and* 2, *respectively, while* $U(\mathbb{Z}_{2^m})$, $m \geq 3$, *is the direct product of a cyclic group of order* 2^{m-2} *and a cyclic group of order* 2.

Proof. (i) By definition, an integer t prime to n has order r modulo n if $|\langle t + n\mathbb{Z}\rangle| = r$, i.e., if $t^k \not\equiv 1 \pmod{n}$ if $k < r$. If $r = \varphi(n)$, we call t a <u>primitive root modulo</u> n. We usually choose t from among the least positive residues $0, 1, \ldots, n-1$ modulo n, but it is not really necessary to fix any particular set of residue class representatives.

According to Theorem 5 §1, the group $\mathbb{Z}_p^* = U(\mathbb{Z}_p)$ is cyclic, i.e., there exists a primitive root a_0 modulo p. Since $a_0^{p^{m-1}} \equiv a_0 \pmod{p}$, it follows that the integer $a = a_0^{p^{m-1}}$ is also a primitive root modulo p. On the other hand, $a^{p-1} = a_0^{p^{m-1}(p-1)} = a_0^{\varphi(p^m)} \equiv 1 \pmod{p^m}$. Hence, the coset $\bar{a} = a + p^m\mathbb{Z}$ generates a cyclic group of order $p-1$ in $U(\mathbb{Z}_{p^m})$.

Furthermore,

$$(1+p)^p = \sum_{i=0}^{p} \binom{p}{i} p^i = 1 + p^2 + \frac{1}{2}(p-1)p^3 + \sum_{i \geq 3} \binom{p}{i} p^i \quad .$$

Since $p > 2$, we have $(1+p)^p \equiv 1 + p^2 \pmod{p^3}$. Making the induction assumption that $(1+p)^{p^j} \equiv 1 + p^{j+1} \pmod{p^{j+2}}$, we find that

$$(1+p)^{p^{j+1}} = [1 + (1+sp)p^{j+1}]^p = \sum_{i=0}^{p} \binom{p}{i} (1+sp)^i p^{(j+1)i} =$$

$$= 1 + (1+sp)p^{j+2} + \frac{1}{2}(p-1)(1+sp)^2 p^{2(j+1)+1} + \cdots \quad .$$

and hence $(1+p)^{p^{j+1}} \equiv 1 + p^{j+2} \pmod{p^{j+3}}$. In particular, $(1+p)^{p^{m-1}} \equiv 1 \pmod{p^m}$, but $(1+p)^{p^{m-2}} \equiv 1 + p^{m-1} \not\equiv 1 \pmod{p^m}$; hence, the coset $\bar{b} = 1 + p + p^m\mathbb{Z}$ with representative $1+p$ generates a cyclic group of order p^{m-1} in $U(\mathbb{Z}_{p^m})$. According to the proposition in Subsection 3 §2 Ch. 4, if elements $\bar{a}$ and $\bar{b}$ have relatively prime orders $p-1$ and p^{m-1}, their product must generate a cyclic group $\langle \bar{a}\bar{b} \rangle$ of order $p^{m-1}(p-1) = \varphi(p^m) = |U(\mathbb{Z}_{p^m})|$.

(ii) The assertion about $U(\mathbb{Z}_2)$ and $U(\mathbb{Z}_4)$ is obvious. If $m > 2$, then starting from the trivial congruence $5 \equiv 1 + 2^2 \pmod{2^3}$ and using induction on j, we easily prove that

$$5^{2^j} \equiv 1 + 2^{j+2} \pmod{2^{j+3}} \quad .$$

In particular,

$$5^{2^{m-3}} \equiv 1 + 2^{m-1} \not\equiv 1 \pmod{2^m}, \qquad 5^{2^{m-2}} \equiv 1 \pmod{2^m} \quad ,$$

so that 5 has order 2^{m-2} modulo 2^m, and the coset $5 + 2^m\mathbb{Z}$ generates a cyclic subgroup of index 2 in $U(\mathbb{Z}_{2^m})$. Note that, since $5^j \equiv 1 \pmod 4$ for all j, the coset $-1 + 2^m\mathbb{Z}$ is not in $\langle 5 + 2^m\mathbb{Z} \rangle$. Since $|\langle -1 + 2^m\mathbb{Z} \rangle| = 2$, we have

$$U(\mathbb{Z}/2^m\mathbb{Z}) = \langle 5 + 2^m\mathbb{Z}\rangle \times \langle -1 + 2^m\mathbb{Z}\rangle$$

is an abelian 2-group of type $(2^{m-2}, 2)$ (see §5 Ch. 7). □

COROLLARY. <u>The group</u> $U(\mathbb{Z}_n)$ <u>is cyclic (equivalently, there exists a primitive root modulo</u> n) <u>if and only if</u> n <u>has the form</u> $2, 4, p^m$ <u>or</u> $2p^m$, <u>where</u> p <u>is an odd prime.</u> □

EXERCISES

1. Prove that a non-zero element p of a unique factorization domain R is prime if and only if R/pR is an integral domain.

2. Prove that, if an integral domain R is not a field, then $R[X]$ is not a principal ideal domain.

3. Prove that the set of elements $x + y\sqrt{-3}$ with either $x, y \in \mathbb{Z}$ or else $x - \frac{1}{2}, y - \frac{1}{2} \in \mathbb{Z}$ is an integral domain R. Show that it is a Euclidean ring with function $\delta = N$ (the norm in $\mathbb{Q}(\sqrt{-3})$). Show that the subring $\mathbb{Z}[\sqrt{-3}] \subset R$ is not even a unique factorization domain.

4. Find all of the prime elements of the ring of Gaussian integers.

5. In the case when R is a unique factorization domain, refine the corollary to Theorem 5 by introducing the elements $\tilde{a}_i = \prod_{j \neq i} a_j$. Find $b_i \in R$ such that $b_i \equiv 1 \pmod{a_i}$ and $b_i \equiv 0 \pmod{\tilde{a}_i}$ for $1 \leq i \leq n$. Let $x_1, \ldots, x_n \in R$. Introduce the element $x = \Sigma\, b_i x_i$, and verify that $x \equiv x_i \pmod{a_i}$, $1 \leq i \leq n$ (this approach is especially convenient when n is large).

6. Apply the previous exercise to $a_1 = 5$, $a_2 = 9$, and to the pairs $(x_1, x_2) = (2, 5), (3, 2)$, and $(3, 5)$. What can you say about x considered modulo 45?

7. Let p be an odd prime. If the congruence $x^2 \equiv a \pmod p$ has a solution, then a is called a <u>quadratic residue modulo</u> p; otherwise it is called a <u>quadratic non-residue.</u> The <u>Legendre symbol</u> (a/p) is defined as follows:

$$\left(\frac{a}{p}\right) = \begin{cases} 0, & \text{if } a \equiv 0 \pmod p \\ 1, & \text{if } a \not\equiv 0 \pmod p \text{ and } a \text{ is a quadratic residue ,} \\ -1, & \text{if } a \not\equiv 0 \pmod p \text{ and } a \text{ is a quadratic non-residue} \end{cases} .$$

Show that $(a/p) = 1$ if and only if $a + p\mathbb{Z} \in \mathbb{Z}_p^{*2}$, and that $(a/p) \equiv a^{(p-1)/2} \pmod p$. Furthermore, $(ab/p) = (a/p)(b/p)$, and the number of quadratic residues in $\{1, 2, \ldots, p-1\}$ is equal to the number of nonresidues. Verify the following so-called <u>quadratic reciprocity law</u> for a few small values of odd prime numbers p and q:

$$\left(\frac{p}{q}\right)\left(\frac{q}{p}\right) = (-1)^{\frac{p-1}{2} \cdot \frac{q-1}{2}} .$$

This law was proved in general by Gauss, who gave several different proofs. Use Example 1 in Subsection 1 to derive the relation $((-1)/p) = (-1)^{(p-1)/2}$.

8. Prove that (in the notation of the previous exercise): $(2/p) = (-1)^{(p^2-1)/8}$, i.e., 2 is a quadratic residue of primes of the form $8k \pm 1$ and a quadratic non-residue of primes of the form $8k \pm 3$.

9. (Supplement to Subsection 5 §2 Ch. 3.) Let $f(X) = f(\ldots, x_{ij}, \ldots)$ be a non-zero polynomial with coefficients in $\mathbb{Z}$ or in a field, where the x_{ij} are n^2 independent variables. We consider f as a function on matrices $X = (x_{ij})$. Prove that if $f(XY) = f(X)f(Y)$ for all $X, Y \in M_n(R)$, where R is a ring, then $f(X) = (\det X)^m$, where m is some non-negative integer. In particular, if $f(\operatorname{diag}(x, 1, 1, \ldots, 1)) = x$, then it follows that $f(X) = \det X$.

10. Show that the ring $\mathbb{Q}_M(Z)$ of all rational numbers a/b with b not divisible by a fixed prime p (see Exercise 6 of §4 Ch. 5) contains a unique maximal ideal

$$J = \{a/b \in \mathbb{Q}_M(Z) \mid p \text{ divides } a\}.$$

A ring which has a unique maximal ideal is called a <u>local ring</u>. (<u>Hint</u>. J is obviously a proper ideal in $\mathbb{Q}_M(Z)$. If $c/d \notin J$, then $c \notin pZ$, and hence $d/c \in \mathbb{Q}_M(Z)$. This means that any ideal L obtained from J by adding just a single element c/d must contain 1, and therefore must coincide with all of $\mathbb{Q}_M(Z)$.)

11. Show that in any local ring R with maximal ideal $\mathbf{M}$, the elements not in M are invertible.

12. Let R be a ring with unit. An ideal P is called a prime ideal if the quotient ring R/P is an integral domain. Every maximal ideal is prime. The complement $M = R - P$ of P in R is a multiplicative subset of R (a monoid not containing zero). The ring $\mathbb{Q}_M(R)$ in this case is usually denoted $\mathbf{M}^{-1}R$ or simply R_P. Show that the ring R_P is always local, and that its maximal ideal M_P consists of fractions of the form a/b, where $a \in P$ and $b \in R - P$. Further show that $M_P \cap R = P$.

The operation of going from R to the local ring R_P is called "localization" of R relative to the prime ideal P.

§3. Modules

The idea of a module incorporates a fundamental principle which developed in algebra about a half century ago: if we are interested in an algebraic system, we should study not only the internal properties of this system, but also all of its representations (in the broadest possible sense of the word).

1. Basic facts about modules. We begin with the classical definition. Let R be an associative ring with unit, and let V be an abelian group written additively. Further suppose that we are given a map $(x,v) \mapsto xv$ from $R \times V$ to V which satisfies the conditions:

$$\text{(M1)} \qquad x(u+v) = xu + xv \ ,$$

$$\text{(M2)} \qquad (x+y)v = xv + yv \ ,$$

$$\text{(M3)} \qquad (xy)v = x(yv) \ ,$$

$$\text{(M4)} \qquad 1 \cdot v = v$$

for all $x, y \in R$ and $u, v \in V$. Then V is called a left R-module (or a left module over R). We similarly define a right R-module. In what follows we shall speak simply of an R-module, even though in some situations we may want to deal with both types of modules at the same time.

Of course, if we want to make this definition when R does not have a unit, then we omit axiom (M4). Moreover, it is possible to modify the axiom (M3) for use with non-associative rings. We shall give an example of a module over a non-associative ring at the end of the section. For now we shall be content with the above definition.

Let V be an R-module. A subgroup $U \subset V$ is called a submodule of V if $xu \in U$ for all $x \in R$ and $u \in U$.

Now suppose that U and V are arbitrary R-modules. By a homomorphism of R-modules (or simply an R-homomorphism) from U to V we mean a map

$\sigma : U \to V$ such that

$$\sigma(u_1 + u_2) = \sigma(u_1) + \sigma(u_2) \quad ,$$

$$\sigma(xu) = x\,\sigma(u)$$

for all $u_1, u_2, u \in U$ and $x \in R$. It is easy to verify that $\operatorname{Ker}\sigma = \{u \in U \mid \sigma(u) = 0\}$ is an R-submodule of U, and $\operatorname{Im}\sigma$ is an R-submodule of V.

Given a submodule $U \subset V$ over R, we define the quotient module $V/U = \{v + U \mid v \in V\}$ to be the quotient group of the abelian group V by U with the action of R defined by the rule:

$$x(v + U) = xv + U \quad .$$

The fundamental isomorphism theorem and the two isomorphism theorems proved first for groups (§3 Ch. 7) and then for rings, carries over exactly, with only small changes in the proofs, to the case of modules.

After §2 Ch. 7, where axioms like (M3) and (M4) were studied, and after Chapter 8 on group representations (with the axioms (M1), (M3) and (M4)), our examples of R-modules will hardly seem very new. Nevertheless, it is worthwhile to discuss and compare these examples.

1) Every abelian group A is a $\mathbb{Z}$-module. Namely, the map $(n, a) \mapsto na$ from $\mathbb{Z} \times A$ to A satisfies all of the axioms (M1)-(M4). It is often very useful to think of an abelian group as a $\mathbb{Z}$-module.

2) Every abelian group A is a module over its endomorphism ring End A. By definition, End A consists of all maps $\varphi : A \to A$ which satisfy the condition $\varphi(a + a') = \varphi(a) + \varphi(a')$. The addition and multiplication operations in End A are introduced in the natural way: $(\varphi + \psi)(a) = \varphi(a) + \psi(a)$, $(\varphi\psi)(a) = \varphi(\psi(a))$, $1(x) = x$, $0(x) = 0$. The map $(\varphi, a) \mapsto \varphi(a)$ from $\operatorname{End} A \times A$ to A clearly gives A the structure of an End A-module.

3) A vector space V over a field K is obviously a K-module. In addition, if we have a fixed linear operator $\mathcal{A} : V \to V$, then we can give V the structure of a $K[X]$-module, which we denote $V_{\mathcal{A}}$, by setting

$$f(X)v = f(\mathcal{A})v = \alpha_0 v + \alpha_1 \mathcal{A}v + \cdots + \alpha_k \mathcal{A}^k v$$

for any $v \in V$ and any polynomial $f \in K[X]$. The axioms (M1)-(M4) hold, since if $\mathcal{A}$ is linear, then so is $f(\mathcal{A})$, and also

$$(f+g)(\mathcal{A}) = f(\mathcal{A}) + g(\mathcal{A}), \quad (fg)(\mathcal{A}) = f(\mathcal{A})\,g(\mathcal{A})$$

(this is the universal property of polynomial rings; see §2 Ch. 5). The submodules of $V_{\mathcal{A}}$ are the $\mathcal{A}$-invariant subspaces. In general, different operators $\mathcal{A}$ give us different (non-isomorphic) $K[X]$-module structures $V_{\mathcal{A}}$ on the same space V.

4) Any left ideal J of a ring R has a natural R-module structure with action $(x,y) \mapsto xy$, for $x \in R$, $y \in J$, which is induced simply by multiplication in R. In the case $J = R$, this means that we are regarding R as a module over itself. This view of R can lead to important results.

5) In the situation of the previous example, we construct the quotient module $R/J = \{y + J \mid y \in R\}$. According to the general definition, $(x, y+J) \mapsto xy + J$ is the action of R on R/J. Note that the canonical epimorphism $\pi : R \to R/J$, which is an R-module homomorphism, satisfies the relation $\pi(xy) = xy + J = x(y+J) = x\pi(y)$. But if J is a two-sided ideal, then R/J is a ring, and π is a ring homomorphism: $\pi(xy) = \pi(x)\pi(y)$.

The intersection $\bigcap_i V_i$ of any family of submodules $V_i \subset V$ is a submodule of V. In particular, the intersection of all submodules containing a fixed set $T \subset V$ gives us the <u>submodule</u> $\langle T \rangle$ <u>spanned by the set</u> T. It consists of all possible elements of the form $x_1t_1 + x_2t_2 + \cdots + x_kt_k$, where $x_i \in R$ and $t_i \in T$. In passing, we note that non-zero elements $t_1, \ldots, t_k \in V$ are called <u>linearly dependent over</u> R if we have $x_1t_1 + \cdots + x_kt_k = 0$, where not all of the $x_i = 0$. The submodule spanned by a family

$\{V_1, \dots, V_m\}$ of submodules is called the sum of the V_i and is denoted in the usual way: $\Sigma V_i = V_1 + \dots + V_m$.

An R-module which is generated by a single element v is called cyclic. It has the form $V = Rv = \{xv \mid x \in R\}$, where $v \in V$; this is the analog of a cyclic group. In particular, R itself is the cyclic R-module $R \cdot 1$ (see Example 4 above).

If $V = Rv_1 + \dots + Rv_n$ is a finite sum of cyclic modules, then V is called finitely generated or an R-module of finite type.

It is easy to check that the map $x \mapsto xv$ is an R-module homomorphism from R to Rv. Its kernel is denoted $\mathrm{Ann}(v) = \mathrm{Ann}_R(v) = \{x \in R \mid xv = 0\}$. $\mathrm{Ann}(v)$ is a left ideal in R, which is called the annihilator of the element v. Thus, $Rv \cong R/\mathrm{Ann}(v)$. An element $v \in V$ with non-zero annihilator is called periodic. A module all of whose elements are periodic is also called periodic. If V does not contain any non-zero periodic elements, then V is called torsion-free.

The annihilator of an R-module V is the set

$$\mathrm{Ann}(V) = \{a \in R \mid aV = 0\} = \bigcap_{v \in V} \mathrm{Ann}(v) .$$

The module is called faithful if $\mathrm{Ann}(V) = 0$.

We can arrive at the same concepts from another point of view. Let $V(x)$ be the set of elements $v \in V$ which are killed by $x \in R$. If R is an integral domain, then $V(x) + V(y) \subset V(xy)$, and it makes sense to speak of the torsion submodule $\mathrm{Tor}(V) = \sum_{x \in R} V(x)$. When $\mathrm{Tor}(V) = V$, we say that V is a torsion module. On the other hand, if $\mathrm{Tor}(V) = 0$, we say that V is torsion-free.

The basic examples of periodic modules are: a) any finite abelian group (considered as a $\mathbb{Z}$-module; its torsion is $m\mathbb{Z}$, where m is the exponent of the group); b) the module $V_{\mathcal{A}}$ over $K[X]$ corresponding to a fixed linear operator $\mathcal{A}$ (see Example 3; the torsion is the principal ideal generated by the minimal polynomial of $\mathcal{A}$).

PROPOSITION 1. $\mathrm{Ann}(V)$ *is always a two-sided ideal of* R. *Setting*

$(x + \mathrm{Ann}(V))v = xv$ <u>gives</u> V <u>the structure of an</u> $(R/\mathrm{Ann}(V))$-<u>module</u>.

<u>Proof.</u> Set $A = \mathrm{Ann}(V)$. A is clearly an additive subgroup of R. Next, $(xax')v = xa(x'v) = (xa)v' = x(av') = x \cdot 0 = 0$ for any $x, x' \in R$, $a \in A$, and $v \in V$. Hence, $RAR \subset A$, i.e., A is a two-sided ideal in R. Now if $x + A = x' + A$, then $x - x' \in A$, so that $(x - x')v = 0$, i.e., $xv = x'v$. Hence, $(x + A)v = (x' + A)v$, i.e., the action of the quotient ring R/A on V is correctly defined. It is not hard to verify that V is an R/A-module under this action. Finally,

$$(x + A)V = 0 \Longrightarrow x + A \in \mathrm{Ann}_{R/A}(V) \Longrightarrow xV = 0 \Longrightarrow x \in A \quad .$$

Consequently, only the zero element in R/A annihilates V. □

Proposition 1 implies that the quotient ring $R/\mathrm{Ann}(V)$ is isomorphic to a subring of the endomorphism ring $\mathrm{End}(V)$ (see Example 2).

If V and W are two R-modules, then the set $\mathrm{Hom}_R(V, W)$ of all R-homomorphisms $\sigma : V \to W$ is an abelian group under the operation of point-wise addition of homomorphisms:

$$(\sigma + \tau)(xv) = \sigma(xv) + \tau(xv) = x\sigma(v) + x\tau(v) = x(\sigma(v) + \tau(v)) = x((\sigma + \tau)(v)) \quad .$$

If V and W are modules over a commutative ring R, then the set $\mathrm{Hom}_R(V, W)$ itself is an R-module, where we define $x\sigma$ for $x \in R$ and $\sigma \in \mathrm{Hom}_R(V, W)$ to be the map $v \mapsto x(\sigma(v))$:

$$(x\sigma)(yv) = x \cdot \sigma(yv) = x(y\sigma(v)) = (xy)(\sigma(v)) = (yx)(\sigma(v)) = y(x\sigma(v)) = y((x\sigma)(v)) \,.$$

If $W = V$, then the set $\mathrm{End}_R(V) = \mathrm{Hom}_R(V, V)$ is a ring, where we take multiplication to be composition of R-homomorphisms $\varphi \circ \psi : (\varphi \circ \psi)(xv) = \varphi(\psi(xv)) = \varphi(x\psi(v)) = x\varphi(\psi(v)) = x((\varphi \circ \psi)(v))$. We should keep in mind that when we regard V simply as an abelian group we write $\mathrm{End}_{\mathbb{Z}}(V)$, and, in general, $\mathrm{End}_R(V)$ is a proper subring in $\mathrm{End}_{\mathbb{Z}}(V)$. When V is a vector space over a field K, we often write $\mathfrak{L}(V)$

for $End_K(V)$ and call this ring the algebra of linear operators.

The ring $End_R(V)$ of R-endomorphisms of a module V is also called the centralizer of R in V. The role of this ring is especially significant in the case of irreducible (also called simple) modules. A module V over a ring R is called irreducible if: a) $V \neq 0$; b) 0 and V are the only submodules of V; and c) $RV \neq 0$ (the third condition automatically holds if R is a ring with unit). It is clear that a non-zero R-module V is irreducible if and only if $V = Rv$ is a cyclic module for any $v \neq 0$ in V.

PROPOSITION 2 (Schur's lemma). If V and W are two irreducible R-modules and σ is a non-zero R-homomorphism from V to W, then σ is an isomorphism. Furthermore, $End_R(V)$ is a division ring (skew field) for any irreducible R-module V.

For the proof see §4 Ch. 8, where the same basic fact (Theorem 1) is proved for irreducible G-spaces. □

2. Free modules. We call an R-module V an (internal) direct sum of the submodules $V_1, \ldots, V_n$ if $V = V_1 + \cdots + V_n$ and $V_i \cap \sum_{j \neq i} V_j = 0$ for $i = 1, \ldots, n$. In other words, we write $V = V_1 \oplus \ldots \oplus V_n$ (denoting the direct sum of submodules) if any element $v \in V$ can be written in one and only one way as a linear combination $v = v_1 + \cdots + v_n$, $v_i \in V_i$. If we are given R-modules $V_1, \ldots, V_n$, then we define their (external) direct sum in the obvious way (just like the rings) with the action of $x \in R$ on a row $(v_1, \ldots, v_n)$, $v_i \in V_i$, defined by: $x(v_1, \ldots, v_n) = (xv_1, \ldots, xv_n)$.

Now suppose that V is an R-module and $\{v_1, \ldots, v_n\}$ is a finite subset in V. We say that $\{v_1, \ldots, v_n\}$ are free generators of V if $V = Rv_1 + \cdots + Rv_n$ and if every map φ from the set $\{v_1, \ldots, v_n\}$ to any R-module W can be extended to an R-homomorphism $\tilde{\varphi}: V \to W$ for which $\tilde{\varphi}(v_i) = \varphi(v_i)$, $i = 1, \ldots, n$.

PROPOSITION 3. The following are equivalent:

(i) the set $\{v_1, \ldots, v_n\}$ is a set of free generators of V;

(ii) the set $\{v_1, \ldots, v_n\}$ is linearly independent, and $\langle v_1, \ldots, v_n \rangle = V$;

(iii) every element $v \in V$ can be written uniquely in the form $v = \Sigma x_i v_i$, $x_i \in R$;

(iv) V is the direct sum $R v_1 \oplus \ldots \oplus R v_n$, and $\mathrm{Ann}(v_i) = 0$, $i = 1, \ldots, n$;

(v) V is isomorphic to the direct sum of n copies of R considered as R-modules, i.e., to the module R^n of rows $(x_1, \ldots, x_n)$ of length n with components $x_i \in R$.

The proof is similar to the argument in Chapter 2 in the case of vector spaces over a field, except that one must be careful not to assume commutativity of R or that every element has an inverse. □

It is possible to construct (rather complicated) examples of non-commutative rings R for which $R^m \cong R^n$ for $m \neq n$, but commutative rings behave well in this respect.

PROPOSITION 4. The rank (number of free generators) in a finitely generated free module over an integral domain R is uniquely determined.

Proof. Let $\{v_1, \ldots, v_n\}$ and $\{u_1, \ldots, u_m\}$ be two bases for a free module V over R. Then

$$v_j = \sum_{i=1}^{m} a_{ij} u_i, \qquad u_i = \sum_{k=1}^{n} b_{ki} v_k .$$

Since R is commutative, we obtain the relations $AB = E_m$ and $BA = E_n$ for the matrices $A = (a_{ij})$ and $B = (b_{k\ell})$, which have dimensions $m \times n$ and $n \times m$, respectively. By imbedding R in its field of fractions $Q(R)$ and using Theorem 4 §4 Ch. 2 (which holds for any field, not just for $\mathbb{R}$), we find that $\min(n, m) \geq m$ and $\min(n, m) \geq n$, so that $m = n$. We note that it is impossible to have $m < \infty$ and

$n = \infty$, since only finitely many basis elements v_k occur in the expression for each u_i, and so that finite set of v_k generates all of V. □

<u>Remark.</u> This proposition can be proved for any commutative ring R by choosing a maximal ideal J in R and passing to the field R/J. We omit the details.

We note that, unlike in the case of vector spaces over a field, it is not true that any set which generates a free R-module necessarily contains a basis. For example, any two primes p and q generate the $\mathbb{Z}$-module $\mathbb{Z}$, but $\{p, q\}$ is not a basis (since the elements have a linear dependence relation $pq - qp = 0$), and $\{p\}$ and $\{q\}$ are not bases, since they do not generate $\mathbb{Z}$.

THEOREM 1. <u>Every R-module of finite type is a homomorphic image of a free R-module of finite type.</u>

<u>Proof.</u> Let $U = \sum_{i=1}^{n} R u_i$ be an R-module which is generated by n elements $u_1, \ldots, u_n$. We take the free R-module R^n (see Proposition 3 (v)), and we let v_i be the standard basis element $(0, \ldots, 0, 1, 0, \ldots 0)$ with 1 in the i-th place. The map $\varphi : V \to U$ given by $(x_1, \ldots, x_n) \mapsto \Sigma x_i u_i$ clearly expresses U as a homomorphic image of the free module V. □

It is not always true that a submodule of a free module is free, even if the submodule is a direct summand. Here is a simple example. Let $R = \mathbb{Z}_6$, $U = R(2 + 6\mathbb{Z})$, $V = R(3 + 6\mathbb{Z})$. Then $R = U \oplus V$ is the direct sum of the R-modules U and V, but neither of these modules is free (since $|R| = 6$, while $|U| = 3$, $|V| = 2$).

THEOREM 2. <u>Let $V = Rv_1 \oplus \ldots \oplus Rv_n$ be a free module of rank n over a principal ideal domain. Then every submodule U of V is a free module of rank $m \leq n$.</u>

Proof. First suppose that $n = 1$, i.e., $V \cong R$. Any submodule $U \subset V$ is isomorphic to an ideal of R, and hence $U \cong (u) = Ru$. If $u = 0$, then $U = 0$ (we consider the zero module to be free of rank zero). But if $u \neq 0$, then $au \neq 0$ for all non-zero $a \in R$, since R is an integral domain. Hence, U is a free (cyclic) module of rank 1. When $n > 1$ we use induction. We consider the free submodule $V' = Rv_2 \oplus \dots \oplus Rv_n$ of rank $n-1$ in V. The quotient module $\overline{V} = V/V'$ is a free cyclic module, with generator $\overline{v}_1 = v_1 + V'$. It contains the submodule $\overline{U} = (U + V')/V'$. If $\overline{U} = 0$, then $U \subset V'$, and then the theorem is true by the induction assumption. But if $\overline{U} \neq 0$, then the case of the theorem which has been proved, tells us that $\overline{U}$ has a cyclic generator $\overline{u}_1 = u_1 + V'$, where $u_1 \in U$. First suppose that $U \cap V' = 0$; then $u \in U \Rightarrow \overline{u} = u + V' \in \overline{U} \Rightarrow \overline{u} = a_1 \overline{u}_1$, $a_1 \in R \Rightarrow$ $\Rightarrow u - a_1 u_1 \in V' \Rightarrow u = a_1 u_1 \Rightarrow U = Ru_1$ is a free module of rank 1. Finally, suppose that $U \cap V' \neq 0$. By the induction assumption, the submodule $U \cap V'$ of V', which has rank $n-1$, has a free basis $\{u_2, \dots, u_m\}$, where $0 < m-1 \leq n-1$. By an argument similar to the previous one, we see that $\{u_1, u_2, \dots, u_m\}$ is a free R-basis for U. Namely, $u \in U \Rightarrow \overline{u} = u + V' \in \overline{U} \Rightarrow \overline{u} = a_1\overline{u}_1$, $a_1 \in R \Rightarrow u - a_1u_1 \in U \cap V' \Rightarrow$ $\Rightarrow u - a_1 u_1 = a_2u_2 + \dots + a_m u_m \Rightarrow u = a_1u_1 + a_2u_2 + \dots + a_m u_m$, $m \leq n$. According to Proposition 3 (ii), we must check that the $u_1, \dots, u_m$ are linearly independent. But $\Sigma x_i u_i = 0 \Rightarrow x_1 \overline{u}_1 = -\sum_{i>1} x_i \overline{u}_i = 0$ in $\overline{V}$. Hence, $x_1 = 0$, since $\overline{u}_1$ is a basis of $\overline{U}$. Since $\{u_2, \dots, u_m\}$ is a free basis in $U \cap V'$, we have: $x_2u_2 + \dots + x_m u_m = 0 \Rightarrow x_2 = \dots = x_m = 0$. □

COROLLARY. Every submodule of a module of finite type over a principal ideal domain is itself a module of finite type.

The proof follows from Theorems 1 and 2 and the second isomorphism theorem (the theorem on the correspondence between submodules). □

It is not very difficult to obtain a complete description of the modules of finite type over a principal ideal domain R. However, the main examples which would interest us in such a description (periodic modules over $\mathbb{Z}$ and over $K[X]$; see Examples 1 and 3) have already been treated (see §5 Ch. 7 and the Appendix). For a unified module-theoretic approach to various problems of this sort, see the list of supplementary reading.

3. Integral elements of a ring. Let R be an integral domain. An element $t \in R$ is called integral (integral over $\mathbb{Z}$) if t is a root of a monic polynomial $X^n + a_1 X^{n-1} + \cdots + a_n \in \mathbb{Z}[X]$. If R is a finite algebraic extension of $\mathbb{Q}$ or when R is the field generated by all complex algebraic numbers, we call such an element t an algebraic integer. Of course, the set of algebraic integers includes $\mathbb{Z}$. Exercise 9 of §4 Ch. 6 shows that a rational number t is an algebraic integer if and only if $t \in \mathbb{Z}$. Next, if we have $a_0 u^n + a_1 u^{n-1} + \cdots + a_n = 0$, $a_i \in \mathbb{Z}$, then $(a_0 u)^n + a_0 a_1 (a_0 u)^{n-1} + \cdots + a_0^n a_n = 0$, and hence any algebraic number can be multiplied by a suitable $a_0 \in \mathbb{Z}$ to obtain an algebraic integer.

Returning to the general case, we note that it is convenient to treat R as a $\mathbb{Z}$-module. Any elements $t_1, t_2, \ldots, t_n \in R$ generate a sub-$\mathbb{Z}$-module $\mathbb{Z}t_1 + \mathbb{Z}t_2 + \cdots + \mathbb{Z}t_n$ of finite type in R. In particular, if t is an integral element, with $t^n + a_1 t^{n-1} + \cdots + a_n = 0$, $a_i \in \mathbb{Z}$, then the subring $\mathbb{Z}[t] \subset R$ is a $\mathbb{Z}$-module of finite type, since $\mathbb{Z}[t] = \mathbb{Z}1 + \mathbb{Z}t + \cdots + \mathbb{Z}t^{n-1}$. Conversely, suppose that $\mathbb{Z}[t]$ is a $\mathbb{Z}$-module of finite type with generators $v_1, \ldots, v_n \in R$. Then the relations

$$(t - a_{11}) x_1 - a_{12} x_2 - \cdots - a_{1n} x_n = 0 \ ,$$
$$\cdots\cdots\cdots\cdots\cdots\cdots$$
$$-a_{n1} x_1 - a_{n2} x_2 - \cdots + (t - a_{nn}) x_n = 0 \ ,$$

considered over the fraction field $Q(R)$, has a non-zero solution $(x_1, \ldots, x_n) = (v_1, \ldots, v_n)$ (not all of the v_i are zero, since $1 \in \mathbb{Z}[t]$). Hence, the determinant of the system is zero (see Chapter 3), and so t is a root of the monic polynomial

$f(T) = \det(TE - A)$. We have proved that <u>an element</u> $t \in R$ <u>is integral if and only if the subring</u> $\mathbb{Z}[t] \subset R$ <u>is a</u> $\mathbb{Z}$<u>-module of finite type.</u>

THEOREM 3. <u>The integral elements of a ring</u> R <u>form a subring of</u> R.

<u>Proof.</u> Let $u, v \in R$ be integral elements. Then $\mathbb{Z}[u,v] = \sum_{1 \leq i, j \leq m} \mathbb{Z} u^i v^j$ is a $\mathbb{Z}$-module of finite type. Since $\mathbb{Z}$ is a principal ideal domain, the corollary to Theorem 2 tells us that the two submodules $\mathbb{Z}[u-v]$ and $\mathbb{Z}[uv]$ are also $\mathbb{Z}$-modules of finite type. By the above criterion, the elements $u - v$ and uv must be integral. □

<u>Example.</u> Any root of 1 is obviously an algebraic integer. By Theorem 3, any integral linear combination of roots of 1 is also an algebraic integer. In particular (see the proof of the proposition in §4 Ch. 8), <u>the values</u> $\chi_\Phi(g)$, $g \in G$, <u>of the character of any complex linear representation</u> Φ <u>of a group</u> G <u>are algebraic integers.</u>

4. <u>Unimodular sequences of polynomials</u>. Let $R = K[X_1, \ldots, X_n]$ be the ring of polynomials in n variables over a field K. A sequence $[f_1, \ldots, f_r]$ of r polynomials $f_i \in R$ is said to be unimodular if $Rf_1 + Rf_2 + \cdots + Rf_r = R$, i.e.,

$$u_1 f_1 + u_2 f_2 + \cdots + u_r f_r = 1 \qquad (1)$$

for some $u_i \in R$, $1 \leq i \leq r$. Further let V be a module of finite type over R. In connection with certain delicate questions in algebraic geometry, the French mathematician J.-P. Serre in 1955 stated the conjecture:

$$V \oplus R^s \simeq R^{s+t} \Longrightarrow V \simeq R^t,$$

which can be given the following elegant form: "every relation (1) can be written in the following form for suitable $u_{ij} \in R$:

$$\begin{vmatrix} f_1 & f_2 & \cdots & f_r \\ u_{21} & u_{22} & \cdots & u_{2r} \\ \cdots & \cdots & \cdots & \cdots \\ u_{r1} & u_{r2} & \cdots & u_{rr} \end{vmatrix} = 1 \text{."} \qquad (2)$$

Despite its apparent simplicity, this conjecture was only proved in 1976, independently by A. A. Suslin (USSR) and D. Quillen (USA) (though the case $n=1$ was done already in 1848 by Hermite). The case $n=1$ is included in the following more general theorem.

THEOREM 4. Let $a_1, a_2, \dots, a_r$ $(r \geq 2)$ be nonzero elements in a principal ideal domain R, and let $d = \text{g.c.d.}(a_1,\dots,a_r)$. Then there exists a matrix $A \in M_r(R)$ with first row $(a_1, a_2, \dots, a_r)$ and with $\det A = d$.

Proof. We use the result at the end of subsection 1 §2. If $r=2$, then, writing d in the form $d = u_1a_1 + u_2a_2$ with $u_i \in R$, we immediately find the required matrix: $A = \begin{Vmatrix} a_1 & a_2 \\ -u_2 & u_1 \end{Vmatrix}$. We now use induction on r. We represent $d' = \text{g.c.d.}(a_1,\dots,a_{r-1})$ in the form $d' = \det A'$, where $A' \in M_{r-1}(R)$ is a matrix with first row $(a_1,\dots,a_{r-1})$. Since $d = \text{g.c.d.}(d',a_r)$, we can write $d = ud' + va_r$. We introduce the matrix

$$A = \begin{Vmatrix} & & & a_r \\ & A' & & 0 \\ & & & \vdots \\ & & & 0 \\ -\frac{va_1}{d'} & \dots & -\frac{va_{r-1}}{d'} & u \end{Vmatrix}.$$

The first row of this matrix is $(a_1,\dots,a_r)$. If we expand $\det A$ along the last column, we find that

$$\det A = u \det A' + (-1)^{r+1} a_r \det A'' = ud' + a_r (-1)^{r+1} \det A'', \tag{3}$$

where A'' is the matrix obtained from A by removing the first row and last column.

On the other hand, if we multiply the first row of A' by $-v$ and then put this row in the last place, preserving the order of the other rows (this can be done by successive transpositions), then we obtain a matrix A''', which is the matrix we get if we multiply the last row in A'' by d'. Thus,

$$d' \det A'' = \det A''' = (-1)^{r-1} v \det A' = (-1)^{r-1} v d'.$$

We substitute this expression in the relation which is obtained from (3) by multiplying both sides by d':

$$d' \det A = u(d')^2 + a_r (-1)^{r+1} d' \det A'' = u(d')^2 + a_r (-1)^{r+1} (-1)^{r-1} d' v =$$

$$= d'(ud' + va_r);$$

canceling d', we arrive at the required relation $d = ud' + va_r = \det A$. □

For $n > 1$ the basic idea of the proof is to study the action of the group $GL(r, K[X_1,\dots,X_{n-1}])$ on the set of unimodular sequences and proceed by induction on n. The reader can read about the proof either in the original article (A. A. Suslin, Projective modules over a polynomial ring are free, DAN SSSR, 229, No. 5 (1976), 1063-1066) or else in the Bourbaki report by D. Ferrand (Sém. N. Bourbaki, 28ème année, 1975/76, juin 1976). The exposition is completely elementary. To appreciate the effort required to attain such a proof, one need only look at the earlier Bourbaki report by H. Bass (Sém. N. Bourbaki, 26ème année, 1973/74, juin 1974). These references contain the statements of some unsolved problems. This whole circle of questions is a good topic for discussion in advanced seminars at the graduate or undergraduate level.

§4. Algebras over a field

1. Definitions and examples of algebras. Until now we have not made much use of the fact that almost all of the rings we know are also vector spaces over a field.

Definition. An algebra over a field K is a pair consisting of a ring $(A, +, \cdot)$ and a vector space A over K (the underlying set A is the same for the ring and the vector space, as are the addition operation and the zero element). Here

$$\lambda(xy) = (\lambda x)y = x(\lambda y)$$

for all $\lambda \in K$, $x, y \in A$. An algebra is called associative if the ring $(A, +, \cdot)$ is associative. By the dimension of the algebra A we mean the dimension of the vector space A over K.

The basic notions of ring theory carry over to algebras. For example, a <u>subalgebra</u> of an algebra A is a subring $B \subset A$ which is also a vector subspace of the vector space A. If T is a subset of A, then the subalgebra $K[T]$ generated by T is defined to be the intersection of all subalgebras of A which contain T. We similarly define ideals and <u>quotient algebras.</u> By an <u>algebra homomorphism</u> we mean a ring homomorphism which is at the same time a K-linear map.

The <u>center</u> $Z(A)$ <u>of an associative algebra</u> A is defined as the set of all elements $a \in A$ which commute with every element of $A : a \in Z(A) \Leftrightarrow ax = xa$, $\forall x \in A$. It is easy to check that the center is a subalgebra of $A : (a - a')x = ax - a'x = xa - xa' = x(a - a')$, $(aa')x = a(a'x) = a(xa') = (ax)a' = x(aa')$, $(\lambda a)x = \lambda(ax) = \lambda(xa) = x(\lambda a)$ for all $a, a' \in Z(A)$, $\lambda \in K$. We have $Z(A) = A$ if and only if A is a commutative algebra.

If A is an associative algebra with unit 1, then we immediately see that $\lambda \cdot 1 \in Z(A)$, and the map $\lambda \mapsto \lambda \cdot 1$, $\forall \lambda \in K$, gives a monomorphism from K to A. Hence, we may consider an algebra to be a ring A together with a specified subfield which is contained in its center $Z(A)$.

Here are some examples of algebras.

1) An extension $F \supset K$ of finite degree $[F:K]$ over a field K is obviously a commutative associative algebra (with unit) having finite dimension $\dim_K F = [F:K]$. We have already studied this example in §1.

2) The polynomial ring $R = K[X_1, \dots, X_n]$ with coefficients in a field K has the structure of an infinite dimensional commutative associative K-algebra. Note that

$$R = R_0 \oplus R_1 \oplus R_2 \oplus \dots$$

is a direct sum of the finite dimensional vector subspaces R_m consisting of homogeneous polynomials of total degree m. We have $R_0 = K$, and $R_i R_j \subset R_{i+j}$. In this situation we call the algebra R <u>graded.</u>

3) The commutative algebra $X_{\mathbb{C}}(G)$ with unit χ_1, which is generated over $\mathbb{C}$ by all of the characters of a finite group G, has dimension r equal to the number of conjugacy classes in G (Theorem 2 §7 Ch. 8).

4) The ring $M_n(K)$ of $n \times n$ matrices with entries in a field K is an algebra of dimension n^2 over K. The basis elements $\{E_{ij} \mid i, j = 1, 2, \dots, n\}$ of the algebra $M_n(K)$ multiply together according to the rule $E_{ik} E_{\ell j} = \delta_{k\ell} E_{ij}$. According to Theorem 3 §3 Ch. 2, we have $Z(M_n(K)) = \{\lambda E\} \cong K$.

We call an associative algebra A with unit <u>central simple over</u> K if $Z(A) \cong K$, and if A has no two-sided ideals other than 0 and A.

PROPOSITION 1. $M_n(K)$ <u>is a central simple algebra.</u>

<u>Proof.</u> It remains to show that any ideal J in $M_n(K)$ which is not the zero ideal must be all of $M_n(K)$. Let

$$0 \neq a = \sum \alpha_{ij} E_{ij} \in J \quad .$$

If $\alpha_{k\ell} \neq 0$, then $E_{st} = \alpha_{k\ell}^{-1} E_{sk} \cdot a \cdot E_{\ell t} \in J$ for any $s, t = 1, \dots, n$; hence, $J = M_n(K)$. □

Proposition 1 also holds for the full matrix algebra $M_n(D)$ over an arbitrary division ring D. The extremely important <u>Wedderburn theorem</u> (which, in a more general context, is known as the Wedderburn-Artin theorem) says that, conversely, <u>every finite dimensional associative simple algebra over a field</u> K <u>is isomorphic to</u> $M_n(D)$, <u>where the natural number</u> n <u>is uniquely determined, and the division ring</u> D (<u>which is a finite dimensional algebra over</u> K) <u>is uniquely determined up to isomorphism.</u>

The matrix algebra $M_n(K)$ also has the following universal property.

PROPOSITION 2. <u>Any</u> n-<u>dimensional associative algebra</u> A <u>over a field</u> K <u>is isomorphic to some subalgebra of</u> $M_k(K)$, <u>where</u> $k \leq n+1$.

Proof. First suppose that A is an algebra with unit 1; we shall imbed it in $M_n(K)$. To do this, we associate to every $a \in A$ the linear operator $L_a : x \mapsto ax$ from A to A. L_a is linear because of the bilinearity of multiplication in A. Since obviously $L_{\lambda a} = \lambda L_a$, $L_{a+b} = L_a + L_b$, $L_{ab} = L_a L_b$ (by associativity!), and $L_1 = \mathcal{E}$, it follows that the map $\varphi : a \mapsto L_a$ is a homomorphism. Injectivity is ensured by the presence of the unit element: $a \neq 0 \Rightarrow L_a(1) = a \cdot 1 = a$, so that $L_a \neq 0$.

Now suppose that A does not have a unit element. Consider the vector space $\widetilde{A} = K \oplus A$, and define a multiplication on $\widetilde{A}$ by setting $(\lambda, a)(\lambda', a') = (\lambda\lambda', aa' + \lambda a' + \lambda' a)$. It is easy to verify that with this multiplication operation $\widetilde{A}$ becomes a K-algebra with unit element $(1, 0)$. Since $\dim_K \widetilde{A} = \dim_K A + 1 = n + 1$, the above argument allows us to imbed $\widetilde{A}$, and thus A, in $M_{n+1}(K)$. $\square$

There is a close resemblance between the proof of Proposition 2 and that of Cayley's theorem for finite groups. In both cases we used the regular representation. More generally, by a representation of a K-algebra A we mean any homomorphism $A \to \mathcal{L}(V) = \mathrm{End}_F(V)$, where $F \supset K$ is a field extension of K. In other words, we supply the F-vector space V with a left A-module structure in the sense of the definition in §3, and we have

$$(\lambda x) \cdot v = x \cdot (\lambda v), \qquad \text{for all} \qquad \lambda \in K, \; x \in A, \; v \in V \quad .$$

If we choose a basis in V, we arrive, as in the case of groups, at a matrix representation $A \to M_r(F)$, where $r = \dim_F V$.

2. Division rings (skew fields). As the above theorem of Wedderburn indicates, the study of division algebras is an important part of the general structure theory for associative algebras. Schur's lemma (Proposition 3 §3) also supports this observation. Before giving some results on division algebras, we take up the following auxiliary fact.

PROPOSITION 3. In an associative algebra A (with unit element 1) having dimension n over a field K, every element $a \in A$ is a root of a polynomial $f_a \in K[X]$ of degree $\leq n$. An element $a \in A$ is invertible if and only if $f_a(0) \neq 0$, where f_a denotes the monic polynomial of least degree. If A has no zero divisors, then A is a division algebra. If K is algebraically closed, then $n = 1$ and $A = K$.

Proof. Since A is finite dimensional, the elements $1, a, a^2, \ldots$ cannot all be linearly independent over K. Hence there exists a monic polynomial $f_a(X) = X^m + \alpha_1 X^{m-1} + \cdots + \alpha_m \neq 0$ of minimal degree $m \leq n$, with coefficients $\alpha_i \in K$, such that $f_a(a) = 0$. If $\alpha_m \neq 0$, then the equality $f_a(a) = 0$, written in the form $[-\alpha_m^{-1}(a^{m-1} + \alpha_1 a^{m-2} + \cdots + \alpha_{m-1})]a = 1$, shows that a is invertible. Conversely, suppose that $a \in A$ is not a zero divisor, but $\alpha_m = 0$. Then

$$(a^{m-1} + \alpha_1 a^{m-2} + \cdots + \alpha_{m-1})a = 0 \Longrightarrow a^{m-1} + \alpha_1 a^{m-2} + \cdots + \alpha_{m-1} = 0 ,$$

which contradicts the minimality of $f_a(X)$. Hence, $\alpha_m \neq 0$. In particular, all elements of A which are not zero divisors are invertible.

If the field K is algebraically closed, then $f_a(X) = (X - c_1) \ldots (X - c_m)$, $c_i \in K$, so that $(a - c_1)b = 0$, where $b = (a - c_2) \ldots (a - c_m) \neq 0$. Since A has no zero divisors, the only possibility is that $m = 1$ and $a - c_1 = 0$, and so $a = c_1 \in K$. Since this is true for any $a \in A$, we have $A = K$. □

We see that the properties of a division algebra depend in an essential way on the ground field K. It is natural that, historically, division algebras over the real numbers $\mathbb{R}$ have aroused special interest. The existence of the field $\mathbb{C} = \mathbb{R} + i\mathbb{R}$ inspired the search for other "hypercomplex systems", i. e., division algebras over $\mathbb{R}$. This search met with success in 1843, when Hamilton constructed his famous algebra of real quaternions.

Example (the quaternion algebra $\mathbb{H}$). Formally, we write

$$\mathbb{H} = \mathbb{R} + i\mathbb{R} + j\mathbb{R} + k\mathbb{R} ,$$

where i, j, k are quantities which are multiplied together according to the rule

$$i^2 = j^2 = k^2 = -1 , \quad ij = k = -ji , \quad jk = i = -kj , \quad ki = j = -ik \quad .$$

An element $x = \alpha_0 + \alpha_1 i + \alpha_2 j + \alpha_3 k \in \mathbb{H}$ is called a quaternion. It can be verified directly that $\mathbb{H}$ is an associative algebra with center $Z(\mathbb{H}) = \mathbb{R}$. But it is more worthwhile first to consider the following model of the algebra $\mathbb{H}$ -- the set

$$\Phi(\mathbb{H}) = \left\{ \begin{Vmatrix} a & b \\ -\bar{b} & \bar{a} \end{Vmatrix} \;\middle|\; a, b \in \mathbb{C} \right\} \subset M_2(\mathbb{C}) \quad .$$

It is an elementary exercise to show that $\Phi(\mathbb{H})$ is a division ring. We did a similar exercise in §1 Ch. 5, when the field $\mathbb{C}$ was introduced. We need only remember that multiplication in $\Phi(\mathbb{H})$ is non-commutative. According to the rule for computing the inverse of a matrix, we have

$$\begin{Vmatrix} a & b \\ -\bar{b} & \bar{a} \end{Vmatrix}^{-1} = \delta^{-1} \begin{Vmatrix} \bar{a} & -b \\ \bar{b} & a \end{Vmatrix} ,$$

where

$$\delta = \det \begin{Vmatrix} a & b \\ -\bar{b} & \bar{a} \end{Vmatrix} = a\bar{a} + b\bar{b} \quad (\neq 0 \quad \text{if} \quad a \neq 0 \quad \text{or if} \quad b \neq 0) \quad .$$

Incidentally, this implies that the multiplicative group $\Phi(\mathbb{H})^* = \Phi(\mathbb{H}) \setminus \{0\}$ contains a subgroup isomorphic to $SU(2)$ (see §1 Ch. 7).

If we set

$$q_0 = \begin{Vmatrix} 1 & 0 \\ 0 & 1 \end{Vmatrix} , \quad q_1 = \begin{Vmatrix} i & 0 \\ 0 & -i \end{Vmatrix} , \quad q_2 = \begin{Vmatrix} 0 & 1 \\ -1 & 0 \end{Vmatrix} , \quad q_3 = \begin{Vmatrix} 0 & i \\ i & 0 \end{Vmatrix} ,$$

we notice that

$$q_s^2 = -q_0 , \quad s \neq 0 ; \quad q_1 q_2 = q_3 = -q_2 q_1 , \quad q_2 q_3 = q_1 = -q_3 q_2 ,$$

$$q_3 q_1 = q_2 = -q_1 q_3 \quad .$$

We see that the map $\Phi : \mathbb{H} \to \Phi(\mathbb{H})$ given by letting $1 \mapsto q_0$, $i \mapsto q_1$, $j \mapsto q_2$, $k \mapsto q_3$, is a two-dimensional complex representation of the quaternion algebra $\mathbb{H}$. The quaternion x corresponds to the matrix

$$\Phi_x = \begin{Vmatrix} a & b \\ -\bar{b} & \bar{a} \end{Vmatrix} = \alpha_0 q_0 + \alpha_1 q_1 + \alpha_2 q_2 + \alpha_3 q_3 \ ,$$

where $a = \alpha_0 + i\alpha_1$, $b = \alpha_2 + i\alpha_3$, $i = \sqrt{-1}$. The quaternion units i, j, k generate the subgroup Q_8 of $\mathbb{H}^*$, which is a group of order 8 we encountered before. The restriction $\Phi|_{Q_8}$ is the irreducible 2-dimensional representation of Q_8 that was constructed at the end of §3 Ch. 7.

To every quaternion $x = \alpha_0 + \alpha_1 i + \alpha_2 j + \alpha_3 k$ we associate the conjugate quaternion $x^* = \alpha_0 - \alpha_1 i - \alpha_2 j - \alpha_3 k$ (this is analogous to complex conjugation). The conjugation operation has the following obvious properties:

$$(x + y)^* = x^* + y^* \ ; \quad x^* = x \iff x \in \mathbb{R} \ ; \quad x^* = -x \iff \alpha_0 = 0$$

(in the latter case we call x a "purely imaginary" quaternion). The product $xx^* = N(x)$ is called the norm of the quaternion x. Using Φ, we can easily show that $(xy)^* = y^*x^*$ and $N(xy) = N(x)N(y)$, where $N(x) = \det \Phi(x) = \alpha_0^2 + \alpha_1^2 + \alpha_2^2 + \alpha_3^2$.

The unique place occupied by the quaternions is clear from the following theorem of Frobenius: There are only three finite dimensional associative division algebras over $\mathbb{R}$, namely, $\mathbb{R}$, $\mathbb{C}$, and $\mathbb{H}$. The essential fact used in the proof is that the minimal polynomial $f_t(X)$ of any non-zero element t in the division algebra D which is not in $\mathbb{R}$ must be quadratic (see Proposition 3 and Theorem 1 §4 Ch. 6). We shall not give the proof here.

Relatively recently a proof using deep topological techniques was given for the fact that any finite dimensional division algebra over $\mathbb{R}$ (not necessarily associative) has dimension 1, 2, 4, or 8. There is actually one example in each dimension.

More than 70 years ago, Wedderburn obtained a beautiful result concerning finite division rings, which is important in geometry. We shall now prove this theorem, which bears a direct relationship to the material in §1.

THEOREM 1 (Wedderburn). Every finite associative division ring is commutative.

Proof. Let D be a finite division ring, and let Z be its center. It is obvious that Z is a field, and that D is a finite dimensional vector space over Z:

$$D = Ze_1 \dotplus Ze_2 \dotplus \cdots \dotplus Ze_n .$$

According to the results of §1, we have $Z = \mathbb{F}_q$ for some $q = p^m$, and hence $|D| = q^n$. Suppose that $x \in D \setminus Z$. The elements of D which commute with x form a set $C(x) = \{y \in D \mid yx = xy\}$, which is closed under addition and multiplication. In other words, $C(x)$ is a subdivision algebra of D which contains Z. If q^d is the number of elements in $C(x)$, then $d = d(x)$ is a divisor of n (with $d < n$), since, if we interpret D as a left vector space over $C(x)$

$$D = C(x)f_1 \dotplus \cdots \dotplus C(x)f_r$$

we have $q^n = |C(x)|^r = q^{dr}$. We now note that Z^* is the center of the multiplicative group D^*, and $(q^n - 1)/(q^d - 1) = (D^* : C(x)^*)$ is the number of elements conjugate to x in D^*. Hence, the formula (2') in §2 Ch. 7 takes the form

$$q^n - 1 = |D^*| = (q - 1) + \sum_d \frac{q^n - 1}{q^d - 1} , \qquad (*)$$

where d runs through some set of divisors of n less than n. The properties of the cyclotomic polynomial $\Phi_n(X)$ proved in §1 show (see Exercise 6 §1) that the integer $\Phi_n(q)$ divides both $q^n - 1$ and $(q^n - 1)/(q^d - 1)$ for $d \mid n$, $d < n$. Thus, by $(*)$, $\Phi_n(q) \mid (q - 1)$, and this means (see Exercise 7 §1) that $n = 1$, and so $D = Z$ is commutative. □

3. Group algebras and modules over them. When studying the regular represen-

tation of a finite group G in §1 of Chapter 8, we introduced the vector space $\langle e_g | g \in G \rangle_K$ over a field K. We now make this space into a K-algebra by setting $e_g e_h = e_{gh}$ and extending this rule by linearity to all vectors $\Sigma \alpha_g e_g$, $\alpha_g \in K$. To simplify notation we shall usually replace e_g by g, and consider the set $K[G]$ of all possible formal sums $\Sigma \alpha_g g$, $\alpha_g \in K$. By definition, $\Sigma \alpha_g g = \Sigma \beta_g g \Leftrightarrow \alpha_g = \beta_g$, $\forall g \in G$. The following operations on formal sums

$$\sum_g \alpha_g g + \sum_g \beta_g g = \sum_g (\alpha_g + \beta_g) g \;,$$

$$\lambda\left(\sum \alpha_g g\right) = \sum_g \lambda \alpha_g g \;, \tag{1}$$

$$\left(\sum_g \alpha_g g\right)\left(\sum_h \beta_h h\right) = \sum_{g,h} \alpha_g \beta_h gh = \sum \gamma_u u \,, \quad \text{where} \quad \gamma_u = \sum_g \alpha_g \beta_{g^{-1}u}$$

give $K[G]$ the structure of an associative algebra. $K[G]$ is customarily called the group algebra of the finite group G over the field K. The elements $1 \cdot g$ of $K[G]$ for $g \in G$ form a basis for $K[G]$ as a K-vector space. Thus, $\dim_K K[G] = |G|$. We consider the group G to be imbedded in $K[G]$ by $g \mapsto 1 \cdot g$. The identity $e \in G$ is the unit element in $K[G]$. If we take K to be a commutative and associative ring with unit (but not necessarily a field), we similarly define the group ring $K[G]$.

Furthermore, a similar construction is possible for a group G which is not assumed to be finite, if we agree only to consider sums $\Sigma \alpha_g g$ having only finitely many non-zero coefficients. It is sometimes convenient to consider such a formal sum $A = \Sigma \alpha_g g$ as a function on the group G (defined by $A(g) = \alpha_g$) with values in K which are non-zero for only finitely many $g \in G$. Then the formulas (1) correspond to the operations of point-wise addition of functions

$$(A_1 + A_2)(g) = A_1(g) + A_2(g)$$

and convolution of functions

$$A_3 = A_1 * A_2 , \qquad A_3(u) = \sum_g A_1(g)\, A_2(g^{-1}u) \quad .$$

The theory of group rings is a vast field of algebra, having its own techniques and problems, but for our purposes $K[G]$ is introduced merely to illustrate some of the general ideas in Chapters 8 and 9 .

THEOREM 2. There exists a one-to-one correspondence between $K[G]$-modules which are finite dimensional vector spaces over the field K and linear representations of the group G .

Proof. Let (Φ, V) be a representation of G. We extend Φ by linearity to the elements of $K[G]$, by defining

$$\widetilde{\Phi}\left(\sum \alpha_g g\right) = \sum \alpha_g \Phi(g) \quad ,$$

and we set

$$\left(\sum \alpha_g g\right) \circ v = \sum \alpha_g \Phi(g) v , \qquad \forall v \in V \quad .$$

The operation $\circ$ gives V a $K[G]$-module structure in the usual sense. We note that

$$\left(\sum \alpha_g g\right) \circ (\lambda v) = \sum \alpha_g \Phi(g)(\lambda v) = \sum \alpha_g \lambda \Phi(g) v =$$

$$= \lambda\left(\sum \alpha_g \Phi(g) v\right) = \lambda\left(\left(\sum \alpha_g g\right) \circ v\right) ,$$

i. e., scalar multiplication in V and $K[G]$ are compatible. It is natural to call the pair $(\widetilde{\Phi}, V)$ a linear representation of the algebra $K[G]$.

Conversely, if V is a vector space over K which is a $K[G]$-module with action $(\Sigma\, \alpha_g g, v) \mapsto (\Sigma\, \alpha_g g) \circ v$, then, setting

$$\widetilde{\Phi}\left(\sum \alpha_g g\right) v = \left(\sum \alpha_g g\right) \circ v \quad ,$$

we define a homomorphism $\widetilde{\Phi} : K[G] \to \mathrm{End}_K(V)$ (i. e., a representation of the algebra

$K[G])$, whose restriction $\Phi = \tilde{\Phi}|_G$ to G gives us a representation of the group G. □

Because of Theorem 1, a representation space V of a group G is often called a G-module. Similar terminology also applies to other concepts in representation theory.

Now let G be a finite group, and let $K = \mathbb{C}$ be the field of complex numbers. According to the results of Chapter 8, every irreducible G-module over $\mathbb{C}$ (i.e., $\mathbb{C}[G]$-module) with character χ_i is isomorphic to some left ideal J_i in the algebra $\mathbb{C}[G]$ (in this connection see Example 4 in §3). If $\dim_{\mathbb{C}} J_i = n_i$, then $\mathbb{C}[G]$ contains a direct sum $A_i = J_{i,1} \oplus \dots \oplus J_{i,n_i}$ of n_i left ideals which are $\mathbb{C}[G]$-isomorphic to $J_i = J_{i,1}$. If we choose one ideal J_i in each isomorphism class of left ideals, we can write the decomposition

$$\mathbb{C}[G] = A_1 \oplus A_2 \oplus \dots \oplus A_r \quad , \tag{2}$$

which corresponds to the decomposition of the regular representation of G. We note that each of the components A_i is uniquely determined.

If J happens to be a minimal left ideal of the algebra $\mathbb{C}[G]$, and if $t \in \mathbb{C}[G]$, then Jt is also a minimal left ideal (possibly the zero ideal). Hence, the map $\varphi : J \to Jt$ given by $v \mapsto vt$ $(v \in J)$ is either the zero map or else a $\mathbb{C}[G]$-isomorphism, since $xv \in J$ for any $x \in \mathbb{C}[G]$ and $\varphi(xv) = (xv)t = x(vt) = x\varphi(v)$. For this reason $J \subset A_i \Rightarrow Jt \subset A_i$ for all $t \in \mathbb{C}[G]$, and hence A_i is a two-sided ideal of $\mathbb{C}[G]$. Since (2) is a direct sum decomposition, we have

$$i \neq j \Longrightarrow A_i A_j \subset A_i \cap A_j = 0 \quad .$$

We would like to obtain more precise information concerning the decomposition (2), using the theory of characters from Chapter 8. We first find the center $Z(\mathbb{C}[G])$ of the group algebra $\mathbb{C}[G]$. By definition,

$$z \in Z(\mathbb{C}[G]) \Longleftrightarrow zg = gz, \quad \forall g \in G \quad .$$

If $z = \sum_{h \in G} \gamma_h h$, then

$$\sum_{t \in G} \gamma_{g^{-1}t} t = g\left(\sum_h \gamma_h h\right) = \left(\sum_h \gamma_h h\right) g = \sum_{t \in G} \gamma_{tg^{-1}} t ,$$

and hence $\gamma_{g^{-1}t} = \gamma_{tg^{-1}}$, $\forall t \in G$. Setting $t = gh$, we obtain $\gamma_h = \gamma_{ghg^{-1}}$. This means that

$$Z(\mathbb{C}[G]) = \langle z_1, z_2, \dots, z_r \rangle_{\mathbb{C}} ,$$

where

$$z_i = \sum_{g \in g_i^G} g ; \qquad i = 1, 2, \dots, r \tag{3}$$

($g_1, g_2, \dots, g_r$ are representatives of the conjugacy classes in G). Clearly, $z_1, z_2, \dots, z_r$ are linearly independent, and hence $\dim_{\mathbb{C}} Z(\mathbb{C}[G]) = r$.

To every element $a \in A_i$ we associate the linear operator $L_a^{(i)}$ which acts on the minimal left ideal $J_i = J_{i,1}$ according to the rule $L_a^{(i)}(v) = av$, $v \in J_i$. Since obviously $L_{\lambda a}^{(i)} = \lambda L_a^{(i)}$, $L_{a+b}^{(i)} = L_a^{(i)} + L_b^{(i)}$, and $L_{ab}^{(i)} = L_a^{(i)} L_b^{(i)}$, it follows that $\varphi : a \mapsto L_a^{(i)}$ is a homomorphism from the algebra A_i to the endomorphism algebra $\mathrm{End}_{\mathbb{C}} J_i \cong M_{n_i}(\mathbb{C})$. Suppose that $0 \neq a \in \mathrm{Ker}\,\varphi$, i.e., $aJ_i = 0$. All of the left ideals $J_{i,j}$ are $\mathbb{C}[G]$-isomorphic, and, if $\varphi_j : J_i \to J_{i,j}$ is an isomorphism, then

$$aJ_{i,j} = a\varphi_j(J_i) = a\varphi_j(eJ_i) = \varphi_j(a \cdot eJ_i) = \varphi_j(0) = 0 .$$

Hence, $aA_i = aJ_{i,1} + \dots + aJ_{i,n_i} = 0$, and in that case we also have $a\mathbb{C}[G] = 0$, since $a \in A_i \Rightarrow aA_j = 0$ for all $j \neq i$. However, $ae = a \neq 0$. This contradiction shows that $\mathrm{Ker}\,\varphi = 0$. Thus, φ is a monomorphism, and, since $\dim A_i = n_i^2 = \dim M_{n_i}(\mathbb{C})$, it follows that φ is an isomorphism from A_i to $M_{n_i}(\mathbb{C})$. Using

Proposition 2, we arrive at the following structure theorem for the group algebra $\mathbb{C}[G]$.

THEOREM 3. <u>The group algebra</u> $\mathbb{C}[G]$ <u>of a finite group</u> G <u>over the field of complex numbers decomposes into a direct sum (2) of simple two-sided ideals which are isomorphic to full matrix algebras:</u>

$$\mathbb{C}[G] \cong M_{n_1}(\mathbb{C}) \oplus M_{n_2}(\mathbb{C}) \oplus \dots \oplus M_{n_r}(\mathbb{C}) \quad .$$

<u>In particular, the group algebra of an abelian group of order</u> n <u>over the complex numbers is isomorphic to the direct sum of</u> n <u>copies of</u> $\mathbb{C}$. □

COROLLARY (Burnside's theorem). <u>Let</u> Φ <u>be an</u> n-<u>dimensional complex irreducible matrix representation of a finite group</u> G. <u>Then there are</u> n^2 <u>of the matrices</u> Φ_g, $g \in G$, <u>which are linearly independent, i.e.</u>, $\langle \Phi_g | g \in G \rangle_{\mathbb{C}} = M_n(\mathbb{C})$. □

The structure of the center $Z(\mathbb{C}[G])$ as a commutative subalgebra of $\mathbb{C}[G]$ is completely determined by the so-called <u>structure constants</u> -- the integers n_{ij}^k in the relations

$$z_i z_j = \sum_{k=1}^{r} n_{ij}^k z_k \quad . \tag{4}$$

Keeping in mind the expression (3) for z_i, we easily see that n_{ij}^k is the number of pairs (g, h), $g \in g_i^G$, $h \in g_j^G$, for which $gh = g_k$.

We choose another basis in $Z(\mathbb{C}[G])$ as follows:

$$e_i = \frac{n_i}{|G|} \sum_{k=1}^{r} \overline{\chi_i(g_k)}\, z_k = \frac{n_i}{|G|} \sum_{g \in G} \overline{\chi_i(g)}\, g \,, \qquad 1 \le i \le r \quad . \tag{5}$$

Here, as in §5 Ch. 8, $\chi_1, \dots, \chi_r$ are the characters of the irreducible representations, and $n_1, \dots, n_r$ are the degrees of those representations. We go backwards in (5) using the formula

$$z_k = |g_k^G| \sum_{i=1}^{r} \frac{\chi_i(g_k)}{n_i} e_i \quad .$$

To see this, one uses relation (4) §5 Ch. 8. That relation also shows that

$$\sum_{i=1}^{r} e_i = \frac{1}{|G|} \sum_{g \in G} g \sum_i n_i \overline{\chi_i(g)} = \frac{1}{|G|} \sum_{g \in G} g \sum \chi_i(e) \overline{\chi_i(g)} = \frac{1}{|G|} e \, |C_G(e)| = e \, .$$

Next, applying the generalized orthogonality relation in Exercise 1 §4 Ch. 8, we find that

$$e_i e_j = \frac{n_i n_j}{|G|^2} \sum_{g,t \in G} \overline{\chi_i(g)}\, \overline{\chi_j(t)}\, gt = \frac{n_i n_j}{|G|} \sum_{h \in G} \frac{1}{|G|} \sum_{g \in G} \overline{\chi_i(g)} \chi_j(hg) \; h^{-1} =$$

$$= \frac{n_i n_j}{|G|} \frac{\delta_{ij}}{n_i} \sum \chi_i(h)\, h^{-1} = \delta_{ij}\, e_i \quad .$$

Thus, the central elements c_i, which are computed by formula (5), satisfy the relations

$$\begin{aligned} e &= e_1 + e_2 + \cdots + e_r \, , \\ e_i^2 &= e_i \, , \qquad e_i e_j = 0 \, , \qquad i \neq j \, , \end{aligned} \tag{6}$$

and for this reason are called the <u>central orthogonal idempotents</u> of the group algebra $\mathbb{C}[G]$. The relation $e = e_1 + \cdots + e_r$ is the condition that this set is complete. If we set $B_i = e_i \mathbb{C}[G]$, we immediately find that B_i is a two-sided ideal in $\mathbb{C}[G]$ with identity element e_i, and that we have the following direct sum decomposition:

$$\mathbb{C}[G] = B_1 \oplus B_2 \oplus \ldots \oplus B_r \quad . \tag{7}$$

It follows directly from (5) that

$$\chi_j(e_i) = n_i \frac{1}{|G|} \sum_g \overline{\chi_i(g)} \chi_j(g) = n_i \delta_{ij} \quad .$$

Hence, B_i contains the minimal left ideal $J \subset A_i$ corresponding to the character χ_i. Since A_i and B_i are two-sided ideals, we have $A_i \subset B_i$. Comparing (2) and (7), we conclude that $A_i = B_i$. We have thereby proved a more complete version of Theorem 3.

THEOREM 4. *The elements* e_i, $1 \le i \le r$, *which are computed using* (5), *form a complete set of central orthogonal idempotents for the group algebra* $\mathbb{C}[G]$ *of a finite group* G. *The simple component* $e_i\,\mathbb{C}[G]$ *in the direct sum decomposition*

$$\mathbb{C}[G] = e_1\,\mathbb{C}[G] \oplus e_2\,\mathbb{C}[G] \oplus \dots \oplus e_r\,\mathbb{C}[G] \quad ,$$

is isomorphic to the full matrix algebra $M_{n_i}(\mathbb{C})$ *and contains all minimal left ideals corresponding to the character* χ_i. □

All of group representation theory can be developed starting from the Wedderburn-Artin theorem (see Subsection 1) and the general structure theory of group algebras (the full story in the case of finite groups is contained in Theorem 4). We have gone in the other direction, essentially using Schur's lemma as our point of departure.

We conclude by proving a useful fact about the degrees of representations.

THEOREM 5. *The degree* n *of an irreducible complex representation* (Φ, V) *of a finite group* G *divides the order* $|G|$.

Proof. Let $\widetilde{\Phi}$ be the corresponding representation of the group algebra $\mathbb{C}[G]$. By Schur's lemma (Proposition 2 of §3), the linear operator $\widetilde{\Phi}(z_i)$, which commutes with all of the $\Phi(g)$, $g \in G$, and hence belongs to $\mathrm{End}_{\mathbb{C}[G]}(V)$, must be a multiple of the identity operator: $\widetilde{\Phi}(z_i) = \omega_i\,\mathcal{E}$. We have

$$n\,\omega_i = \mathrm{tr}\,\omega_i\,\mathcal{E} = \mathrm{tr}\,\widetilde{\Phi}(z_i) = \sum \mathrm{tr}\,\Phi(g_i^h) = |g_i^G|\,\chi_\Phi(g_i) \quad ,$$

and hence

$$\omega_i = \frac{|g_i^G|\,\chi_\Phi(g_i)}{n} \quad .$$

Applying $\widetilde{\Phi}$ to the relations (4), we obtain

$$\omega_i\,\omega_j = \sum_{k=1}^{r} n_{ij}^k\,\omega_k \quad .$$

Hence, $\mathbb{Z}[\omega_i]$ is a submodule of the $\mathbb{Z}$-module of finite type $\mathbb{Z}[\omega_1, \ldots, \omega_r]$, and so, by the results of Subsection 3 of §3, ω_i is an algebraic integer. Using the same results, we find that

$$\frac{|G|}{n} = \frac{|G|}{n}(\chi_\Phi, \chi_\Phi)_G = \frac{1}{n}\sum \chi_\Phi(g)\overline{\chi_\Phi(g)} = \frac{1}{n}\sum_{i=1}^{r} |g_i^G| \cdot \chi_\Phi(g_i)\overline{\chi_\Phi(g_i)} = \sum \omega_i \overline{\chi_\Phi(g_i)}$$

is an algebraic integer. Thus, $|G|/n \in \mathbb{Z}$. □

4. Non-associative algebras. Let A be any (not necessarily associative) algebra of arbitrary dimension over a field K. To every triple of elements $x, y, z \in A$ we associate their associator $(x, y, z) = (xy)z - x(yz)$. Depending on the identities satisfied by the associators and other expressions, we obtain various types (called primitive classes or varieties) of algebras. Examples of such classes of algebras are:

1) associative algebras: $(x, y, z) = 0$;
2) elastic algebras: $(x, y, x) = 0$;
3) alternative algebras: $(x, x, y) = (y, x, x) = 0$;
4) Jordan algebras: $(x, y, x^2) = 0$ and $xy - yx = 0$.

Of course, we can use this axiomatic procedure endlessly. But what is remarkable is that many classes of non-associative algebras have arisen naturally in fields far removed from the science of algebra per se. The most notable examples are the Jordan algebras, which arose from quantum mechanics, and the Lie algebras, which were originally designed to describe (under certain conditions) the local structure of topological groups (Sophus Lie was a XIX century mathematician). We have alluded to Lie algebras before in this book, and so now will devote a brief discussion to them.

In a Lie algebra L over a field K the product of two elements x and y is customarily denoted $[xy]$. By definition, in a Lie algebra the bilinear map $(x, y) \mapsto [xy]$ must satisfy the following two properties:

(i) $[xx] = 0$ ($[xy] = -[yx]$, the property of anti-commutativity);

(ii) $[[xy]z] + [[yz]x] + [[zx]y] = 0$ (the Jacobi identity).

Example 1. Let A be an associative algebra over a field K. We can give the vector space A the structure of a Lie algebra, which we denote $L(A)$, by setting $[xy] = xy - yx$. Obviously, $[xx] = 0$. Furthermore,

$$[[xy]z] = (xy - yx)z - z(xy - yx) = xyz - yxz - zxy + zyx \quad ,$$

$$[[yz]x] = (yz - zy)x - x(yz - zy) = yzx - zyx - xyz + xzy \quad ,$$

$$[[zx]y] = (zx - xz)y - y(zx - xz) = zxy - xzy - yzx + yxz \quad .$$

By simply adding these expressions, we obtain the Jacobi identity.

For example, let $A = \mathrm{End}_K(V) = \mathcal{L}(V)$ be the algebra of all linear operators on a finite dimensional K-vector space V. Any homomorphism φ from a Lie algebra L to the Lie algebra $L(\mathcal{L}(V))$ is called a representation of the Lie algebra L. The representation space V is also called an L-module (or a module over the Lie algebra L). Formally, an L-module is given by three axioms:

$$\text{(L1)} \qquad x(\alpha u + \beta v) = \alpha x u + \beta x v\,;$$

$$\text{(L2)} \qquad (\alpha x + \beta y)v = \alpha x v + \beta y v\,;$$

$$\text{(L3)} \qquad [xy]v = x(yv) - y(xv) \quad .$$

Example 2. If A is any (not necessarily associative) algebra over a field K, by a differentiation $\mathcal{D}$ of the algebra A we mean a differentiation of the ring A (see the definition in Subsection 3 of §1 Ch. 6) which commutes with scalar multiplication, i.e., $\mathcal{D}(\lambda a) = \lambda\mathcal{D}(a)$ for $\lambda \in K$ and $a \in A$. Example 1 and Exercise 8 in §1 Ch. 6 show that the formula $[\mathcal{D}_1 \mathcal{D}_2] = \mathcal{D}_1\mathcal{D}_2 - \mathcal{D}_2\mathcal{D}_1$ gives the K-vector space $\mathrm{Der}(A)$ a Lie algebra structure. For example, if $A = K[X]$ is a polynomial algebra, then $\mathrm{Der}(A)$ consists of the differentiations $\mathcal{D}_u$, $u \in A$, which act according to the rule: $\mathcal{D}_u(f) = u\, df/dX = uf'$. By definition, we have: $[\mathcal{D}_u \mathcal{D}_v](f) = \mathcal{D}_u(\mathcal{D}_v f) - \mathcal{D}_v(\mathcal{D}_u f) =$

$= \wp_u(vf') - \wp_v(uf') = u(vf')' - v(uf')' = u(v'f' + vf'') - v(u'f' + uf'') = (uv' - u'v)f'$. Consequently, $[\wp_u \wp_v] = \wp_{uv' - u'v}$, and we see that the algebra $\mathrm{Der}(A)$ is isomorphic to the infinite dimensional Lie algebra $(A, [\ \ ])$ having A as its underlying space and multiplication law $[uv] = uv' - u'v$. Setting $A_{(i)} = \langle X^{i+1} \rangle_{\mathbb{C}}$, we obtain a direct sum decomposition of A

$$A = A_{(-1)} \oplus A_{(0)} \oplus A_{(1)} \oplus A_{(2)} \oplus \cdots ,$$

which shows that A is a graded Lie algebra: $[A_{(i)} A_{(j)}] \subset A_{(i+j)}$ (compare with Example 2 in Subsection 1). The Lie algebra $(A, [\ \ ])$ acts on the vector space A in two ways: 1) $(a, f) \mapsto af'$ (the natural action); 2) $(a, f) \mapsto af' - a'f$ (the action by adjoint endomorphisms). These two actions give two non-isomorphic $(A, [\ \ ])$-modules.

Example 3. The skew-hermitian matrices K_1, K_2, K_3 with non-zero trace that were constructed in Exercise 3 of §1 Ch. 7 for the group $SU(2)$, satisfy the relations

$$[K_1 K_2] = K_3, \qquad [K_2 K_3] = K_1, \qquad [K_3 K_1] = K_2 ,$$

which are exactly the rules for the cross-product of vectors in $\mathbb{R}^3$ ($[K_s K_t] = K_s K_t - K_t K_s$ is the commutator of the matrices in $M_2(\mathbb{C})$; see Example 1). Hence, the three-dimensional real space $\langle K_1, K_2, K_3 \rangle_{\mathbb{R}}$ has been given a Lie algebra structure.

The general theory of representations of compact groups tells us that there is a one-to-one correspondence between the irreducible representations of $SU(2)$ and those of its Lie algebra $\underline{su}(2) = \langle K_1, K_2, K_3 \rangle_{\mathbb{R}}$. Intuitively, we can see this by taking into account the continuity of the group representation and considering the linear operator $\lim_{t \to 0} \frac{1}{t} \Phi(g_t)$ in the linear span of the operators $\Phi(g_t)$ (where g_t is an element of $SU(2)$ which depends in a differentiable way on t and for which $g_0 = e$); this limit operator is actually in the algebra $\underline{su}(2)$. In order to see that the list of irreducible representations of $SU(2)$ in §6 Ch. 8 is complete, we must verify that, for any natural number n, there is up to isomorphism exactly one irreducible $\underline{su}(2)$-module of dimension n over $\mathbb{C}$. To do this, it is useful from the very beginning to pass from the real Lie algebra $\underline{su}(2)$ to its "complexification", which is the Lie algebra

$$L = sl(2) = su(2) \otimes_{\mathbb{R}} \mathbb{C}$$

of all complex 2×2 matrices having zero trace. The basis elements

$$e_{-1} = -iK_1 + K_2, \qquad e_0 = -2i K_3, \qquad e_1 = -iK_1 - K_2$$

of the Lie algebra L multiply together according to the rule

$$[e_1\, e_{-1}] = e_0, \qquad [e_0\, e_{-1}] = -2e_{-1}, \qquad [e_0\, e_1] = 2e_1 \quad . \tag{8}$$

Forgetting for a moment how L originated, we may take $L = \langle e_{-1}, e_0, e_1 \rangle_{\mathbb{C}}$ to be the abstract three-dimensional Lie algebra over $\mathbb{C}$ with multiplication table (8). It is easy to verify that L is a simple Lie algebra. Hence, any irreducible L-module of dimension > 1 is faithful.

First suppose that $V \neq 0$ is an arbitrary L-module of finite dimension over $\mathbb{C}$, and let E_{-1}, E_0, E_1 be the linear operators on V corresponding to e_{-1}, e_0, e_1, respectively. The representation theory of Lie algebras has its own terminology, which we shall adhere to. If $V^\lambda = \{v \in V | E_0 v = \lambda v\}$ is the eigen-subspace of E_0 in V with eigen-value $\lambda \in \mathbb{C}$, then the vectors in V^λ are customarily called vectors of <u>weight</u> λ. The dimension of V^λ is called the <u>multiplicity of the weight</u> λ.

LEMMA 1. <u>If</u> $v \in V^\lambda$, <u>then</u> $E_1 v \in V^{\lambda+2}$ <u>and</u> $E_{-1} v \in V^{\lambda-2}$.

<u>Proof.</u> By the axiom (L3) we have

$$E_0(E_1 v) = [E_0\, E_1]v + E_1(E_0 v) = 2E_1 v + E_1(\lambda v) = (\lambda+2) E_1 v \quad ,$$

so that, by definition, $E_1 v \in V^{\lambda+2}$. We similarly show that $E_{-1} v \in V^{\lambda-2}$. □

We know from linear algebra that vectors corresponding to different eigen-values are linearly independent. Hence the sum $W = \sum_\lambda V^\lambda \subset V$ is a direct sum. It further follows from Lemma 1 that $W = \sum_{\lambda} V^{\lambda}$ is an L-submodule of V. Since $W \neq 0$, we must have $W = V$ if V is an irreducible L-module.

A vector $v_0 \in V$ is called a <u>highest weight vector</u> of weight λ if $v_0 \neq 0$ and $E_1 v_0 = 0$, $E_0 v_0 = \lambda v_0$.

LEMMA 2. <u>Any finite-dimensional L-module V has a highest weight vector.</u>

<u>Proof.</u> Take an arbitrary non-zero vector v of weight μ, and construct the sequence of vectors $v, E_1 v, E_1^2 v, \ldots$ with weights $\mu, \mu+2, \mu+4, \ldots$ (see Lemma 1). Since $\dim V < \infty$, it follows that $E_1^{m+1} v = 0$ for some m. If we take m to be the least integer with this property, we can set $v_0 = E_1^m v$, $\lambda = \mu + 2m$. □

As an example, we consider the $(n+1)$-dimensional $\mathbb{C}$-vector space V_n with fixed basis $v_0, v_1, \ldots, v_n$. We define the operators E_{-1}, E_0 and E_1 by the formulas

$$\begin{aligned} E_{-1} v_m &= (m+1) v_{m+1} , \\ E_0 v_m &= (n-2m) v_m , \\ E_1 v_m &= (n-m+1) v_{m-1} , \end{aligned} \tag{9}$$

where we set $v_{-1} = 0 = v_{n+1}$. A direct computation shows that we have:

$$\begin{aligned} E_1(E_{-1} v_m) - E_{-1}(E_1 v_m) &= E_0 v_m , \\ E_0(E_{-1} v_m) - E_{-1}(E_0 v_m) &= -2 E_{-1} v_m , \\ E_0(E_1 v_m) - E_1(E_0 v_m) &= 2 E_1 v_m , \end{aligned}$$

which corresponds to the multiplication table (8) and the axioms of an L-module. Since $E_1 v_0 = (n+1) v_{-1} = 0$ and $E_0 v_0 = n v_0$, it follows that v_0 is a highest weight vector of weight n, and the entire space V_n can be written as the direct sum

$$V_n = V^n \oplus V^{n-2} \oplus \ldots \oplus V^{-n} \tag{10}$$

of one-dimensional weight subspaces $V^{n-2m} = \langle v_m \rangle$ (each weight has multiplicity 1). If we had a non-zero submodule U in V_n, we could take any eigen-vector $u \in U$ of E_0. Then, by (10), $u = \lambda v_m$ for some m. By successively applying E_1 (see (9)), we would obtain $v_{m-1} \in U, \ldots, v_0 \in U$, and by successively applying E_{-1} to v_0 we would obtain all of the other v_j. Hence, $U = V_n$, and V_n is an irreducible L-module.

We note that V_0 is the trivial (one-dimensional) module, and V_1 is the module corresponding to the natural definition of the Lie algebra L: in the basis $\{v_0, v_1\}$ the operators E_{-1}, E_0, E_1 have the matrices

$$\begin{Vmatrix} 0 & 0 \\ 1 & 0 \end{Vmatrix}, \quad \begin{Vmatrix} 1 & 0 \\ 0 & -1 \end{Vmatrix}, \quad \begin{Vmatrix} 0 & 1 \\ 0 & 0 \end{Vmatrix} .$$

The following theorem answers the remaining question before us.

THEOREM 6. <u>Every</u> $(n+1)$-<u>dimensional irreducible</u> L-<u>module</u> V <u>over</u> $\mathbb{C}$ <u>is isomorphic to</u> V_n.

<u>Proof.</u> By Lemma 2, our module V has some highest weight vector v_0 of weight λ. We set $v_{-1} = 0$ and $v_m = \frac{1}{m!} E_{-1}^m v_0 = \frac{1}{m!} E_{-1}(E_{-1}(\ldots(E_{-1}v_0)\ldots))$ for $m \geq 0$. We claim that the following formulas hold for any $m \geq 0$:

$$\begin{aligned} E_{-1} v_m &= (m+1)\, v_{m+1} , \\ E_0 v_m &= (\lambda - 2m)\, v_m , \\ E_1 v_m &= (\lambda - m + 1)\, v_{m-1} . \end{aligned} \tag{10'}$$

In fact, for $m = 0$ the formulas (10') reduce to the definitions of the highest weight vector v_0 and the vector v_1. We now use induction on m: a) the vector v_{m+1} is defined by the formula $E_{-1}v_m = (m+1)v_{m+1}$; b) the formula $E_0 v_m = (\lambda - 2m)v_m$ follows from Lemma 1; c) if we already know that $E_1 v_{m-1} = (\lambda - m + 2)v_{m-2}$, then we obtain the last formula in (10') by dividing both sides of the following equality by m :

$$m E_1 v_m = E_1(E_{-1} v_{m-1}) = [E_1 E_{-1}] v_{m-1} + E_{-1}(E_1 v_{m-1}) =$$
$$= E_0 v_{m-1} + (\lambda - m + 2) E_{-1} v_{m-2} = \{(\lambda - 2m + 2) +$$
$$+ (\lambda - m + 2)(m - 1)\} v_{m-1} = m(\lambda - m + 1) v_{m-1} \quad .$$

If the vectors $v_0, v_1, \dots, v_r$ are non-zero for a certain r, then they must be linearly independent, since they have different weights. On the other hand, since V is irreducible, the submodule generated by the vector v_0 is all of V, and, since $\dim V = n + 1$, it follows that $V = \langle v_0, v_1, \dots, v_n \rangle$ and $v_{n+1} = v_{n+2} = \dots = 0$. In particular,

$$0 = E_1 v_{n+1} = (\lambda - n) v_n = 0 \Longrightarrow \lambda = n$$

(note the interesting fact that $\dim V < \infty$ implies that λ is a non-negative integer).

Substituting $\lambda = n$ in (10') and taking into account the notation we are using, we arrive at the formulas (9) which define the L-module V_n. Hence, $V \cong V_n$. □

EXERCISES

1. How many solutions does the equation $x^2 + 1 = 0$ have in the quaternion algebra $\mathbb{H}$?

2. The algebra of generalized quaternions over $\mathbb{Q}$. Show that the multiplication table

	1	e_1	e_2	e_3
1	1	e_1	e_2	e_3
e_1	e_1	n	e_3	ne_2
e_2	e_2	$-e_3$	m	$-me_1$
e_3	e_3	$-ne_2$	me_1	$-nm$

with $n, m \in \mathbb{Z}$, m and n non-zero, gives the structure of an associative algebra with unit to the four-dimensional vector space $H(n, m) = \langle 1, e_1, e_2, e_3 \rangle_{\mathbb{Q}}$ over $\mathbb{Q}$. To do this, use the representation

$$x = x_0 + x_1 e_1 + x_2 e_2 + x_3 e_3 \longmapsto A_x = \begin{Vmatrix} x_0 + x_1\sqrt{n} & x_2\sqrt{m} + x_3\sqrt{nm} \\ x_2\sqrt{m} - x_3\sqrt{mn} & x_0 - x_1\sqrt{n} \end{Vmatrix} .$$

The determinant $\det A_x = x_0^2 - x_1^2 n - x_2^2 m + x_3^2 nm = N(x)$ is called the norm of the element x. Prove that $H(n, m)$ is a division algebra provided that the norm of any non-zero $x \in H(n, m)$ is non-zero. Using the ideas and results of Exercise 7 §2, show that, if p is a prime congruent to $\pm 3 \pmod 8$, then $H(2, p)$ is a division algebra.

3. Consider $\mathbb{F}_{2^n}$ as an n-dimensional vector space V over $\mathbb{F}_2$. Along with the addition operation coming from $\mathbb{F}_{2^n}$, introduce a multiplication operation on V by letting $(x, y) \mapsto x \circ y = \sqrt{xy}$. Here $x \mapsto \sqrt{x}$ is the automorphism of $\mathbb{F}_{2^n}$ which is inverse to $x \mapsto x^2$; thus, $\sqrt{x+y} = \sqrt{x} + \sqrt{y}$. Show that $(V, +, \circ)$ is a commutative, non-associative algebra over $\mathbb{F}_2$ with the properties: a) V has no zero divisors and does not have a unit element; b) the equation $a \circ x = b$ with $a \neq 0$ has a unique solution; c) the automorphism group $\mathrm{Aut}(V)$ acts transitively on $V \setminus \{0\}$.

4. Show by a direct computation that the following identity holds in any algebra:

$$t(x,y,z) + (t,x,y)z = (tx,y,z) - (t,xy,z) + (t,x,yz) \quad .$$

Prove that, if an algebra A with unit element 1 over a field K has the property that $(x,y,z) \in K \cdot 1$ for all associators (x,y,z), then A is an associative algebra.

Appendix. The Jordan Normal Form of a Matrix

We are including here a discussion of this particular corner of linear algebra in order to emphasize its similarity with §5 Ch. 7, where we have the classification of finite abelian groups. We decided not to insist in §3 Ch. 9 on the unified point of view of modules over principal ideal domains, since different readers may find it more convenient to have direct proofs of the necessary facts concerning groups or concerning linear operators.

1. If we want to understand the action of a given linear operator $\mathcal{A} : V \to V$, it is natural for us to try to find a basis of V which is most compatible with $\mathcal{A}$. In other words, in the class of similar matrices $C^{-1} A C$ corresponding to the operator $\mathcal{A}$, we would like to find a matrix with the simplest possible form. Solving this problem depends on the nature of the field K over which the vector space V is defined. In what follows we shall assume that K is the complex number field $\mathbb{C}$ or any other algebraically closed field.

Let $n = \dim V$, and let $\lambda_1, \ldots, \lambda_n$ be the roots of the characteristic polynomial

$$f_{\mathcal{A}}(t) = f_A(t) = \det(tE - A) = t^n + a_1 t^{n-1} + \cdots + a_n = \prod_{i=1}^{n} (t - \lambda_i) ,$$

$$a_1 = -\operatorname{tr} A = -(\lambda_1 + \cdots + \lambda_n) ,$$

$$a_n = (-1)^n \det A = (-1)^n \lambda_1 \ldots \lambda_n .$$

The complex numbers λ_i are also the eigen-values of the linear operator $\mathcal{A}$: the subspaces

$$V^{\lambda_i} = \{v \in V \mid \mathcal{A} v = \lambda_i v\}$$

are non-zero, and the non-zero vectors in these subspaces are called the eigen-vectors of $\mathcal{A}$. The set $\operatorname{Spec}(\mathcal{A})$ of all pair-wise distinct eigen-values (characteristic roots) of the operator $\mathcal{A}$ is called its spectrum. We similarly speak of the spectrum $\operatorname{Spec}(A)$ of the matrix A.

We note the following facts.

(i) Eigen-vectors having different eigen-values are linearly independent. The sum $\sum_{\lambda \in \operatorname{Spec}(A)} V^\lambda$ is a direct sum (in general, ΣV^λ does not necessarily coincide with V).

(ii) The matrix of a linear operator $\mathcal{A}$ can always be reduced to triangular form (within the class of similar matrices).

The simplest way to see these facts is to use induction. Take a one-dimensional $\mathcal{A}$-invariant subspace $\langle e_1 \rangle$ (where $\mathcal{A} e_1 = \lambda_1 e_1$), pass to the quotient space $\overline{V} = V/\langle e_1 \rangle = \{\overline{v} = v + \langle e_1 \rangle \mid v \in V\}$, which has dimension $n - 1$, and to the quotient operator $\overline{\mathcal{A}}$, defined by $\overline{\mathcal{A}}\,\overline{v} = \overline{\mathcal{A} v}$. In $\overline{V}$ choose a basis $\overline{e}_2, \ldots, \overline{e}_n$ which brings $\overline{A}$ to triangular form (using the induction assumption), and then return to V to obtain:

$$A = \begin{Vmatrix} \lambda_1 & & & * \\ & \lambda_2 & & \\ & 0 & \ddots & \\ & & & \lambda_n \end{Vmatrix} . \qquad \square$$

(iii) (The Hamilton-Cayley theorem). A linear operator $\mathcal{A}$ and the corresponding matrix A (in any basis) are annihilated by their characteristic polynomial.

Since this assertion does not depend on the choice of basis, to prove it it is useful to make use of property (ii). We consider the chain of $\mathcal{A}$-invariant subspaces $V = V_0 \supset V_1 \supset \supset \dots \supset V_{n-1} \supset 0$, where $V_k = \langle e_1, e_2, \dots, e_{n-k-1}, e_{n-k}\rangle$. Since $(\mathcal{A} - \lambda_{n-k}\mathcal{E})e_{n-k} \in V_{k+1}$, it follows that $(\mathcal{A} - \lambda_{n-k}\mathcal{E})V_k \subset V_{k+1}$, and hence

$$f_{\mathcal{A}}(A)V = \prod_{i=1}^{n} (\mathcal{A} - \lambda_i\mathcal{E})V = (\mathcal{A} - \lambda_1\mathcal{E}) \dots (\mathcal{A} - \lambda_n\mathcal{E})V_0 \subset (\mathcal{A} - \lambda_1\mathcal{E}) \dots (\mathcal{A} - \lambda_{n-1}\mathcal{E})V_1 \subset$$

$$\subset (\mathcal{A} - \lambda_1\mathcal{E}) \dots (\mathcal{A} - \lambda_{n-2}\mathcal{E})V_2 \subset \dots \subset (\mathcal{A} - \lambda_1\mathcal{E})V_{n-1} = 0 .$$

But $f_{\mathcal{A}}(\mathcal{A})V = 0$ if and only if $f_{\mathcal{A}}(\mathcal{A}) = 0$. □

(iv) The minimal polynomial $h_{\mathcal{A}}(t) = h_A(t)$ of an operator (i.e., the monic polynomial of minimal degree $m \leq n$ which annihilates $\mathcal{A}$ and A) is a divisor of the characteristic polynomial $f_{\mathcal{A}}(t)$ and is divisible by all linear factors $t - \lambda$, $\lambda \in \operatorname{Spec}(\mathcal{A})$.

The division algorithm, which gives $f_{\mathcal{A}}(t) = q(t) \cdot h_{\mathcal{A}}(t) + r(t)$, $\deg r(t) < \deg h_{\mathcal{A}}(t)$, along with the fact that $f_{\mathcal{A}}(\mathcal{A}) = 0 = h_{\mathcal{A}}(\mathcal{A})$, allow us to conclude that $r(\mathcal{A}) = 0$, and so $r(t) = 0$. Next, if λ is an eigen-value of $\mathcal{A}$, then, choosing v so that $\mathcal{A}v = \lambda v$, we have: $0 = h_{\mathcal{A}}(\mathcal{A})v = h_{\mathcal{A}}(\lambda)v$, so that $h_{\mathcal{A}}(\lambda) = 0$, and hence $(t - \lambda) \mid h_{\mathcal{A}}(t)$. □

Example. A linear operator $\mathcal{A} : V \to V$ is called nilpotent if $\mathcal{A}^m = 0$ for some positive integer m; m is called the index of nilpotence if $\mathcal{A}^{m-1} \neq 0$. We claim: if $\mathcal{A}^{m-1}v \neq 0$, then the vectors $v, \mathcal{A}v, \dots, \mathcal{A}^{m-1}v$ are linearly independent. In fact, any non-trivial linear dependence relation has the form

$$\mathcal{A}^k v + \alpha_1 \mathcal{A}^{k+1} v + \dots + \alpha_{m-1-k}\mathcal{A}^{m-1} v = 0, \qquad 0 \leq k \leq m-1 .$$

Applying the operator $\mathcal{A}^{m-k-1}$ to both sides of this equation gives us $\mathcal{A}^{m-1}v = 0$, which contradicts the choice of v.

Thus, the index of nilpotence m of $\mathfrak{a}$ does not exceed $n = \dim V$. Suppose that $m = n$, and that $\mathfrak{a}^{n-1} v \neq 0$. We introduce the following notation for the basis vectors: $v_1 = \mathfrak{a}^{n-1} v$, $v_2 = \mathfrak{a}^{n-2} v, \ldots, v_{n-1} = \mathfrak{a} v$, $v_n = v$. Then $\mathfrak{a} v_k = v_{k-1}$ for $k > 1$, and $\mathfrak{a} v_1 = 0$, so that the matrix of $\mathfrak{a}$ in the basis $\{v_1, \ldots, v_n\}$ is the so-called "Jordan cell"

$$J_{n,0} = \begin{Vmatrix} 0 & 1 & 0 & \ldots & 0 & 0 \\ 0 & 0 & 1 & \ldots & 0 & 0 \\ 0 & 0 & 0 & \ldots & 0 & 0 \\ \cdot & \cdot & \cdot & \cdot & \cdot & \cdot \\ 0 & 0 & 0 & \ldots & 0 & 1 \\ 0 & 0 & 0 & \ldots & 0 & 0 \end{Vmatrix} .$$

For example, if $V = \langle 1, X, X^2, \ldots, X^{n-1} \rangle_{\mathbb{C}}$ is the space of polynomials of degree $< n$ over $\mathbb{C}$, and if $\mathfrak{a} = d/dX$ is the differentiation operator, then the matrix of $\mathfrak{a}$ in the basis $\{v_i\}$, where $v_i = \frac{1}{i!} X^i$, is precisely $J_{n,0}$.

More generally, by the $m \times m$ (upper) Jordan cell corresponding to the eigen-value λ, we mean the matrix

$$J_{m,\lambda} = \begin{Vmatrix} \lambda & 1 & 0 & \ldots & 0 & 0 \\ 0 & \lambda & 1 & \ldots & 0 & 0 \\ 0 & 0 & \lambda & \ldots & 0 & 0 \\ \cdot & \cdot & \cdot & \cdot & \cdot & \cdot \\ 0 & 0 & 0 & \ldots & \lambda & 1 \\ 0 & 0 & 0 & \ldots & 0 & \lambda \end{Vmatrix} .$$

We note that $J_{m,\lambda} - \lambda E = J_{m,0}$ is a nilpotent matrix with index of nilpotence m. We further see that $(t - \lambda)^m$ is the minimal polynomial of the Jordan cell $J_{m,\lambda}$, and λ is the only eigen-value: $\mathrm{Spec}(J_{m,\lambda}) = \{\lambda\}$.

If $u(t)$ is any polynomial, then we have

$$u(J_{m,\lambda}) = \begin{Vmatrix} u(\lambda) & u'(\lambda)/1! & u''(\lambda)/2! & \ldots & u^{(m-1)}(\lambda)/(m-1)! \\ 0 & u(\lambda) & u'(\lambda)/1! & \ldots & u^{(m-2)}(\lambda)/(m-2)! \\ \cdot & \cdot & \cdot & \cdot & \cdot \\ 0 & 0 & 0 & & u(\lambda) \end{Vmatrix} ,$$

so that it is much easier to operate with $J_{m,\lambda}$ than with arbitrary matrices.

FUNDAMENTAL THEOREM. *Every* $n \times n$ *matrix* A *over an algebraically closed field* K *(for example, over* $\mathbb{C}$*) is similar to a direct sum of Jordan cells. That is, there exists a non-singular matrix* C *such that*

$$C^{-1}AC = J_{m_1,\lambda_1} + \cdots + J_{m_s,\lambda_s} = \begin{Vmatrix} J_{m_1,\lambda_1} & & & 0 \\ & J_{m_2,\lambda_2} & & \\ & & \ddots & \\ 0 & & & J_{m_s,\lambda_s} \end{Vmatrix}$$

(this is called the Jordan normal form $J(A)$ *of the matrix* A*). The Jordan normal form is unique except for the order of the Jordan cells.*

The fundamental theorem is proved in three steps in Subsections 2, 3, and 4.

Since the minimal polynomials of similar matrices are the same, it follows from the fundamental theorem and the above remarks on Jordan cells that

$$h_A(t) = (t - \lambda_{i_1})^{m_{i_1}} \cdots (t - \lambda_{i_p})^{m_{i_p}},$$

where $\{\lambda_{i_1}, \ldots, \lambda_{i_p}\} = \mathrm{Spec}(A)$, and m_{j_k} is the maximal order of the Jordan cells corresponding to the eigen-value λ_{j_k}.

Clearly, a matrix A is diagonalizable (i.e., similar to a matrix of the form $\mathrm{diag}\{\lambda_1, \ldots, \lambda_n\}$) if and only if there are no Jordan cells of order greater than 1 in $J(A)$. Hence, we have the following useful criterion.

COROLLARY. *A square matrix* A *over* $\mathbb{C}$ *is diagonalizable if and only if its minimal polynomial* $h_A(t)$ *has no multiple roots.*

Note that to apply this corollary, i.e., to find $h_A(t)$, we do not have to reduce A to Jordan normal form.

2. The set of vectors

$$V(\lambda) = \{v \in V \mid (\mathcal{A} - \lambda\mathcal{E})^k v = 0 \quad \text{for some} \quad k\}$$

is called the root space corresponding to the eigen-value $\lambda \in \operatorname{Spec}(\mathcal{A})$. It is easy to check that $V(\lambda)$ is really a subspace. Namely, suppose that $u, v \in V(\lambda)$ with $(\mathcal{A} - \lambda\mathcal{E})^s u = 0$ and $(\mathcal{A} - \lambda\mathcal{E})^t v = 0$. If $m = \max\{s, t\}$, then

$$(\mathcal{A} - \lambda\mathcal{E})^m (\alpha u + \beta v) = \alpha(\mathcal{A} - \lambda\mathcal{E})^m u + \beta(\mathcal{A} - \lambda\mathcal{E})^m v = 0 ,$$

and so $\alpha u + \beta v \in V(\lambda)$ for any $\alpha, \beta \in \mathbb{C}$. Since $V(\lambda)$ contains an eigen-value corresponding to λ, it follows that $V(\lambda) \neq 0$. Furthermore, we have $V^\lambda \subset V(\lambda)$, but the two spaces do not necessarily coincide, as we see from the example of a nilpotent operator $\mathcal{A}$ having index of nilpotence n. In that case $\lambda = 0$ is the only eigen-value, and $\dim V^0 = 1$, but $V(0) = V$.

Since $\dim V(\lambda) \leq n$ and the restriction of $\mathcal{A} - \lambda\mathcal{E}$ to $V(\lambda)$ is a nilpotent operator, it follows that

$$V(\lambda) = \{v \in V \mid (\mathcal{A} - \lambda\mathcal{E})^n v = 0\} .$$

THEOREM 1. <u>Let</u> $\mathcal{A} : V \to V$ <u>be a linear operator with characteristic polynomial</u>

$$f_{\mathcal{A}}(t) = \prod_{i=1}^{p} (t - \lambda_i)^{n_i} \quad (\lambda_i \neq \lambda_j \quad \text{for} \quad i \neq j) .$$

<u>Then</u> $V = V(\lambda_1) \oplus \ldots \oplus V(\lambda_p)$ <u>is the direct sum of the root spaces</u> $V(\lambda_i)$, <u>each of which is invariant with respect to</u> $\mathcal{A}$ <u>and satisfies</u> $\dim V(\lambda_i) = n_i$. <u>The operator</u> $\mathcal{A} - \lambda_i\mathcal{E}$, <u>which is nilpotent on</u> $V(\lambda_i)$, <u>is non-singular on the subspace</u>

$$V_i = V(\lambda_1) \oplus \ldots \oplus V(\lambda_{i-1}) \oplus V(\lambda_{i+1}) \oplus \ldots \oplus V(\lambda_p) .$$

Finally, λ_i is the only eigen-value of the operator $\mathcal{A}|_{V(\lambda_i)}$.

Proof. None of the prime factors $t - \lambda_k$ can be a divisor simultaneously of all of the polynomials

$$f_i(t) = \prod_{j \neq i} (t - \lambda_j)^{n_j}, \qquad i = 1, 2, \ldots, p ,$$

and hence we have g. c. d. $(f_1(t), \ldots, f_p(t)) = 1$. Thus, we can find polynomials $g_1(t), \ldots, g_p(t) \in \mathbb{C}[t]$ for which

$$\sum_{i=1}^{p} f_i(t) g_i(t) = 1 \quad . \tag{1}$$

The subspaces

$$W_i = f_i(\mathfrak{A}) g_i(\mathfrak{A}) V = \{f_i(\mathfrak{A}) g_i(\mathfrak{A}) v \mid v \in V\}, \qquad 1 \leq i \leq p ,$$

are invariant under $\mathfrak{A}$:

$$\mathfrak{A} W_i = f_i(\mathfrak{A}) g_i(\mathfrak{A}) \mathfrak{A} V \subset f_i(\mathfrak{A}) g_i(\mathfrak{A}) V = W_i \quad .$$

In addition,

$$(\mathfrak{A} - \lambda_i \mathcal{E})^{n_i} W_i = f_{\mathfrak{A}}(\mathfrak{A}) g_i(\mathfrak{A}) V = 0$$

(since $f_{\mathfrak{A}}(\mathfrak{A}) = 0$ by the Hamilton-Cayley theorem), so that

$$W_i \subset V(\lambda_i) \quad . \tag{2}$$

The relation (1), re-written in the form

$$\mathcal{E} = \sum_{i=1}^{p} f_i(\mathfrak{A}) g_i(\mathfrak{A}) ,$$

gives us:

$$V = \sum_{i=1}^{p} W_i$$

and so all the more (because of the inclusion (2)):

$$V = \sum_{i=1}^{p} V(\lambda_i) \quad .$$

Suppose that $v \in V(\lambda_i) \cap V_i$, where, as in the statement of the theorem, $V_i = \sum_{j \neq i} V(\lambda_j)$. Then $(\mathcal{a} - \lambda_i \mathcal{E})^n v = 0$, and since $v = \sum_{j \neq i} v_j$ and $(\mathcal{a} - \lambda_j \mathcal{E})^n v_j = 0$, it follows that we also have $\{\prod_{j \neq i} (\mathcal{a} - \lambda_j \mathcal{E})^n\} v = 0$. But, because the polynomials $(t - \lambda_i)^n$ and $c(t) = \prod_{j \neq i} (t - \lambda_j)^n$ are relatively prime, there exist $a(t)$ and $b(t)$ for which

$$a(t)(t - \lambda_i)^n + b(t)c(t) = 1 \quad .$$

We obtain

$$v = a(\mathcal{a})(\mathcal{a} - \lambda_i)^n v + b(\mathcal{a})\{\prod_{j \neq i} (\mathcal{a} - \lambda_j \mathcal{E})^n\} v = 0 \quad ,$$

i.e., the spaces $V(\lambda_i)$ and V_i have intersection zero. Thus, we have the direct sum decomposition

$$V = V(\lambda_1) \oplus \ldots \oplus V(\lambda_p) \tag{3}$$

into $\mathcal{a}$-invariant subspaces.

The inclusion (2) and the decomposition (3) immediately imply that $W_i = V(\lambda_i)$. We have thereby obtained the following explicit expression for $V(\lambda_i)$:

$$V(\lambda_i) = f_i(\mathcal{a})\, g_i(\mathcal{a}) V \quad ,$$

where $f_i(t)$ and $g_i(t)$ are the polynomials in (1). In particular, we have

$$(\mathcal{a} - \lambda_i)^{n_i} V(\lambda_i) = 0 \quad .$$

The minimal polynomial for $\mathcal{a}$ on $V(\lambda_i)$ must be a divisor of the polynomial $(t - \lambda_i)^{n_i}$. This implies, first of all, that λ_i is the only eigen-value for the operator $\mathcal{a}\big|_{V(\lambda_i)}$. Furthermore, if we take a basis for V which is a union of bases for the $V(\lambda_i)$, then the operator $\mathcal{a}$ has matrix

$$A = \begin{Vmatrix} A_1 & & 0 \\ & \ddots & \\ 0 & & A_p \end{Vmatrix},$$

where A_i is an $n'_i \times n'_i$ matrix (with $n'_i = \dim V(\lambda_i)$) whose only eigen-value is λ_i and whose characteristic polynomial is $f_{A_i}(t) = (t - \lambda_i)^{n'_i}$, where $n'_i \le n_i$. Since $f_A(t) = \prod_{i=1}^{p} f_{A_i}(t)$, it follows that $n = n'_1 + \cdots + n'_p$ and $n'_i = n_i$.

It remains to prove that the restriction $(\mathcal{A} - \lambda_i \mathcal{E})\big|_{V_i}$ is non-singular. But this is clear: otherwise we would have $\{\operatorname{Ker}(\mathcal{A} - \lambda_i \mathcal{E})\} \cap V_i \neq 0$ and $\mathcal{A}v - \lambda_i v = 0$ for some non-zero $v \in V_i$. But the characteristic polynomial for $\mathcal{A}$ on V_i is $f_i(t) =$ $= \prod_{j \neq i} (t - \lambda_j)^{n_j}$, and λ_i cannot be an eigen-value. □

3. Theorem 1 reduces the problem of choosing the simplest possible matrix for $\mathcal{A} : V \to V$ to the case when $\mathcal{A}$ has only one eigen-value λ, and $(\mathcal{A} - \lambda\mathcal{E})^m = 0$, $m \le \dim V$. If we set $\mathcal{B} = \mathcal{A} - \lambda\mathcal{E}$, we obtain a nilpotent operator with index of nilpotence m and with matrix B.

THEOREM 2. <u>The Jordan normal form</u> $J(B)$ <u>exists for the nilpotent matrix</u> B (here the ground field K can be arbitrary).

<u>Proof.</u> We must show that the vector space V on which the nilpotent operator $\mathcal{B}$ acts with matrix B splits into a direct sum of so-called <u>cyclic subspaces</u> $K[\mathcal{B}]v_i =$ $= \langle v_i, \mathcal{B}v_i, \dots, \mathcal{B}^{m_i - 1} v_i \rangle$ with $\mathcal{B}^{m_i} v_i = 0$. We would like to use induction on the dimension of the space. Suppose that the theorem holds for all pairs $(V', \mathcal{B}')$, where $\dim V' < \dim V$ and $\mathcal{B}'$ is a nilpotent operator on V'.

Suppose that $\mathcal{B}^m = 0$, $\mathcal{B}^{m-1}u \neq 0$. We introduce the cyclic subspace $U = \langle u, \mathcal{B}u, \dots, \mathcal{B}^{m-1}u \rangle$ and the quotient space $\overline{V} = V/U$, and we define the quotient operator $\bar{\mathcal{B}}$ on $\bar{V}$ in the usual way: $\bar{\mathcal{B}}\bar{v} = \overline{\mathcal{B}v}$. Here $\bar{v} = v + U$ is the coset with representative v. Since $\bar{\mathcal{B}}^m \bar{v} = \overline{\mathcal{B}^m v} = 0$, it follows that $\bar{\mathcal{B}}$ is a nilpotent operator with index of nilpotence $\bar{m} \leq m$. In other words, $\mathcal{B}^{\bar{m}-1} V \not\subset U$ while $\mathcal{B}^{\bar{m}} V \subset U$.

Since $\dim \bar{V} < \dim V$, by the induction assumption we have

$$\bar{V} = \bar{U}_1 \oplus \dots \oplus \bar{U}_{s-1}, \qquad \bar{U}_i = K[\bar{\mathcal{B}}]\bar{u}_i \quad .$$

We obtain the decomposition of V

$$V = U_1 \oplus \dots \oplus U_{s-1} \oplus U \quad , \tag{4}$$

where

$$U_i = \langle u_i, \mathcal{B}u_i, \dots, \mathcal{B}^{m_i-1} u_i \rangle, \quad \mathcal{B}^{m_i} u_i \in U, \quad m_i \leq \bar{m} \leq m \quad .$$

The subspaces U_i are not $\mathcal{B}$-invariant, since, in general, $\mathcal{B}^{m_i} u_i \neq 0$.

For convenience, for fixed i we set $w = u_i$, $\ell = m_i$, $W = U_i = \langle w, \mathcal{B}w, \dots, \mathcal{B}^{\ell-1} w \rangle$. By assumption,

$$\mathcal{B}^{\ell} w = \alpha_k \mathcal{B}^k u + \alpha_{k+1} \mathcal{B}^{k+1} u + \dots + \alpha_{m-1} \mathcal{B}^{m-1} u, \qquad \alpha_k \neq 0$$

(if all the α_j are zero, we have nothing left to do). Applying the operator $\mathcal{B}^{m-1-k}$ to this equality, we obtain $\mathcal{B}^{m-1-k+\ell} w = \alpha_k \mathcal{B}^{m-1} u \neq 0$. Since $\mathcal{B}^m = 0$, this is only possible if $\ell \leq k \leq m-1$. Setting

$$v = w - \alpha_k \mathcal{B}^{k-\ell} u - \alpha_{k+1} \mathcal{B}^{k-\ell+1} u - \dots - \alpha_{m-1} \mathcal{B}^{m-1-\ell} u \quad ,$$

we find that $\mathcal{B}^{\ell-1} v = \mathcal{B}^{\ell-1} w + u' \neq 0$, but

$$\mathcal{B}^{\ell} v = \mathcal{B}^{\ell} w - \alpha_k \mathcal{B}^k u - \dots - \alpha_{m-1} \mathcal{B}^{m-1} u = 0 \quad .$$

The cyclic space $\langle v, \mathcal{B}v, \dots, \mathcal{B}^{\ell-1} v \rangle$ with $\mathcal{B}^{\ell} v = 0$, together with U, generates the subspace $U_i \oplus U$.

This argument holds for any i, $1 \leq i \leq s-1$, so that in (4) we can replace each subspace U_i by $V_i = \langle v_i, \mathcal{B} v_i, \dots, \mathcal{B}^{m_i - 1} v_i \rangle$, $\mathcal{B}^{m_i} v_i = 0$. Further setting $v_s = u$, $m_s = m$, and $V_s = U$, we obtain the decomposition

$$V = V_1 \oplus \dots \oplus V_s ,$$

which has all the required properties. □

4. We now prove uniqueness. At the same time we give a practical method for reducing an arbitrary $n \times n$ matrix A to Jordan normal form.

To do this we must be able to find the number $N(m, \lambda)$ of Jordan cells $J_{m,\lambda}$ of order m corresponding to the eigen-value λ of A. As usual, we let the matrix A correspond to an operator $\mathcal{A}$ which acts on an n-dimensional vector space V. We decompose V into the direct sum

$$V = V(\lambda) \oplus V' , \tag{5}$$

where

$$V(\lambda) = \bigoplus_{j=1}^{s} \langle v_j, (\mathcal{A} - \lambda \mathcal{E}) v_j, \dots, (\mathcal{A} - \lambda \mathcal{E})^{m_j - 1} v_j \rangle , \qquad V' = \sum_{\lambda' \neq \lambda} V(\lambda') .$$

We shall compute the rank $r_t = \operatorname{rank}(A - \lambda E)^t$ of the matrix $(A - \lambda E)^t$, or, equivalently, the dimension of the space $(\mathcal{A} - \lambda \mathcal{E})^t V$. Of course, this dimension does not depend on the choice of basis in V. Each of the spaces in (5) is invariant relative to $(\mathcal{A} - \lambda \mathcal{E})^t$; hence

$$\dim (\mathcal{A} - \lambda \mathcal{E})^t V = \sum \dim (\mathcal{A} - \lambda \mathcal{E})^t \mathbb{C}[\mathcal{A}] v_j + \dim (\mathcal{A} - \lambda \mathcal{E})^t V' .$$

To be definite, suppose that $m_1 \le m_2 \le \dots \le m_s$. If $m_j \le t$, then $(\mathcal{A} - \lambda \mathcal{E})^t \mathbb{C}[\mathcal{A}] v_j = 0$. For $m_j > t$ we have

$$(\mathcal{A} - \lambda \mathcal{E})^t \mathbb{C}[\mathcal{A}] v_j = \langle (\mathcal{A} - \lambda \mathcal{E})^t v_j , (\mathcal{A} - \lambda \mathcal{E})^{t+1} v_j, \dots, (\mathcal{A} - \lambda \mathcal{E})^{m_j - 1} v_j \rangle ,$$

so that $\dim (\mathcal{A} - \lambda \mathcal{E})^t \mathbb{C}[\mathcal{A}] v_j = m_j - t$. The operator $\mathcal{A} - \lambda \mathcal{E}$ is non-singular on V' (by Theorem 1), so that $\dim (\mathcal{A} - \lambda \mathcal{E})^t V' = \dim V'$. We obtain

$$r_t = \sum_{m_j > t} (m_j - t) + \dim V' ,$$

so that

$$\begin{aligned} r_t - r_{t+1} &= \sum_{m_j > t} (m_j - t) - \sum_{m_j > t+1} (m_j - t - 1) = \\ &= \sum_{m_j > t} (m_j - t) - \sum_{m_j > t+1} (m_j - t) + \sum_{m_j > t+1} 1 = \\ &= \sum_{m_j = t+1} 1 + \sum_{m_j > t+1} 1 = N(t+1, \lambda) + N(t+2, \lambda) + \cdots . \end{aligned}$$

Hence, $r_{m-1} - r_m - (r_m - r_{m+1}) = \{N(m, \lambda) + N(m+1, \lambda) + \cdots\} - \{N(m+1, \lambda) + N(m+2, \lambda) + \cdots\} = N(m, \lambda)$, and we finally obtain the formula

$$N(m, \lambda) = r_{m-1} - 2r_m + r_{m+1} , \tag{6}$$

$$m \geq 1, \quad r_t = \text{rank}\,(A - \lambda E)^t , \quad r_0 = n .$$

We note that r_t is an invariant of the matrix A (i.e., a number that depends only on the class of matrices similar to A). Hence, <u>the uniqueness of the Jordan normal form</u> $J(A)$ <u>is also established by the formula (6)</u>.

So far we have said nothing about the matrix C which realizes the reduction

$$J(A) = C^{-1} A C .$$

But since we now know the matrices A and $J(A)$, we can find $C = (c_{ij})$ from the homogeneous system of n^2 linear equations

$$CJ(A) - AC = 0 .$$

Let $C_1, \ldots, C_r$ be a fundamental system of solutions. In general, not all of the C_i are non-singular matrices, but, since the Jordan normal form $J(A)$ exists, it follows that

$\det(t_1 C_1 + \cdots + t_r C_r) \neq 0$ with indeterminate coefficients $t_1, \ldots, t_r$, and it is possible to choose $\alpha_1, \ldots, \alpha_r \in \mathbb{C}$ for which $\det(\alpha_1 C_1 + \cdots + \alpha_r C_r) \neq 0$. Then $C = \alpha_1 C_1 + \cdots + \alpha_r C_r$ can serve as the change of basis matrix which reduces A to $J(A)$. Of course, C is by no means uniquely determined, even if we normalize by requiring that $\det C = 1$.

Hints to the Exercises

Page

58 1. For any natural number n, the number $4n! - 1$ has at least one prime divisor p of the form $4k - 1$, where $p > n$.

58 2. Set $n = 2$ and $m = p_1 p_2 \cdots p_s$, where $p_1, \ldots, p_s$ are different primes of the form $p_i = 4k_i + 1$. Then every prime divisor p of the odd number $n^2 + m^2$ has the form $4k + 1$, where p does not belong to the set $\{p_1, p_2, \ldots, p_s\}$.

59 4. In case of difficulty, the reader can turn to §4 of Chapter 4, where this is proved using more sophisticated considerations.

77 1. Let $\dim V_h(A) = r$, $\dim V_v(A) = s$. Choose r rows which form a basis; without loss of generality we may assume that they are the first r rows $A_1, A_2, \ldots, A_r$. Consider the shortened $r \times n$ matrix $\widetilde{A} = [A_1, A_2, \ldots, A_r]$ made up of the first r rows of A. Choose t columns of $\widetilde{A}$ which form a basis, where $t = \dim V_v(\widetilde{A})$. Without loss of generality we may assume they are $\widetilde{A}^{(1)}, \ldots, \widetilde{A}^{(t)}$. Since $V_v(\widetilde{A}) \subset \mathbb{R}^r$, we have $t \leq r$. Next, prove that $s \leq t$ as follows. For every column $A^{(k)}$, $k > t$, we must find scalars $\lambda_1, \ldots, \lambda_t \in \mathbb{R}$ such that $A^{(k)} =$

$= \lambda_1 A^{(1)} + \cdots + \lambda_t A^{(t)}$, i.e., $a_{ik} = \sum_{p=1}^{t} \lambda_p a_{ip}$, $1 \leq i \leq m$. Choose $\lambda_1, \ldots, \lambda_t$ so that $\widetilde{A}^{(k)} = \lambda_1 \widetilde{A}^{(1)} + \cdots + \lambda_t \widetilde{A}^{(t)}$ in the shortened matrix. Then $a_{ik} = \sum_{p=1}^{t} \lambda_p a_{ip}$ for $i \leq r$. For $i > r$, use the expression $A_i = \mu_1 A_1 + \cdots + \mu_r A_r$ for the i-th row as a linear combination of the first r rows. We then have:

Page

$$a_{ik} = \sum_{\ell=1}^{r} \mu_\ell a_{\ell k} = \sum_{\ell=1}^{r} \mu_\ell \sum_{p=1}^{t} \lambda_p a_{\ell p} = \sum_{p=1}^{t} \lambda_p \sum_{\ell=1}^{r} \mu_\ell a_{\ell p} =$$

$= \sum_{p=1}^{t} \lambda_p a_{ip}$. Thus, $s \leq t$, and, since $t \leq r$, we have $s \leq r$. Next, consider the so-called transpose matrix

$$^tA = \begin{Vmatrix} a_{11} & a_{21} & \cdots & a_{m1} \\ a_{12} & a_{22} & \cdots & a_{m2} \\ \cdots & \cdots & \cdots & \cdots \\ a_{1n} & a_{2n} & \cdots & a_{mn} \end{Vmatrix}$$

which is the $n \times m$ matrix whose rows are the columns of A and whose columns are the rows of A. We have: $r_h({}^tA) = r_v(A)$, $r_v({}^tA) = r_h(A)$, and so, applying the inequality we proved to ${}^tA$, we obtain $r \leq s$. Thus, $r = s$.

104 2. Notice that, if a matrix B has basis columns with indices $j_1, \ldots, j_r$, then all of the columns in AB can be expressed as a linear combination of the columns $(AB)^{(k)}$, $k = j_1, \ldots, j_r$. The same can be said for the transpose matrix ${}^t(AB) = {}^tB\,{}^tA$.

104 3. Consider the restriction $\bar{\varphi} = \varphi|_V$ of the map φ to an arbitrary subspace $V \subset \mathbb{R}^m$. It is obvious that $\operatorname{Ker}\bar{\varphi} \subset \operatorname{Ker}\varphi$. Hence, by Theorem 1, which is applicable, because we already know that V can be interpreted as $\mathbb{R}^k$ for some $k \leq m$, we have: $\dim V - \operatorname{rank}\bar{\varphi} = \dim \operatorname{Ker}\bar{\varphi} \leq \dim \operatorname{Ker}\varphi$. Thus, $\dim V - \dim \varphi(V) \leq \dim \operatorname{Ker}\varphi$. Setting $V = \psi(\mathbb{R}^n) = \operatorname{Im}\psi$, we finally obtain: $\dim \operatorname{Ker}\varphi\psi = n - \operatorname{rank}\varphi\psi = (n - \operatorname{rank}\psi) + (\dim V - \operatorname{rank}\bar{\varphi}) \leq \dim \operatorname{Ker}\psi + {}$ $+ \dim \operatorname{Ker}\varphi$.

105 5. Show that $A = [x_1, \ldots, x_n](y_1, \ldots, y_n)$.

Page

129 3. Use the equality $\det C = \det C'$ and relations (4) and (5).

130 6. Recall Example 3 of Subsection 3 §3 Ch. 2, and take note of the fact that

$\det C_n(1, \ldots, 1) = (-1)^n \det C_n(-1, \ldots, -1)$.

137 5. If this is not the case, use the criterion in Exercise 3. Namely, if $[x_1^o, \ldots, x_n^o]$ is a non-trivial solution of the linear system $AX = 0$ and if x_k^o is a component with maximal absolute value, then the k-th equation $a_{kk} x_k^o + \sum_{j \neq k} a_{kj} x_j^o = 0$ gives us the estimate $(n-1)|a_{kk}|\,|x_k^o| =$

$= (n-1)\left|\sum_{j \neq k} a_{kj} x_j^o\right| < (n-1)\,|a_{kk}|\,|x_k^o|$, which is a contradiction.

137 6. Since $c_{ij} = \Sigma\, a_{ik} b_{kj}$, where $C = (c_{ij})$, repeated application of the rule for expanding a determinant along a row (Theorem 1' of §2 Ch. 3) gives

$$\det C = \sum_{k_1, \ldots, k_n = 1}^{n} \begin{vmatrix} a_{1k_1} & a_{1k_2} & \cdots & a_{1k_n} \\ a_{2k_1} & a_{2k_2} & \cdots & a_{2k_n} \\ \cdots & \cdots & \cdots & \cdots \\ a_{nk_1} & a_{nk_2} & \cdots & a_{nk_n} \end{vmatrix} b_{k_1 1} b_{k_2 2} \cdots b_{k_n n},$$

where the summation is over all pair-wise distinct $k_1, \ldots, k_n$. If $m < n$, there are no such indices, and hence $\det C = 0$. But if $m \geq n$, then $k_1, \ldots, k_n$ is a choice of numbers $\{j_1, \ldots, j_n\}$ from $1, 2, \ldots, m$, taken in some order. Then collect all terms corresponding to a fixed set $\{j_1, \ldots, j_n\}$, and, using the complete expansion of a determinant, obtain

Page

$$\sum \begin{vmatrix} a_{1k_1} & \cdots & a_{nk_1} \\ \cdot & \cdots & \cdot \\ a_{1k_n} & \cdots & a_{nk_n} \end{vmatrix} b_{k_1 1} \cdots b_{k_n n} = \begin{vmatrix} a_{1j_1} & \cdots & a_{nj_1} \\ \cdot & \cdots & \cdot \\ a_{1j_n} & \cdots & a_{nj_n} \end{vmatrix} \cdot \sum_{\pi} \epsilon_{\pi} b_{k_1 1} \cdots b_{k_n n} =$$

$$= \begin{vmatrix} a_{1j_1} & \cdots & a_{nj_1} \\ \cdot & \cdots & \cdot \\ a_{1j_n} & \cdots & a_{nj_n} \end{vmatrix} \cdot \begin{vmatrix} b_{j_1 1} & \cdots & b_{j_1 n} \\ \cdot & \cdots & \cdot \\ b_{j_n 1} & \cdots & b_{j_n n} \end{vmatrix}, \quad \text{where } \pi = \begin{pmatrix} j_1 & \cdots & j_n \\ k_1 & \cdots & k_n \end{pmatrix}.$$

163 5. Consider the partition of G into pairs $\{g, g^{-1}\}$.

164 12. Use the Fundamental Theorem of Arithmetic in §8 Ch. 1. Does this group have any finite set of generators?

181 1. Find an upper estimate for the number of different Cayley tables of order n. Use Theorem 2 to show that $\rho(n)$ is at most equal to the number $\binom{n!}{n}$ of different subsets of n elements in S_n. Actually, $\rho(n)$ is much less, but no good estimate, i. e., close to a best possible estimate, has been found for $\rho(n)$.

181 3. Partition G first into left and then into right cosets with the same representatives.

182 6. If $x^2 = e$ for all $x \in G$, then $abab = c \Rightarrow ab = b^{-1}a^{-1} =$
$= b(b^{-1})^2(a^{-1})^2 a = beea = ba$.

205 8. Use induction on m together with the fact that the binomial coefficient $\binom{p}{k}$ is divisible by p if $0 < k < p$.

206 12. Use proof by contradiction. Let $N = \{a_1, \ldots, a_n\}$ be the set of all non-zero non-invertible elements. The map $\rho_x : a_i \mapsto xa_i$ is a bijection $N \to N$ for any non-zero $x \in R\backslash N$. The kernel of the map $\rho : x \mapsto \rho_x$ is infinite.

Page

221 6. Answer. The identity automorphism and the map $a + b\sqrt{d} \mapsto a - b\sqrt{d}$.

235 4. After proving that

$$\sum_{k=0}^{m} \binom{k+s}{k} = \binom{m+s+1}{m},$$

use induction on m.

252 7. Suppose the contrary: $\det(x_{ij}) = g_1(\dots, x_{ij}, \dots)\, g_2(\dots, x_{ij}, \dots)$. Since $\det(x_{ij})$ is a linear homogeneous polynomial in the variables in a fixed column, it follows that one of the g_1, g_2 is a linear homogeneous polynomial in the x_{ij}, $1 \le i \le n$, for a fixed j, while the other factor does not involve those x_{ij}. Apply the same argument to the variables in a fixed row. Now suppose, for example, that x_{11} appears in g_1. Then g_2 does not contain x_{1j}, $1 \le j \le n$, and then g_2 does not contain x_{ij}, $1 \le i, j \le n$, i.e., g_2 is a constant.

261 2. Answer. No.

277 4. Since f is homogeneous, clearly $f(0, \dots, 0) = 0$. Suppose the theorem is false, i.e., $(a_1, \dots, a_n) \neq (0, \dots, 0) \Rightarrow f(a_1, \dots, a_n) \neq 0$. Using Exercise 3 and Fermat's Little Theorem, show that the reduced polynomial for $g(X_1, \dots, X_n) = 1 - f(X_1, \dots, X_n)^{p-1}$ must then be $g^*(X_1, \dots, X_n) = (1 - X_1^{p-1}) \dots (1 - X_n^{p-1})$. But

$$\deg g = (p-1)\deg f = (p-1)r < (p-1)n = \deg g^*$$

This contradiction proves the theorem.

279 11. Apply the corollary to Gauss's Lemma, in §3 Ch. 5, and the previous exercise, and use the fact that $\mathbb{Z}_2[X]$ is a unique factorization domain.

Page

296 3. Consider f as a polynomial in X_n with coefficients in $K[X_1, \ldots, X_{n-1}]$. Notice that, since f is skew-symmetric, $f = 0$ for $X_n = X_{n-1}$, and hence, f is divisible by $X_n - X_{n-1}$.

317 2. For $x > 1$ start with the inequality

$$f(x) \geq a_0 x^n - B \frac{x^{n-m+1} - 1}{x - 1} > \frac{x^{n-m+1}}{x - 1} [a_0 x^{m-1}(x - 1) - B] \quad .$$

318 3. Differentiate the formal expression $f(X) = \sum b_i (X - a)^i$ k times and set $X = 0$.

318 4. Use Exercise 3.

318 7. It is useful to consider the "dual" polynomial $X^5 f(1/X)$ and use formulas (12) of §1 and (9) of §2.

318 8. Use the fact that $f(X) = (X - c) g(X) \Rightarrow g(X) \in \mathbb{Z}[X]$. Now use these facts to find the integer roots of the polynomial $X^4 + X^3 - X^2 + 40X - 100$ (answer: $c = 2$).

318 9. If $c = a/b$ is a fraction in lowest terms, then $a^n/b = -a_1 a^{n-1} - a_2 a^{n-2} b - \cdots - a_n b^{n-1}$.

318 10. Using Theorem 1, factor $f(X)$ into polynomials of the form $(X + a)^2 + b^2$, and then apply the formal identity

$$(p^2 + q^2)(r^2 + s^2) = (pr + qs)^2 + (ps - qr)^2 \quad ,$$

which comes from the relation

$$|p + iq|^2 \, |r + is|^2 = |(p + iq)(r + is)|^2 \quad .$$

Page

319 11. $f(X) = X^3 + aX^2 + bX + c = (X^2 + \alpha X + \beta)(X + \theta)$, where $a = \alpha + \theta$, $b = \beta + \alpha\theta$, $c = \beta\theta$, with $\alpha, \beta, \theta \in \mathbb{R}$. Stability of $f(X)$ is equivalent to stability of the two polynomials $X^2 + \alpha X + \beta$, $X + \theta$, i.e., to the inequalities $\alpha > 0$, $\beta > 0$, $\theta > 0$. It is easy to check that this system of inequalities is equivalent to: $a > 0$, $b > 0$, $c > 0$, $ab - c > 0$. Similar considerations apply to a fourth degree polynomial in $\mathbb{R}[X]$.

341 1. Take H to be the stationary subgroup G_1 of a point $1 \in \Omega$, use (3), and set $\sigma(i) = g_i G_1$.

341 3. Notice that all elements of P have the form

$$g = A^i B^j C^k, \quad \text{where} \quad A = \begin{Vmatrix} 1 & 1 & 0 \\ 0 & 1 & 0 \\ 0 & 0 & 1 \end{Vmatrix}, \quad B = \begin{Vmatrix} 1 & 0 & 0 \\ 0 & 1 & 1 \\ 0 & 0 & 1 \end{Vmatrix}, \quad C = \begin{Vmatrix} 1 & 0 & 1 \\ 0 & 1 & 0 \\ 0 & 0 & 1 \end{Vmatrix};$$

if $g \notin Z(P)$, then $C_P(g) = \langle g \rangle Z(G)$, $|C_P(g)| = p^2$.

341 4. If $\sigma \in S_n$ and $\pi = \pi_1 \dots \pi_m$, then $\sigma\pi\sigma^{-1} = \sigma\pi_1\sigma^{-1} \dots \sigma\pi_m\sigma^{-1}$; in addition, $\sigma \cdot (i_1 i_2 \dots i_k) \cdot \sigma^{-1} = (\sigma(i_1)\sigma(i_2) \dots \sigma(i_k))$ for any cycle $(i_1 i_2 \dots i_k)$ of length k.

342 8. In $\sum N(g)$, each element $x \in \Omega$ is counted $|St(x)|$ times. Thus, the elements in the same orbit as x give a contribution to the sum equal to $(G : St(x)) \cdot |St(x)| = |G|$.

365 11. Count the number of elements of order 2, or make use of the result in Exercise 10.

366 14. $aba = ba^2b = ba^{-1}b \Rightarrow ab^2 = aba \cdot a^{-1}b = ba^{-1}b \cdot a^{-1}b = ba^{-1} \cdot aba = b^2a$. Conclude from this that $ab = ba$, and so, if we take into account the other relations, we have $b = e$.

Page

384 1. The beginning of the proof is the same. After taking a cyclic group $\langle a \rangle$ of maximal order m in A, take a subgroup B such that $\langle a \rangle \times B$ is maximal. If this direct product is A, then we are done. Otherwise, consider an element $c \in A$ not in $\langle a \rangle \times B$ but such that $c^p \in \langle a \rangle \times B$ (where p is a prime). Now work in the group $\langle c, \langle a \rangle \times B \rangle$, and try to write it in the form $\langle a \rangle \times B'$, where $B' \supset B$.

412 1. Differentiate the equation $e^{i\alpha t}\,\overline{e^{i\alpha t}} = 1$ with respect to t and then set $t = 0$.

420 1. See the proof of Theorem 5 in §3 Ch. 7.

420 2. Using the fact that all elements of order 2 are conjugate, show that they form a "bouquet" (Fig. 21) of five pair-wise disjoint (except for the identity e) conjugate Sylow subgroups of order 4. The group I acts on the "bouquet" by conjugation. This action is faithful, since I is a simple group (see Exercise 1).

420 3. Apply the homomorphism theorem to $\Phi : SU(2) \to SO(3)$.

420 7. Use the ideas in the computation of the number of necklaces (Problem 2 at the beginning of the chapter).

432 1. Rewrite (4) and (5) in the form

$$|G|^{-1} \sum_g \psi_{jj_0}(g)\,\varphi_{i_0 i}(g^{-1}) = \delta_{\Phi,\Psi} \frac{\delta_{ji}\,\delta_{j_0 i_0}}{\chi_\Phi(e)} .$$

Multiply both sides by $\psi_{kj}(h)$ and sum over j, taking into account the equality $\sum_j \psi_{kj}(h)\,\psi_{jj_0}(g) = \psi_{kj_0}(hg)$. In the resulting relation

$$|G|^{-1} \sum_g \psi_{kj_0}(hg)\,\varphi_{i_0 i}(g^{-1}) = \delta_{\Phi,\psi} \frac{\psi_{ki}(h)\,\delta_{j_0 i_0}}{\chi_\Phi(e)}$$

Page

set $j_0 = k$ and $i_0 = i$, and then sum over i and k to obtain the characters.

432 3. Let Φ be the irreducible representation, and let h be an element of G. Since G is commutative, we have $\Phi(g)\Phi(h) = \Phi(h)\Phi(g)$, $\forall g \in G$. Setting $\sigma = \Phi(h)$ in Schur's lemma, we obtain: $\Phi(h) = \lambda_h \mathcal{E}$. This holds for any $h \in G$. Since Φ is irreducible, the only possibility is that Φ is one-dimensional.

433 5. By assumption, $C\Phi_g C^{-1} = \Psi_g$ for some matrix $C = (c_{ij}) \in GL(n, \mathbb{C})$. The operation $A \mapsto A^* = {}^t\overline{A}$, applied to $C\Phi_g = \Psi_g C$, gives $\Phi_g^{-1} C^* = C^* \Psi_g^{-1}$, and hence $\Phi_g^{-1} C^* C = C^* C \Phi_g^{-1}$. By Schur's lemma, we have $C^* C = \lambda E$. Furthermore, $\lambda = \sum_{k=1}^{n} |c_{ki}|^2 = \mu\bar{\mu}$, $\mu \in \mathbb{C}$, and $U = \mu^{-1} C$ is the desired matrix.

445 2. Since $a^{\tau}(\chi_1\chi_2) = a^{\tau}(\chi_1)\, a^{\tau}(\chi_2)$, it follows that a^{τ} is a character of the group $\hat{A}$. Since $(a\,a')^{\tau} = a^{\tau}(a')^{\tau}$, it follows that τ is a homomorphism from A to $\hat{\hat{A}}$. Furthermore,

$$\operatorname{Ker}\tau = \{a \in A \mid a^{\tau}(\chi) = \chi(a) = 1, \ \forall \chi \in \hat{A}\} \implies \operatorname{Ker}\tau = e \quad ,$$

and so $|\hat{\hat{A}}| = |\hat{A}| = |A|$ implies that τ is an isomorphism.

446 7. Answer: 5, 1, -1, 0, 0.

453 2. Comparing dimensions, obtain a direct sum decomposition

$$P_m = H_m \oplus (x^2 + y^2 + z^2) H_{m-2} \oplus (x^2 + y^2 + z^2)^2 H_{m-4} \oplus \ldots \quad .$$

453 4. Since $SO(3)$ is simple, if τ were non-trivial it would mean that τ is a faithful representation of degree 2. But we can see from Example 3 of Subsection 4 §5 or from the description of the finite subgroups of $SU(2)$ in §3

Page

that even the restriction of τ to $S_4 \cong \mathbf{O}$ cannot be faithful.

471 3. There exist non-singular matrices C and D such that

$$A' = CAC^{-1} = \begin{Vmatrix} \alpha_1 & & * \\ & \ddots & \\ 0 & & \alpha_n \end{Vmatrix}, \quad B' = DBD^{-1} = \begin{Vmatrix} \beta_1 & & * \\ & \ddots & \\ 0 & & \beta_m \end{Vmatrix}.$$

Hence

$$A' \otimes B' = (C \otimes D)(A \otimes B)(C^{-1} \otimes D^{-1}) = (C \otimes D)(A \otimes B)(C \otimes D)^{-1}$$

is a triangular matrix with diagonal entries $\alpha_i \beta_j$. These diagonal entries are the eigen-values of the matrix $A' \otimes B'$, and hence of $A \otimes B$. We have:

$$\det(A \otimes B) = \prod_{i,j} \alpha_i \beta_j = \left(\prod_i \alpha_i\right)^m \left(\prod_j \beta_j\right)^n = (\det A)^m (\det B)^n .$$

493 4. Consider the splitting field F of the polynomial $X^p - a$. Let $\theta \in F$ be one of the roots, so that $a = \theta^p$ and $X^p - a = (X - \theta)^p$. If it is possible to write $X^p - a = u(X)\, v(X)$, where $u(X)$ is a monic polynomial over K having positive degree $m < p$, then, since $F[X]$ is a unique factorization domain, we must have $u(X) = (X - \theta)^m$. In particular, $\theta^m \in K$; since $\theta^p \in K$, we have $\theta \in K$.

494 5. By the preceding exercise, it suffices to verify that the equality

$$X^p - Y = \left(X - \frac{g(Y)}{h(Y)}\right)^p \quad \text{with } g, h \in \mathbb{Z}_p[Y] \text{ is impossible.}$$

494 6. By (8), we have $X^d - 1 = \prod_{e \mid d} \Phi_e$. Hence

$$X^n - 1 = (X^d - 1) \prod_{s \mid n;\, s \nmid d} \Phi_s(X) = (X^d - 1)\, \Phi_n(X) \prod_{s \mid n;\, s \nmid d;\, s \neq n} \Phi_s(X) .$$

It remains to refer to (10).

Page

494 7. Since $\Phi_n(X) = \Pi(X - \varepsilon)$, where ε runs through the primitive n-th roots of 1, it follows that when $n > 1$ all of the ε are $\neq 1$, and so the distance in the complex plane from q to any ε is greater than the distance from q to 1. Hence, $|\Phi_n(q)| = \Pi |q - \varepsilon| > q - 1$, and there's no way $\Phi_n(q)$ can divide $q - 1$.

494 11. Work with the equation $x^2 + y^2 - z^2 = 0$ with $x, y, z \in \mathbb{F}_p$. According to Chevalley's theorem (see Exercise 4 §1 Ch. 6), the total number N of solutions of this equation is divisible by p. Suppose that there is no solution with $xyz \neq 0$. Compute N by considering two cases separately. If no $a \in \mathbb{F}_p$ exists for which $a^2 + 1 = 0$, then the only solutions are $(0, 0, 0)$, $(0, n, \pm n)$, $(n, 0, \pm n)$, $n = 1, 2, \ldots, p - 1$, and hence $N = 4p - 3 \equiv 0 \pmod p \Rightarrow p = 3$. If $a^2 + 1 = 0$ for some $a \in \mathbb{F}_p$, then $N = 6p - 5 \equiv 0 \pmod p \Rightarrow p = 5$.

494 12. Answer: not in general.

510 8. Using Exercise 9 of §1, consider a primitive 8-th root of one α in the algebraic closure Ω_p of the field $\mathbb{F}_p$. Since $\alpha^4 = -1$, we have $\alpha^2 + \alpha^{-2} = 0$; in addition, $\alpha^5 = -\alpha$ and $\alpha^{-5} = -\alpha^{-1}$, so that $\alpha^5 + \alpha^{-5} = -(\alpha + \alpha^{-1})$. Setting $\beta = \alpha + \alpha^{-1}$, we have $\beta^2 = \alpha^2 + \alpha^{-2} + 2 = 2$, so that $p \equiv \pm 1 \pmod 8 \Rightarrow \beta^p = \alpha^p + \alpha^{-p} = \alpha + \alpha^{-1} = \beta \Rightarrow 1 = \beta^{p-1} = (\beta^2)^{\frac{p-1}{2}} = 2^{\frac{p-1}{2}} \Rightarrow (\frac{2}{p}) = 1$. Similarly, $\beta \equiv \pm 5 \pmod 8 \Rightarrow \beta^p = \alpha^p + \alpha^{-p} = \alpha^5 + \alpha^{-5} = -(\alpha + \alpha^{-1}) = -\beta \Rightarrow -1 = \beta^{p-1} = 2^{\frac{p-1}{2}} \Rightarrow (\frac{2}{p}) = -1$.

Page

510 9. If $n = 1$ and $f(x) = \sum_{i=1}^{m} a_i x^i$, $a_m \neq 0$, then

$$f(xy) = \sum_{i=1}^{m} a_i x^i y^i = f(x) f(y) = f(x) \left(\sum_{i=1}^{m} a_i y^i \right) ,$$

where x and y are independent variables. Equating coefficients of y^m, we see that $a_m x^m = f(x) a_m$, and so $f(x) = x^m$. Now suppose $n > 1$. Setting $g(x) = f(x \cdot E)$, we have $g(xy) = g(x) g(y)$. This, along with the fact that we proved our claim when $n = 1$, imply that $g(x) = x^s$. Since $X \cdot X^{\vee} = (\det X) \cdot E$, it follows that

$$f(X) f(X^{\vee}) = f((\det X) E) = g(\det X) = (\det X)^s .$$

But $f(X)$, $f(X^{\vee})$ and $\det X$ are polynomials in the x_{ij}, $1 \leq i, j \leq n$, and $\det X$ is an irreducible polynomial (see Exercise 7 §3 Ch. 5). By Theorem 2, which says that a polynomial ring in any number of variables has unique factorization, we have $f(X) = c(\det X)^m$, where c is a constant, and $f(XY) = f(X) f(Y) \Rightarrow c^2 = c$, and, since $c \neq 0$, we have $c = 1$.

Index

Universitext

Editors: F.W. Gehring, P.R. Halmos, C.C. Moore

Chern: Complex Manifolds Without Potential Theory
Chorin/Marsden: A Mathematical Introduction to Fluid Mechanics
Cohn: A Classical Invitation to Algebraic Numbers and Class Fields
Curtis: Matrix Groups
van Dalen: Logic and Structure
Devlin: Fundamentals of Contemporary Set Theory
Edwards: A Formal Background to Mathematics 1: Logic, Sets and Numbers
Edwards: A Formal Background to Mathematics 2: A Critical Approach to Elementary Analysis
Frauenthal: Mathematical Modeling in Epidemiology
Fuller: FORTRAN Programming: A Supplement for Calculus Courses
Gardiner: A First Course in Group Theory
Greub: Multilinear Algebra
Hájek/Havránek: Mechanizing Hypothesis Formation: Mathematical Foundations for a General Theory
Hermes: Introduction to Mathematical Logic
Kalbfleisch: Probability and Statistical Inference I/II
Kelly/Matthews: The Non-Euclidean, Hyperbolic Plane: Its Structure and Consistency
Kostrikin: Introduction to Algebra
Lu: Singularity Theory and an Introduction to Catastrophe Theory
Marcus: Number Fields
Meyer: Essential Mathematics for Applied Fields
Moise: Introductory Problem Courses in Analysis and Topology
Oden/Reddy: Variational Methods in Theoretical Mechanics
Reisel: Elementary Theory of Metric Spaces: A Course in Constructing Mathematical Proofs
Rickart: Natural Function Algebras
Schreiber: Differential Forms: A Heuristic Introduction